Springer Series in Operations Research and Financial Engineering

More information about this series at http://www.springer.com/series/3182

John Celona

Winning at Litigation through Decision Analysis

Creating and Executing Winning Strategies in any Litigation or Dispute

Springer

John Celona
San Carlos, CA, USA

ISSN 1431-8598 ISSN 2197-1773 (electronic)
Springer Series in Operations Research and Financial Engineering
ISBN 978-3-319-30038-2 ISBN 978-3-319-30040-5 (eBook)
DOI 10.1007/978-3-319-30040-5

Library of Congress Control Number: 2016940906

Printed on acid-free paper

This Springer imprint is published by Springer Nature
The registered company is Springer International Publishing AG Switzerland

This book is dedicated to
Professor Ronald A. Howard
and Dr. Peter McNamee, who taught me
decision analysis and how to write about it;
to our son Trey, whose love lights up
our lives in ways we never could have
predicted; and to my spouse and partner
in life Bubba Gong, without whose constant
love, support, belief and counsel this work
would never have come to fruition.

Foreword

Over the course of more than half a century, my main professional concern has been improving decision making. The importance of decisions should be self-evident: Decisions are the only means you have to change your future life.

Most of us consider ourselves pretty good at making decisions despite having little or no formal training in the subject. Yet the principles of good decision making have been known for generations. Unfortunately, it is all too easy to demonstrate how frequently people make decisions that seem unwise to them upon more careful review. Hence my personal commitment in spreading decision analysis is to provide anyone who wishes the competence to attain clarity of action in any decision they face. My latest book with Ali Abbas, *Foundations of Decision Analysis*, presents the field at levels that progress from novice to professional.

In this work John Celona applies decision analysis to decisions in litigation or, more generally, to disputes of all sorts: an area in which improved decision competence can greatly help people to clarify what they want, what it is worth to them, and how best to achieve it. Litigation decisions are especially difficult because they involve high stakes, great uncertainty, and because they require close cooperation between the principal (a client) and a highly trained professional (the attorney). Decision analysis should be a part of every attorney's training. Unfortunately, it is not. Further, among the many works on decisions in litigation, none tackle in depth the application of decision analysis to litigation. This book will help to fill that gap.

A further challenge arises when an analyst (possibly trained in decision analysis, management science, or operations research) is called in to support the legal decision-making process. Legal terminology, issues, and how to structure and analyze them are critical, yet foreign to an analyst—and most attorneys are not trained to provide what the analyst needs to know about legal proceedings. This book will be of great assistance.

For over three decades, I have known, taught, and worked with the author at Stanford University, at Strategic Decisions Group, and during his subsequent independent consulting practice. I have always found him to be an exceptionally clear thinker and commend his contributions to the field of decision analysis. His first book on decision analysis, *Decision Analysis for the Professional* (coauthored with

Peter McNamee), played a key role in developing the professional practice of decision analysis. This work, first published in 1986 and presently in its fourth edition, continues in use at Stanford University and around the world.

The author's combination of clear thought, acumen, ability, training, and experience in both the law and decision analysis make him uniquely qualified to author this work. I applaud his newest contribution to wider decision competence and recommend it highly.

Stanford, CA Ronald A. Howard

Production Credits

Apple OS Support: Adam Rosen, Oakbog.com.

Preface

In 2014 I attended the 50th anniversary celebration of the founding of decision analysis by Ronald A. Howard of Stanford University and Howard Raiffa of Harvard.

Some at the gala lauded the accomplishments of the last 50 years, while others lamented the past, present, and future of decision analysis, opining that the fervor for its widespread adoption had slipped away.

I remain optimistic.

Decision analysis has become standard practice in some industries, such as oil and gas development and pharmaceuticals. Before a well is drilled at Chevron or Exxon, before a drug development project is funded at Pfizer, Merck, or Genentech, before an airplane is developed at Boeing, there is a decision analysis to figure which course of action is best and why.

Likewise, there have been many applications of decision analysis to litigation and legal problems, and many attorneys have some level of training in how to apply it. It has now been 25 years since, while a law student at Hastings College of the Law, I was asked to deliver a lecture in the corporate finance class on decision analysis. Other panicked students asked afterwards "Will this be on the final?!"

Since then, as a professional decision analyst and attorney, I've applied decision analysis to innumerable cases—some with potential liabilities in the billions of dollars. They have run the gamut from shareholder derivative suits to environmental liability to patent litigation to medical malpractice. I've also had the privilege of implementing routine decision analysis for all cases and matters at Stanford Medical Center, possibly the first such implementation ever.

Still, among the many books aimed at decision analysis or litigation management, there appears not to be any other books on applying decision analysis to litigation. Hence this book.

This book is aimed at both decision analysts (who typically hail from the Operations Research and Management Science community) with need to help on legal matters, and at attorneys and law students seeking to understand why decision analysis can be such a powerful addition to their arsenal of professional skills.

For both audiences, and for the clients who must make key decisions and insurers responsible for covered losses, the message is simply this: legal matters are rife with decisions, ranging from where to direct expensive legal resources to whether to litigate or settle and for how much. Insurers face those same decisions, plus how much to set aside in reserves. For all these decisions, there is a better way.

Private and business clients yearn for concrete guidance on how to economically and ethically manage the myriad legal issues which arise, whether they have yet given rise to litigation or not. Using decision analysis, it is entirely feasible for analysts and attorneys to give them the quantified and defensible insight they are looking for.

It is my earnest hope that this book will assist in making that happen.

Acknowledgements

Conversations and work with many people contributed to this book. The experience of coauthoring *Decision Analysis for the Professional* with Peter McNamee shaped my thinking in how to write a book on decision analysis. Work with Mark Victor at Litigation Risk Analysis, Inc., Bruce Beron at the Litigation Risk Management Institute, Michael Pierrantozzi in the Intellectual Property Licensing group at Hewlett Packard, and Jeff Driver and Brad Briegleb in the Risk Management organization at Stanford Medical Center was key in honing how to apply decision analysis to litigation.

To all the attorneys and clients I've worked with over the years who are too numerous and too confidential to mention, allow me to offer a collective thanks.

Lastly, much thanks goes to my editor at Springer, Donna Chernyk, without whose belief in this work it would not be before you today.

Contents

1 Introduction ... 1
1.1 Winning at Litigation ... 1
1.2 Decision Analysis: Conceptual Technology to Make Better Use of Your Brain ... 2
1.3 Graphical Presentation of Theory ... 4
1.4 A Note on Ethics ... 5
1.5 How to Use This Book ... 5
1.6 How Not to Use This Book ... 6
1.7 How to Use the Problems in this Book ... 6
1.8 A Note on the Use of Software ... 7
1.9 Problems ... 8
Reference ... 9

Part I Why Should You Do a Decision Analysis?

2 The Nature of the Beast ... 13
2.1 Uncertainty and Complexity ... 14
2.2 Cognitive Problems ... 15
2.2.1 Missing Issues ... 15
2.2.2 Adjustment and Anchoring ... 16
2.2.3 Representativeness and Availability ... 17
2.2.4 Miscommunication ... 19
2.2.5 Misvaluing Combinations of Uncertainties ... 20
2.3 Motivational Problems ... 21
2.4 Summary ... 22
2.5 A Solution Template ... 23
2.6 Problems ... 24
Reference ... 29

3 Why Decision Analysis Is the Right Solution ... 31
3.1 Relevance Diagrams Catch Issues ... 31
3.2 Decision Trees Take Care of Complexity ... 33

3.3 Probability Assessments Capture Judgment 36
3.4 Explicit Values Test Intuition and Create Insight as to Which Alternative Is Best and Why 38
3.5 Decision Analysis Creates a Robust Understanding the Best Decision—Regardless of How Things Turn Out 42
3.5.1 The Difference Between Good Decisions and Good Outcomes 42
3.5.2 The Elements of a Good Decision 43
3.6 The Results Should Make Intuitive Sense 45
3.7 How to "Sell" Your Case 46
3.7.1 Why Is Decision Analysis Different from Other Methods? 46
3.7.2 What Benefits Does Decision Analysis Offer? 47
3.7.3 Where Have These Kinds of Techniques Been Used? 47
3.8 A Note on Game Theory 48
3.9 Problems 49

Part II How To Do a Decision Analysis

4 Structure the Problem 55
4.1 Draw the Relevance Diagram 56
4.1.1 Introduction to Relevance Diagrams 56
4.1.2 Howitzer v. Crabgrass et al. 62
4.1.3 Procedure for Relevance Diagrams 64
4.1.4 Hints for Drawing Better Relevance Diagrams 67
4.2 Draw the Decision Tree 77
4.2.1 Procedure for Drawing the Decision Trees 77
4.2.2 Common Questions in Drawing the Decision Tree 78
4.3 Determine the Value Formulas 105
4.4 Problems 108
4.4.1 Draw the Relevance Diagram 108
4.4.2 Draw the Decision Tree 111
4.4.3 Determine the Value Formulas 116

5 Elicit Probabilities 121
5.1 Background on Probability Theory 121
5.1.1 Subjective Probabilities Represent a Person's State of Knowledge 121
5.1.2 Why Do We Need Subjective Probabilities? 122
5.1.3 Probability Arithmetic 126
5.1.4 Mutually Exclusive and Collectively Exhaustive Events..... 128
5.2 Procedures for Assessing Probabilities 130
5.2.1 Procedure for Assessing Discrete or "Yes/No" Variables.... 131
5.2.2 Introduction to Continuous Variables 135
5.2.3 Procedures for Assessing Continuous Variables 139
5.2.4 When to Revise Your Assessments 148

5.3 A Note on Monte Carlo Simulation Methods 149
5.4 Problems 150
5.4.1 Background on Probability Theory 150
5.4.2 Assessment Calibration Exercises 155
5.4.3 Procedures for Assessing Probabilities 157

6 Integrate All Information, Judgment, and Consequences 161
6.1 Calculate Scenario Values 162
6.2 Calculate the Mean Value 164
6.2.1 Calculating the Mean Value with the Roll-Forward 164
6.2.2 Rollback Mean Value Calculations 167
6.3 Cumulative Probability Distribution and Histogram 169
6.4 More on Subtrees 172
6.4.1 Definition and Purposes of Subtrees 173
6.4.2 Requirements for Subtrees 174
6.4.3 Procedure for Subtrees 177
6.5 Problems 180
6.5.1 Calculating Scenario and Mean Values 180
6.5.2 Cumulative Probability Distribution and Histogram 182

7 Analyze the Results 183
7.1 Do the Results Make Sense? 183
7.2 Have You Left Out Anything Important? 188
7.2.1 Issues Omitted from the Analysis 188
7.2.2 Issues Handled Grossly 189
7.2.3 Other Strategy Considerations 190
7.3 Settlement Strategy 190
7.3.1 Risk-Neutral Decision-Maker 191
7.3.2 Risk-Averse Decision-Maker 192
7.4 Discovery Strategy 199
7.4.1 Values of Clairvoyance (Information) and Wizardry (Control) 199
7.4.2 Probability Sensitivity 213
7.5 Trial Strategy 218
7.6 Crafting Winning Legal Strategies 222
7.7 Problems 223
7.7.1 Do the Results Make Sense? 223
7.7.2 Settlement Strategy 224
7.7.3 Discovery Strategy 230
7.7.4 Trial Strategy 232

Part III Advanced Decision Analysis Topics

8 Difficult Questions to Structure 239
8.1 Complicated Damages 239
8.1.1 Sum of Two Kinds of Damages 239
8.1.2 A Share of the Verdict 241

8.1.3 A Multiplier on Damages ... 243
8.1.4 An Offset to the Award ... 244
8.1.5 Insurance or Reimbursement ... 246
8.2 Legal Fees and Costs ... 248
8.2.1 Don't Include Sunk Costs ... 249
8.2.2 Add or Subtract Legal Fees and Costs to an Assessment from Total Case Value ... 252
8.2.3 Add Legal Fees and Costs to the Distribution on Case Value ... 253
8.2.4 Make Multiple Assessments If Potentially Large and Variable by Scenario ... 253
8.2.5 Include in the Analysis If Potentially Large, and Variable, and Uncertain ... 254
8.2.6 Use Discounting Where Necessary ... 255
8.2.7 Handling Contingent Legal Fees ... 255
8.3 Fund Flows Over Time ... 256
8.3.1 Converting Fund Flows Over Time to a Single Value ... 256
8.3.2 Time Preference: Choice of a Discount Rate ... 257
8.3.3 No Prejudgment Interest and No Continuing Injury ... 258
8.3.4 Prejudgment Interest ... 260
8.3.5 Injuries and Damages Occurring Over Time ... 260
8.4 Precedence ... 261
8.5 Valuing Nonmonetary Consequences ... 267
8.5.1 Nonmonetary Consequences in Business Cases ... 268
8.5.2 Nonmonetary Consequences in Governmental Problems ... 279
8.5.3 Decision-Making Where Increased Chance of Death Is a Possibility ... 284
8.6 Uncertainty as to Which Forum ... 286
8.7 Cases Where You Have Almost No Information ... 287
8.8 Problems ... 288
8.8.1 Complicated Damages ... 288
8.8.2 Legal Fees and Costs ... 294
8.8.3 Prejudgment Interest and Discounting ... 296
8.8.4 Precedence ... 298
8.8.5 Valuing Nonmonetary Consequences ... 300

9 Special Probability Approaches ... 307
9.1 Many Possible Arguments ... 308
9.1.1 Focus on How the Judge or Jury Will Think About It ... 308
9.1.2 Don't Use More Detail Than Is Helpful ... 313
9.1.3 Look for "Clincher" Arguments ... 317
9.2 Handling "Spill-Over" ... 319
9.2.1 Argument Synergies and Order of Proof ... 319
9.2.2 Jury (or Judge) Synergy ... 320
9.2.3 Jury Consideration of an Issue in Stages ... 322

9.3 Working with a Group ... 324
9.3.1 Problem Structuring in a Group ... 325
9.3.2 Probability Assessment in a Group ... 325
9.3.3 Working with Experts ... 327
9.4 Tracking Estimates Over Time ... 328
9.5 Constructing and Reading Probability Distributions ... 331
9.5.1 Histograms ... 331
9.5.2 Cumulative and Excess Probability Distributions ... 337
9.5.3 Piecewise-Continuous Distributions ... 341
9.5.4 Contingent Probability Distributions ... 342
9.6 A Closer Look at Dependencies ... 347
9.6.1 Dependent Probabilities and Dependent Outcomes ... 347
9.6.2 Order of dependencies ... 348
9.6.3 Jury Science and the Value of Imperfect Information ... 354
9.7 Problems ... 361
9.7.1 Many Possible Arguments ... 361
9.7.2 Handling "Spill-Over" ... 362
9.7.3 Working with a Group ... 364
9.7.4 Tracking Estimates Over Time ... 364
9.7.5 Constructing and Reading Probability Distributions ... 364
9.7.6 A Closer Look at Dependencies ... 367

10 Multiple Parties and Claims ... 375
10.1 Codefendants ... 375
10.1.1 How to Determine Your Value of Litigating the Case with a Codefendant ... 375
10.1.2 How to Allocate a Settlement Between Yourself and a Codefendant ... 377
10.1.3 How to Figure a Settlement Value Where Codefendant Has Already Settled ... 378
10.1.4 Splitting a Settlement Between Insurer and Insured ... 379
10.2 Multiple Defendants ... 382
10.3 Counter-Claims ... 389
10.4 Two Opposing Motions for Summary Judgment ... 392
10.5 Problems ... 392
10.5.1 Codefendants ... 392
10.5.2 Multiple Defendants ... 394
10.5.3 Counter-Claims ... 396
10.5.4 Two Opposing Motions for Summary Judgment ... 396

11 Pretrial and Trial Strategy Planning ... 399
11.1 Updating Your Probabilities and Analysis ... 399
11.2 Contingent Strategy Tables ... 401
11.3 Claim Inclusion or Exclusion ... 404
11.4 Analyzing from Your Opponent's Point-of-View ... 406
11.5 Applying Decision Analysis to Alternative Dispute Resolution (ADR) ... 409

11.5.1 Why Do Trial Risks Need To Be Evaluated? ... 409
11.5.2 How Does Decision Analysis Improve the Traditional Settlement Process? ... 411
11.5.3 What Benefits Does Decision Analysis Bring to ADR? ... 413
11.6 Problems ... 415
11.6.1 Updating Your Probabilities and Analysis ... 415
11.6.2 Contingent Strategy Tables ... 416
11.6.3 Claim Inclusion or Exclusion ... 416
11.6.4 Analyzing from Your Opponent's Point-of-View ... 419
11.6.5 Pretrial and Trial Strategy ... 420
11.6.6 Applying Litigation Risk Analysis to Alternative Dispute Resolution (ADR) ... 425

12 Review and Communication ... 429
12.1 Reviewing Your Analysis ... 429
12.1.1 Relevance Diagrams ... 429
12.1.2 Assessments and Decision Trees ... 433
12.1.3 Probability Distributions ... 438
12.1.4 Sensitivity to Probability ... 439
12.1.5 What to Do and Why ... 442
12.2 Discussing the Case ... 442
12.2.1 Purposes of Discussion with Client or Co-counsel ... 443
12.2.2 Discussing the Analysis with an Opponent ... 444
12.3 Documenting Your Work ... 445
12.4 Problems ... 446

Bibliography ... 447

Index ... 449

Chapter 1
Introduction

1.1 Winning at Litigation

Legal disputes are complex, time-consuming, extremely expensive, and common. Matters can take years to resolve.

For individuals, being a party to a legal dispute is hopefully a rare—though often painful—experience. For entities of any size, managing legal disputes is a constant task. Large organizations typically have multiple matters pending at any time, some ripened to a filed lawsuit and some not. For insurance companies, managing many claims and lawsuits is a key and central part of the business.

Counsel bring their considerable training, talents, and expertise to doing the best possible job for clients. Their work is critical and appreciated.

For counsel and clients, there is a powerful way to build on their respective contributions to consistently win at litigation. By winning, I mean creating a sustained strategic and tactical advantage to achieve the most favorable possible result in any case. This method is Decision Analysis.

Decision Analysis doesn't replace the respective roles of clients (determining their interests) and counsel (doing the best possible job of meeting the client needs). Rather, it builds on their respective contributions to:

- Create a better understanding the key issues in a case, including potential damages exposure.
- Apply these insights to crafting the best possible case strategies, including negotiation and settlement strategy.
- Update case strategy as a matter progresses.

This superior understanding of what's truly important in a case, what it's worth, and what you could do creates a pivotal advantage over your opponent. As in other conflicts, it won't always be possible to achieve an easy and lopsided victory

J. Celona, *Winning at Litigation through Decision Analysis*,
Springer Series in Operations Research and Financial Engineering,
DOI 10.1007/978-3-319-30040-5_1

(sometimes you will be lucky) but, relative to what the stakes, issues, and exposures really are, you will consistently come out ahead. I've seen this borne out time and again in decades of applying decision analysis to litigation.

With the aid of decision analysis, either counsel or analysts assisting counsel or clients can craft and execute strategies to win at litigation.

1.2 Decision Analysis: Conceptual Technology to Make Better Use of Your Brain

Over the years, I have noted a number of typical reactions when people are first introduced to the theory, methods, and tools of Decision Analysis. Occasionally someone gobbles up the message whole, wondering why he or she hasn't already been doing something as intuitively sensible as Decision Analysis. Others listen and suspend their judgment until they've seen all the pieces laid out before them and then put together. Others balk and reject it outright, sometimes with an almost instinctive negative reaction to structuring a decision and a refusal to use numbers instead of words to think about possibilities.

The last reaction is perhaps most unfortunate because I believe that Decision Analysis offers a unique way to make better use of the two modes of thinking all people innately possess, in the process creating powerful and defensible insights as to which course of action is best and why.

All people possess a fast, intuitive mode of thinking which I like to call the intuitive supercomputer. It's the one people naturally use most of the time to make decisions. It's fast, easy, intuitive, and works well most of the time. It uses subtle and complex reasoning processes that are difficult to trace. It governs emotions, empathy, trust, and—most importantly—action. It is highly developed for dealing with other people. Daniel Kahneman refers to this mode of thinking as the Affective Decision System or Type 1 thinking.

Unfortunately, the fast, easy, and intuitive heuristics and methods employed by the Intuitive Supercomputer reliably go astray when dealing with complexity and uncertainty. They are further subject to motivations, willpower, emotions and stress which all too predictably drive demonstrably unreasonable action. These failure modes are discussed in further detail in Chapter 1.

Thankfully, we are equipped with a second mode of thinking which I call the Logical Co-Processor or slow thinking. It is indeed slower and literally more effort to invoke (measurable in calories burned). It requires concentration and focus, and uses transparent and logical reasoning methods.

For example, consider the purchase of a car. Whether you like a car or not is an instant, intuitive determination. Whether you actually purchase that particular car or another (or particular options on that car) brings in slow thinking. What's the cost of a particular configuration? Does it have enough power, gas efficiency, or room? Here one typically begins to logically puzzle out the alternatives and their respective costs and benefits.

Your logical co-processor likewise isn't omniscient. It requires considerable concentration and focus—at the risk narrowing the focus to where you try to solve the wrong problem. In Decision Analysis terms, we call this Frame Blindness: setting up the wrong framework for the decision. Going back to our car example, are you considering a car for every day use for yourself, your entire family, or for taking to the race track? Each different frame for the problem drives a different preferred choice.

Further, a logical argument by itself doesn't lead to action unless trusted and persuasive. This is self-evident to people trained in formal logic. Change the premises or the conditional ("if-then") and the conclusion changes. It is very easy to manipulate a logical argument or a quantitative analysis to lead to the desired conclusion. Attorneys, in particular, are trained to put forth the best possible arguments for their clients regardless of their own thoughts as to the intuitive merits of the case.

Decision analysis frames the interchange between your intuitive supercomputer and your logical co-processor to develop solid and defensible insights as to which course of action is best and why. Intuition creates the alternatives and your judgments on the possibilities for uncertainties that affect which alternative will turn out to be best in the end. Logic structures these into a quantitative model to generate results and create insight as to which course of action is best and why. This interchange is illustrated in Figure **1.1**.

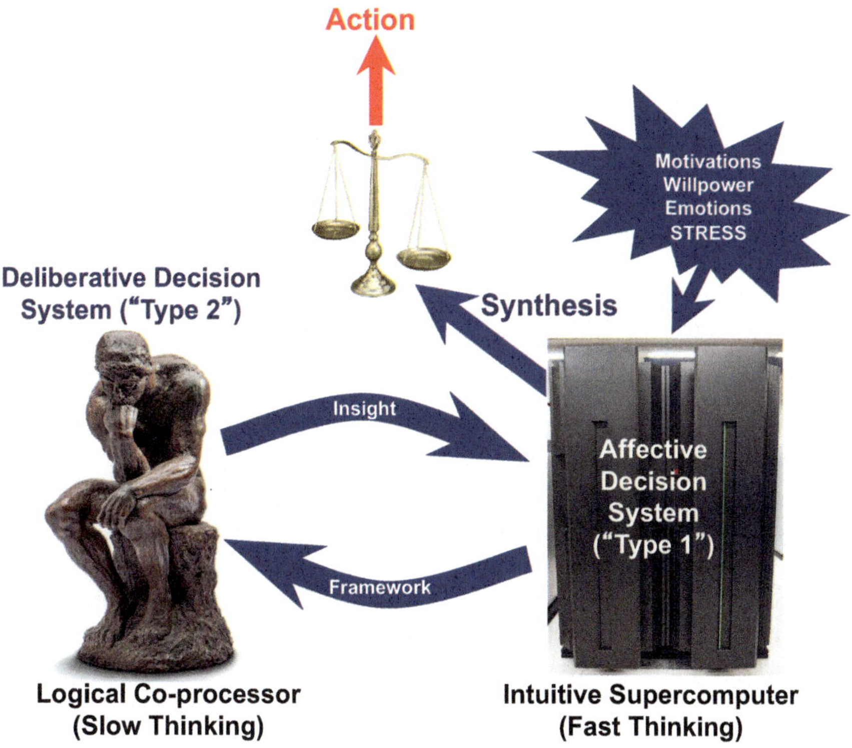

Figure 1.1 Decision analysis frames the interchange between intuitive and logical thinking

Done well, the insight created by the interchange framed by Decision Analysis creates insights—a qualitative synthesis—which can indeed drive well-considered action.

The process is similar to the one Einstein used to develop the theory of relativity. When he first published it, he was asked "Where is your lab work?" He had none. He used the now-famous thought experiments documented in his book on relativity to develop the theory (the trains traveling toward each other and so forth).

In the same way, decision analysis creates thought experiments expressed in quantitative models to explore which alternative might be best in various conditions and why.

These thought experiments and they insight they create furnish a very robust and defensible basis for action. The insights and synthesis yield the two- or three-sentence description of what should be done and why that senior decision makers seek. The analysis behind it elaborates the rationale for the decision. Done well, it addresses pretty much any "Yeah but," "What if," or "Did you consider…"

In summary, Decision Analysis provides:

- A philosophy for how to make decisions employing your intuitive and logical thinking modes;
- A framework for breaking decisions down into components: alternatives, information, preferences, and logic;
- A language for communicating clearly about the decision and its possible consequences;
- A systematic process for making decisions;
- A methodology employing specific tools and analyses.

This book is neither a complete exposition on Decision Analysis nor on the law. For the former, please see *Decision Analysis for the Professional.*

Rather, it aims to equip people with some analytical background (e.g., management science and operations research) with methods and examples to tackle legal problems effectively, and people with legal training with enough analytical methods and guidance to do the same.

1.3 Graphical Presentation of Theory

The selections from Decision Analysis theory, philosophy, and methods in this book are presented with largely graphical means, omitting possible accompanying derivations and representations with mathematical formulas.

For example, in Chapter 5, we present graphically the equal-areas method for discretizing a continuous probability distribution. A presentation on this same topic in *Foundations of Decision Analysis* concludes with the following formula:

$$\int_{x_1}^{x_2} \left(p_2 - F(x)\right) dx = \left(< x \mid x \in [x_1, x_2], \& > - x_1\right)\left(p_2 - p_1\right)$$

Our aim with employing graphical rather than formula presentations is twofold:

- For those with a technical background (e.g., mathematics, operations research, or management science), to present a complete set of methods and calculations and how they apply to analyzing complex and uncertain legal situations.
- For those with a legal background (e.g., law students and attorneys) to likewise introduce them to applying these methods to legal situations without needing to learn the underlying mathematical representations.

For those desiring to supplement this book with a formula presentation, we refer you to *Foundations of Decision Analysis*.

1.4 A Note on Ethics

Decision Analysis works regardless of the motives of a party. It can be used to plan the best strategy to prosecute or defend a meritorious suit as well as a frivolous one.

Naturally, we hope that, as in all matters in life, parties act ethically. What one considers "ethical" could obviously be the subject of many books and a lifetime of study. In the interest of keeping the length of this book finite, we omit a discussion.

1.5 How to Use This Book

- Read Part I for a qualitative explanation of Decision Analysis.
  - For the novice, Chapter 2 provides a minimally quantitative first pass through the analysis.
- Read Part II for a complete explanation of the rules and procedures of a Decision Analysis.
- Refer to Part III as you confront the advanced topics discussed there (or as curiosity moves you). It is, however, worthwhile at least to skim through the Advanced Topics to acquaint yourself with the material discovered there—otherwise, you may find it difficult even to *recognize* when you're faced with one of those situations or problems.
- You may read this book without doing any of the problems at the end of each chapter. However, the problems do range from simple to complex and thereby provide a way to test and develop your comprehension and skills before being confronted with the unstructured complexity of a case of your own.
- Use the Index to find any of the example trees used throughout the book (such as the tree for Two Arguments to Recover a Single Loss), as well as for a more detailed topic breakdown than is presented in the table of contents.

- Refer to the References for more advanced discussions of various topics, and for other applications and more general discussions of decision analysis.
- Refer to the Note on Software (below) for information on the decision analyis software used in this book.

1.6 How Not to Use This Book

- This book is *not* a guide to the substantive law used in examples throughout it. Much of the law will be generally applicable to your jurisdiction, but you should not use this book as a legal reference to either the substantive law or to legal terminology.
- Nor does this book present such complete examples of sample analysis that you should merely pick the one closest to your problem and use it. Certainly, many Relevance Diagrams and trees illustrate commonly encountered types of cases and occasionally one may almost exactly capture your particular problem. However, your analysis would be far more illuminating and valuable if you used the illustrated diagrams and trees as ideas for structuring your own representation of your case, rather than simply copying a diagram that may not reveal the important uncertainties and relationships in *your* case.
- Don't put this book away after a first reading. It contains a distillation of years of experience in decision analysis, and the importance of many hints and discussions in it won't be apparent to you after a first reading. Use the Contents and Index and keep referring back to it. You'll be surprised at how you continually find helpful new ideas and information as you analyze your own cases.

1.7 How to Use the Problems in this Book

Each chapter concludes with a variety of problems organized by chapter section. It is not intended that an instructor would assign students every problem to complete or that a reader would complete them all. A variety is included so that instructors and readers can select which ones to assign or complete. Depending on the class focus, different instructors may choose different problems to assign, or different ones each time a class is offered. To test understanding, tackling one or more problem in each section would be recommended.

The problems in each section progress from simple discussion or calculation problems to case studies. Though the case studies may appear complex, they are far simpler than real-world situations in which one of the major challenges is to define the problem framework, as opposed the frameworks defined for each case study. As such, they hopefully furnish an appropriate intermediate exercise between a simple calculation and tackling a real world problem.

Likewise, the verbal response questions are intended to provoke thought rather than directing the reader to repeat something stated in the text. As an example, we offer an answer to question 1.1 below as follows:

- The amounts and types of the damages at issue (general, special, and punitive damages at a broad level, each of which has many types under it).
- The theories of liability and the prospects for the plaintiff prevailing on each one.
- Whether the trier-of-fact is a judge or a jury, and whether they are known to be favorable or unfavorable to plaintiffs in the venue in question.
- Evidentiary issues bearing on the ability to demonstrate particular assertions.
- Factors driving whether the plaintiff or defendant would be particular appealing/unappealing or credible/not credible to the trier-of-fact.
- Whether the dispute involves well-settled or novel legal theories.
- The cost and duration of litigation in terms of legal fees, experts, discovery expenses, etc.

This list is more extensive than what might be expected from a typical student, and could easily be even longer. The objective of teaching is creating learning and understanding in the student. These questions aim to test what has been achieved.

In sum, it is hoped that the combination of reading the text and working some of the problems in fact equips the student or reader to successfully tackle real cases and disputes.

1.8 A Note on the Use of Software

Use of software to evaluate decision trees is critical to making decision analysis practical for real-world problems. Nonetheless, when learning decision analysis, it is best to take a "paper and pencil" approach and structure and evaluate problems on paper and maybe with a simple calculator (your phone or Excel). This is the approach taken in this text.

All too often, people dive into using software and literally can't see the forest for the tree. They make the analysis too complex and can't tell if it's correct, what it means, or explain it to someone else—much less it being a compelling basis for a particular course of action. Best to start simple and only add complexity as your ability and the problem merit it. As Einstein said, "Everything should be made as simple as possible, but not simpler."

Software is introduced when the complications become too cumbersome for paper-and-pencil methods.

The software used in this text is Supertree for three reasons: (1) a student version of it is available for free at www.supertree.net; (2) it does everything you need it to do; (3) it does a lot of things few other programs do and almost none do everything that Supertree does.

There are many other decision analysis and decision tree program available. OR/MS Today regularly publishes a survey of them.

1.9 Problems

1. Given your present knowledge and experience, consider how you would formulate a recommendation on settling or continuing a case, and for how much.
    1.1. What matters would you consider?
    1.2. How would you relate them to each other?
    1.3. How would you create and support a specific recommendation?
2. Prior to reading this chapter, how would you define "winning at litigation?" Would you now define it differently, and, if so, how?
3. In each of the following cases, list at least two or three of the roles or responsibilities of the analyst or attorney, whether any formal analysis would be useful in performing that role (whether tax analysis, decision analysis, etc.) and why.
    3.1. Defending a corporate client in an antitrust suit.
    3.2. Prosecuting a patent infringement case against a particular defendant.
    3.3. Estate planning for a client, including preparing a will and an inter vivos trust.
    3.4. Representing one spouse in a contested divorce suit.
    3.5. Representing both spouses in an uncontested, no-fault divorce.
    3.6. Mediating a labor contract dispute.
    3.7. Working as a US prosecutor with the responsibility of deciding whether a particular narcotics trafficking case would be suitable for testing the use of the R.I.C.O. statute to impound the defendant attorney's fees.
    3.8. Rendering an opinion on whether a two clients should form their business as a partnership or as a corporation.
    3.9. Advising a client on whether to accept a particular settlement offer in a case you're prosecuting for him, and in which you are being paid hourly.
    3.10. Working as a public defender with the responsibility of advising a client on whether to accept a plea bargain for 6 months of jail plus probation.
    3.11. Considering how resolution of a particular case might affect future cases (either in terms of setting a precedent or the perceived attractiveness of future cases).
4. Consider how you presently use your affective and deliberative systems on a case.
    4.1. What do you use intuition for?
    4.2. What do you use logical reasoning for?
    4.3. How do you relate them together, if at all?
    4.4. Where do you consider any gaps to be?
5. Consider how you've faced other unfamiliar decisions in the past, like buying your first car or house, or accepting your first career-oriented job (or even the summer job at Burger King). How did you handle the decision, and were you satisfied with it? Would you feel comfortable facing the same decision again? In what respects do you feel your decision making fell short?

6. The analysis techniques presented in this book are generally applicable to any area of law, and, indeed, were originally developed outside of the legal area. Are there areas of law you feel these techniques would be particularly useful in, or not applicable, and why?

Reference

1. Kahneman, D.: Thinking, fast and slow. Farrar, Straus and Giroux, New York (2011)

Part I
Why Should You Do a Decision Analysis?

Chapter 2
The Nature of the Beast

For all my mind is clouded with a doubt

—Alfred, Lord Tennyson, The Passing of Arthur

The sun…
In dim eclipse disastrous twilight sheds
On half the nations, and with fear of change
Perplexes monarchs.

—Milton, Paradise Lost

After reading Chapter 1, you have hopefully begun pondering how in the past you've employed your Affective/Intuitive and Deliberative/Logical thinking styles, and possibly even how you've related them to each other. In this book, you will see how to use Decision Analysis structure your internal "conversation" between these two thinking modes to make better use of both, to make better decisions, and to use this superior decision making to win at litigation.

Decision Analysis can and should also be used to structure the conversation with other people. You will see in this book the concepts to structure both your internal and external conversations and how to apply them to making better decisions.

Certainly, it would be easier to stick with the way you currently make decisions, which, for most people, relies heavily on Affective/Intuitive thinking. That works fine for simple or familiar decisions where uncertainty is not a concern.

Unfortunately, a large body of cognitive psychological work has definitely shown that where either complexity or uncertainty are present (and especially if both are present), intuition reliably fails. By failing, we mean that people intuitively make decisions which, upon further scrutiny, they no longer believe are the best decisions. Our aim is to avoid these failings and create a decision and basis for the decision which look good at the time and remain good (meaning it still looks like a good decision) as events unfold.

We begin by exploring exactly how our trusty intuition fails us.

J. Celona, *Winning at Litigation through Decision Analysis*,
Springer Series in Operations Research and Financial Engineering,
DOI 10.1007/978-3-319-30040-5_2

2.1 Uncertainty and Complexity

The only certainty is that nothing is certain.

—Pliny the Elder, Historia Naturalis

The uncertainties and complexities in planning and valuing a case are myriad to consider and mystifying to reconcile.

- What evidence will surface during discovery?
- What will be the results of various pretrial and trial motions?
- Whose witnesses will be more believable?
- What instructions will the judge give the jury?
- What conclusions will the jury reach on each of the individual elements that make up a finding of liability?
- What damages will be awarded? Or what will be the value (cost) of injunctive relief?
- How long will it be until final judgment? What is the client's time value of money?
- What will the litigation cost?
- Will there be indirect financial consequences to winning or losing this case? Will there be procedural ramifications, or an effect from the precedent set in this case?
- What will be the effect on the client's stock price?
- What is the client's attitude towards risk-taking?

This uncertainty coupled with the complexity makes it extremely difficult for counsel and clients to properly weight and balance all of the important factors and arrive at the best pretrial, trial, and settlement strategies. And with millions of dollars at stake in many lawsuits, the consequences of imperfect decisions can be disastrous—for companies and for their lawyers.

Furthermore, the more complex the problem becomes—including more uncertainties—the more difficult it becomes to think about it and the more you need assistance.

These same characteristics—uncertainty and complexity—also make it hard for counsel to communicate clearly and confidently with each other and with the client. It is usually hard to explain convincingly why particular settlements should be accepted or rejected, why particular pretrial and trial strategies should be pursued or abandoned, or why particular financial and legal resources should be invested in or withheld from a case.

Consider one of your more complicated cases. Did you consider all these questions? Did you consider how they all interact to effect the settlement value of the case? Did you finally come up with a "safe" settlement figure that explicitly reflected your answers to all these questions? Certainly, you can consider these questions in an abstract fashion and then intuit an answer. But can you justify and explain why your answer isn't $10,000 higher or lower? Uncertainty and complexity make this process exceedingly difficult—even if your intuition is working correctly. The next section discusses why it frequently does not.

2.2 Cognitive Problems

For now we see through a glass, darkly...

—1 Corinthians 13:12 (King James Bible)

Now we see only puzzling reflections in a mirror...

—1 Corinthians 13:12 (New English Bible)

Throughout their lives, people develop habitual and almost unconscious methods of evaluating information and basing decisions on it. In their simplest form, these methods help to cope with the physical environment, such as using haze and clarity to help estimate the distance of objects, assuming that glowing red objects are hot, or that larger objects are heavier than small ones.

However, even these simplest methods break down in some situations. For instance, people often drastically underestimate the distance of objects in the desert because of the exceptionally clear air. Would you think that a glowing red fluorescent light tube is hot? Would you say a washing machine is heavier than a full-size refrigerator-freezer?

These breakdowns in intuition become even more acute in dealing conceptualizations like uncertainty and complexity. Psychological studies have identified a number of specific intuitive breakdowns,[1] to which I add problems particular to risk analyses.

2.2.1 Missing Issues

Alas for you, lawyers and Pharisees, hypocrites! You pay tithes of mint and dill and cummin; but you have overlooked the weightier demands of the Law, justice, mercy, and good faith.

—Matthew 23:23

Perhaps the single most serious shortfall in evaluating uncertain and complex situations is simply not thinking of all the issues relevant to the final result (the value of a case, for instance). Omitting issues is often a reaction to a case that already seems to complex too think about, and would become even more so if you added more issues to the pot. Consider the generic list of issues given above. That list becomes even longer when the issues specific to the relevant substantive law is considered.

Another reason issues are omitted is the lack of any method to relate together their effects when they have been all listed out. How do you think about some of the possible permutations of evidentiary rulings? How do you relate them to the types of damages awarded?

[1] A. Tversky, and D. Kahneman, *Judgment Under Uncertainty: Heuristics and Biases,* Science 185 (September 27, 1974): 1124–31.

A final reason for missing issues is the lack of a formal process to reveal as many of the important issues as possible. Unless you've dealt with a particular kind of case at least several times in the past, it's very difficult to simply sit down and list all the relevant issues. Further, even for familiar types of cases, the tendency is to ignore the unique aspects of the case that might make new issues relevant. Have you ever been surprised by a new and critical issue raised in the pleadings or discovery, and lambasted yourself for not having thought of it sooner? (Certainly, the purpose of discovery is exactly revealing the surprises, but hasn't your surprise been to your opponent's advantage nonetheless?) For cases with novel elements or with significant uncertainties (Decision Analysis-kind of cases), this phenomena is perhaps more the rule than the exception.

2.2.2 *Adjustment and Anchoring*

Matters of fact, which as Mr. Budgell somewhere observes, are very stubborn things.

—Will of Matthew Tindal (1733)

Try filling in the following chart. For each city listed, estimate the probability (a percentage between 0% and 100%) that the actual annual average precipitation in a year is *greater* than the number shown, then make your best guess as to what the average annual rainfall actually is (Table 2.1).

Have you filled it in yet? At least do a few before reading on.

How well do you think you did? Look at the last column. The truth is that all the cities listed receive between 21 and 24 inches on average per year except Tokyo, which receives 61.6![2]

Table 2.1 Exercise for estimating annual rainfall

	Number	Probability that **actual annual average precipitation** is greater than this number	Your best guess as to actual rainfall
Canberra, Australia	25		
Dakar, Senegal	6		
Godthaab, Greenland	40		
London, UK	45		
Mexico City, Mexico	15		
Tokyo, Japan	35		
Warsaw, Poland	30		

[2] Average precipitation figures courtesy of the American Express Appointment Book. You might also note that statistical figures for the middle column would accordingly all be around .50 except for Tokyo, which might be around .90.

Did your answers tend to follow the numbers given in the first column? Is so, then you demonstrated the adjustment and anchoring bias. In the form aimed for by the table exercise, people will typically start with some number they know or is given to them (the numbers in the first column), and then make a mental adjustment to arrive at the desired number (the numbers in the third column). The adjustment is almost always inadequate and, by following this kind of method, people will ignore other information they possess that might lead to a different guess. Adjustment and anchoring is thus a cognitive bias (a fallacious way of thinking) because the adjustment is inadequate to compensate for the focus on the initial anchor. In other words, your guess is stuck on the first number you think of.

2.2.3 Representativeness and Availability

The best way to suppose what may come, is to remember what is past.

—George Savile, Marquis of Halifax (1633–1695), Miscellaneous

Consider the following questions.

- Robert is aggressive, well-educated, and well-paid, and always wears wool suits and carries a leather briefcase. What is the chance that Robert is a lawyer?
- The victim was shot with a .38 caliber weapon. The defendant owns a .38 caliber revolver. What is the chance that the defendant's weapon shot the victim?

The representativeness bias describes when people judge according to a stereotype, or misconstrue the mathematics of chance. Thus, in the first question, the description of Robert might correspond to that of the stereotypical lawyer, leading you to think it quite likely that Robert is a lawyer. The fallacy in this reasoning is more obvious if it's broken down as follows.

- All lawyers are aggressive, well-educated, well-paid, wear wool suits, and carry leather briefcases.
- Robert is aggressive, well-educated, well-paid, wears wool suits, and carries a leather briefcase.
- Therefore, Robert is a lawyer.

One might object that the description fits many other professionals, and that given, for instance the large number of MBAs in the population, it's more likely that Robert is a some other kind of professional or businessman.

The second question raises a more subtle kind of misestimation. You might think it more likely the defendant used his weapon and is guilty because the caliber of his weapon matches the caliber used on the victim. However, given the *millions* of .38 caliber weapons in circulation,[3] the likelihood that any particular weapon was used,

[3] U.S. Department of Justice, Office of Justice Programs, NCJ-148201, July 1995.

solely based on it being a .38 caliber, is quite small (1 divided by the number of .38 caliber weapons). One interesting example of such misunderstanding and misuse of probability is in *People v. Collins* (68 C.2d 319; 1968), in which a prosecutor attempted to conclude that there was only a 1-in-12 million chance that the defendant was innocent, although, even using the prosecutor's own estimates of the probability of a man with a mustache being at the scene of the crime, etc., the math actually worked out to a probability of .40 that someone else matching the defendant's description was at the scene of the crime.[4]

Consider the following questions.

- On the average, what are the chances of punitive damages where a suit has successfully challenged denial of benefits under a disability insurance policy (in your jurisdiction)?
- What are chances of being mugged while walking through the business district of your city after 11 p.m.?
- What are the chances of a palimony suit against a 4-year live-in succeeding in California?

Differences between your answers and those of somebody else likely illustrate the availability bias. For instance, the answers to all these questions are nearly a compilation of statistical data (since you don't know any more details), yet answers to the first question would probably differ between an insurance lawyer and a criminal lawyer. An insurance lawyer would likely give a higher estimation of the chances of punitive damages because he can more easily recall instances of it happening; that is, he judges the chances higher of easily available occurrences. Similarly, the criminal lawyer would likely give higher chances of being mugged than the insurance lawyer (because the criminal lawyer can more easily recall specific muggings and can recall more of them), while a police officer's probability might be even higher: he can recall the muggings for which there were no arrests or no prosecutions.

The third question illustrates another aspect of the availability bias: the greater ease of recalling more dramatic and vivid instances. After the famous Mitchelson palimony suit, the average lawyer might judge the chances of obtaining palimony in California higher than does a lawyer practicing family law in the state (and thereby knows more of the unfamous and unsuccessful cases).

---

[4]The prosecutor's actual error was in multiplying probabilities together without accounting for the overlap between them. Thus, the suspect was a white man with a mustache, as was the defendant. If 35% of the population is white males and 15% of the people have mustaches, you can't just say that the probability of being a white male with a mustache is $.35 \times .15 = .05$, and there is therefore only a 5-in-100 chance the defendant is innocent on this basis alone. You need the probability of a mustache given you already know that that defendant is a white male to account for the overlap and make the calculation. For instance, the 15% of the population with mustaches would clearly give a wrong answer if the defendant was a white female.

2.2.4 Miscommunication

If you try to equate it by a scale in a civil case, preponderance of the evidence is just a little more than, say 50 percent, 51 percent, 90 percent; anything over that might be a preponderance of the evidence. As I have defined a reasonable doubt, that is more or less a greater degree of proof is required, so that you don't have a reasonable doubt. What is that if you put it on a 1 to 100 scale? I don't know. It's above 50 percent.

—jury instructions that were the basis for overturning a criminal conviction, from The National Law Journal

Unlike the problems discussed above, miscommunication problems don't have their origin in the subtle and complicated operation of the mind. They simply stem from the fact that people seldom think hard enough about what they mean when they use qualitative (verbal) descriptions of uncertainty. As a result, two people may use the same words and mean quite different chances, or vice versa—making the words actually misleading! This unsuspected difference in numerical odds ascribed to the same words can mask disagreement or indicate disagreement when there is none. For example, try filling in the chart below with percentages ranging from 0 to 100 (Table 2.2).

Was there any overlap in the numerical ranges you assigned to different phrases? Do you think the ranges you assigned would match those of other lawyers in your firm?

After conducting this experiment with many groups, the same result has held true in every test: *qualitative* evaluations are extremely misleading. For example, we present attorneys with several phrases commonly used to express the likelihood of winning or losing, like "it is very likely we will win (or lose)." They were then asked to use a numerical scale (0–100) to indicate the range of probabilities they mean when they use each phrase. The differences among attorneys have been frightening. In each group, there are always several attorneys who use the words "very likely" to convey a 60–75% chance, while others use "very likely" to convey an 85–100% chance. On several occasions, these widely divergent views have been held by two partners from the same law firm who work together on the same cases. And they admit they never realized their colleague meant something very different than they did.

Unfortunately, these results repeat themselves continually during actual case evaluations. While working on one lawsuit, the author heard an attorney for the company describe his thinking on a key legal issue and conclude with the comment,

Table 2.2 Exercise for quantifying verbal descriptions of uncertainty

	Minimum probability	Maximum probability
There is a good possibility that ...		
It is very likely that ...		
It is likely that ...		
There is a distinct possibility that ...		

"And that's why I think there is a 'good possibility' the court will rule in our favor." The second attorney, having liked the reasons advance by the first one, said, "I'll agree with you on that one." Before letting them continue, however, the author asked each to write down his probability of a favorable ruling. The first had meant 80% when he concluded "good possibility;" the second meant only 35%! In another situation, the author asked outside counsel to translate into a probability his appraisal of a "distinct possibility" that the judge would let a key issue go to the jury. It was only then that the client's assistant general counsel learned that his outside lawyer was trying to convey by these words chances in the 5–10% range, not the 30–40% range.

2.2.5 *Misvaluing Combinations of Uncertainties*

A further psychological difficulty posed by uncertainty is that untrained people seldom know how to evaluate combinations of uncertainties. In one experiment, Tversky and Kahneman[5] found that people regarded the coin flip sequence H–T–H–T–T–H as more likely than the sequence H–H–H–T–T–T (which did not appear random to them), or the sequence H–H–H–H–T–H (which did not appear to represent the fairness of the coin). These beliefs are not borne out by the mathematics of a fair coin toss, which would only predict an equal number of heads and tails with a great many coin tosses. The fact that any particular coin toss may have come up heads or tails, or even that five in a row have come up tails, still leaves you with 50/50 odds of tails on the next flip.

Even if you manage to overcome the psychological problems and to agree on what the words for uncertainty mean, a further difficulty is the lack of any consistent way to combine qualitative descriptions of various uncertainties to reach a correct conclusion. For instance, consider the graphic depiction below for a hypothetical product liability claim in which punitive damages hinge on whether the harm from the product defect was reasonably foreseeable, and whether the company's defense that it used state-of-the-art safety design is successful (Note that there is a circle below the heading for each uncertainty and a branch for each possible outcome—making this a "tree" depiction of all the possible scenarios.) (Figure 2.1).

How would you assess your chances of the court finding the harm reasonably foreseeable, but then finding for you on your state-of-the-art defense? How about finding the harm not foreseeable, and but then for you on your state-of-the-art defense? And how should these estimations affect the amount your client should be willing to settle for?

[5] See reference, *supra* note 1.

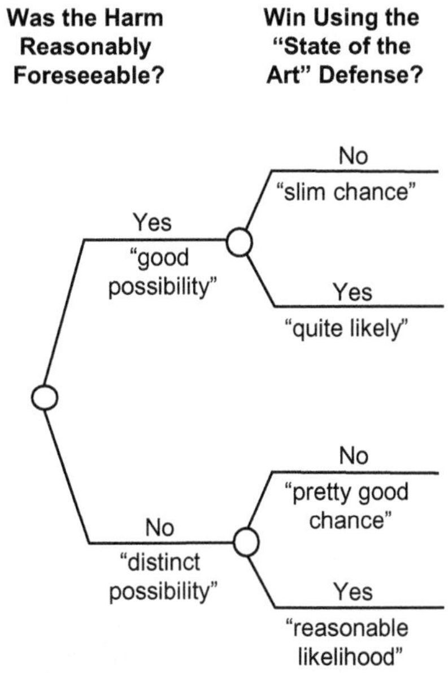

Figure 2.1 Tree representation of possibilities in a product liability case

2.3 Motivational Problems

While the world's tide is bearing me along;
Other desires and other hopes beset me,
Hopes which obscure...

—Emily Brontë, Wuthering Heights, ch. 9

Estimations of uncertainty may also be affected by the motives of the estimator. Counsel, for instance, may feel that putting too low a likelihood on the chances of winning a case somehow indicates lack of diligence or enthusiasm on his or her part. Counsel may unconsciously compensate for unduly optimistic odds of winning by then shifting the uncertainty to a factor not under her control, such as whether the court admits a "crucial" piece of evidence.

Or counsel may feel that estimates of the chances of winning should be based on "objective data" of how prior cases have turned out, rather than also on her own judgment and intuition—thus ignoring the unique aspect of the instant case. Another aspect of the desire to be "accurate" can be seen when, for instance, counsel gives an unduly small range of possible awards in case because she feels that an expert should be "sure."

Finally, counsel may suppress her own doubts and judgment to be consistent with the opinion of, for instance, a senior partner in her firm. Undoubtedly, experienced counsel can quickly size up a familiar type of case, but it may be too easy and inaccurate to assume that a new case has no significant differences from prior cases.

2.4 Summary

Hopefully, at this point you're convinced that there are serious issues to be addressed in improving how people think about uncertainty and complexity. For an extensive discussion, see Thinking, Fast and Slow (reference listed at the end of Chapter 1). Figure 2.2 presents a graphical summary of the salient issues.

A skeptic might ask: how could this be? How could mankind have progressed this far and be such a wreck at a very key skill for thriving in the modern world?

Indeed, mankind is a wreck at dealing with uncertainty in the modern world. Documented failures abound, ranging from the Columbia disaster to housing and financial meltdown to the Asian flu pandemic that never happened.

I believe the answers lies in the modernity of the modern world. Over the millennia, mankind became quite well adapted, for example, to not being eaten by lions, to determining whether an unfamiliar person intends to kill you or not, and to surviving interruptions in food supply.

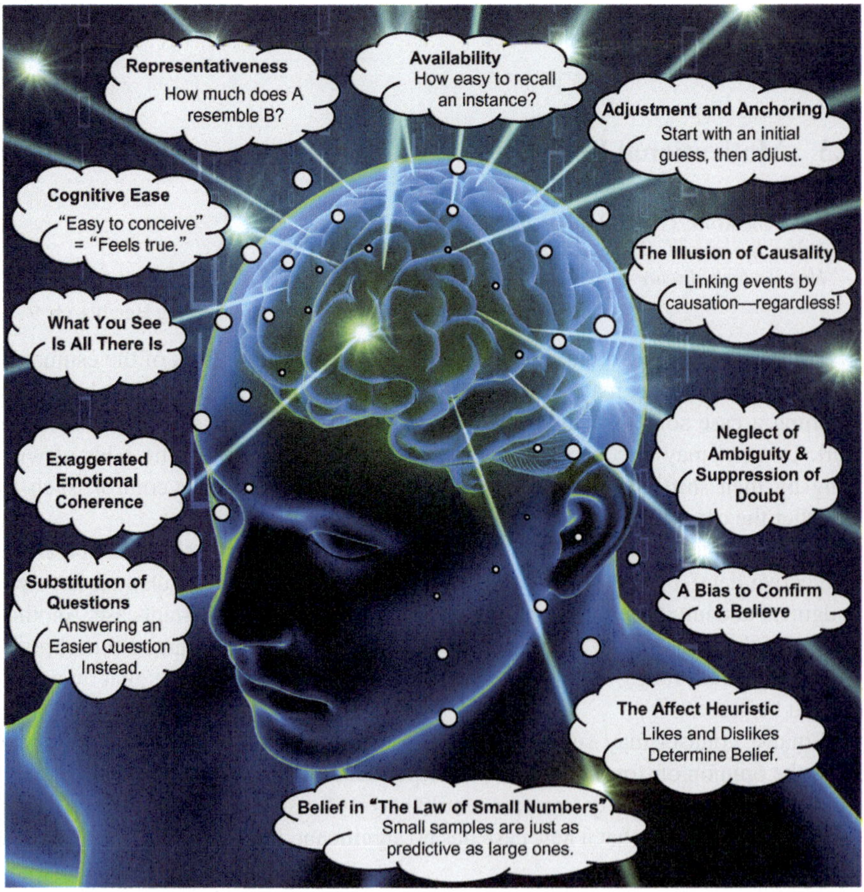

Figure 2.2 Many issues cloud our thinking about uncertainty

However, we are not well adapted to the changes introduced by civilization in the last few thousand years. Evolution simply hasn't had enough time to catch up. We are not adapted at all to having too much food to consume and, with our ready capacity to accumulate the excess for the lean times that never come, that excess kills us.

Similarly, pilots readily become disoriented and crash in low visibility conditions unless rigorously trained to rely on instruments, and sometimes even then. For the preceding few million years, there was never an evolutionary advantage to always knowing which way is down, so we never developed it. In contrast, birds fly at night, in storms, and into clouds without issue. Humans lack a basic ability present in every "bird brain."

Even in areas where we might be expected to know better, such as building houses in flood plains or coastal areas periodically ravaged by hurricanes, we are quite able to convince ourselves that the inevitable disaster will strike somewhere else or at some other time.

Most importantly for our quest, though, is how few people recognize the fundamental uncertainty in predicting how *other* people will react to uncertainty. It is the uncertainty in human behavior that drives the uncertainty in financial markets. Markets go up when people believe they will, and down when the consensus sentiment changes. It is the battle between greed and fear, with greed definitely being more popular. It is highly uncertain when that sentiment will change, and it does so without respective to "market fundamentals" (price-earning ratios and the like). Indeed, changes in sentiment drive changes in those fundamentals. If people believe housing prices will go down, they are reluctant to buy them and that drives the prices down—regardless of other economic trends.

It is the same for legal disputes. If your opponent strongly believes his or her case is well merited (or is simply angry enough), that suit will go on regardless of what you may think the "reasonable" value of the case is.

Hence the need for Decision Analysis. In the same way that instruments allow pilots to fly when their intuition would cause them to crash, Decision Analysis forces you to think better about uncertainty and complexity and explicitly relate that to the decision you make. It is a prescription for the uncertainty and complexity introduced by modern civilization, just as a healthy diet and exercise are a prescription for our modern abundance of food and lack of physical activity.

2.5 A Solution Template

A solution to these issues to create winning litigation strategies would:

- Make full use of counsel, analyst, and client's Affective/Intuitive thinking (experience, judgment, and intuition) and Deliberative/Logical thinking (creating thought models to quantify possible outcomes).
- Identify which alternative is best and why for the many decisions faced in litigation, such as:
  - Whether to continue a case or settle it, and a minimum or maximum settlement value as an alternative to going to trial;

  - Which issues are most and least important, why, and whether they merit further exploration (e.g., legal research, discovery, expert testimony);
  - Providing tactical guidance for specific case management decisions, such as moving for summary judgment, changing venue, stipulating as to a legal or factual issue, etc.

- Create intuitively sensible and logically consistent rationales for action which are readily explainable, defensible, and can be updated as a case progresses.

To accomplish these goals, we specifically need to:

- **Identify the Right Issues and Their Relationships to Each Other**. The method should neither overlook any important issues, nor get bogged down in myriad details that aren't that significant to the final result (like the chances of winning or losing, or the value of the case).
- **Articulate Clearly Counsel's Assessment of Each Issue**. The method should clearly communicate which issues are felt important, and how uncertain those issues are in unambiguous terms.
- **Evaluate Each Issue Correctly**. The method should avoid all the distorting cognitive and motivational problems in evaluating uncertainties, and thereby produce accurate representation's of counsel's true judgment.
- **Draw the Right Conclusion**. The method should produce a clear, defensible, and logical conclusion from the assessments of the uncertainties, avoiding the problems of misvaluing combinations of uncertainties and of inconsistency in reaching conclusions.

The following chapters lay out the philosophy, theory, and methods of Decision Analysis which accomplish these aims.

A skeptical reader might ask: are there alternative methods to achieve these aims? Why bother with decision analysis?

Indeed, I have worked with many attorneys who initially question both the validity and necessity for employing decision analysis to craft winning legal strategies. They have thus far relied on intuition, experience, and professional competency and it has been "just fine."

So far, even the greatest skeptics (of which there have been many) have been convinced of merits, strengths, and necessity of Decision Analysis methods in addition to their own skills and experience. It is simply a demonstrably better way, borne out in superior results.

2.6 Problems

1. In each of the following situations, list some of the issues at stake. Which can be more readily quantified, and might the object of numerical analysis be? Is it clear how the issues relate to each other, or would an explicit structure for those relationships be helpful?

Table 2.3 Sample solution to Problem 1.1

Issue	Object of analysis
Negligence Rule	Likelihood of the court picking either negligence rule
Plaintiff's Fault	Likelihood of the trier-of-fact finding the plaintiff contributorily negligent
Damages	Estimating possible awards if the plaintiff prevails
Sympathy Factor	Considering how much sympathy with the jury the plaintiff's condition will evoke (qualitative factor to be considered in other analysis)
Plaintiff's Time	Impact on the plaintiff's future earnings. Loss of consortium claim?
Value of Money	The plaintiff would be willing to take by, for example, receiving a settlement now instead of a judgment in the future
Credibility of Witnesses	Considering how they will help or hurt the contention of contributory negligence
History of Judge	Considering whether he ever chosen either negligence system in the past
Policy Arguments	Considering whether policy arguments favor either negligence system

1.1. **Example**: The plaintiff was struck by the defendant's car and seriously injured while making a left-hand turn in an intersection. The defendant's car was approaching the intersection at an excessive rate of speed (by the length of the skid marks) and entered the intersection during a yellow light. The defendant contends that he should not be held liable because of the plaintiff's contributory negligence. This jurisdiction currently has conflicting decisions making their way through appeal on whether "pure" comparative negligence is the rule (damages apportioned strictly according to the percent at fault), or a modified comparative negligence system applies with the plaintiff only recovering if he is less than 50% at fault. This particular plaintiff will not be able to have rehabilitative surgery unless he recovers (Table 2.3).

1.2. The president of a company solicited a tender offer for the company, then obtained board approval without soliciting any further offers or undertaking studies of the company's value. A number of shareholders initiated a derivative suit, contending that the president and other directors breached their fiduciary duty and were negligent by not making an informed determination of the value of the company before accepting the offer. The directors respond that the offer was fair because it was higher than the current market value, and that the business judgment rule only requires that they act in good faith. This jurisdiction has not yet determined whether the business judgment rule applies to sale of the company (as opposed to normal operation of the business), or whether the rule is one of loyalty (requiring only good faith and no self-interest), or of a duty of care (requiring that the action not be either negligent or grossly negligent).

1.3. The defendant conveyed land to the plaintiff's predecessors with an easement reserved in the deed for access via a driveway on the south side of the defendant's property. The defendant always mistakenly used a way on the north side, which is also on the plaintiff's land. The plaintiff built a house on the south side that, along with shrubbery, partially blocks the easement.

The plaintiff now sues to extinguish the easement, contending it has been extinguished by nonuse, that it is not now physically usable, and that the defendant has plenty of space to build a new access entirely on his own land. The defendant responds that he never intended to abandon the easement, and that he should at least retain usage rights to the north driveway. It is unclear whether the court will require affirmative intention to abandon to extinguish the easement, or whether it would apply equity jurisdiction to rewrite the easement.

1.4. The Commissioner contends that a taxpayer's gift of property to a non-revocable trust and lease back from the trust is a mere sham because the taxpayer retained informal control over the property through his influence with trustees, and the assignment had no purpose other than its tax consequences. The taxpayer responds that the assignment was legally effective, he has no reversionary interest in the trust, and it is unfair to allow the IRS to reach beyond the letter of the law and pierce an otherwise valid transaction. The property in question is a country club with an 18-hole golf course, tennis courts, and three pools, and the taxpayer is attempting to stipulate the identity of the property, while the Commissioner wants to introduce detailed evidence on the value of the property and the amount of income from the club.

1.5. The plaintiff and the defendant jointly founded and ran a corporation. The corporation is now dissolved, and the parties dispute certain payments made to and from the corporation while it was in business, and certain dissolution payments. Throughout the case, the judge has been giving the jury his own conclusions and comments on both issues of fact and of law, though he has been careful to warn the jury that theirs is the final judgment. The comments have been generally unfavorable to your case. This jurisdiction has no provision for raising objections out of the hearing of the jury in civil cases of this type. You are unsure whether your case will be harmed more by allowing the judge to continue commenting, or by objecting in front of the jury and being almost inevitably overruled. Precedents are not conclusive on whether you must object at trial to preserve the issue for appeal.

2. For each of the following situations, tell whether you think the person is reasoning correctly, and, if not, whether the problem stems from cognition problems, or biases, or both, and why.

    2.1. Millie Hopeful buys two lottery tickets and each pays $5. She buys ten more, figuring she's "on a roll."

    2.2. Bill Candy hates having his cavities filled, and has therefore been brushing and flossing diligently. He therefore decides that if his dentist still finds cavities, he's going to a new dentist.

    2.3. Horatio Chevy's bonus is determined by how much he exceeds his sales targets. He always sets those targets slightly below the sales last year at this time to allow for a downturn in the economy.

2.4. Bulldog Trucker noted how clearly he could see the mountains as he drove across the desert. "Must be close." he thought.
2.5. Irving Quiver heard that there's a one-in-one-hundred-thousand chance that a bomb is on any particular flight, but a one-in-ten million chance of two bombs. To be safe, he therefore always flies with a bomb of his own.
2.6. Steven Estoppel was trying to decide how much to offer to settle a case. At a staff meeting, a senior partner who had worked in related but different areas remarked "I'd offer about $40,000." Steven offered that amount.
2.7. Mario Presto in general drives over the speed limit (using a radar detector). He has nonetheless been ticketed once in a particular 35 mph zone, and, as a result, always drives the speed limit in that zone.
2.8. Bill Practical subscribes to a service that provides statistical analyses of the awards various judges make in different kinds of cases. He uses those averages to help advise clients on the values of their cases.
2.9. Roger Doubter's associate Bill Brash asserted that they could get $1 million for a case—almost four times Roger's highest possible estimate. Roger was dubious at first, but, as Bill explained the different rulings that could combine to produce that result, he became more and more convinced.
2.10. Melvin Dutiful was very worried that his requests to a foreign government for various papers wouldn't be fulfilled in time for trial. However, he felt more confident after a senior partner told him to "Make it your top priority."
2.11. You've retained an expert to advise you on how the seawater intake pipe of a local power plant has affected the local population of bay crabs. "It's been reduced by 2/7." he told you.
2.12. In the last two reckless driving cases Cy Presser defended, putting the client on the stand to testify about his careful driving gave the prosecutor the opportunity to introduce a raft of prior moving violations. Cy therefore resolved not to put reckless driving defendants on the stand any more.
2.13. Curt Class took his last big, multi-party product liability case to trial and lost, angering the clients and looking bad in his firm. He resolves to make greater efforts to settle his next such case before trial.
2.14. The defendant succeeded in joining Archibald Gavel's plaintiff and another separate plaintiff in a single trial, contending the cases involved the same issues and facts. The other plaintiff settled for $35,000. Archibald figured that his case was worth about $3,000 more than the other and settled for $38,000.

3. Below are several phrases commonly used to express the likelihood that a particular event will occur. Indicate the range of probabilities that you mean when you use each phrase (using numbers between 0 and 100 percent). Compare your answers with those of several classmates, or tabulate the answers in class. Are there any ranges for the same phrase that don't even overlap? (Table 2.4).

Table 2.4 Table for problem 3

	Maximum probability	Minimum probability
There is a good possibility …		
It is very likely …		
It is likely …		
There is a distinct possibility …		
It is unlikely that …		
It is extremely unlikely that …		
We're dead.		
I am confident that …		
I am hopeful that …		

4. For each of the following situations, list some of the relevant uncertainties and note how you think they might interact. Can you tell what the best course of action is from this level of analysis? (That is, is the best alternative clear given the combination of uncertainties?)

   4.1. Your client was hit in a crosswalk by a car running a red light, and it is the eve of trial. The defendant contends he can prove your client was drunk (he had just left a bar, and no blood alcohol test was done), and that he was involved in a near-accident a year ago. The defendant has offered to settle for 75% of your client's claim. The case is being tried before a judge who has a reputation for being very tough on damages claims.

   4.2. Your client is a renowned hospital being sued by a plaintiff who tripped and fell in the parking lot on the way in. The parking lot is a temporary one (during construction), several lights were not working, and concrete car stops were irregularly arranged in it. The plaintiff was treated and released from the Emergency Room for scrapes and bruises. The plaintiff contends she is severely injured and will need surgery. The plaintiff lives in a three-story house on the beach. You are considering hiring a private investigator to video her comings and goings from the house.

   4.3. Your presentation to the jury has gone very well so far, including very convincing testimony by your client, and you are confident of winning your case and getting the prayed-for relief. However, you've just learned your opponent will move to introduce evidence that your client lied on his income tax return 2 years ago, and you doubt you can succeed in excluding it. At the same time, your opponent has offered to avoid the embarrassment of the revelation and settle for half the relief you asked for.

   4.4. Your client is a doctor and is being sued for malpractice for failing to use a fairly new medical technique that, in hindsight, would have been appropriate in the plaintiff's case. The technique only emerged from clinical trials a year ago, and, so far, has only been used in 5% of the cases where it could potentially be applied. Your client has read about the technique, but has not seen it applied. The damages claimed are $1 million, and the defendant has offered to settle for $250,000.

4.5. Your client runs a small but successful chain of pasta shops which offers takeout and cafe-style sit-down service. Two years ago, you drafted a license agreement for the client's trademark and methods ("Noodle Mania") and the client licensed others to use them. Business is booming. You now discover, to your horror, that the client has not been doing any quality control on the licensees—contrary to the terms of the agreement—and thus risks having his trademark rights voided for abandonment (that is, licensing without quality control) if someone brings a suit. The client counters that the specified inspections would be expensive and time-consuming, and that he would either have to take too much time away from his own business or would have to hire a full-time employee to do it.

4.6. Your client owns a trademark, has been using it continuously in a small business which includes worldwide sales via the Internet, and has filed for a received incontestability in use of the mark. Revenues from this business are less than $50,000 per year. A large company with total revenue of several billion dollars per year is using the mark as part of a product line. Your client is considering whether to pursue action for infringement. You don't know how much revenues are specifically in the infringing product line, or whether your client's field of use is broad enough to cover the infringing product.

Reference

1. Hubbard, D.W.: The failure of risk management. Wiley, NJ (2009)

Chapter 3
Why Decision Analysis Is the Right Solution

If a man begin with certainties, he will end with doubts; but if he will be content to begin with doubts, he shall end in certainties.

—Francis Bacon, Advancement of Learning

In Chapter 2, you saw why uncertainty and complexity make it difficult to value cases and plan case strategy. The objective of a Decision Analysis is overcome these difficulties and, thereby, to provide the counsel and the clients with carefully thought-out, unambiguous descriptions of the full risks of a lawsuit, allowing confident litigation, settlement, and strategy decisions. Decision Analysis does this by:

- Providing a structured process for unambiguously eliciting intuition and judgement and incorporating them into quantitative models to create insight as to which course of action is best and why;
- Imposing a discipline on case planning that both requires and assists the counsel to think more systematically and carefully through any issue;
- Introducing tools that allow for the explicit evaluation and correct integration of each of the relevant factors;
- Offering a language to improve communications between the counsel and the clients, and for the clients to make economic and prudent case decisions.

The following sections go through the steps of a Decision Analysis, showing at each stage how the methods and tools deal with the problems of uncertainty and complexity to achieve these goals.

3.1 Relevance Diagrams Catch Issues

What is the answer? ...In that case, what is the question?

—Gertrude Stein, last words as quoted by Donald Sutherland

J. Celona, *Winning at Litigation through Decision Analysis*,
Springer Series in Operations Research and Financial Engineering,
DOI 10.1007/978-3-319-30040-5_3

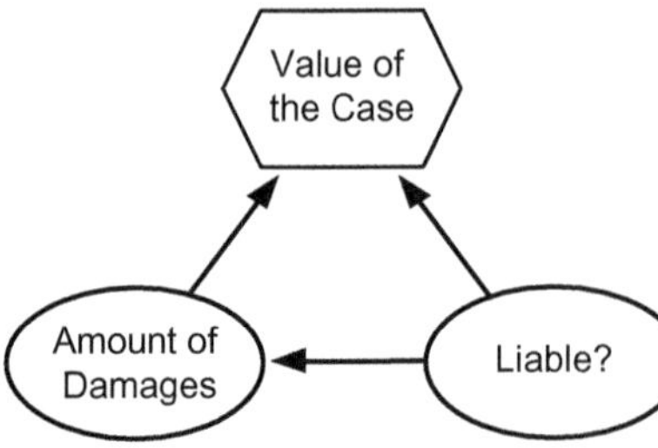

Figure 3.1 Simple relevance diagram to show the relationships between uncertainties and the value measure

The first step in a Decision Analysis to begin structuring the problem by drawing a relevance diagram. A relevance diagram is a simple method of showing the important issues in a case and how they're related. It consists of ovals for the uncertainties, boxes for decisions, a hexagon for the value measure, and arrows to show what is relevant to what. The diagram below shows the value of a generic case, with the value (cost) of the case depending on the amount of damages awarded, and both the ultimate value and the damages awarded depending on whether the defendant is found liable or not (Figure 3.1).

Relevance diagrams are crucial for dealing effectively and rigorously with uncertainty and complexity.

- **Relevance Diagrams Help Elicit Issues**. Using a relevance diagram makes it easy to put down lots of issues by prompting you to start at the final result you're interested in (such as the value of the case), and then working logically back through all the uncertainties that affect the final result—reducing the chances of missing any important issues.
- **Relevance Diagrams Help Disaggregate the Problem**. You've already seen how the mind is particularly bad at valuing combinations of uncertainties. A relevance diagram helps you to break down a complicated problem (like figuring the value of a case) into easier-to-understand pieces.
- **Relevance Diagrams Fight Biases**. A relevance diagram helps you to combat cognitive and motivational problems by breaking the value of the case, for example, down into smaller pieces on which the problems are either not so strong or easier to spot.
- **Relevance Diagrams Highlight Crucial Issues**. Helps you to find which issues are important by asking only for those critical issues that are both *uncertain* and have a *significant effect* on the final result (such as the chances of winning or losing, or the financial consequences of losing), and by showing the relationships between them;
- **Relevance Diagrams Aid Communication About Your Case**. Using a relevance diagram makes it easy to show all the issue you think are important in a case and how they interact.

By using relevance diagrams, you can quickly and clearly lay out the important issues in the case, making easy to explain those issues to others and to see if any important issues have been missed. By sketching out a relevance diagram for each new or unfamiliar case, you'll gain a better and clearer understanding your case—improving your case—even if you don't analyze the case any further. And though the diagram shown above may seem simple enough to be unnecessary, you need only begin considering the damages issues in a simple contract case (expected benefit, reliance, or restitution measures of damages; foreseeability, certainty, and mitigation limits on damages, etc.) to see how quickly the diagram can become complicated—without even yet considering the evidentiary level!

3.2 Decision Trees Take Care of Complexity

The second step in a Decision Analysis is to complete the problem structuring by using the relevance diagram to draw a decision tree. Below is the decision tree corresponding to the relevance diagram shown before (Figure 3.2).

Each circle in the decision tree corresponds to an oval in the relevance diagram, while each branch coming out of the circle represents a different possible outcome for that uncertainty. The decision tree reads from left to right, following the sequence of events and the direction of the arrows in the relevance diagram. Each path through the tree (combination of outcomes) thus describes a complete scenario, such as being found liable and the court awarding low damages.

These tree attributes of dividing the uncertainties into possible outcomes and relating them together to form scenarios enable the tree to perform a variety of functions:

- **Decision Trees Check Your Issue Identification**. By relating all the issues together into scenarios, the tree helps and prompts you to think about whether each scenario is completely described, or whether there are important issues you've left out.
- **Decision Trees Check Your Understanding of Complexity**. Similarly, drawing out the tree to form scenarios probes your understanding of the relationships between the uncertainties. Does each scenario make sense? Are the factors that influence the outcome in each scenario consistent with the way you think about the case?
- **Decision Trees Provide Scenarios for Probability Assessments**. The scenarios described by the tree will enable more meaningful probability assessments for the uncertainties in each scenario. Thus, although it is fairly obvious that

Figure 3.2 Simple decision tree

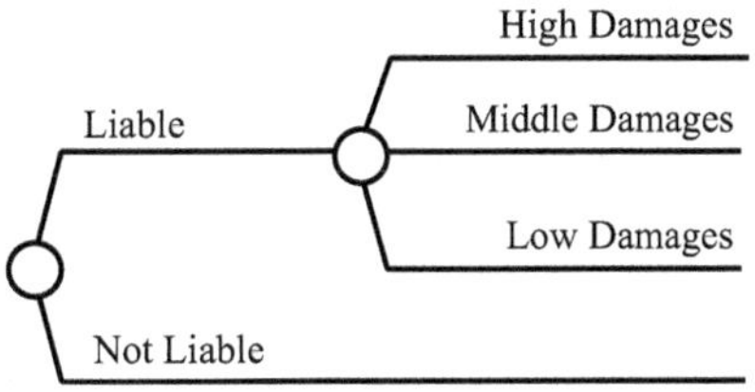

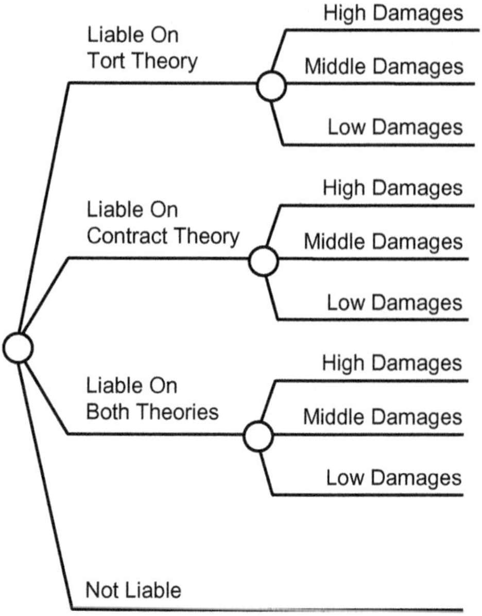

Figure 3.3 Decision tree for multiple theories of liability

you would only consider damages if you've already been found liable (reading the tree from left to right), the tree below shows how quickly the tree becomes useful for identifying the scenario that the uncertainty must be considered in (Figure 3.3).

There are many correct (and innumerable wrong!) ways the draw decision tree for a particular situation. The criteria for a correct tree are:

- It makes sense to everybody involved (analyst, client, counsel).
- It accurately captures their intuition and judgement.
- It correctly follows the rules of probability described in Chapter 5.

The tree above could be correctly completed (probabilities and outcomes filled in), but it would require one to be very careful about assessing probabilities and outcomes for all the scenarios. In particular, getting the probabilities right for liability on the tort theory only versus on contract only versus on both theories can be tricky.

An alternative way of drawing the tree is shown below. Although possibly more difficult to draw, it is easer to keep straight exactly what is going on with the respective theories of liabilities and damages awards (Figure 3.4).

Forcing you to think harder about what could happen and providing methods to help you think clearly is precisely the aim of analysis and why it works. Sometimes the best relevance diagram and tree aren't the simplest ones, but the ones that are just complex enough and no more so. A useful rule (and quote) are shown below (Figure 3.5).

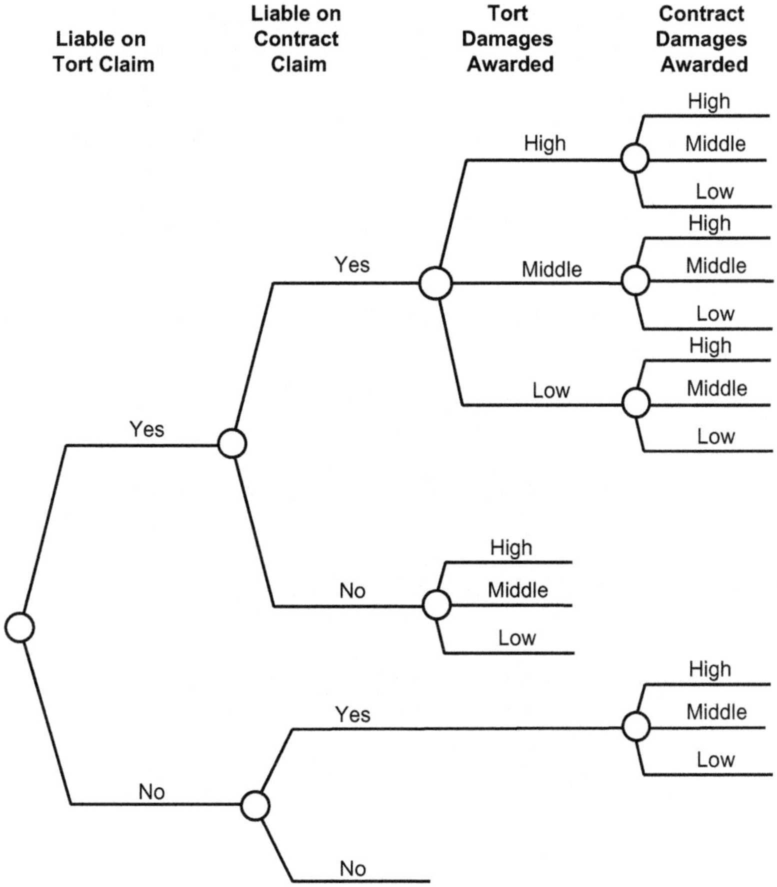

Figure 3.4 Alternative tree for multiple theories of liability

Everything should be made as simple as possible, but not simpler.
--Photograph by Orren Jack Turner, Princeton, N.J. Modified with Photoshop by PM_ Poon and later by Dantadd. - This image is available from the United States Library of Congress's Prints and Photographs division under the digital ID cph.3b46036.

- **Decision Trees Enable Numerical Analysis**. The division of the uncertainties into outcomes enables later calculation of the value of the case and the probability distribution on the value of your case—so you can draw consistent, logical, and defensible conclusions even though you're faced with a combination of uncertainties or even a single uncertainty.

Decision trees thus further aid you in breaking an uncertain and complex problem down into easily digestible pieces and in developing your thinking about your case. Drawing the relevance diagram and decision tree typically require an even clearer understanding of your case than is necessary for writing a brief.

Figure 3.5 A useful rule for figuring how complex to make the analysis

3.3 Probability Assessments Capture Judgment

But to us, probability is the very guide of life.

—Bishop Butler (1692–1752), Introduction

Certainty is never true and truth is never certain.

—Charles Smith, Sensism: the Philosophy of the West

After you've described the uncertainties in the relevance diagram and then tested and clarified your understanding with the decision tree, probability assessments capture your judgment on the different possible outcomes for each uncertainty in the tree. You've already seen why qualitative, verbal descriptions of uncertainty are unsatisfactory. But why are numerical probabilities (that is, numbers from 0 to 100 percent, or from 0.0 to 1.0) the right way to describe uncertainty?

- **Probabilities Make Intuitive Sense**. For any uncertainty, it's easy to see that any possible outcome can be rated anywhere from 0 percent (a certainty the outcome won't happen) to 100 percent (a sure thing it will). The numbers are so natural a way of expressing the odds that, for example, people never question describing a coin flip as a 50/50 chance of heads or tails.

- **Probabilities Capture All Your Information and Judgment**. Subjective probabilities simply ask: Given all your experience, information, and judgment, what do you think the odds are? They thus allow you to express all your intuition and experience in one, clear number.
- **Probability Assessments Avoid Biases**. The procedures for assessing probabilities (discussed in Part II) are specifically designed to avoid the cognitive and motivational problems discussed above, and thereby to capture a true representation of your judgment.
- **Probabilities Bring Out Why Issues Are Important**. Assessing probabilities points out which issues are most important by revealing which ones have the greatest effect on the value of your case, either through affecting the chances of winning or losing or through affecting the value (cost) of winning or losing.
- **Probabilities Communicate Unambiguously**. There is no doubt about what .72 means—in contrast to "quite likely." Skeptics should note that .72, for instance, does not imply more or less precision: just clarity and the best representation of a person's judgment.
- **Probabilities Allow Logical Combination of Judgment and Conclusions**. Unlike with verbal descriptions of uncertainties (as in the tree below, repeated from before), probabilities allow you to combine your judgments on different uncertainties to reach conclusions on ultimate issues, such as the overall chances of winning or losing a case, or the minimum (maximum) amount you should be willing to settle for. Probabilities do this via the mean value (next topic) and via roll-forward calculations (Part II) (Figure 3.6).

Figure 3.6 Hard-to-calculate decision tree

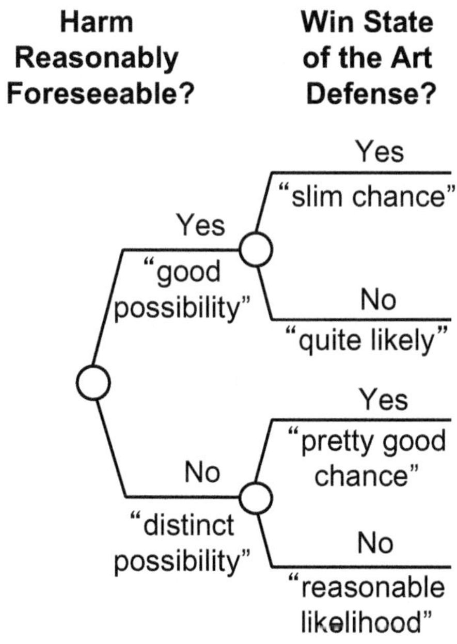

- Probabilities Are Instrumental to Calculating the Value of Your Case. Without subjective probability, there is no theoretically rigorous and consistent way to calculate the value of a single case whose outcome is uncertain. You need probabilities to calculate a mean value.[1]
- Probabilities Work Best. Besides their intuitive appeal, it can be mathematically proven that, over the long run, careful subjective probabilities and the mean value (next topic) are the best available basis for assessing future, uncertain risks like litigation risks.

3.4 Explicit Values Test Intuition and Create Insight as to Which Alternative Is Best and Why

Once you've drawn the decision tree and assessed probabilities, the first step in analyzing the tree is calculating the mean value of litigating your case. However, before giving you the simple calculation procedure (if you don't already know it), you should think about why the mean value is the right way to look at the uncertain value of your case.

Below is a decision tree on whether to litigate or settle a case. Note that one new wrinkle is the decision in the front of the tree, indicated by a box rather than the circle for uncertainties. It simply says that you get to decide whether to litigate or settle your case, and, reading the tree from left to right, you have to do so before you find out whether you'd be held liable or not (of course, but the entire order of the tree should make sense like this). We employ the following definitions for distinguishing between decisions and uncertainties.

- A decision is a choice between two or more alternatives which involves an allocation of resources that is either irrevocable or only revocable at a cost.
- An uncertainty is anything you're not sure how it will or did turn out.

Decisions are under your control. Uncertainties are not under your control, though you may be able to influence how they turn out.

For simplicity, this tree indicates that you'll pay $50k in damages if you lose, instead of the previous uncertainty on damages. This tree shows the value of the case for the plaintiff or the cost of the case for the defendant.

Consider the $35,000 settlement offer. Would you rather settle or take your chances at trial? Can you explain why you would rather litigate or settle? (Figure 3.7).

A defendant might be willing to pay $20k to settle the case rather than risking an 80% chance of losing and paying $80k at trial. A plaintiff would probably be more

[1] There are methods of directly assessing the value of a case, some with a structuring process similar to the one used in assessing probabilities. However, as you might imagine, such methods run smack into the cognitive and motivational biases that problem dis-aggregation through relevance diagrams and decision trees helps to combat.

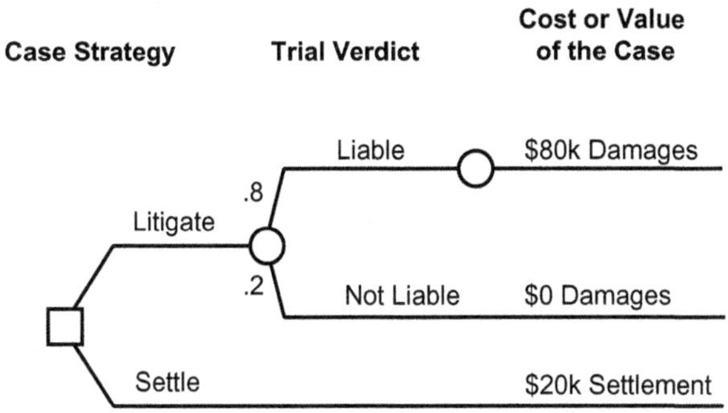

Figure 3.7 Simple decision tree for case strategy

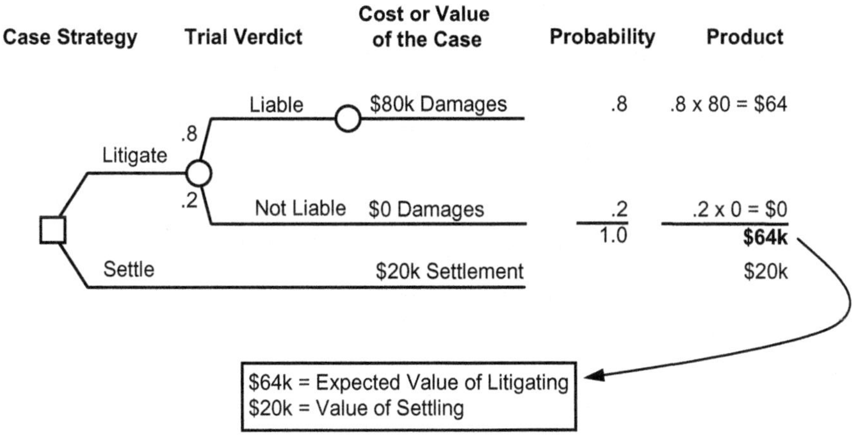

Figure 3.8 Roll forward calculation for simple case strategy tree

inclined to refuse the settlement offer and go to trial (assuming that, for both parties, the stakes are small compared to their total assets).

Doing a roll-forward calculation to find the mean value (that is, simply multiplying the probabilities by the outcomes and then summing the resulting products) confirms that the mean value is better for litigating than for settling. The mean value is simply a probability weighted average of all the possible scenarios. It allows you represent many possible scenarios with a single number valuation of an alternative for comparison to other alternatives (Figure 3.8).

Note that the probabilities for all the individual scenarios for how an alternative could turn out add up to 100%.

You would perhaps agree, then, that the lowest-expected-value choice matches what you would pick in a simple situation like the one above. But suppose that, even though you believe there is an 80% chance of liability, one element of the damages claim is highly speculative, and there is only a 10% chance of the court awarding

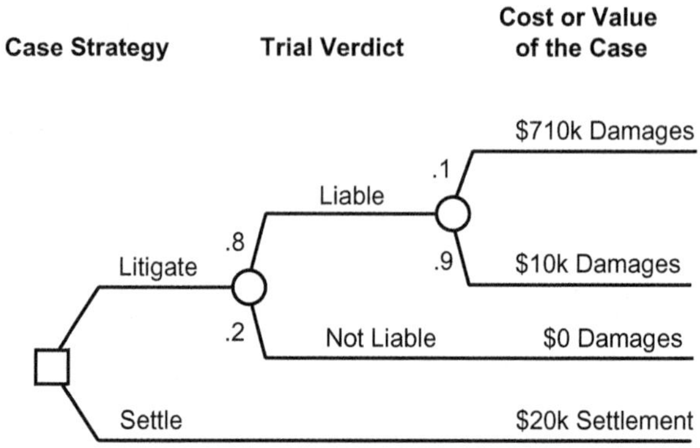

Figure 3.9 Case strategy tree with "speculative" damages element uncertainty

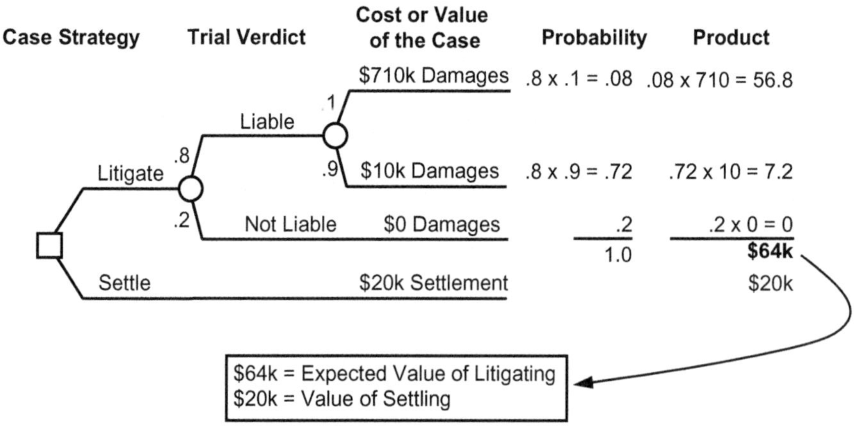

Figure 3.10 Case strategy tree with roll forward calculation on speculative damages element

this. That claim amounts to $700k. Otherwise, the court will only award minimal damages of $10k (700 + 10 = 710 in the worst case scenario). This tree is shown below (Figure 3.9).

Now what does your intuition tell you? Would you rather settle or litigate the case, and can you explain your choice?

The best choice is looking less sure, but hopefully this layout makes the tradeoffs much clearer: there's a large chance of being found liable at trial, but only a small chance of being hit with major damages. The roll forward values are the same (Figure 3.10).

For the defendant, the best choice is clearly to pay $20k to settle the case. For the plaintiff, they would have to think long and hard about not taking the settlement when there is only an 8% chance of doing better at trial—and a 92% chance of doing worse ($10K or zero).

Is this result in accord with your intuition? Or are there just too many combinations of uncertainties to intuit the answer? The aim of the analysis is to relate together all of your probability and outcome information (from previous steps in the Decision Analysis) and produce a recommendation that matches what you'd intuitively choose in a simple enough situation. However, the great advantage of using decision analysis is the ease of applying it in situations where uncertainty and complexity would frustrate your intuition.

The intuitive reasons for using the mean value, then, can be summed up as follows.

- The best expected-value choice matches your intuition in simple situations.
- The mean value extends that same method of choosing to more complicated problems with consistency, ease, and simplicity.
- The mean value is the only measure that relates together all your probability and outcome information.
- The simple math of the mean value (multiplication and addition) lets you balance probabilities and damages and avoid a one-sided focus.

However, as mentioned in the previous section on probabilities, there is also a formal, mathematical reason for using the mean value to make your decisions. It can be mathematically shown that, for situations where the outcomes aren't potentially ruinous, using the mean value will maximize wealth over the long run. Thus, although the example above had only a $44 thousand difference between litigating or settling the case, if you faced a hundred of lawsuits like that one, or if you consistently spurned the lowest expected-value alternative over a long time, then you would most likely be paying more money than you needed to. It would be bad business.

Further, the existence of the insurance industry is proof that the mean value works over the long term for cost-effectively managing many risks transferred by other parties who don't have the "luxury" (or the problem!) of balancing many similar risks. The modus operandi of the insurance industry is simply making sure that premiums are higher than the mean claims cost on a particular policy. Certainly, some policies will have much higher claims costs, and some much lower. But as long as the insurer is careful enough about estimating the mean and charges more, it will make money on its portfolio of policies as a whole. And the insurers that analyze the risk better to make better estimates of the mean claims cost—or even to identify risks and *reduce* the mean claims cost—can price its policies lower and gain market share. The mean value is big business.

Of course, using the mean value does not guarantee good results in any one particular case. The essence of an uncertainty is that you really don't know exactly how it will turn out (hence the estimates with probabilities). However, you can be confident that, over the long run, you will be better off using the mean value to make decisions.

Finally you should note that other valuation techniques, such as return on investment or using a risk-adjusted discount rate, don't work for a variety of reasons.[2]

---

[2] See the discussion in *Decision Analysis for the Profession* (listed in the Bibliography) for more details.

3.5 Decision Analysis Creates a Robust Understanding the Best Decision—Regardless of How Things Turn Out

The conduct of a losing party never appears right: at least it never can possess the only infallible criterion of wisdom to vulgar judgments—success.

—Edmund Burke, Letter to a Member of the National Assembly

People commonly classify a decision as good or bad according to how things turn out in the end (what the outcome is). An investment was a good decision if it made you money. A decision to litigate a case was a bad one if you ended up paying a judgment greater than the settlement offer.

However, when you begin to look at the picture created by a Decision Analysis, you realize that this perspective is too short-sighted. The distinction between decisions and outcomes means that decisions are under your control, uncertainties are not. You can only do the best possible job of making the choices under your control given the uncertainties not under your control.

An informed choice takes into account the possibility that damages may be higher—it doesn't mean they won't be. Making a well-informed choice doesn't ensure you'll be lucky in a particular case; only that you'll be better off over the long run when all the case results are averaged out. With these points in mind, how should you think about good decisions and good outcomes?

3.5.1 *The Difference Between Good Decisions and Good Outcomes*

In a nutshell,

- Good outcomes are when the uncertainties turn out the way you'd like.
- Good decisions are when you make decisions that are logically consistent with all the information available to you at the time.

This is a strategy both for making better decisions and for living with their consequences.

It helps you make better decisions by more carefully considering them—using all your intuitive and logical faculties—and by documenting the basis on which you made a choice. After the fact, you can look back and examine whether there was an error in your thinking (such as a bias corrupting a probability assessment) or whether the basis for the decision still makes sense.

The explicit basis for the decision also makes it much easier to deal with fortunate or unfortunate outcomes. Fortunate outcomes are seldom a problem, but realizing you got lucky is helpful in avoiding hubris souring future decisions.

If you are unfortunate, the possibilities of unfortunate outcomes are explicitly laid out for all to see. "We knew it could turn out this way and we showed that possibility." tends to be a very effective way to cover one's backside.

Accordingly, a decision is either good or bad when it's made—and this doesn't change according to how the uncertainties turn out. Sure, things always seem much clearer with hindsight. Remember how they were not clear at the time. You wrote it down in a tree just to be sure.

Certainly, you may know more tomorrow than you do today, and that additional information may or may not be helpful. In Part II, we'll show you how to actually calculate what additional information would be worth—before you get it. Often the answer is "not much." Means to accurately predict or control uncertainties are seldom available, but we can nonetheless calculate what the worth would be ahead of time. This is especially helpful in looking at which legal or factual issues are worth further investigation.

With this background, we can lay out the elements of a good decision.

3.5.2 The Elements of a Good Decision

Separated out, then, making a good decision logically based on all the information available to you at the time means (Figure 3.11):

- The inputs to the decision are the alternatives under consideration, your information and beliefs as expressed in the probability assessments, and your preferences as to the time value of money and attitude towards risk taking.
- The logic to relate it all together should be rigorous, consistent, reasonable, and include explicit valuation to rate the prospects for each decisions.
- Your decision, although the best one possible, may or may not be followed by a good outcome of the uncertainties.

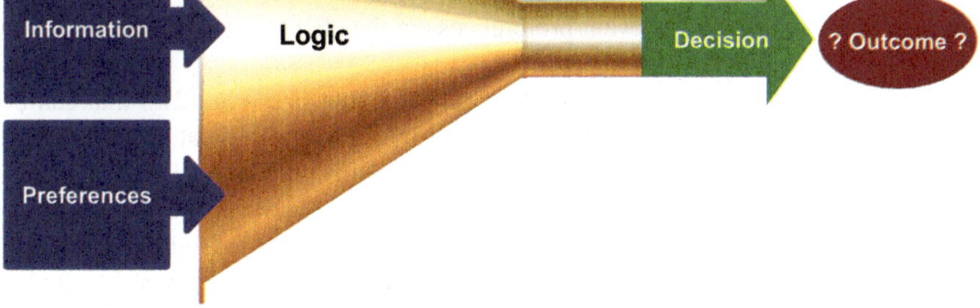

Figure 3.11 The elements of a good decision

3.5.2.1 Alternatives

The alternatives you evaluate should be meaningfully different from one another. Otherwise, there's no real decision. In thinking about whether the alternatives make a difference, it's sometimes useful to think of a decision as an allocation of resources that's only revocable at some cost (that is, no going back for free). In most litigation risk analyses, the main decision is between settling a case and litigating it, though you can also evaluate other decisions (pursuing discovery, getting a teardown analysis in a patent infringement, deposing additional experts, etc.).

To come up with meaningfully different alternatives (such as when pondering different discovery strategies), remember that generating alternatives is a creative exercise and creativity should be encouraged at that stage. Save criticizing alternatives for when you're assessing probabilities and calculating mean values. One alternative you should always include, though, is "Do Nothing." For example, this applied when a matter has come to you attention, but a lawsuit has not yet been filed. Sometimes they never ripen to a suit, and sometimes the statute of limitations runs in the interim. This alternative will help to highlight exactly what you're gaining from the proposed activities and potentially save a lot of client time and money.

3.5.2.2 Information

Information consists of data and probabilities. Data includes whatever you may have (contracts, court decisions, depositions, medical records, etc.) In most cases, that data won't supply the answers you're looking for, such as how a particular uncertainty (a court finding liability or the amount of damages awarded) will turn out.

Accordingly, a key part of decision analysis is going beyond looking at the data to creating probabilities which describe a person or person's opinion and judgment as to how a particular uncertainty may turn out. This critical assessment process will be covered in Part II.

3.5.2.3 Preferences

For decision making, the relevant preferences are the clients's time value of money and attitude towards risk-taking.

The time value of money is expressed as a discount rate (e.g., 2%) that can be used to convert funds being paid or received over time into what the equivalent single sum today is. This single sum today is called a net present value.

These kinds of calculations are required, for example, when you need to convert a future stream of lost income into a sum the court would award today, or likewise to convert a future stream of medical costs into an equivalent amount today. It also applies to calculating prejudgment interest.

The client's attitudes towards risk-taking is expressed as a risk preference (covered in more detail in Chapter 7). If the amounts at stake are not potentially ruinous or a life-changing windfall, we use a risk-neutral attitude which just looks at the mean values (probability weighted averages) for each alternative. If attitude towards risk does need to be considered, using decision analysis you can show the client exactly how the alternative valuation changes with differing attitudes towards risk. Typically, this discussion changes both how the client perceives risk and what they feel is the best alternative, and is thus one of the ways an analyst or the counsel can be especially valuable to the client.

3.5.2.4 Logic

The logic employed should match what the decision-maker would intuitively do in simple situations, and extend that same reasoning consistently to situations with more complexities and uncertainties where his or her intuition would fail. As you've seen, the mean value fits the bill for risk-neutral decision-making. Accordingly, the logic employed should be simply picking the highest mean value alternative, with the possible proviso that the client's attitude towards risk-taking may require a further step.

We cannot emphasize enough that making a good decision doesn't guarantee a good outcome. However, the analysts and the counsel employing decision analysis have a huge advantage over their opponents from far clearer understanding of the case which can almost always be turned to great advantage. Possibly this asymmetric advantage will go away when both sides routinely employ decision analysis but, for now, you have satellite photos and precision guidance and they don't. Make the best of it.

3.6 The Results Should Make Intuitive Sense

After analyzing the decisions and alternatives, the next step is to draw conclusions. Those conclusions may be as simple as "We should settle because the settlement offer is $50 thousand less than the mean value of litigating the case.", or, as you'll see later on, as sophisticated as "It is not worthwhile to conduct any further discovery on the issue of (pick one)." However, those conclusions should always make intuitive sense if you think carefully about them because:

- They draw from the best of your Affective/Intuitive thinking expressed in a Deliberative/Logical thought model to explore the possibilities.
- They follow transparently from the pieces of the decision basis (the elements of a good decision).
- They tell a story you understand, believe, and can relate.

Put another way, a Decision Analysis don't tell you what your opinion *should* be. Rather, it helps you to improve, clarify, and communicate yours' and others' information, and to draw consistent and correct conclusions from that information.

3.7 How to "Sell" Your Case

Any intelligent fool can make things bigger, more complex, and more violent. It takes a touch of genius—and a lot of courage—to move in the opposite direction.

E.F. Schumacher, The Radical Humanist

There are two parts to "selling" your case:

- Communicating the results to the client (if the counsel is doing the analysis) or to the counsel and the client (if an analyst).
- Selling the need to do a decision analysis in the first place.

For the first part of communicating the result, remember that the purpose of decision analysis is to create insight as to which course of action is best and why. If you understand that, you can express in 2–3 sentences that even the busiest and most senior executive will want to hear. The presentation consists of a few summary pages and 50 pages of appendices detailing how you got there. More on communicating results is found in Chapter 12.

To get there, though, you may need to convince someone that an analysis is merited in the first place, and that undertaking one does not denigrate or question counsel's wisdom, competence, and experience in any way.

Possibly you can lend this book to someone to read just the first part. Or you may need to pitch it.

The pitch is straightforward: Decision Analysis is fundamentally simple and sensical. It does exactly what you would do intuitively in simple situations, and is necessary to extend that same kind of judgment to more uncertain and complications situations where your intuitive fails. The following sections offer more detail to motivate this conclusion.

Doubters might also be assured that, although this procedure will help you exercise better judgment, all the usual aspects of good lawyering are still required: diligence, specific legal knowledge, attention to client's needs, etc.

3.7.1 Why Is Decision Analysis Different from Other Methods?

- Decision Analysis offers clear language and concepts for dealing with uncertainty and complexity.
- Decision Analysis offers a transparent process and symbols for capturing judgments about uncertainties and the relationships between complexities.
- Decision Analysis uses simple and transparent logic to relate everything together in a way that yields often surprising but ultimately explicable and sensible results.

3.7.2 *What Benefits Does Decision Analysis Offer?*

Decision Analysis is a disciplined approach that can assist analysts, attorneys, and clients in planning their litigation and settlement strategies. The techniques of carefully disaggregating a case and logically capturing the key factors in relevance diagrams and decision trees, and then of realistically and explicitly evaluating each of the key factors with probabilities, offer tremendous benefits:

- Individual judgments can be properly integrated to determine the overall value of a case for settlement purposes.
- Through sensitivity analyses, the counsel can efficiently manage cases during the pretrial and trial phases.
- As new information is developed in pretrial discovery and counsel's judgments change, the case value can be rapidly updated and any shifts in strategy that are indicated can be quickly identified.
- The client can easily understand the logic of counsel's recommendations because the thought process and underlying judgments have been unambiguously set forth.
- Cases can be settled more quickly and more favorably because the counsel understands the case better then their opponent and can use that understanding to drive negotiations.

3.7.3 *Where Have These Kinds of Techniques Been Used?*

Decision Analysis is standard and required in the pharmaceutical industry for drug development projects and in oil and gas exploration. In more than three decades as a professional decision analyst, I have actually done A (automobiles) to Z (zinc mine) in addition to the many applications in litigation. Here are some of my legal applications:

- Multibillion dollar environmental liability
- Shareholder derivative suits
- Patent litigation
- Software licensing and copyright litigation
- Medical malpractice
- Privacy breach
- Pension funds
- Workers' compensation
- Construction disputes

Whatever legal area you practice or are assisting in, Decision Analysis is applicable. It applies to any decision. Some can made with quick application of the key concepts (no analysis required). Others may required more detailed work and even computer modeling.

Here's wishing you all the success in applying Decision Analysis to make better decisions, drive better outcomes, and win at litigation.

3.8 A Note on Game Theory

I am frequently asked about the possible application of game theory or negotiation analysis to litigation problems rather than the decision analysis approach presented in this book. Here is how Howard Raiffa summarizes the various approaches[3]:

- Decision analysis (a prescriptive approach—how an analytically inclined individual *should* and *could* make wise decisions)
- Behavioral decision making (a descriptive approach—the psychology of how ordinary individuals *do* make decisions)
- Game theory (a normative approach—how groups of ultra-smart individuals *should* make separate, interactive decisions)
- Negotiation analysis (how groups of reasonably bright individuals *should* and *could* make joint collaborative decisions)

Certainly, the final outcome of almost all civil cases is the result of a negotiation. Almost all cases settle before trial. Even after a trial proceeds to verdict, there is usually a negotiated settlement to end the appeal process.

Negotiation strategy is a primary focus of this text: how much should each party be willing to settle for versus proceeding? A decision analysis furnishes the basis for developing a quantified negotiation strategy.

What about a game theory approach instead? To answer this, we need to focus on the necessary elements for a game theory analysis:

- Multiple parties make separate decisions.
- The payoffs for each player are dependent on their own decisions and on the decisions of the other players.
- The rules of the game (the possible alternatives and their joint impact on respective payoffs) are set and cannot be changed during the game.

Thus, for example, game theory works quite well for online auctions (Google ads etc.).

However, though there are certainly game-like aspects, disputes and litigation do not seem to quite fit the game theory model. The parties can and almost always do negotiate an agreed settlement rather than causing a result through their separate decisions. Decisions by the court greatly affect the "game," though the court is not a player in the sense having a stake in the outcome other than a general interest in upholding the law.

Most importantly, the rules of the game aren't set but, rather, evolve throughout the process. Alternatives change as the case develops and valuations change continually (new evidence develops, new settlement demands are made, claims and parties are added or dropped). The legal process and the actions of the parties continually reshape the case and change what is at issue and for how much.

---

[3] Howard Raiffa with John Richardson and David Metcalf, Negotiation Analysis: The Science and Art of Collaborative Decision Making. The Belknap Press of Harvard University Press. Cambridge, Massachusetts, and London, England: 2002.

The game-like aspect that should be considered and is covered in this text is looking at a case from your opponent's perspective, and that is indeed very valuable. However, I've not yet seen a formal, analytical game theory application to a lawsuit or dispute. I would be very curious to see what one looked like.

Lastly, I would note that, in the one litigation example in Dr. Raiffa's book (in Chapter 8), he applies a decision analysis approach exactly along the lines presented in this book.

3.9 Problems

1. Explain why relevance diagrams are similar to or different from other diagramming techniques you've used. Give at least one example of another problem diagramming technique you've learned or used.
2. List at least one decision-making technique (besides relevance diagrams) that breaks the problem up into smaller pieces (A sophisticated example is the management science technique of dynamic programming which, for instance, can determine an "optimum" inventory policy by starting with the desired final inventory levels and working backwards through the steps to reach it.).
3. Consider and list at least one or two other decision-making aids you've encountered (a horoscope is perhaps the most widespread). Does the aid tell you what to do, and, if so, is it clear how the recommendation is reached? Or does the aid only furnish information to aid in making your own decision and, if so, is it clear how the information should be incorporated into your decision-making process?
4. The decision analysis techniques used in decision analysis have been widely applied, in areas ranging from new product introduction to space exploration, and many decision analysts use the techniques to make decisions in their personal lives. List at least three areas in your personal life where you think decision analysis might be of use to you, and three areas where you think it wouldn't be. Either discuss the list with your instructor, or keep it on hand as you go through this book. You may surprise yourself.
5. List at least one or two personal decisions where the numbers of details and factors make it difficult to make the best choice. One common example is choosing where to live. In the case you listed, how would you reconcile and relate the factors together to make a decision? Are you satisfied with the method and results?
6. How would you explain to someone else what a decision tree is and how it's used?
7. Decision trees explicitly calculate different values for different scenarios. How is this the same or different than how you would typically look at a prospective decision?
8. Decision trees typically calculate scenario values in terms of dollars. What do you think are the advantages and disadvantages of this method?
9. Discuss how a mean value is different from an arithmetic average. Are these differences advantages or disadvantages?

10. Suppose someone offered you a chance at a coin flip where you'd win $10 if it came up heads and nothing if it came up tails. How much would you be willing to pay for a chance at this flip? Is the answer intuitively obvious to you, and why or why not?
11. Suppose you were playing blackjack and had been dealt an eight and a four (both face down). You now have a chance to double your $5 bet. Is it intuitively obvious what you should do? Can you explain your answer? Sketch a decision tree for this problem (you don't have to fill in all the numbers). Does the decision tree help you without all the numbers filled in and the mean value solved for?
12. How does the method used by Decision Analysis for creating an understanding of which alternative is best and why differ from how you would usually justify a proposed plan of action?
13. Pick a particular kind of legal problem you've had occasion to face, such as responding to a motion for summary judgment, or, if you're a student, perhaps deciding which classes to take next semester.

    13.1 How do you go about solving the problem? Discuss whether and how you do such steps as establishing a criteria for evaluating alternatives, generating alternatives, and evaluating them. Would anything more formal seem helpful, and why?
    13.2 Contrast the solution method you described in part (a) with the five steps of a decision analysis. Do you think the LRA methodology is applicable (or would be helpful), and why or why not?
    13.3 List and describe briefly the kinds of concerns you'd like a decision-making method to address. How does decision analysis do so, and how does your method from part (a) compare? What has been left out that you'd like addressed?

14. Imagine that you're an English lord or lady with an opportunity to sail on the maiden voyage of the Titanic. The voyage has been sold out for months, but a friend has offered you two tickets. You know that the voyage will be the social event of the season and offers a rare opportunity to be in with the uppermost crust. You suspect that profitable business deals may be hatched over late-night brandy. You know that the ship does not carry enough life boats for all the passengers, but it is not standard practice to do so, and the Titanic boasts a new system of waterproof compartments, prompting the claim that it's "unsinkable." Because of your membership in the local Royal Geographic Society, you also know that icebergs drift in the North Atlantic at that time of year, but you've never heard of a ship being sunk by one. Assume that history will not necessarily repeat itself, and thus that the odds of the Titanic striking an iceberg are the same as any ship randomly hitting one.

    14.1 Describe a process of reasoning by which it would be a good decision to sail on the Titanic.
    14.2 Describe a process of reasoning by which it would be a bad decision to sail on the Titanic.

14.3 How confident are you in your decision? List the most significant uncertainties and values at stake. Would you feel more confident if you assessed numerical uncertainties and values and saw how the decision came out? Would it further help you to see over what ranges of values and probabilities the decision is to sail? (This is what sensitivity analysis does.)

15. Consider some of the decisions you've made in your professional or personal life, and some of the twists of fate you've endured.

    15.1 List one or two bad outcomes you've suffered. Did the outcomes result from a decision you made? With hindsight, can you tell if it was a good decision or a bad one?
    15.2 List one or two good outcomes you've enjoyed. Again, can you tell if they followed a good decision or a bad one?
    15.3 List one or two bad decisions you've made. What information did you ignore or not properly take account of?
    15.4 List one or two good decisions you've made. Were they really good decisions because you logically considered your alternatives, information, and values, or because a good outcome followed? How would you make it differently if you were to make the exact same decision again?

16. How easy or difficult do you think it is to tell after the fact that a decision was good or bad without having explicitly laid out the basis for the decision?
17. Consider the reward structure at your workplace or school. Does it reward good decisions or good outcomes? List at least two advantages and disadvantages to rewarding good decisions, and the same for rewarding good outcomes. What implications would a switch from one system to the other have for your own behavior?
18. Is it clear how the results of the individual pieces combine to produce a good decision? What shortcomings do you see in the technique?
19. The traditional view of the split of authority between lawyer and client is that the client makes decisions in matter dispositive of the case (with counsel's advice), while the lawyer makes decisions on how to proceed with the case. Classify each of the decisions below according to whether they fall more to the lawyer or to the client, and list for each some of the issues and sources of information in making the decision. Also, for each client decision, describe how the relevant information can be communicated to client. Do you think the communication means you listed are adequate?

    19.1 Whether or not to settle a case for a specific amount (for the plaintiff).
    19.2 Whether or not to settle a case for a specific amount (for the defendant).
    19.3 Whether or not to depose the opposing party's mother.
    19.4 Whether to appeal a decision.
    19.5 On what grounds to appeal a decision.
    19.6 Which legal claims to include in an initial case filing.
    19.7 Which legal claims to move for summary judgment on.
    19.8 Whether your client-defendant should testify in a criminal case.

19.9 Whether your client should test a grey area in I.R.S. regulations that could save him $15,000 on his tax return.

19.10 Whether your client should breach a contract with a long-time customer where the client would lose $50,000 in performing as specified, and the contract contains a liquidated damages for breach clause of $10,000.

20. For each part of problem 19 that you listed issues for, try relating those issues together in a simple relevance diagram (You'll get more rules for doing this in the next chapter.). Do you find that drawing the diagram requires more thought than just listing the issues? Did any new issues occur to you?
21. For each of the decisions in problem 19, list the remaining elements of a good decision (you've already got the uncertainties). Does doing so force you to think even harder about the problem? What are some potential sources for the elements?
22. A probability of 0.0 or 1.0 can be interpreted as meaning you're so sure that you're willing to stand up and be shot dead if you're wrong.

22.1 List at least three events or conditions you're that sure about. Now come up with at least two reasons why you could be wrong (There is always at least one reason even if it seems utterly fantastic to you now.) Are you really that sure, or would a .98, for instance (two chances in a hundred) better describe your (un)certainty?

22.2 Does it make any difference that an event is in the past or in the future? List at least two pairs of past and future events where you're more certain of the future event than of the past one. (Example: an asteroid crash wiped out the dinosaurs, and the sun will rise tomorrow.)

23. Make a list of reasons why somebody else (anybody else) could be in your home right now, and a list why nobody could be home (this process will be covered in more detail in the chapter on probability assessment.) Write down a number between 0 and 1.0 for the probability that someone else is home right now. Now classify each reason according to whether it's a fact you know, a conjecture based on your intuition and experience or an opinion. Is it easy to divide up the reasons this way? Does the classification make a difference in arriving at your probability? Is there a way other than the probability to relate together all the reasons in a clear estimate?
24. Many government agencies (EPA, OSHA, etc.) are required to do cost–benefit analysis on proposed regulations before promulgating them. Do you think that they should be comparing mean-value costs and mean-value benefits, and why or why not? Do you see any problems with this kind of comparison, or anything that's been left out?
25. Suppose you've been asked to help on a case that's similar to one a senior partner handled a number of years back. You're in charge of the case on a day-to-day basis, with that senior partner supervising. What factors would lead you to think that a decision analysis would be helpful? How would you put it to the senior partner? (If you're a senior partner, assume you're talking to another senior partner.)

Part II
How To Do a Decision Analysis

Chapter 4
Structure the Problem

You have seen in Part I why something more than intuition is needed for dealing with decision making under uncertainty and complexity, and why Decision Analysis is a great solution to the problem. Other great solutions include omniscience and omnipotence. Unfortunately, they tend to be in short supply.

In seeing why Decision Analysis is a great solution, you have also seen a general explanation of the steps in a Decision Analysis. As the heading indicates, Part II goes through that same process step-by-step, but in more detail than in Part I. Those steps (the chapter headings) are:

1. Structure the Problem;
2. Elicit Probabilities;
3. Integrate All Information, Judgment, and Consequences;
4. Analyze the Results.

Structuring the Problem then further breaks down into three steps that, together, determine the important uncertainties, their interrelationships, and their consequences:

- Draw the Relevance Diagram;
- Draw the Decision Tree;
- Write Value Formulas;

If you're really unsure of how all the steps fit together, you can either go through the detailed discussion in this part, or refer back to Part I for a briefer and more general discussion. You may also find it useful occasionally to refer to the cognitive and motivational problems discussed in Part I to help you understand why the procedures detailed in this part are necessary, and how they deal with those problems.

J. Celona, *Winning at Litigation through Decision Analysis*,
Springer Series in Operations Research and Financial Engineering,
DOI 10.1007/978-3-319-30040-5_4

4.1 Draw the Relevance Diagram

Part I has given a general introduction to relevance diagrams. This section expands on that background to make you a pro diagram-er, first, with a closer look at what the diagram aims at and what the symbols mean; second, with the procedure for drawing relevance diagrams; and third, with hints for drawing better relevance diagrams.

4.1.1 Introduction to Relevance Diagrams

4.1.1.1 What's the Problem?

The starting point in drawing any relevance diagram is deciding what the problem is; or, more specifically, what are the ultimate results you want to understand better. For most legal problems, this step is usually not difficult because that legal nature tends to focus on the likelihood of winning or losing a case and the value of the case.[1]

It is important, though, to focus on what's going to happen to *you*—win, lose, the amount of the judgment, etc.—and *not* on how the complaint is laid out. The complaint is your opponent's attempt to frame the case in a legal structure most favorable to his client. Your object is to decide what to do according to the chances of different outcomes happening, and those outcomes are dependent on what you, your opponent, the trier-of-fact, the witnesses, etc. do along the way. You don't want to lose that focus on better understanding the end result, nor do you want to get lost in the details of the complaint. As you'll see later, you usually don't need the complaint's level of detail to draw meaningful conclusions (you can look ahead to the discussion of Focusing on What's Most Important if you're really curious.)

Once you've decided what the problem is, the next step is to draw a relevance diagram for it. This section explains what goes into a relevance diagram and how they work, before part B gives you the detailed procedure for drawing a relevance diagram.

4.1.1.2 Showing Dependencies

The ovals in a relevance diagram represent uncertainties. Those uncertainties are labeled by putting a question in the oval, like "Liable?" or "Conduct found reckless or intentional?" Every question should be answerable by yes, no, or a number. This test is called the ***Clairvoyant Test*** because it asks whether a ***clairvoyant***

[1] Business problems, in contrast, usually focus on the net present value of various business opportunities—but they also may have unusual focuses, like determining the demand of some product to help plan capacity for producing that product.

(someone who sees all) could give you a simple answer to your question, or would have to ask you what you meant.

"Our first theory." is obviously not a proper question, nor is "codefendant." (Something like "Win on first theory?" or "Co-defendant held liable?" would do the trick.) This yes, no, or number requirement must be strictly followed to figure out what happens in the different possible scenarios.

An arrow indicates dependencies among the uncertainties. For instance, in the diagram below, the arrow indicates that your chances of being held liable depend on whether or not particular evidence is let in (Figure 4.1).

You can tell whether you've put down the questions and arrows correctly by whether you can draw the decision tree for them. For this diagram, you start at the bottom with whether or not the evidence is allowed in, then follow the arrow to the question of whether you're held liable. The combinations of answers to the two questions produce four scenarios, as you can see in the tree (Figure 4.2).

Decisions are indicated in the diagram with boxes. Thus, if you had a chance to settle a case, you'd indicate that decision with a box and arrows as in the diagram below. Here, the value of the case depends on whether you litigate or settle (the value is the settlement amount of you settle), and on whether you're held liable if you litigate. Note the arrow from the decision to the liability uncertainty. It's there because you never find out if a court would find you liable or not unless you go to trial. The arrows should correspond to the order things will happen, or the order you'll find out how uncertainties turn out.

The hexagon shows the value measure you'll use to quantify outcomes and compare alternatives (Figure 4.3).

Again, a little tree sketch will help you decide if you've made reasonable sense out of your problem. Below, you start with the litigate or settle decision, then find out your liability only if you decide to litigate. But what happened to the value of the case? (Figure 4.4).

Figure 4.1 Simple start to a relevance diagram

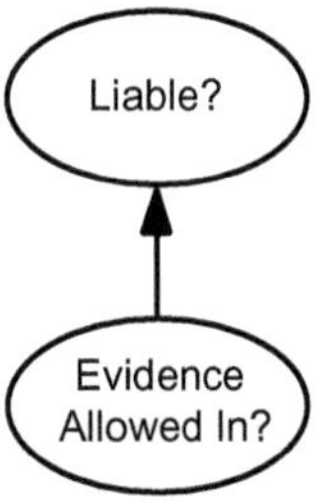

Figure 4.2 Simple relevance diagram drawn as a decision tree

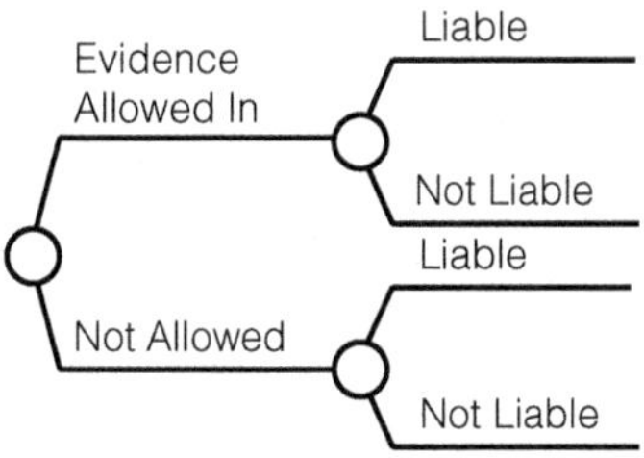

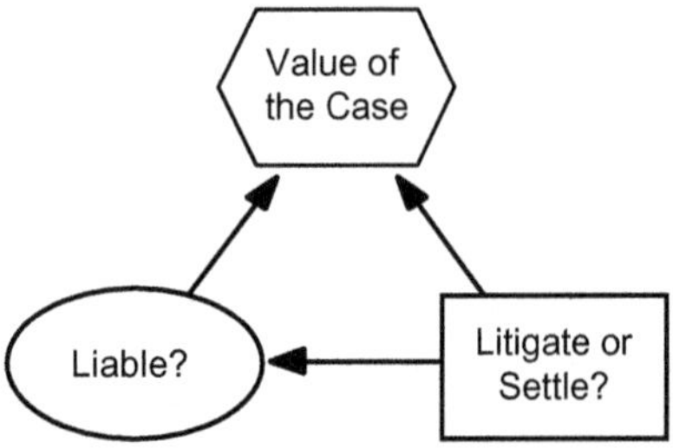

Figure 4.3 Adding a decision on litigating for settling, plus a value measure

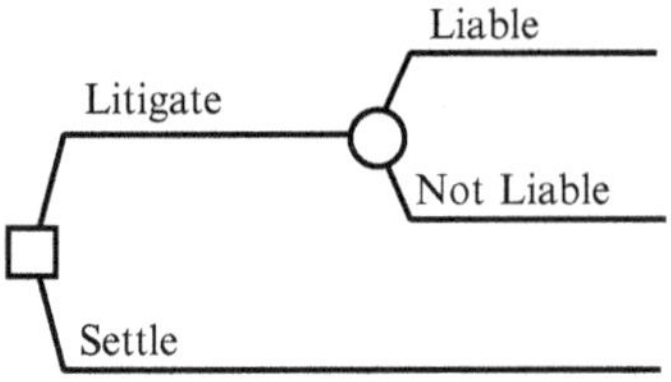

Figure 4.4 Decision tree for litigating or settling a case

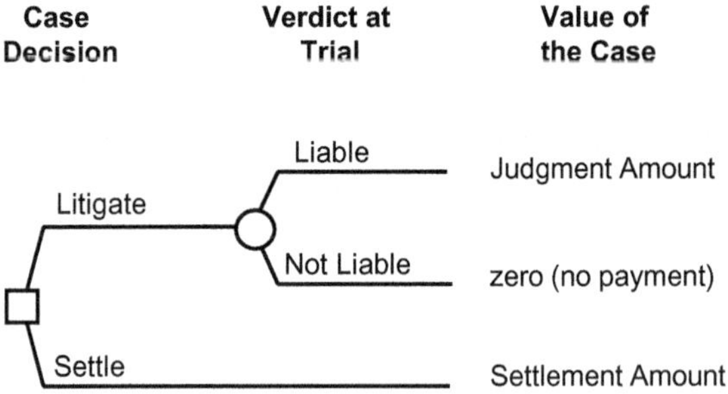

Figure 4.5 Decision for litigating or settling with value calculation

The value of the case is determined by your decision and the answers to the uncertainties. It's a calculating (possibly a simple one) for each of the three scenarios shown in the tree. If you settle, the value is the settlement amount, and the value if you litigate and lose is the judgment, as shown below (Figure 4.5).

Note that this value calculation is only for indemnity payments (potential damages). It does not include costs (legal fees, expert witnesses, etc.). We will include those, but are starting simply and building the analysis as we need to. This is also a very good recommendation for how to start your own analyses. Recall the admonition to make the analysis as simple as possible—but no more so!

4.1.1.3 Finding Dependencies

Drawing the arrows is easy to do and often harder to explain, but thinking about the dependencies is one of the most important steps in the whole process (part of breaking the baffling combinations down into tractable factors). Here are two tests to help you find dependencies.

- **Intuitive Test**: does knowing how one uncertainty turned out help you in thinking about the other?
- **Numerical Test**: do your probabilities for one uncertainty change depending on how the other turns out (on how a prior issue turns out)?

A "no" answer to both these questions means you don't think there's any dependency; that is, the factors are independent. If you really think so, then you can treat the factors separately and not worry about possible interactions.

But do think hard about the question. Although it is easier to treat factors as independent of each other, recognizing and analyzing dependencies often provide some of the most helpful and meaningful insights in the whole analysis. In my experience, dependencies are almost always present, and almost always of critical importance to the best decision when they are.

4.1.1.4 Testing Dependencies

If you have indeed posited a dependency (drawn an arrow), you should make sure you understand clearly what you mean. Start with the clairvoyance test (described above). Does the uncertainty really pass the clairvoyance test on its own? For instance, "win on a motion to exclude evidence?" and "win on motion for summary judgment?" are two separate questions that the clairvoyant could answer yes or no on separately.

In contrast, people occasionally write down as two separate uncertainties questions that are vague whacks at the same log, like "Try a conspiracy theory?" and "Use conspiracy evidence?" (If you asked what these questions meant, that's the right answer.)

The most common effect of a dependency is it changes the odds, and this is also often helpful in thinking about it. For instance, in the very first relevance diagram shown (repeated below), the arrow could indicate that your chances of being held liable go down (or up) if the evidence is allowed in (Figure 4.6).

4.1.1.5 Special Dependencies

Beside increasing or decreasing the odds of an uncertainty coming out particular ways (making various possible outcomes more or less likely), the dependency arrow can represent other kinds of relationships. For instance, an arrow may only apply if

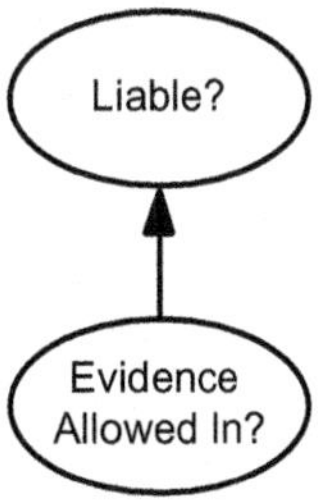

Figure 4.6 Chances of liability depend on evidence allowed in

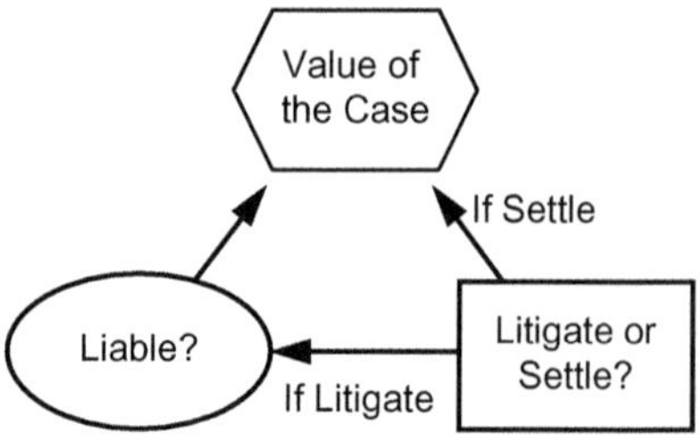

Figure 4.7 Labels to clarify dependencies

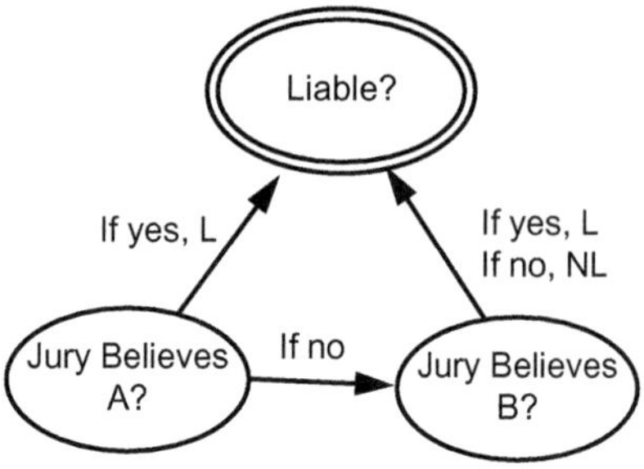

Figure 4.8 Uncertainty determined by other factors shown with a double line

the answer to a question is "yes" (or "no"). The first relevance diagram with a decision sneaked one of these by you. The diagram appears below with the extra labels on the arrows (Figure 4.7).

Of course, if the question were simply "Litigate?" the labels would only then need by "If yes" going to "Liable?" and "If no" going to "Value of Case."

You could also have an uncertainty that's only uncertain until you know the answers to the questions it depends on. In the diagram above, the value of the case was only this kind of uncertainty because there was no damages uncertainty (a rare case). Below is a more realistic example. Suppose there are two bad witnesses that the jury may or may not believe, but you'll be liable if the jury believes either one of them. Thus, once you know the answers to which witnesses were believed (A, B, or neither), you know whether you are liable. The relevance diagram for this situation appears below, with the extra clarifiers of "L" for liable and "NL" for not liable added after the if yes' and if no's (Figure 4.8).

Note that this diagram doesn't include the possibility that the jury believes both A and B. That's because, in this example, the jury won't ever get to the question of B's credibility if they believe A: they will conclude you're liable on that basis. Accordingly, the additional complication of both believed isn't necessary because it doesn't affect what happens to *you*. This is the ***Trier-of-Fact Rule***, which will be emphasized throughout this book. The Trier-of-Fact Rule says do your analysis the way you think the trier-of-fact will think about your case. You don't care about including every nuance you can think of, just like you don't care about analyzing every detail in the complaint. Consequently, the tree below is all you need (Figure 4.9).

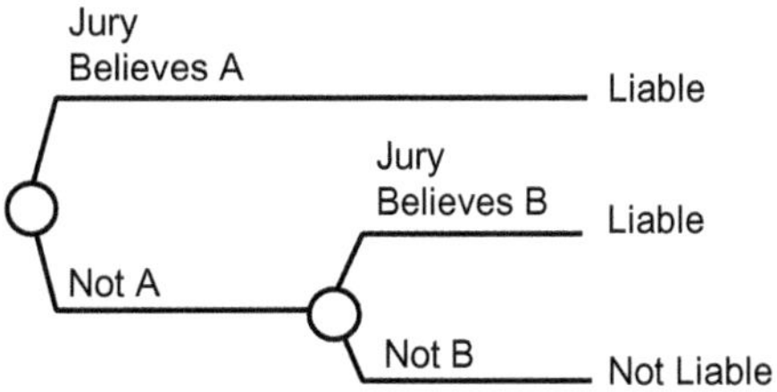

Figure 4.9 Tree for two ways to lose (either or)

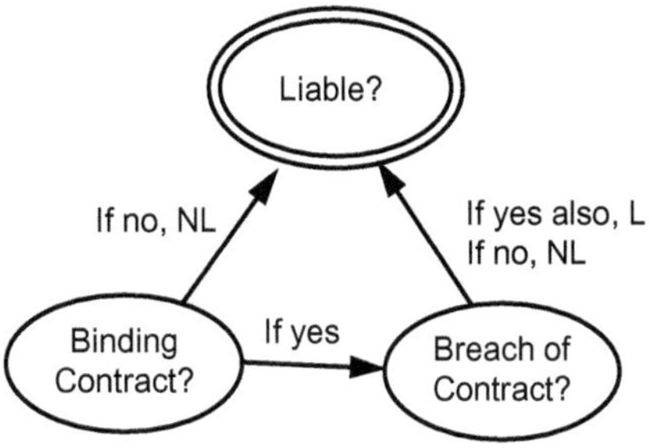

Figure 4.10 Diagram for need both to lose (an "and" dependency)

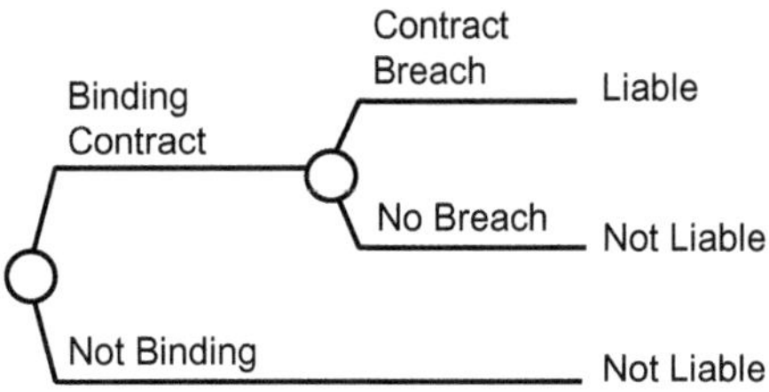

Figure 4.11 Tree for need both to lose

Another common possibility is the case were you only lose if both issues are found against you, such as the existence of a binding contract and a breach of that contract in a contract case (assuming no excuse, etc.). The relevance diagram for this situation is found below (Figure 4.10).

Similarly, you don't even get to the question of whether there was a contract breach if the contract is found not binding, and the tree below is all you need (Figure 4.11).

Note that the labels you place on the arrow depend on how you frame the question. For example, if the question in the above diagram was "Not Liable?" rather than "Liable?" it would describe a situation where you could win if either question was answered in your favor, rather than only losing of both are answered against you (Figure 4.12).

Whether you should write down the question as "Liable?" or as "Not Liable?" is a ***Framing Question***. A framing question arises when you have a choice between representing the problem two different ways that are formally equivalent

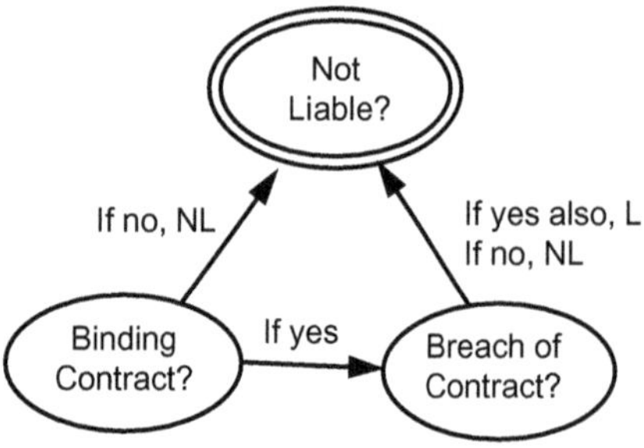

Figure 4.12 Changing the framing changes the dependency

(that is, they're the same as far as the diagram is concerned). The trier-of-fact rule would apply to say that you should frame the question the way the jury would think about it. The way the jury considers the issues often depends on who has the burden of proof. For instance, in a contract case, you might have to prove "Excuse?" or "Lack of Excuse?" Thus, both the burden of proof and the way the jury will think about the issue (when not specifically instructed) should be reflected in the way you frame the questions in the relevance diagram.

The same decision tree can still be used to represent the relevance diagram. Do you understand why this is true?

4.1.1.6 Key Influencing Facts

Strictly speaking, your relevance diagram should only include uncertainties and decisions. However, you may want to include key influencing facts (not uncertain) in the diagram, either to remind yourself where they've been considered or to show someone else you've taken them into account. For instance, if the plaintiff in an unfortunate accident was blinded for life at the age of 6 years, that fact may influence so many other uncertainties that it will help your thinking to include it in the diagram. You can include it, as below, likewise with a double line around it to show it's not an uncertainty (Figure 4.13).

These influencing facts may also appear in your tree drawing in the list of reasons under each branch explaining how that outcome could have happened (see section on drawing decision trees, below).

4.1.2 Howitzer v. Crabgrass et al.

Throughout the remainder of Part II, a hypothetical case will be used to illustrate each Decision Analysis step as it's discussed. The case contains elements from a number of actual cases (as well as pure fiction)—all added to illustrate various aspects of a Decision Analysis. Henceforth, all discussion of the case will appear in italic type to distinguish it from the general discussion.

Figure 4.13 Relevance diagram with key influencing fact

Dominguez Howitzer is a veteran and six months ago bought a house in the town of Sulphur Springs from Wilbur Crabgrass. He financed his purchase with a loan provided by the Veteran's Administration under a home-purchase-for-veterans plan. Zelda Suave of Throckmorton, Plimpton Real Estate represented the seller Crabgrass. Dominguez was not represented by a broker.

It was a first-time home purchase for Dominguez, and, although he was somewhat handy with home repairs and inspected the property himself, he was leery about buying without having a professional contractor inspect the property for him. However, he claims he was dissuaded from doing so by Zelda, who assured him that "The V.A. has to inspect anyway, and any problems they find will be fixed by Crabgrass before closing." An inspector from the V.A. did, in fact, look over the property, and found no problems with it.

After living in the house for several months, Dominguez claimed there were problems with the house, and retained a lawyer to pursue damages. After Crabgrass balked at Dominguez's demand for $100,000 to remedy the "defects'" in the house, Dominguez' s lawyer filed suit against Crabgrass, Zelda, and Throckmorton, Plimpton, alleging intentional misrepresentation, negligent misrepresentation, and violation of the Consumer Fraud Act. He is demanding $200,000 in compensatory damages and is also asking for punitive damages. You represent Throckmorton, Plimpton.

You note first of all that the general law regarding house purchases is caveat emptor, and there is no general duty to disclose. Accordingly, Dominguez's cause of action focuses on misrepresentation in Zelda's statement about the V.A. inspector. You assume that Throckmorton will be held liable for any wrongdoing of Zelda via respondeat superior.

As for the alleged misrepresentation, you're not sure if the V.A. is actually required by regulation to inspect any property they're financing, or if doing so is merely a custom (Or is there some unique reason that the V.A. inspected this house?). The fact that a V.A. inspector did inspect and didn't find any defects is a curious twist, raising the questions of whether the property does in fact have any defects, and of whether you can be misrepresenting if what you say turns out to happen—even if you were lying at the time. On Zelda's side, she has been selling real estate for fifteen years and has never been sued before. Could it be that Zelda knew of these "defects" that eluded both Dominguez and the V.A. inspector and tried to dissuade Dominguez from inspecting to conceal them and make the sale? Would she have done so by lying about the V.A. inspecting, or was she only mistaken? Does the house actually have any defects? And what's the scoop on private causes of action under this Consumer Fraud Act? You glance up at the clock, see that it's 6:45 p.m., and decide to call it a day.

4.1.3 Procedure for Relevance Diagrams

Now that you've seen how a relevance diagram works, the following steps provide a general way to build your diagram while maintaining the proper focus. The section after this one then gives more hints for drawing better relevance diagrams.

- Ask Yourself: What is the ultimate outcome you're most interested in? How would you like to value scenarios and evaluate alternatives?
  - The value measure goes in a hexagon at the very top of your relevance diagram.
  - Typically, you'd most like to know the value of the case, or the chances that a defendant will be found liable or not.
  - Each relevance diagram must focus on *one* value measure or uncertainty you're trying to understand.
  - Do a separate relevance diagram if there's another uncertainty you'd also like to study. For instance, you might do separate diagrams for the value of a case and the chance of the plaintiff developing a particular medical condition 20 years hence—although the conclusion from the medical diagram may go into the case diagram.
- Can you answer that question directly, or do you need to know more?
  - If you can answer the question directly, leave the hexagon or oval alone.
  - You don't have to know what the answer to the question will be. What the answer will be is uncertain; you should just be able to say that the answer will be a simple yes or no.
  - Note that this is another application of the clairvoyant test: can the clairvoyant give you a straight yes or no or numerical answer, or is he going to ask you another question because you've left out something important?
  - If you can't answer the question directly, then write underneath the further ovals and arrows for the other questions you need the answers to (thus adding another level to the diagram).
  - Generally, you should focus on issues that are important because they greatly affect the overall chance of winning or losing, or the financial consequences of losing.
  - Your questions can include legal or factual questions, or issues of liability or of damages.

- Repeat the step above until you've answered all the questions.
  - You will often get to the evidentiary level before deciding you have information to answer the questions directly.
  - Again, remember that being able to say what the possible answers are doesn't mean you know the answers: the answers (outcomes) are still uncertain. You've just arrived at the level of detail where you no longer need to ask about other factors the question depends on.
  - You diagram may include key influencing facts: not uncertain, but included because they greatly affect other uncertainties.
  - You may find more influencing factors when you draw the decision tree or assess the probabilities.

You arrive in the office to find a settlement demand from Howitzer's lawyer for $100,000. What could be wrong with that house? You decide it's time to start analyzing the case.

You pull out your Decision Analysis book and begin with a relevance diagram: what would you most like to know? Hmm... Whether the house has any defects and whether Zelda was lying certainly are important questions, but maybe you'd rather know whether you'd win the case. But wait—you glance through the book and realize you really want to know the value of the case so you can decide whether or not to settle it. Better to start there.

You decide you're not comfortable assessing the value of the case directly (step 2). The clairvoyant would want to know if you were held liable or not, and what the damages were. The value of the case, then, depends on whether you're held liable or not, and what the damages are. It occurs to you that you might get some apportionment of the award between Throckmorton (your client) and Zelda as an individual. You're not sure whether you should bother with that in your first-pass analysis, so put it off to the side, producing the first level of the relevance diagram as below (Figure *4.14*).

You then begin to think about the question of liability. Are you comfortable assessing directly whether you'd be held liable or not? Not with those three different theories of liability floating around. But how to draw them in? The intentional representation turns on the infamous "Five Fingers of Fraud" (a false statement, of fact, intentionally or recklessly made, that the plaintiff reasonably relied on, causing

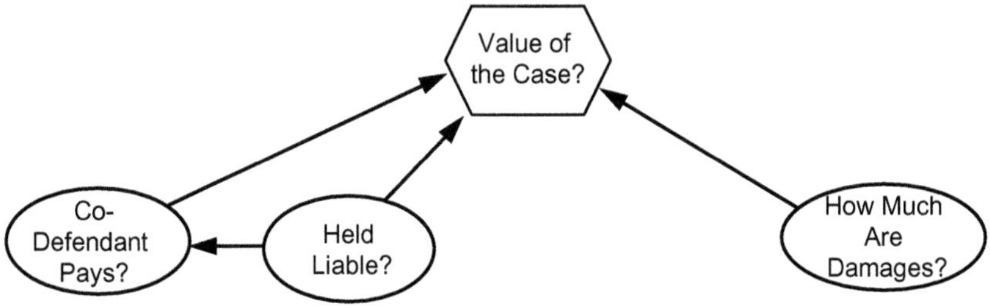

Figure 4.14 Start of relevance diagram for *Howitzer v Crabgrass et al.*

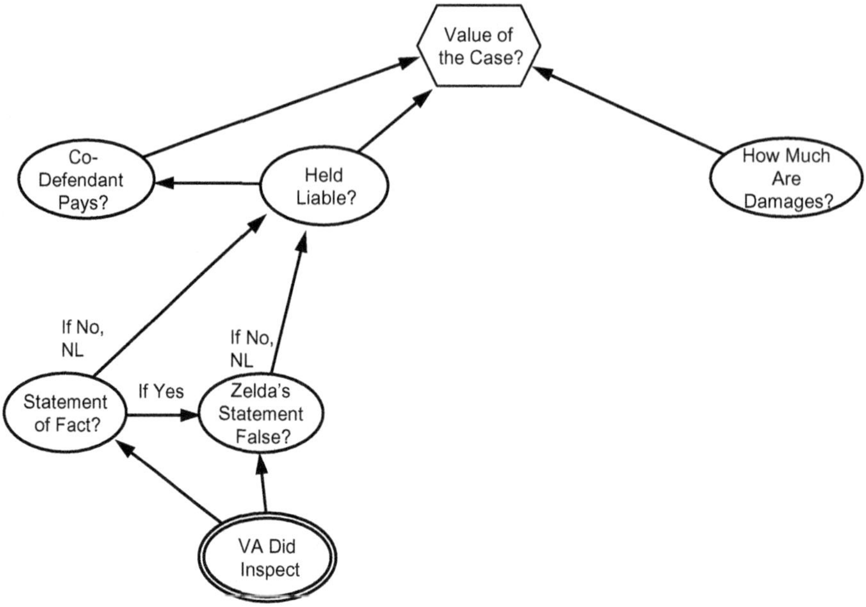

Figure 4.15 Adding Zelda's false statement and VA inspection

damages as a result).[2] *The negligent misrepresentation charge also requires falsity, reasonable reliance, causation and damages, but the misrepresentation need only have been negligently made. And the Consumer Fraud Act, you discover, only requires that the agent have been a conduit of false information from seller to buyer.*

You turn these considerations over for a few minutes and decide that the first question in determining liability is whether Zelda's statement was false. You decide to put that down, including the key influencing fact (the double oval) that the VA did inspect. You remember to put down the label "If No, NL" to indicate there's no liability under any theory if Zelda's statement was true. You then apply some hornbook law: actions for misrepresentation do not lie in misrepresentations of opinion or law. You consider whether this question is too technical for the trier-of-fact, and whether they will just look focus on the falsity question. You decide it has to be included because it's a potentially dispositive issue you need to consider and one you might even get summary judgment on (Figure *4.15*).

You next consider the liability theories themselves. You decide that causation is not going to be an issue (defects in the house is cause enough), the damages, if any, are a separate question, and the reasonable reliance will be subsumed under the scienter considerations. You are then left with the questions of intention, negligence, or merely acting as a conduit, all conditioned on a prior finding of a false statement (Figure *4.16*).

You then consider what else you'd like to know to decide the misrepresentation questions, and the issue of defects comes to mind. A colleague had argued to you that

[2] Check these.

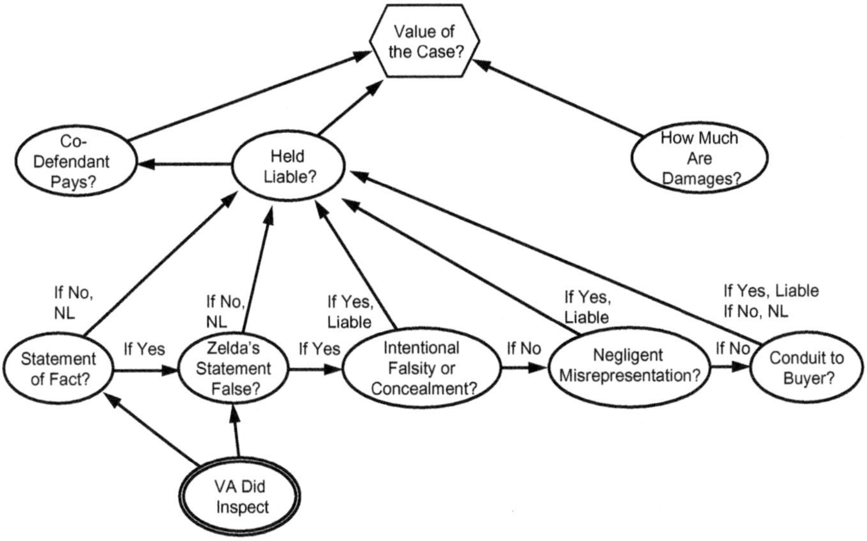

Figure 4.16 Adding alternative theories of liability in the order you feel the trier-of-fact will consider them

you could defeat the whole case on the defects question. That is, there is technically only liability if the property has latent defects (defects not readily apparent). There is no liability for any misrepresentation if there are no defects because there are then no damages, and there's no liability it the defects are patent because then it wouldn't be reasonable for the plaintiff to have relied on the misrepresentation (causation would be superseded by a kind of contributory negligence on defendant's part).

However, you think those arguments are splitting the hairs too finely for the jury. Considering that this plaintiff's home is at issue, and that he's a veteran, you think it will be sufficient for the jury if the house has demonstrable defects, regardless of arguments about patency or latency. Of course, no defects would cut off liability (no damages). You therefore incorporate the defects question into your diagram as below (Figure *4.17*).

Time to break for the weekly staff meeting and to assign the next week's research for the law student interns (heh, heh, heh).

4.1.4 Hints for Drawing Better Relevance Diagrams

4.1.4.1 Remember the Focus

The main object of the relevance diagram is to disaggregate the problem into the factors that determine the ultimate result (what will happen to your client). You need to stay focused on those important and operative factors, and not make your analysis an overly technical recitation of legal nuances.

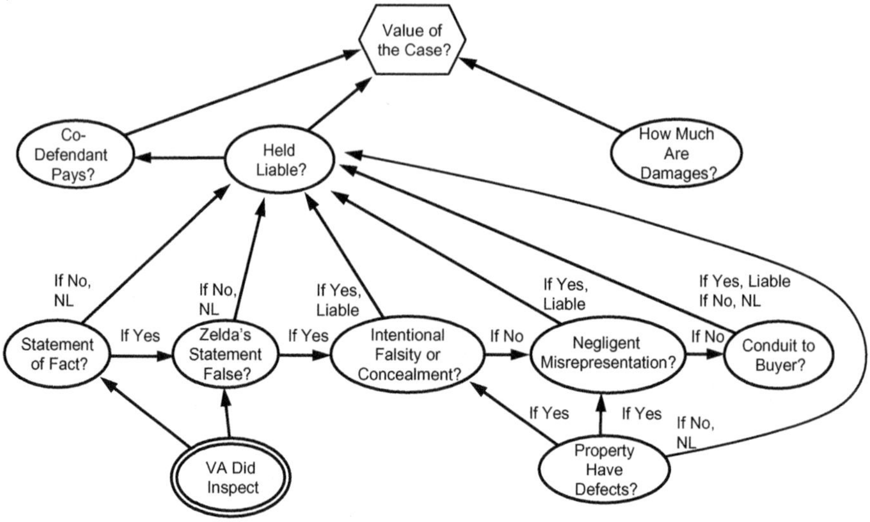

Figure 4.17 Adding the question of whether the property has defects

Those important factors will typically fall in one of the categories listed below. If a factor doesn't fall into one of these categories, you should think hard about why and how it's important before including it in your diagram.

- **Ultimate Issues**: like liable or not.
- **Dispositive Issues**: like negligent or not. Other factors, like whether the jury likes the witnesses or not, influence dispositive factors but are not themselves dispositive.
- **Key Influencing Uncertainties**: factors that are uncertain and important because they greatly affect the odds of a dispositive or ultimate issue turning out one way or the other. These uncertainties are often legal issues and thus have a close and direct effect on ultimate or dispositive issues, as opposed to the global issues (next heading) that you can think of as underlying the legal dispute. Possible examples of key influencing uncertainties include:
  - Will the jury conclude that there was a product defect?
  - Will the jury conclude that the product defect was a cause of damages?
  - Will the memo destroy our credibility on all points, or only on the matters referred to in it?
- **Global Issues**: includes factors that underlie the legal dispute. At trial, these issues are often a matter of evidence. Examples of global issues include:
  - Did the bulldozer experience a mechanical failure or not?
  - Does the basement experience flooding?
  - Was the floor wet and slippery?
  - Did the defendant bus company know there had been other muggings at the bus stop after late-night drop-offs?

- **Important Influencing Facts**: as previously discussed, these facts are not uncertain, but strongly influence the factors and/or ultimate result. Possible examples include:
  - A harmful memo exists and is admissible.
  - The plaintiff will be a very good witness.
  - The opposing counsel is inexperienced.
  - The defendant never signed the formal agreement enclosed with the cover letter.
  - **The Client's Time Value of Money**. Although this issue seldom affects the question of liability and usually doesn't have to go in the relevance diagram, it will affect settlement amounts where possible damages extend for years in the future. You can find further discussion of discounting future sums in the discussions on ascertaining management values in this part, and in Part III on prejudgment interest and discounting.

4.1.4.2 Look at Issues from the Trier-of-Fact's Point-of-View

You've already seen the Trier-of-Fact Rule: always look at your case from the trier-of-fact's point of view. In drawing your relevance diagram, this means framing the issues the way the trier will consider them, including not making your analysis any more complicated than the trier will understand and consider and especially including phrasing your question to reflect who the trier gives the burden of proof to. Thus, your questions should not look like a highly nuanced law review analysis when the jury simply won't consider 12 micro-factors, for instance, in arriving at their verdict. Similarly, whether the right question for your relevance diagram is "Basement Floods?" or "No Basement Flooding?" depends on who has the burden of proof.

Beside how you frame the questions, using the trier-of-fact's perspective also means putting the dependencies in your diagram (and thereby determining the later order in the tree), according to the order the jury will consider the issues in. When you are unsure about which way the arrows should go in your relevance diagram (that is, which issue comes later and depends on an earlier issue), start with the issue that logically comes first in the jury's mind.

After the staff meeting and the usual delicious roast beef and horseradish grinders, you get back to your relevance diagram for Howitzer v. Crabgrass (which you kept pondering during the meeting). You decide that the Consumer Fraud Act claim is just bogus. To sustain a claim under the Act, the plaintiff must prove that Zelda was the conduit from seller to buyer of false information, and there isn't a shred of evidence pointing to Zelda's statement about the V.A. inspection originating with Crabgrass. Accordingly, you erase the issue from your diagram (Figure *4.18*).

It does occur to you that the bogus Fraud Act claim might have a "spill-over" effect by making the rest of plaintiff's case look bad, but you decide to put that down as a reason to keep in mind, rather than as an uncertain question to address directly. You begin to build up your reasons as follows.

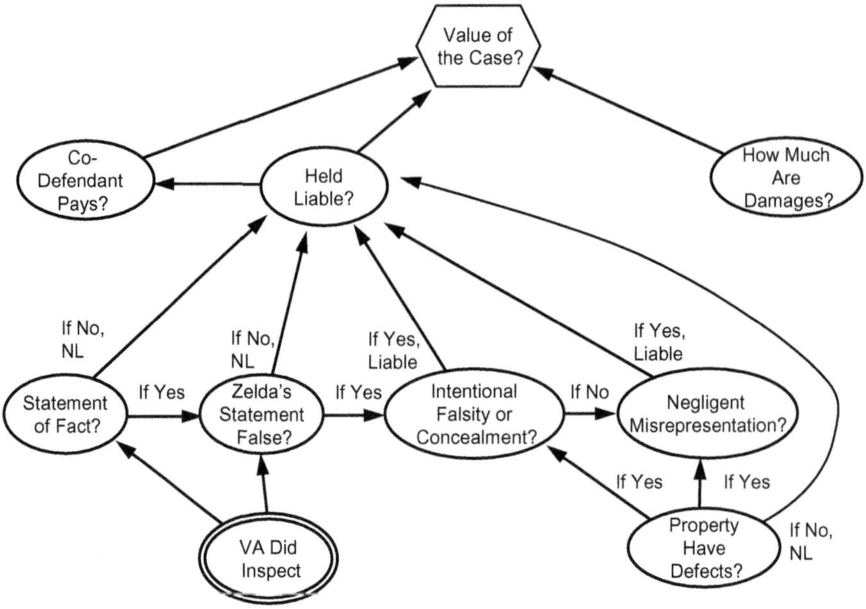

Figure 4.18 Deleting the conduit to the buyer theory of liability

	Reasons for Answers
Liable	• Howitzer is a veteran and the jury feels sorry for him.
	• Zelda looks like a really bad witness and the jury thinks Throckmorton and she are trying to get away with something.
	• The jury thinks Throckmorton is a heartless, big business with deep pockets.
	• Howitzer hired a very experienced attorney.
Not Liable	• The claim looks bad because of bogus Consumer Fraud Act claim.
Statement of Fact	• The trier-of-fact refuses to get into the wrangle of whether the statement was of fact, opinion, or law.
Not a Statement of Fact	• The trier-of-fact decides that Zelda was trying to be helpful about V.A. regulations, and Howitzer should have checked himself.
Statement False	• Zelda thought V.A. wouldn't inspect and was trying to cover up something.
Statement Not False	• V.A. did inspect.
	• Zelda had no reason to lie.
Defects	• The guy's house is at stake.
	• **Question**: what does proof of defects look like?
No Defects	• The plaintiff inspected it himself and didn't find anything wrong.
	• The V.A. inspector didn't find any defects.
Zelda Intentional	• Effective cross makes Zelda look like she's lying.
	• The plaintiff comes up with some real corker defects in the property.

	Reasons for Answers
	• The plaintiff shows that Zelda has handled V.A.-financed sales before, and therefore should have known the process.
Zelda Not Intentional	• Difficult to meet the plaintiff's clear and convincing evidence burden-of-proof.
Zelda Negligent	• The plaintiff shows that other agents knew how V.A. sales worked, so Zelda failed to meet standard of care.
Zelda Not Negligent	• The jury impressed by her professionalism.

As you work on your reasons, you realize that the questions of defects and of damages are the ones you know the least about. What defects is Howitzer claiming, and what proof does he have? And how did he come up with his compensatory damages number? You also note that, at some point, you'll need to figure how to work the punitive damages question into your analysis.

4.1.4.3 Play Devil's Advocate

As you draw your relevance diagram, play devil's advocate (or, better yet, have someone else do it) for each oval as you write it down and before you move on to the next oval. The idea is to examine carefully the reasons for each oval so you don't leave out something important. The following questions are often useful in probing your understanding of the uncertainty and its dependencies.

- How might a bad outcome occur at a lower level in the diagram?
- How might a good outcome occur at a higher level in the diagram?

If you think carefully (or have a good questioner to help you), playing devil's advocate should help you to clarify what you were thinking of when you wrote down the oval, and what the oval means in relation to the rest of the uncertainties in the diagram. You may find that you left out important issues, that you no longer agree with the dependencies you previously wrote down, or that you've overlooked subtle interactions among the issues (This last phenomena is sometimes called "spill-over," and is covered in more detail in Part III.).

Following are some of the more common problems you should look for in playing devil's advocate.

- Are interactions are pointed out?
- Is there unintentional overlap among the uncertainties?
- Are two closely related issues the same, but with a different label, or two different issues?

For instance, on the last point, consider the relevance diagram below. Is there any difference between being found negligent or not, versus being found liable or not? Can you ever be found negligent and held not liable? (The answer will vary with the type of case.) (Figure 4.19).

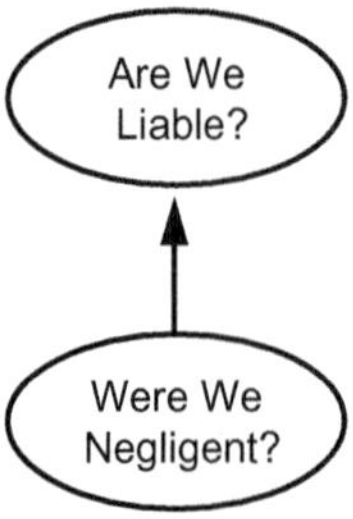

Figure 4.19 Relevance diagram of questionable clarity

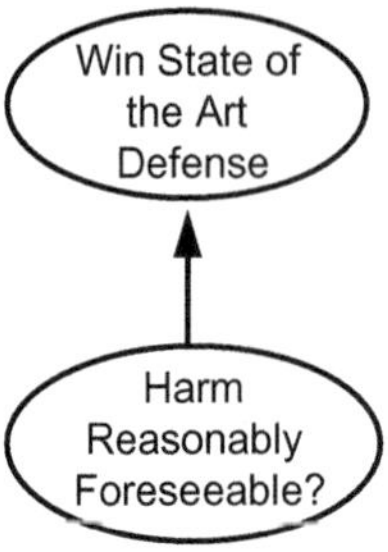

Figure 4.20 Two separate questions or only one?

Likewise, the following diagram came up in a product liability case. The question was whether the manufacturer would win on its defense that the contested harm from the product was not reasonably foreseeable and that the manufacturer used the state-of-the-art safety technology (Figure 4.20).

Although the two questions were originally written down separately in the relevance diagram, playing devil's advocate revealed that the two questions would be resolved together—that is, the defendant would never win the state-of-the-art defense if the harm were found foreseeable, and vice versa. Although this kind of insight is very valuable, it should be carefully tested by considering exactly what it implies, as illustrated by the dDecision tree below. Is there really a zero probability of winning the defense if the harm is found foreseeable? And really a sure chance (100%) of winning if the harm is found not foreseeable? As one decision analyst memorably put it, "One-hundred percent means you're willing to stand up and be shot if you're wrong." Are you? (Figure 4.21).

4.1.4.4 Use the Relevance Diagram to Clarify Your Case

To sum up, then, the relevance diagram should help you both to disaggregate the complexity and uncertainty and catch issues and to clarify your understanding of your case. Use it as a thinking tool to help you straighten things out. For instance, if you're

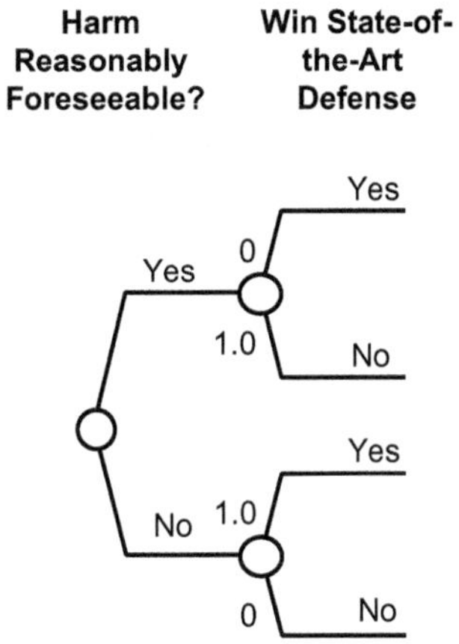

Figure 4.21 Probabilities test whether it's one question or two

not sure how a particular issue affects your case, draw the issues you're comfortable with and use the relevance diagram to help decide where the problem issue fits in. Play around with different representations until you arrive at the picture that best captures how you think the trier-of-fact will view your case. Experience has shown that the relevance diagram is the best way to explain your case both to yourself and to someone else.

When you become more familiar with using relevance diagrams, you can also use it to identify potential subtrees that you can evaluate separately and thereby keep the main decision tree small. The section on subtrees in Part III discusses this process in more detail.

You start to think about drawing the decision tree for the Howitzer case, but decide to work a little more on your relevance diagram first. After a phone call to Howitzer's lawyer and an off-hand remark about avoiding an avalanche of subpoenas duces tecum, you managed to get some information from him.

Apparently, Howitzer's house was built in an area that was a prehistoric flood plain, so it (like all the other houses in the neighborhood) sits on mud. The mud expands when it gets wet, and contracts when it dries. Howitzer's house is securely bolted to the perimeter foundation walls, but the interior support pillars only rest on concrete footings without any attachment. As a result, the mud expansion and contraction has caused the house to shift, so that the interior support pillars are titled on angles ranging from 5 degrees to 15 degrees.

You also obtained an engineer's report from Howitzer's lawyer, and two V.A. inspection reports (the one done at time of sale and the other after the suit was filed). As you expected, the first V.A. report is clean. However, it indicates that the crawl-space under the house was not inspected, and the reason checked is "inadequate clearance." The second V.A. report states that no problems are visible in the perimeter foundation wall from outside the house (which was inspected the first time), but that inspection of the crawl-space revealed the tilted pillars. You note that this second report describes the clearance in the crawl-space as ranging from 12 to 18 inches, and that the space is littered with "concrete fragments, lumber, nails, and other construction debris." You have little doubt that Zelda did not inspect the crawl-space. The second report also notes that a number of the doorways in interior walls are out-of-square, and that several doors bind on closing as a result. Is this something that Zelda could or should have noticed?

The engineer's report notes that the construction techniques used complied with all building codes at the time of construction, some forty years ago, although they would not comply with current code. The report further states that the settling seems to have occurred within the first few years after the house was built (judging from the deformation of the stressed pillar edges, dust deposits, and nail corrosion), and the house appears to have been stable for the last few years. However, the report does conclude that the house is subject to "possible floodwater or seismic dislocation," and recommends the remedial steps of raising the house off the foundation, pressure-driving pilings down into the mud, and then securing the house and foundation to the pilings—at a cost of $100,000. You note that the total sale price for the house was only $205,000. You also discover from a quick check of city engineer's maps that the house is located in a 100-year floodplain, and that the seismic hazard rating is only 2 on a scale of 1 to 5. You therefore see a number of ways the jury could come out on the defect question.

Finally, you checked into the V.A. inspection regulations. The V.A. only requires an inspection of the property when they are financing more than 80 percent of the appraised value (as they were for Howitzer), and the inspection can be by an employee or by one of a number of certified inspectors. Further, the V.A. has two classifications of inspection problems and different ways of handling them: class 1 problems must be fixed prior to loan funding, while class 2 problems need only be noted in the inspection report unless the sale price minus the estimated repair cost puts the buyer in a different percent financing category—in which case the buyer must either qualify under the stricter financing standards or the problem must be fixed prior to loan funding. In short, you can see about twelve different ways the jury could approach the question of whether Zelda's statement was true or not.

Thankfully, now that you know how damages were arrived at, you have a clearer sense of how damages will work. You think the jury will only consider punitive damages if they find intentional misrepresentation, and you still think the amounts awarded of compensatory or punitive damages could vary.

You put all this together by revising your relevance diagram and list of reasons (Figure *4.22*).

You note that you could have put down arrows connecting the defect and compensatory damages questions, but figure that interaction will be handled by whether you're held liable or not.

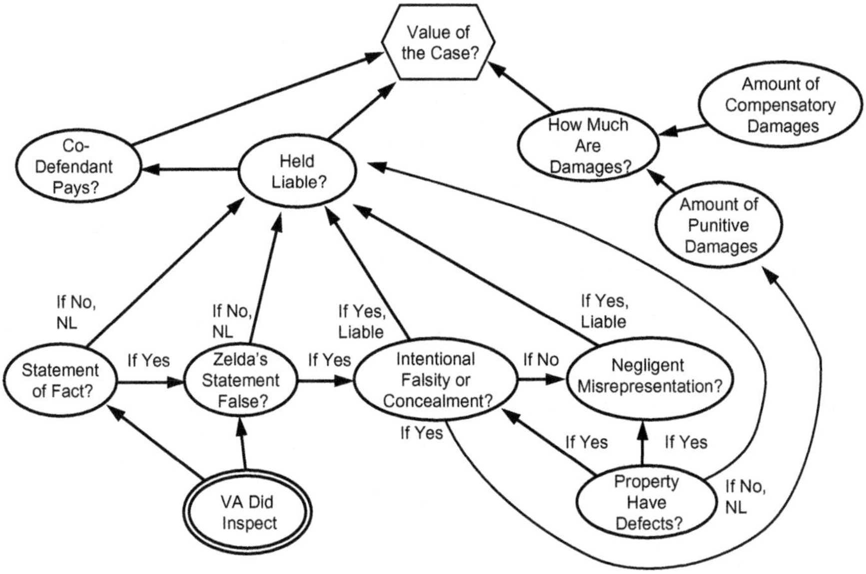

Figure 4.22 Adding factors driving total damages awarded if found liable

	Reasons for Answers
Liable	• Howitzer sounds pitiful on the stand, is a veteran, and the jury feels sorry for him.
	• Zelda looks like a really bad witness and the jury thinks Throckmorton and she are trying to get away with something.
	• The jury thinks Throckmorton is a heartless, big business with deep pockets.
	• Howitzer hired a very experienced attorney.
Not Liable	• The claim looks bad because of bogus Consumer Fraud Act claim.
	• The jury doesn't think Howitzer should have a beef because all the other houses in the neighborhood are in the same boat.
	• The jury feels Howitzer is trying to unfairly skewer Zelda.
Statement of Fact	• The judge wants the case to get to the jury, rather than dismissing on summary judgment.
Not a Statement of Fact	• The judge thinks the claim is junk and wants to get rid of it on summary judgment.
Statement False	• The jury thinks Zelda thought V.A. wouldn't inspect and was trying to cover up something.
	• The jury is too confused by the actual V.A. regulations and buys the opposing counsel's arguments.
Statement Not False	• V.A. did inspect.

	• Zelda had no reason to lie.
	• The jury rejects even trying to figure out the V.A. regulations and goes on gut feeling that what Zelda said was not too far off.
Defects	• The guy's house is at stake.
	• The jury is scared by possible risk to house of tilted pillars (Maybe the plaintiff presents dramatic photos.).
	• The jury wants the plaintiff to recover.
	• The jury feels there were a number of slipshod actions and doesn't want the plaintiff to end up holding the bag.
No Defects	• The plaintiff inspected it himself and didn't find anything wrong.
	• The V.A. inspector didn't find any defects.
	• The jury unimpressed after photos of crack-free exterior wall and age of house.
	• Seems silly to the jury that a house where people have been living in for 25 years suddenly needs $100,000 in repairs when nothing has changed.
	• The jury thinks this 100-year flood plain and 2-rating on seismic hazard are really out to lunch.
	• The jury thinks the engineer is looking for work.
Zelda Intentional	• Effective cross makes Zelda look like she's lying, or trying to hid something.
	• The plaintiff shows that Zelda has handled V.A.-financed sales before, and therefore probably knows the process.
	• The jury decides a realtor would have noticed crooked doorways.
	• The jury decides Zelda understood the V.A. regulations and exploited Howitzer's naïveté.
Zelda Not Intentional	• Difficult to meet the plaintiff's clear and convincing evidence burden-of-proof.
	• The jury decides Zelda didn't know about the pillars.
	• The jury decides Zelda didn't know or misunderstood the regulations.
	• The jury is confused about what the different liability theories mean and doesn't want to pin Zelda with something that sounds "criminal."
Zelda Negligent	• The plaintiff shows that other agents knew how V.A. sales worked, so Zelda failed to meet standard of care.
	• The jury decides a realtor has a duty to be more careful about what she says.
	• The jury finds Zelda was trying to pressure Howitzer into the sale and take advantage of his not having an agent.
Zelda Not Negligent	• The jury is impressed by her professionalism.
	• The jury feels no one would make sense of the V.A. regulations.
	• The jury finds it ridiculous that Zelda should have crawled under the floor.
	• The jury finds second V.A. report too convenient and thinks they're trying to pawn the problem off on the defendants.
	• The jury feels no one should be held liable for failing to notice such a subtle problem.

As you write out the reasons, you realize that, because the jury will never be able to sort out the question, whether the alleged misrepresentation is fact or law/opinion is really a question of whether or not the judge is so unimpressed with Howitzer's case that he's willing to dismiss it on summary judgment. You make a note of this on your relevance diagram.

4.2 Draw the Decision Tree

4.2.1 Procedure for Drawing the Decision Trees

Once you've completed the relevance diagram, the final step in problem structuring is drawing the decision tree. This step may be nearly routine if you already have a very clear understanding of your case, but, more typically, the process of drawing the tree raises more questions about your case. The most frequently encountered questions are discussed in more detail in the next section. You will probably want to begin drawing your tree according the procedure below and deal with these questions and considerations as they arise.

The steps for drawing the tree, then, appear below.

- ***Advanced Step***. Once you're familiar with subtrees, the first step is to divide the relevance diagram up into the main tree and the various subtrees (whose mean values will then go into the main tree). Don't worry about this step until you're familiar with the basic process and with how subtrees are used.
- **Pick Where to Start**. Start at the lowest level of questions in the relevance diagram that will be included in the tree (See the next section for which influencing factors should go into the tree.).
- **Draw the Current Level of the Tree (with Reasons)**. Draw the chance and/or decision nodes for the questions on the current level of the tree.
  - Each node should include a branch for each possible outcome, and a list of reasons under that branch why the question might turn out that way.
  - See the next section for a discussion of how to order questions on the same level of the diagram (if not already indicated by "if yes" arrows, for instance).
  - Remember that double-ovals (things not uncertain) should not appear in the tree (Recall that double-ovals are used for key influencing facts or questions whose answers are no longer uncertain once you know how all the factors influencing them turned out.). The possible outcomes of a double-oval just tell you what happens (what the branch labels are) for different combinations of its influencing factors, as illustrated in the diagrams repeated below (Figures **4.23** and 4.24).
- **Go Up to the Next Level**. Go up to the next level in the relevance diagram until you get to the ultimate issue (such as the question of liability or the value of the case). Now go back to Step ①.

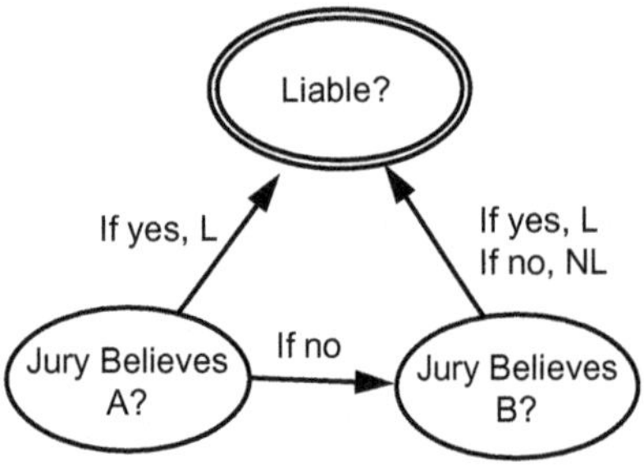

Figure 4.23 Relevance diagram with factor no longer uncertain once preceding questions are answered

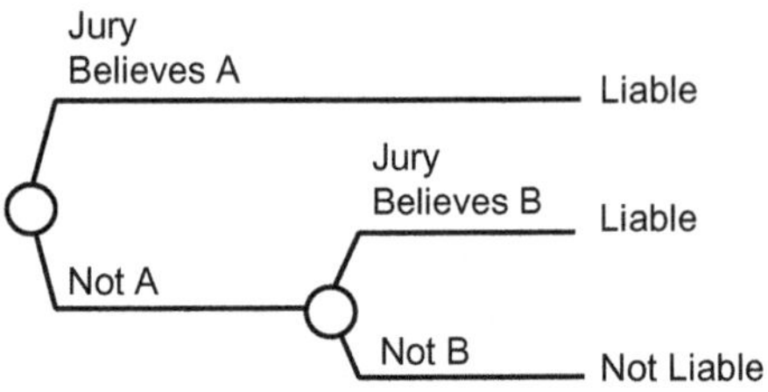

Figure 4.24 Tree showing factor determined by answers to preceding questions

- **Assess Probabilities and Redraw the Tree**. Once the decision tree is completely drawn with all reasons under the branches, you need to assess the probabilities (Chapter 4), and then make a new tree drawing with just the probabilities. Two separate tree drawings are usually more convenient, but, for small trees, you may find it helpful in thinking about your case to include the reasons and probabilities on the same tree drawing.

4.2.2 *Common Questions in Drawing the Decision Tree*

You may have found in the procedure above all the information you need to draw your decision tree. More typically, at least one (and usually more than one) of the following questions arises in drawing the decision tree. You should read through all the topics to familiarize yourself with the problems, then refer back to the discussions as you encounter the questions in your practice.

4.2.2.1 How Do I Order the Decision Trees?

The basic sequencing rule for decision trees is that you read them from left to right, and that the time sequence likewise goes from earlier on the left to later on the right until, at the far right of the tree, you've arrived at the ultimate question you were interested in (liability or value of the case) and know what the answers to the uncertainties or decisions were in any particular scenario.

However, where you have a number of legal issues/uncertainties that the jury will consider together, there is often no obvious way to order them by time sequence. In those cases, one of the following methods will help you put them in an order that makes the most sense to you (and thus facilitates your analysis).

By Trier-of-Fact Order

In another example of the Trier-of-Fact Rule, you should order the issues according to the sequence that you think the trier-of-fact will logically consider them in—as shaped both by the judge's instructions and by the burden of proof. Thus, if you followed the Trier-of-Fact Rule when drawing your relevance diagram, the order in the tree from left to right should be consistent with the direction of arrows in the relevance diagram: no arrows from the relevance diagram pointing to the left from issue-to-issue in the tree ("backwards" in time).[3]

If the arrows from the relevance diagram would point "backwards" in your tree, you should reexamine both your relevance diagram and your tree. Which (or neither) is the best representation of how you think the trier will decide your case?

By Dependencies

Where there are dependencies between the questions (arrows from one question to another), the arrows give you the order for the decision tree. Thus, influencing facts (lower in the relevance diagram) go to the left in the tree of those factors they influence. Ordering this way will also give you the different scenarios for later assessing the dependent probabilities (probabilities that differing according to which influencing scenario you're in).

By Order Logical to You

If neither the trier-of-fact order nor the dependencies in your diagram give you the order of questions for the tree, put the questions in whatever order seems logical to you to consider this issues in. In most cases, this order *should* be the same as the trier-of-fact order because you're trying to look at the case from the judge or jury's perspective.

---

[3] People with prior exposure to probability theory will note that any dependency (the direction of any arrow in the dependency diagram) can be reversed to a mathematically equivalent form. If you're curious, you can examine the section on flipping dependent probabilities.

4.2.2.2 How Do I Draw Dependencies in the Decision Tree?

In drawing relevance diagrams, you saw a number of different meanings for the dependency arrows between questions. Each different meaning leads to a different representation in the decision tree.

Dependencies from Asymmetry

In almost every relevance diagram, you'll have questions where there's a different dependency arrow depending on the answer to the question. Thus, for a claim with several necessary elements, as in the vicarious liability example below, answering "no" to any of the necessary elements leads the jury straight to a finding of no liability, while a "yes" answer brings them to the next necessary element. This is a common situation for claims with multiple hurdles where the plaintiff must get over each (The alternative, of course, is a "factors" kind of case, such as many negligence cases, where the jury considers all the factors together and a "no" or "yes" answer on any single factor is not dispositive.) (Figure 4.25).

The tree for this diagram appears below. Thus, you see that a "no" answer to any question leads to no liability, while a "yes" answer leads to the next question, or, ultimately, to liability (Figure 4.26).

The fact that each branch leads to a different question or result is what's referred to as the "asymmetry" in the tree. Contrast the diagram and tree below, where, regardless of how the jury comes out on the preliminary questions, they still face the same question of liability (Figures **4.27** and 4.28).

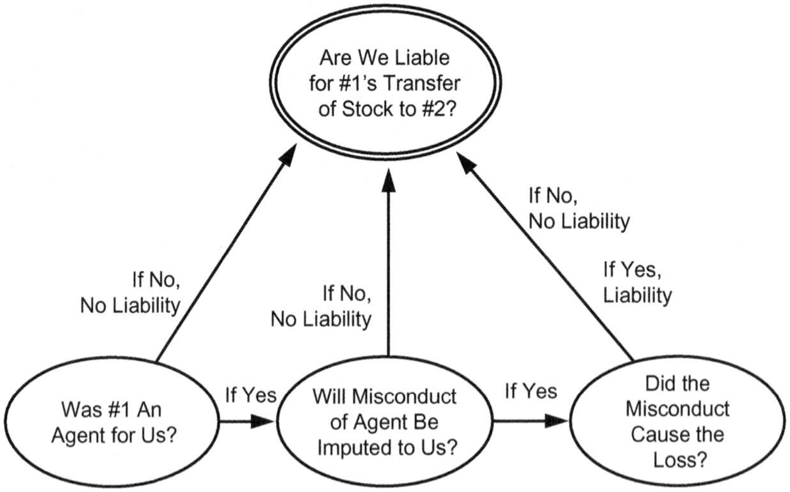

Figure 4.25 Relevance diagram for dependency leading to asymmetric tree

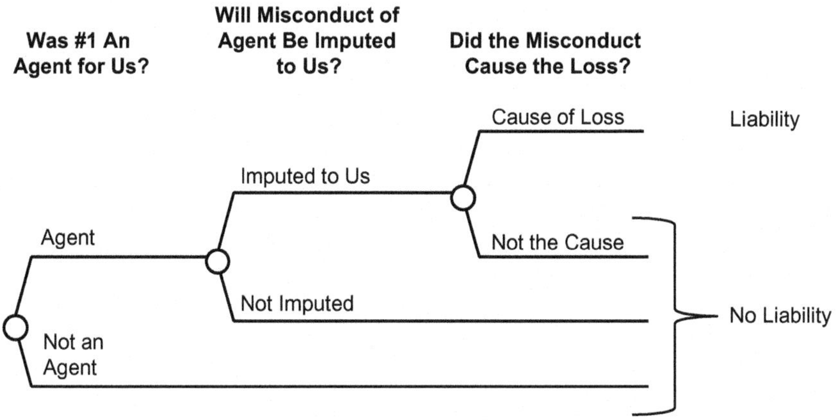

Figure 4.26 Asymmetric tree for vicarious liability

Figure 4.27 Relevance diagram leading to a symmetric tree

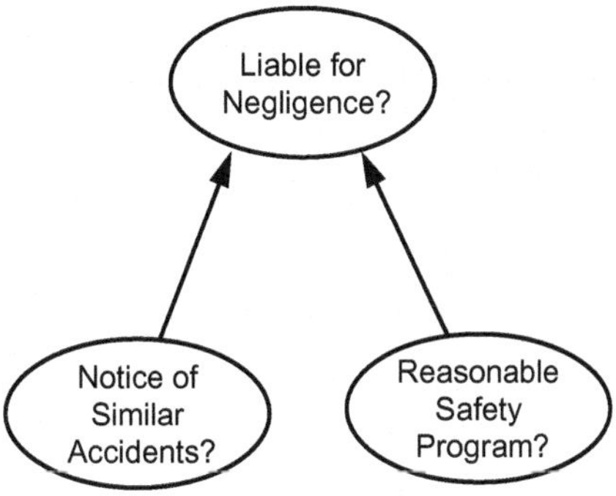

Figure 4.28 Symmetric tree for liability with multiple factors

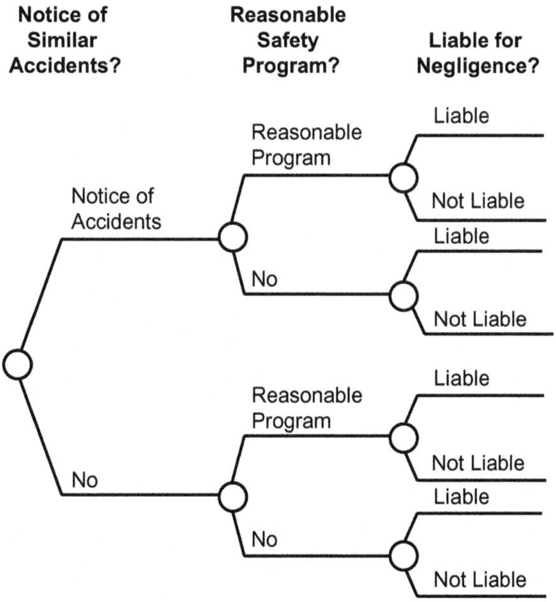

Dependent Probabilities

Another meaning a dependency arrow can have is changing the odds of various possible answers to a question (dependent probabilities). For instance, in the relevance diagram above for negligence liability, there are arrows from the two factual questions to liability. Those arrows mean that the factual questions have *some* affect on the determination of liability. Thus, at the least, there should be different probabilities for liability in the decision tree depending on what the answers to those two questions were, as illustrated below (Figure 4.29).

However, because the reasonable program question follows either finding on notice, this tree should present a further question to you: are the odds of the jury finding the safety program reasonable also affected by whether they find notice of other similar accidents? That is, is there any spill-over? If you thought so, the first step would be to add another arrow to the relevance diagram, as below (assuming you thought the trier-of-fact would first consider the notice question) (Figure 4.30).

A tree with this latest set of dependent probabilities added, then, might resemble the one below (Figure 4.31).

In contrast, if there were no arrow in the dependency diagram from "Notice of Similar Accidents?" to "Reasonable Safety Program?" (e.g., the answers to one question didn't depend on another), the probabilities for "Reasonable Safety Program?" would be the same, as shown below (Figure 4.32).

Figure 4.29 Probabilities depend on a preceding factor

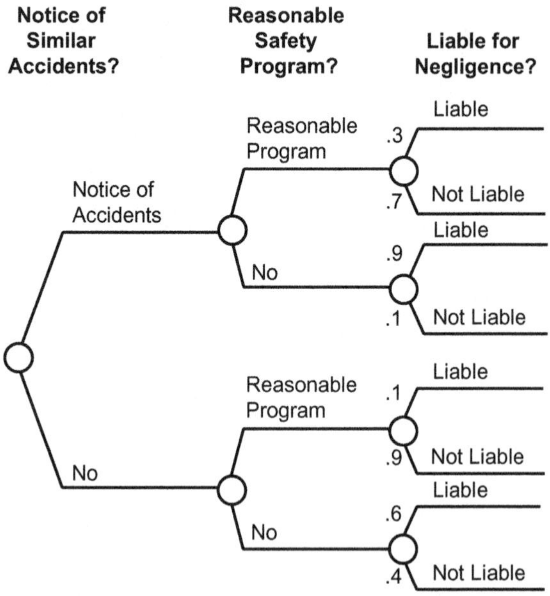

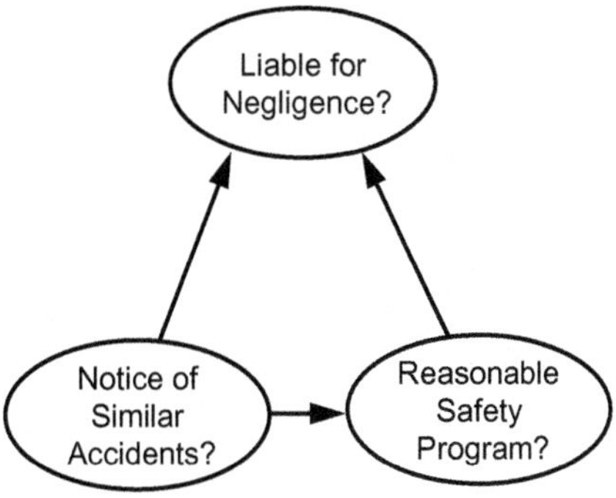

Figure 4.30 Adding an arrow to show "spill-over" between factors

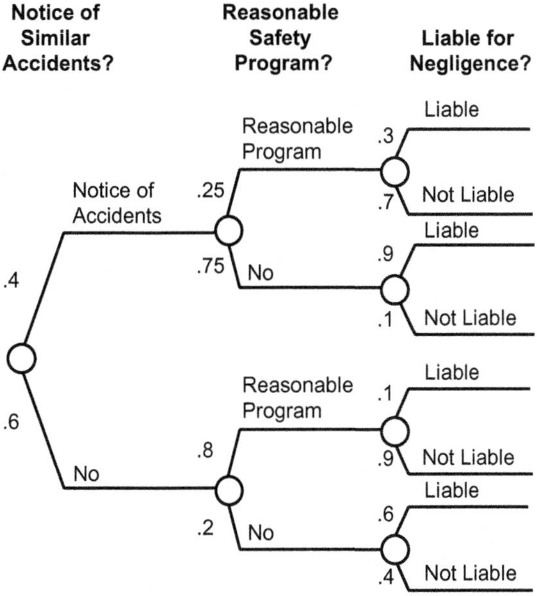

Figure 4.31 Tree showing two levels of dependent probabilities

Uncertainties Determined by Other Questions

This final type of dependency may have snuck by you in the discussion of dependencies from asymmetry. The double-oval around liability in the vicarious liability relevance diagram (repeated below) indicates that the question of liability is only uncertain until all the questions in the level below have been answered (as properly indicated by the arrows) (Figure 4.33).

Thus, the kind of the dependency indicated by the labeled arrows and the double-oval is reflected in the decision tree (below) simply by which scenarios result in liability or no liability. There is no need for dependent probabilities or for a chance node for the liability question (Of course, the situation would be different if you didn't think these three questions were dispositive on liability.) (Figure 4.34).

Figure 4.32 Tree without an arrow from "Notice of Similar Accidents?" to "Reasonable Safety Program?"

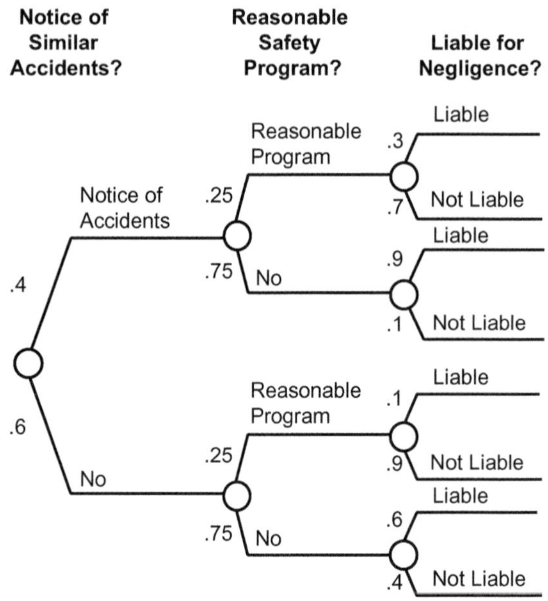

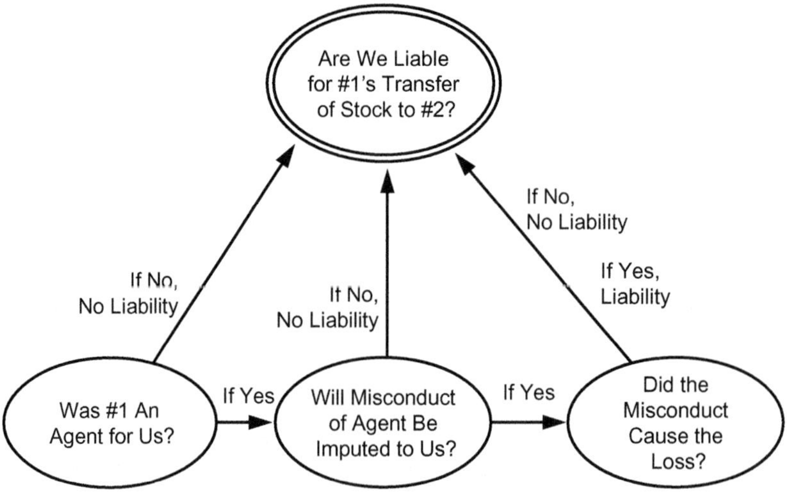

Figure 4.33 Uncertainty determined by other uncertainties

4.2.2.3 Which Influencing Factors Should Go into the Tree?

As discussed in the section on drawing relevance diagrams, your diagram generally gets more detailed as you add levels to the bottom of it, often reaching the evidentiary level before you decide the questions can be simply answered with yes, no, or

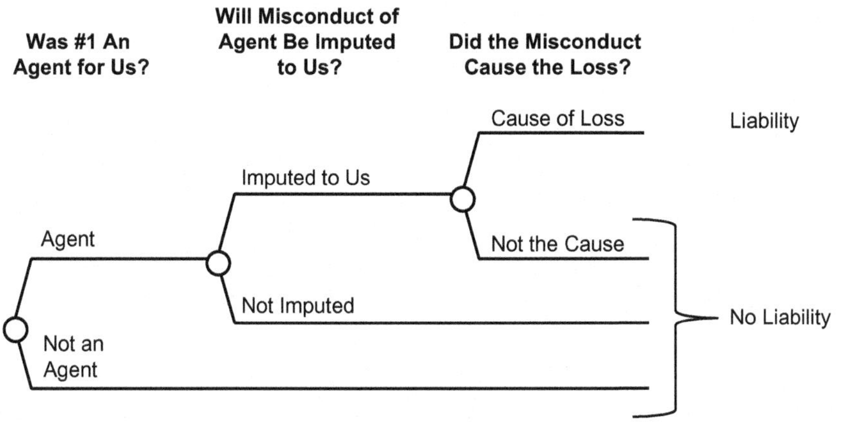

Figure 4.34 Uncertainty determined by other uncertainties shown with an asymmetric tree

a number (the clairvoyance test). However, although it is important for your understanding and insight to develop your diagram to this level of detail, it is usually not necessary or desirable to include all the questions in the decision tree. For one thing, your objective is to identify and to focus on those uncertainties that make the most different to what will happen to you, and, second, simply putting all the questions in the tree quickly produces a tree too large for the largest computers, much less your legal pad.

Accordingly, only those questions that are both *significant* and *uncertain* should go into the decision tree.

- **Significant**: those questions whose resolution has a substantial impact on the ultimate issue (or, in more complex problems, on another influencing issue which in turn has a significant impact on an ultimate issue); and
- **Uncertain**: those questions whose resolutions are uncertain (such as a .98 chance or better of one outcome and a .02 or less chance of the other.)

One way to help decide which factors are significant is to consider whether you need the factor in question to facilitate your thinking about the ultimate issue. Can you think about your chances of winning or losing the case without knowing how that particular question turned out? Or is it so important that you must know the answer to make any guess?

Based on the author's experience (and considering the number of cases you have to analyze), shoot for the two or three "most important" influencing uncertainties. One may be okay.[4] Don't force yourself to come up with another, but if in thinking about the ultimate issue you say to yourself "It really depends on…," then that factor should be drawn in also.

[4] See Section X in the Advanced Topics section for how you should handle those cases where you have so little information that you can't identify even one influencing factor.

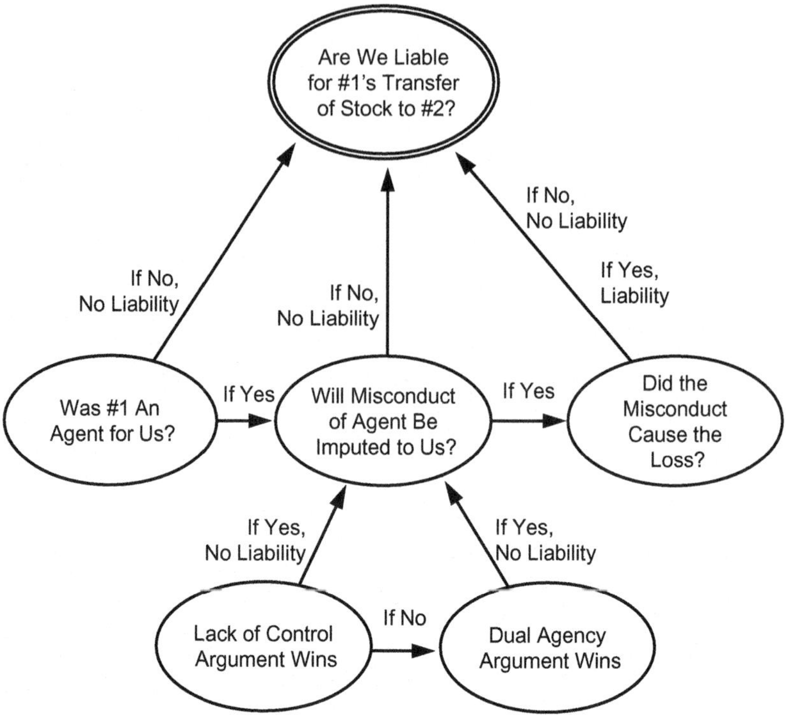

Figure 4.35 Pondering additional uncertainties for vicarious liability

You may also recall from the prior discussion of relevance diagrams that key influencing facts from the relevance diagram (the double ovals) may be listed among the reasons in the decision tree. Similarly, questions that don't have a direct and important enough effect on the ultimate issue may also have their answers listed among the reasons in the tree—rather than including those questions in the tree as chance nodes.

For instance, another level of detail could be added to the vicarious liability diagram to examine the question of whether the misconduct of your agent will be imputed to you, as below (Figure 4.35).

In drawing the tree, then, your first reaction might be to include these two new questions in the decision tree, producing the enlarged tree below (Figure 4.36).

Alternatively, if you decided to use the answers to these two questions as reasons in the tree, your tree would look like the one below (Figure 4.37).

However, you may not be comfortable simply subsuming these two clear legal questions in the issue of imputation when you can address them more carefully (remember the cognitive problems in evaluating combinations of uncertainties.) There is an alternative.

Note how in the tree the reasons are listed with "and" and "or." You won't lose unless you lose on dual agency and control, but can win with either one of them.

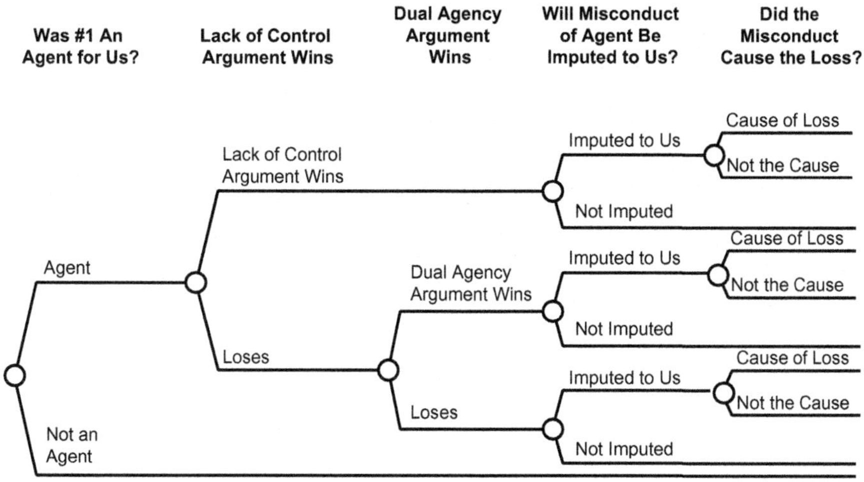

Figure 4.36 Adding additional questions to the decision tree

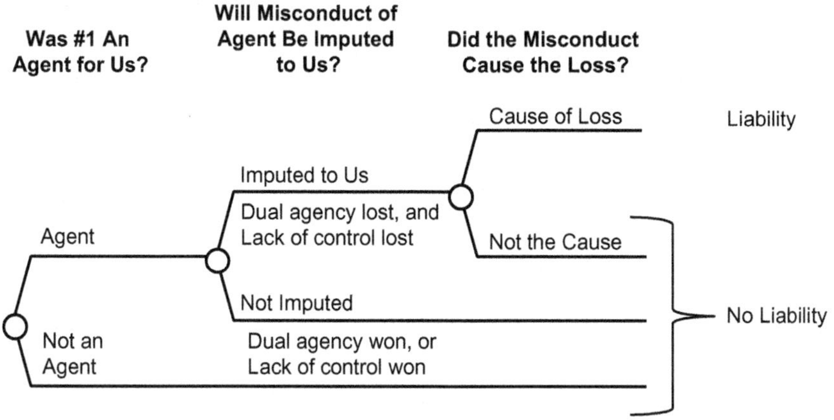

Figure 4.37 Additional questions handled as supporting reasons for assessed probabilities

Thus, liability will be imputed to you *unless* you win on the lack of control or the dual agency argument. In that case, you needn't include these arguments in your main tree. Only the odds of being held liable need be included in your tree. You can solve for the probability that liability will be imputed to you in a separate tree, as illustrated below, and then plug the appropriate probabilities into the main tree. This separation of the tree into its independent parts to solve in pieces, known as subtrees, will be covered in more detail in Advanced Topics, but you can see already that use of subtrees helps keep the main decision tree small and forces you to think more carefully about how the uncertainties are related (Figure 4.38).

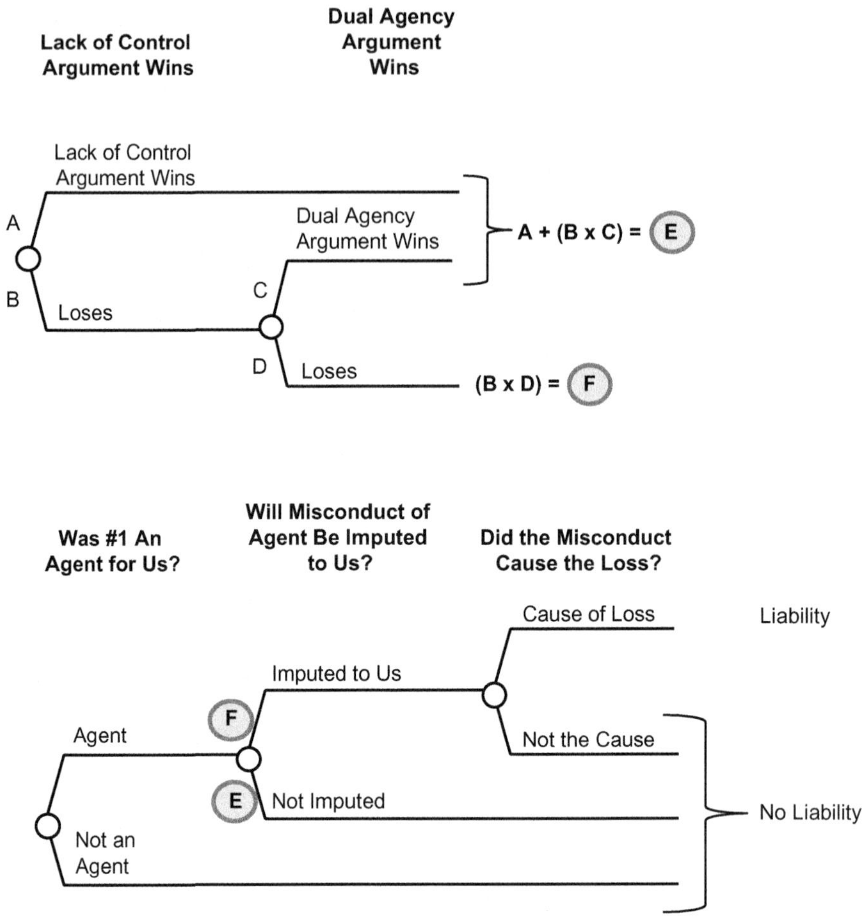

Figure 4.38 Using a subtree to calculate probabilities for the main tree

4.2.2.4 What Do I Do If the Decision Tree Gets Too Large?

It quickly becomes evident how large decision trees become as you add questions. Most of the time, it's not necessary to have a tree that extends across expanses of pages taped together (as many people construct when they begin drawing trees). The following suggestions will help you to keep your tree manageable without sacrificing insight or significant.

Focusing on What's Most Important

The first suggestion is to reconsider whether you've only included that questions that are significant to the ultimate issue. Are you only including what you think ***really*** affects winning or losing and the amount of damages? Or are you including too many issues, making the analysis more "academic" and not really reflecting how the case will be resolved?

Making your analysis too technical and over-inclusive will actually hinder your analysis because that's not the way the jury will think about your case.

Limiting Your Analysis

There are a number of reasons for simply omitting issues from the decision tree. You might, for instance, simply ignore issues that could be important, but don't seem to you as important as the main issues of the case (like the effect on precedence, for example).

Or you could be omitting a level of detail in your first-pass analysis, like omitting the last level of detail added to the influence diagram above. There are a number of reasons for simply omitting details from your first-pass analysis.

- Lack of time (some analysis is a lot better than none);
- You decide to include the possible details in your lists or reasons (either in the tree, as you've seen, or, as you'll see later, in your probability assessments list);
- The details are key influencing facts so that you *should* put them in the list of reasons and *not* in the tree (there's no "uncertainty" at 100% probability);
- You're actually applying the Trier-of-Fact Rule in deciding that the jury won't separately focus on the issues; thus, that much detail would actually make your analysis worse.

The main point is that you can and should limit your first-pass analysis of any case so that it doesn't get unmanageably complicated and so that you can complete the analysis and think about the results—the most important part. You can then consider whether the results make intuitive sense to you and whether you should now include some of the omitted issues (a consideration discussed at greater length in Chapter 6).

Subtrees and Lists of Reasons

You've already seen how the detail of a relevance diagram can be handled in a subtree, or in your list of reasons in the tree why an issue came out a certain way. To add one more example, consider the relevance diagram below for a case in which you could be liable on either a contract theory or on a tory theory (Figure 4.39).

Because the questions affecting contract liability don't affect anything else in the tree (a determination discussed at greater length in the section on subtrees in Part III), you could do a subtree for contract liability as below (Figure 4.40).

If you felt more comfortable separately assessing the probabilities for each of these questions and then plugging the probabilities for contract liability or no liability into the main tree, you should proceed that way. Alternatively, you could assess the contract liability issue directly, and merely list these sub-questions as possible reasons for coming out one way or another in your tree (Even if you decide to do the direct contract assessment method, you might still want to list the reasons in the tree to remind yourself that these sub-questions were considered.) (Figure 4.41).

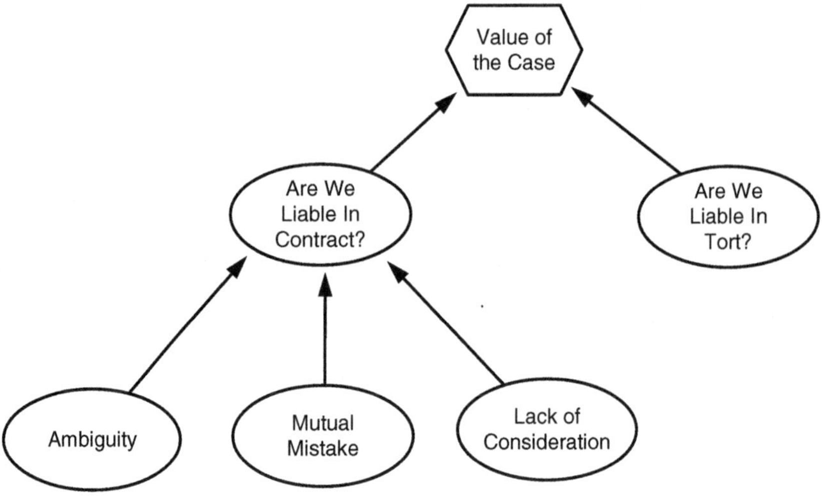

Figure 4.39 Relevance diagram for two theories of liability

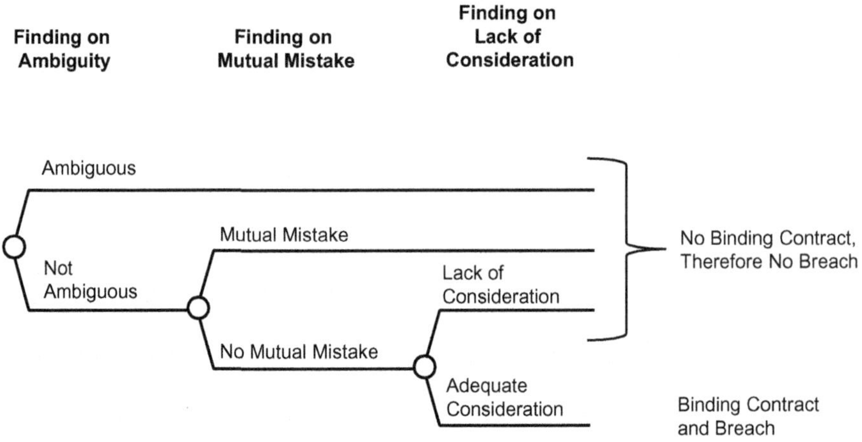

Figure 4.40 Subtree to calculate probabilities of liability under contract

The Problem with Pruning

Another method of limiting the size of the tree is carefully thinking about when questions are relevant, and not drawing them in the tree when they're not (otherwise known as "pruning the tree"). For instance, in the subtree for contract liability shown above, you never get to the other contract questions if the trier-of-fact has already found the contract ambiguous (that is, you go directly to no contract liability).

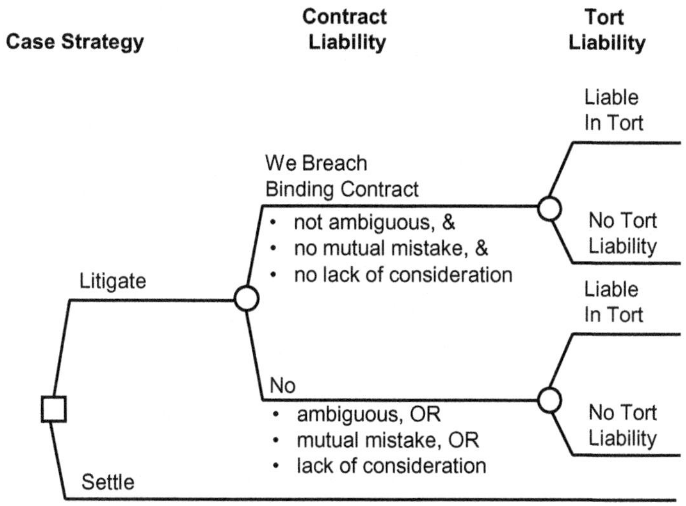

Figure 4.41 Reasons included in lieu of or in addition to subtree analysis

The problem with pruning is that it tempts you to remove issues from the tree where they might still be relevant, thus missing possible spill-over effects. Thus, pruning is rather a two-edged sword:

- On the pro-side, pruning makes you play devil's advocate again. Are you really sure the issue is not relevant in certain scenarios?
- On the con-side: if you don't prune irrelevant issues, you end up with much larger trees.

One way to test if you're appropriately pruning an issue is to ask yourself if it would also make sense to prune the issue in your relevance diagram by placing "if yes" or "if no" on the appropriate arrows in your relevance diagram. The two methods are equivalent, and an appropriate pruning should make sense either way (In terms the earlier dependency discussion, you are in effect introducing an asymmetric dependency into your relevance diagram and tree.).

To make this discussion a little more concrete, consider the tree below for product liability. The plaintiff is asking for compensatory and punitive damages, and the key questions (besides the liability and damages uncertainties) are whether the harm from using the product was foreseeable, and whether the product testing program was adequate.[5]

The question, then, is whether it even matters what the testing was if the harm is found foreseeable. Similarly, punitive damages may not even be an issue if the harm is found not foreseeable (no egregious behavior, etc.). These potential prunes to the tree are indicated by the X's and question marks in the tree below (Figure 4.42).

[5] To test your understanding of dependency diagrams and trees, you should try to draw the dependency diagram that corresponds to this tree.

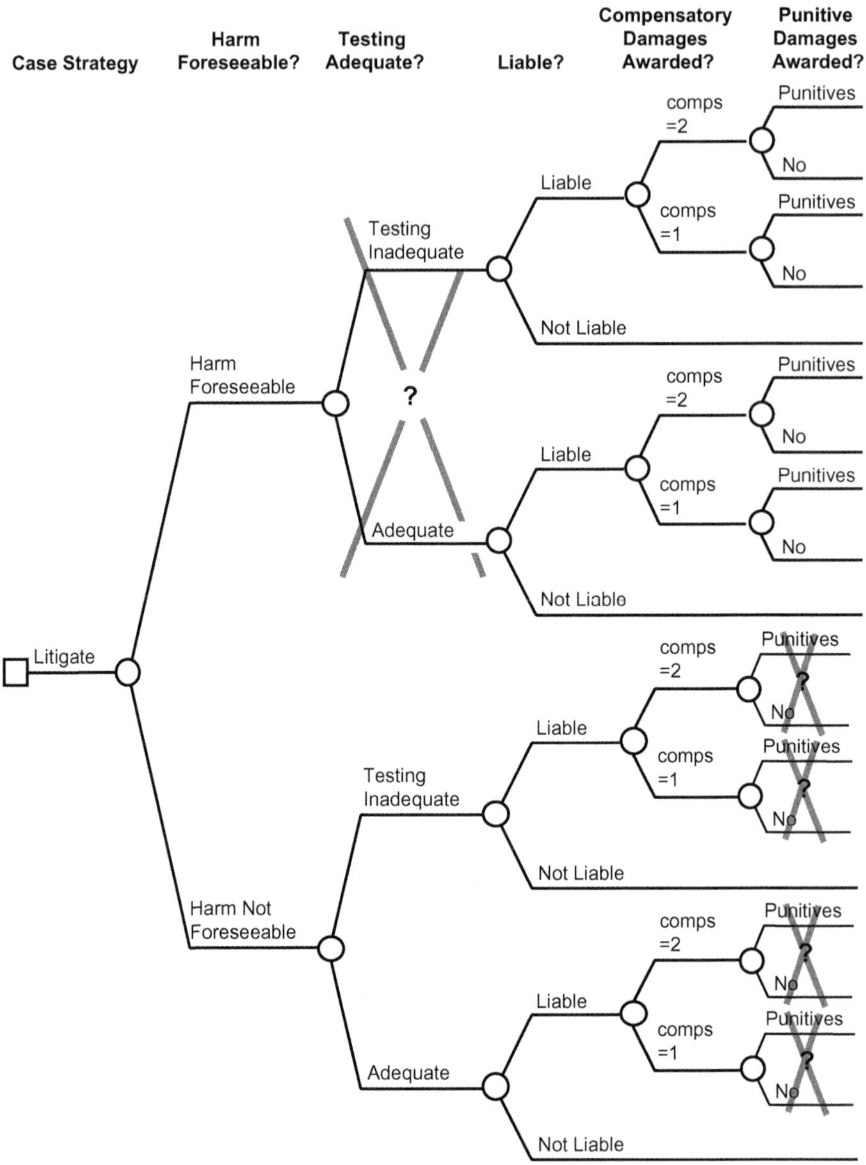

Figure 4.42 Pondering whether to prune nodes from the tree

The resolution of these questions depends on the results of playing devil's advocate and testing your understanding of the case (in both relevance diagram and decision tree form).

However, if you do prune the tree, you should at least still have each issue appear in its own column, as below, so that it's clear where issues have been omitted (longer branch lines where appropriate does the trick). This simple technique will help remind you of your earlier carpentry (Figure 4.43).

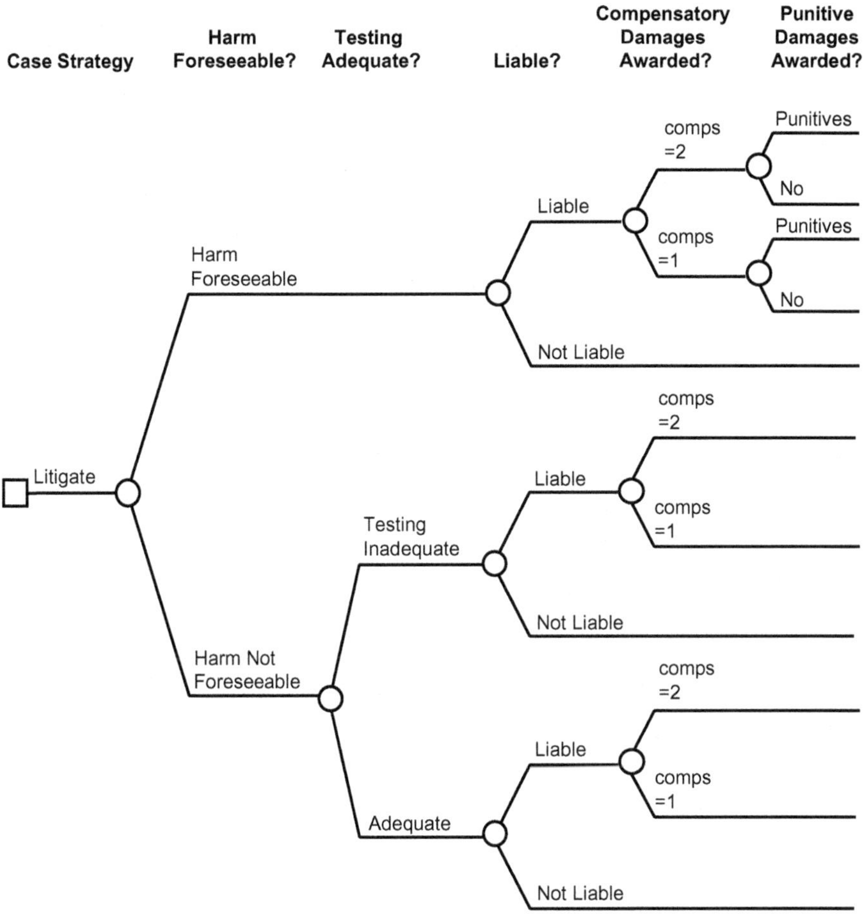

Figure 4.43 Pruned tree with columns preserved to remind of removed nodes

An Alternative Approach to Managing Tree Size: Deterministic Sensitivity Analysis

There is an alternative approach to keeping tree size manageable: it's called deterministic sensitivity analysis.

This approach required you to create a model of the value you're interested in (such as total exposure in the case) using all the factors you've identified in the relevance diagram. The model is frequently created in Excel. Instead of calculating the value for each scenario in the tree (as in the examples in this section), you would build a model which takes the outcomes of the decisions and uncertainties as inputs and calculates the value for a particular scenario.

Using software, you then vary the outcome of each uncertainty one at a time and see which ones make the most difference in the value of the case. Typically, five to eight uncertainties account for 90% or more of the variance in value of the case. Those uncertainties are candidates for inclusion in the decision tree.

For cases in which potential damages are very complex (e.g., patent litigation with the need to project past and future revenues covered by the patents and potential license rates), a model is required. Our aim here is to get you started with analysis that can largely be completed without software, although we do introduce software in Part III. For a complete exposition on deterministic sensitivity analysis, its use, and the modeling methods required to use it, please see *Decision Analysis for the Professional.*

As you begin to think about the decision tree, you consider the hints for drawing trees and how they apply. The first question is how much of the relevance diagram you should put into the tree. You decide to leave out the codefendant (Zelda or Crabgrass) apportionment question. To begin with, it's doubtful that Throckmorton wouldn't get stuck with the whole judgment under respondeat superior. You also think it's very doubtful that Crabgrass would be held liable for Zelda's possible unless some sort of collusion were shown, and you don't think there's any evidence of that. Besides, it's just too complicated a wrinkle to put in until you've got some idea of settlement strategy.

The next question is how much detail you want to go into with damages. Your first reaction is to do a subtree for damages, as you've seen in many examples, but then you realize that the amount of compensatory damages depends on which basis you're held liable on (no comps unless liable for intentional misrepresentation). You decide to wait and see how big the tree gets. You could combine comps and punitives together into a single question that has different possible answers depending on how you're held liable, but it may not be such a big deal to have two damage nodes (comps and punitives) in the intentional misrepresentation scenario instead of one.

Next question is what order the trier-of-fact will logically decide the issues in (as indicated by the arrows in your relevance diagram). Oh yeah, the first node in the tree is your own decision on whether to litigate or settle the case. Then, whether Zelda's statement is held a statement of fact or otherwise (law or opinion) will be determined on your motion for summary judgment (whether the judge thinks the case is junk and wants to dispose of it).

Looking at the arrows, you see that the next issue for the trier could be whether Zelda's statement is false, or whether the property has defects. You could do either issue and still have the tree order from left-to-right consistent with the arrows. You decide that the jury will first decide whether the property has any defects, because that will be their way of considering whether Howitzer even has a legitimate beef. So that uncertainty will go after the summary judgment question.

Having decided that much, you realize that the arrows give you the rest of the ordering. And, thankfully, the "if yes" and "if no" labels help you by showing where you'll all done (no liability) and where you go on the next question. Thus, it's not too much work to have separate nodes for compensatory and punitive damages, and you end up with a first drawing for the tree (Figure *4.44*).

In drawing the tree, you employed a shorthand of drawing the punitive damages node after but not connected to the compensatory damages node. This indicates that the amount of punitive damages is uncertain regardless of how the compensatory damages turns out, but you just didn't draw it three times (Figure *4.45*).

However, as you thought about the tree further, you decided that the jury wasn't going to award punitives automatically if they found intentional misrepresentation.

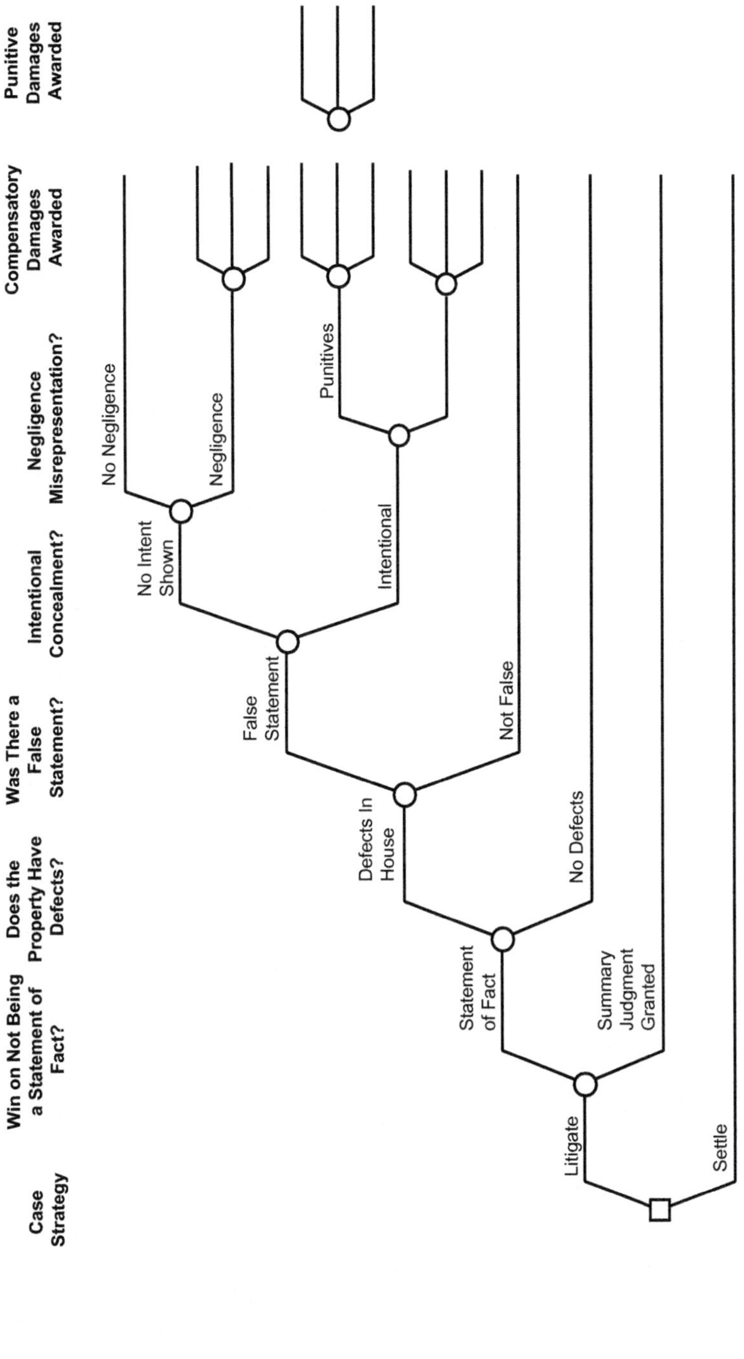

Figure 4.44 First pass at the Howitzer v. Crabtree decision tree

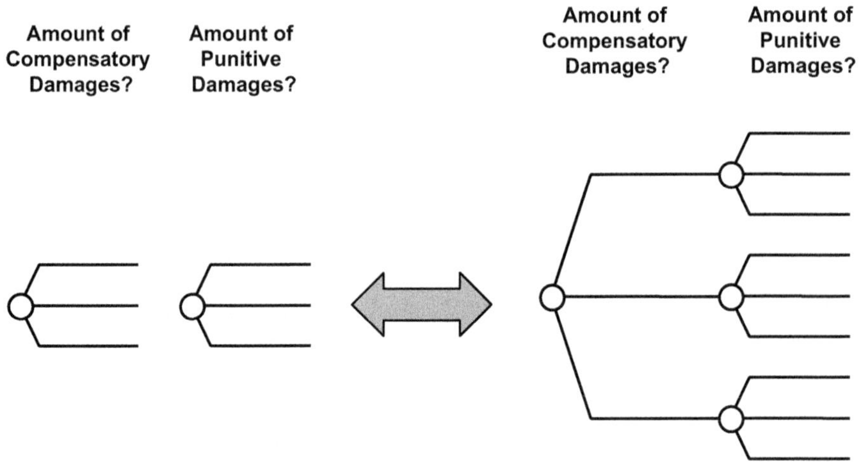

Figure 4.45 Equivalence of shorthand convention for drawing decision trees

Figure 4.46 Subtree to calculate damages awarded if found liable

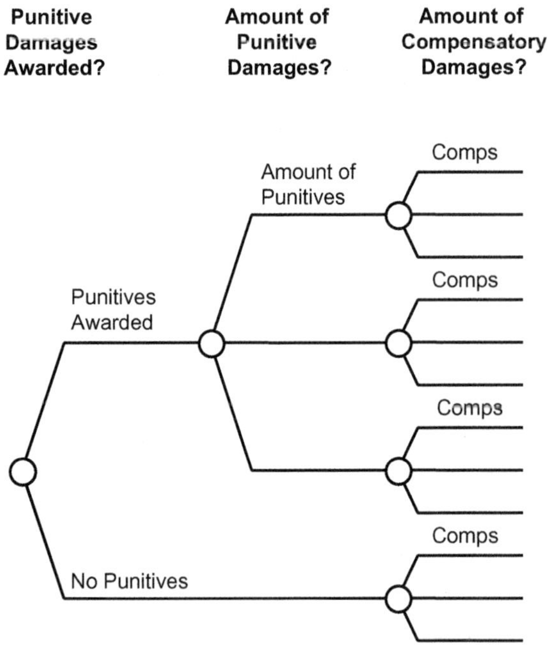

Rather, you thought they'd first decide whether punitives were merited at all, according to how egregious they felt Zelda's behavior was, and would then go on to fix an amount of punitives if merited. However, you also thought that the compensatory damages would be whatever the plaintiff could demonstrate, regardless of whether punitive damages were awarded. This allowed you to draw a subtree to calculate the damages with the order reversed (i.e., the numbers on the branches are the same regardless of the order drawn) (Figure *4.46*).

Then, as you reviewed your list of reasons for different possible answers in the tree, you added the reasons for a punitives award, for high and low punitives, and for high and low damages—remembering that the reasons for punitives amounts were conditioned on the jury already having decided to award some punitives.

Punitives Awarded	• The jury getting mad at Zelda after having found against her on all the preliminary issues.
	• The jury wants to "get back" at realtors. **Note**: can you test for attitude towards realtors in *voir dire*?
	• The jury decides Zelda was lying after all.
	• The jury buys the opposing counsel's argument that Zelda, Throckmorton, and Crabgrass were trying to scam Howitzer.
	• The opposing counsel gets in particularly favorable jury instructions.
Punitives Not Awarded	• The jury doesn't think Zelda's behavior egregious enough to punish.
	• The jury thinks Howitzer well-enough compensated for damages without punitives.
	• The jury balks at awarding Howitzer more than the purchase price in damages.
	• The jury feels forced by the legal definition to find Zelda's statement intentional, but still doesn't think her really culpable.
	• The jury's instructions favorable to our side make the jury feel really antsy about awarding punitives.
High Punitives	• Same kind of reasons as for awarding punitives get the jury angry and on a roll.
	• Zelda looked terrible on the stand, Howitzer looked great, and the judge gives the jury a whole spiel about "punitives don't depend on actual damages" to make the jury feel free to make a high award.
Low Punitives	• The jury wants to put a little sting in the award to punish Throckmorton, but doesn't want to give Howitzer a windfall.
	• The jury gets worried they're going too far after having decided on punitives and wants to pull back a little.
	• The jury gets bogged down in punitives deliberations and begins to reconsider, then settles on a low punitives award.
	• New instructions or rereading the transcript during punitives deliberations causes the jury to cool down after they already decided on punitives.
	• The judge invokes judgment *non obstante verdicto* standard to overrule the jury's decision and lower punitives.
High Comps	• The jury buys that the foundations *must* be fixed or the house will fall down and wants to make sure Howitzer has the money to do it.
	• We put in our own damages theory on just doing a little reinforcement, and the jury thinks we're trying to bilk Howitzer and goes for big comps.
	• The jury sees it as little, unrepresented Howitzer against big, insured Throckmorton and wants Howitzer to get paid.
	• The opposing counsel comes up with some really dramatic damages evidence, like a big crack in the foundations or the house next door falls down.
	• The opposing counsel effectively plays on the jury's anxiety about the same thing happening to *their* houses.

Low Comps	• The jury decides there is enough to find liability, but that the repair bill is overblown and just wants Howitzer to have some compensation.
	• We come up with our own engineers that look great on the house only needing lesser repairs.
	• We manage to contend that the condition of Howitzer's house is "normal" for the neighborhood and construction methods used at the time without looking like slime buckets.
	• We weaken the causation angle by reraising the issues of whether anyone would have noticed the condition, and by arguing that Howitzer didn't really rely on Zelda's assurance because he got a V.A. report before buying.
	• We argue that yeah, even if Zelda should have been more careful in what she said, she wasn't trying to cheat Howitzer, he got a report, and it would be unjust to award him a windfall when he has a liveable house.

Lastly, you note that it may seem curious to have the jury decide on punitives in two passes, and that you'll have to keep that in mind and think hard about whether they'll actually do it that way. You also now count five different compensatory damages scenarios (one in the main tree and four in the subtree), and you may or may not be able to think meaningfully about how the compensatory damages vary in each of those scenarios.

4.2.2.5 How Do I Do Multiple Arguments?

One of the most common sources of difficulty in drawing the tree is how to handle multiple arguments, such as when you have more than one argument you can win with (by itself), or when you can only make one argument (usually for some procedural or "spill-over" reason). The following topics show how to deal with these difficulties in their most common form. There's also a discussion in Part III on how to handle many possible arguments without having to draw a chance node for each one, as is done in this section.

Alternative Arguments

When people have more than one argument they can win the case with, they often come up with a drawing like the one below (Figure 4.47).

Unfortunately, it doesn't work to put down alternative arguments as the branches of a single chance node. It doesn't work on a qualitative level because it's not clear whether, for instance, the first or the second chance node/uncertainty tells if you first argument won, and also because combinations like losing on the first argument but going on to win on the second don't appear anywhere. It doesn't work on a quantitative level because people figure they're going to try both arguments and omit probabilities on the first node, so the total probabilities add up to 200%!

A correct drawing for two alternative arguments to win on appears below. This tree includes all the possibilities of winning and losing on the second and first

Figure 4.47 Incorrect tree drawing

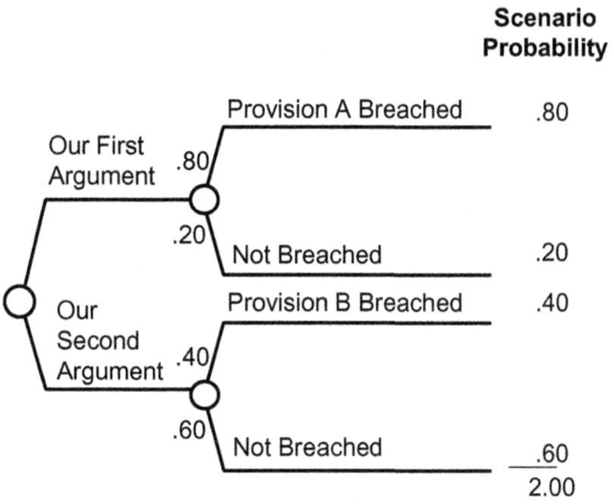

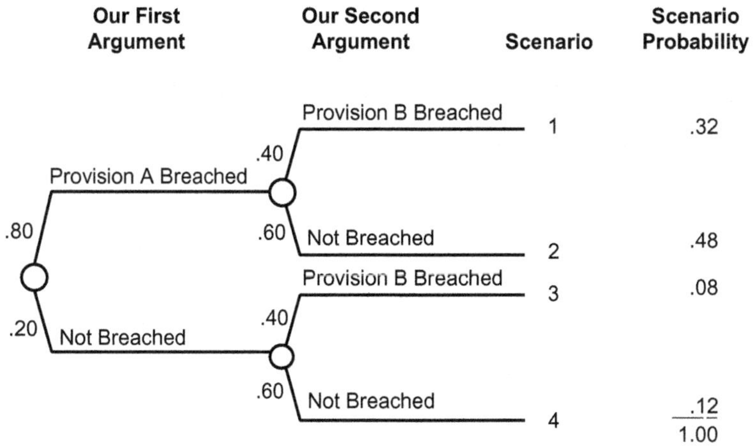

Figure 4.48 First way of drawing two alternative and independent arguments

arguments, and the probabilities come out to a proper 100% when you multiply and add them together.

Note that this tree is drawn with the second argument coming first. If the two arguments really independent of each other (that is, you win or lose on either one without that success or failure affecting the other argument), then it doesn't matter which order you draw them in. The second tree drawing is equivalent to the first.

You will note, however, the order of the two middle scenarios switches when you reverse the order of nodes. This doesn't matter because the probabilities still come out correct (Figures 4.48 and 4.49).

As you can see, the probabilities of particular scenarios don't change with the order of the arguments. It's just a matter of which order makes more sense to you. If the probabilities did depend on each other, there is still a way of changing the

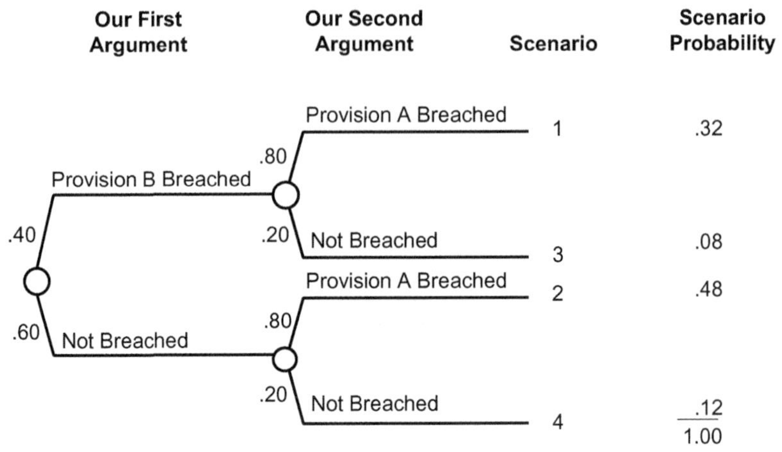

Figure 4.49 Second and equivalent way to draw two independent arguments

Figure 4.50 Relevance diagram for two independent arguments

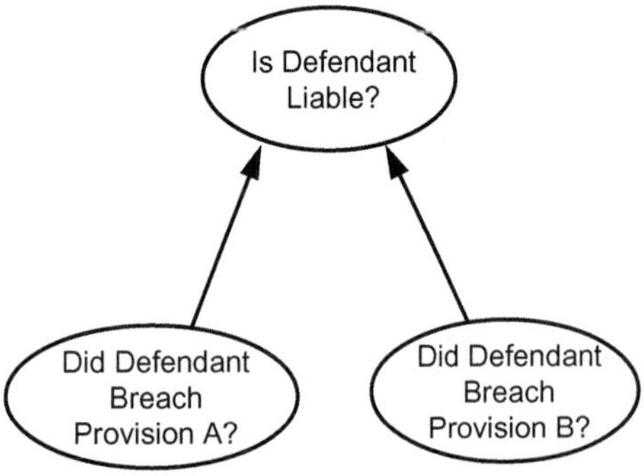

ordering which is covered in the section in Chapter 9 on A Second Look at Dependent Probabilities.

If you thought that your chances of winning on your last argument did change according to whether you won or lost on the first one you presented, then you'd be positing a "spill-over" effect. See the discussion of "spill-over" in Part III for details on how to handle it.

Lastly, the relevance diagram for two alternative arguments appears below. Does this diagram correspond to the first tree shown or the second one (or both), and how do you know? (Figure 4.50).

Another difficulty with alternative arguments is when they both aim to recover the same loss, just by different legal theories. People commonly draw each argument as a separate tree (here, only a one-node tree), and then add up the mean values, as in this case where there are two theories for recovering a single $5 million loss (Figure 4.51).

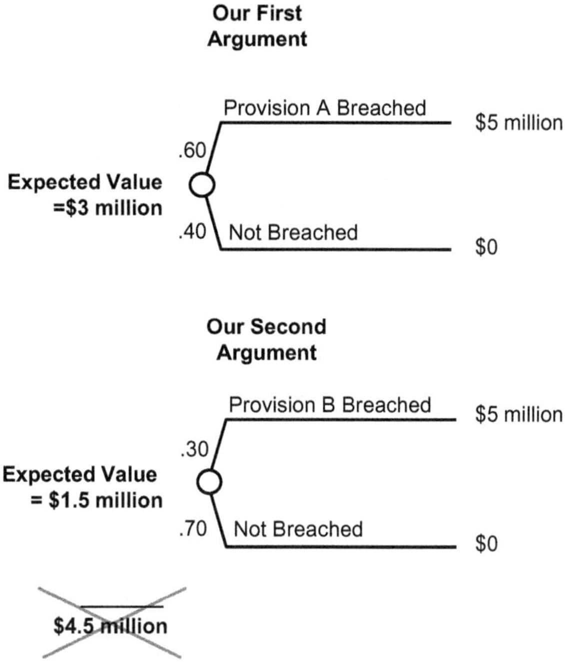

Figure 4.51 Wrongly adding up two independent arguments for achieving the same damages

This method produces a result mean value of the case of \$4.5 million. Sound plausible? Try using this method to solve for the mean value of the case if your chance of winning the second argument is .40, then again if it's .60? Now what is the "mean value" of the case? How do you explain the results?

A correct drawing for two alternative methods of recovering a single loss appears below. This tree correctly handles the fact that if you've already won on Provision A (a .60 probability), it doesn't buy you anything extra to win on Provision B. Your B argument only buys you an extra $.4 \times .3 = .12$ chance at the \$5 million in the case where you already lost on A—not an additional .3 chance at the \$5 million as the two separate trees above assumed. Now do you see where the previous tree over-counted and came up with a too-high mean value?[6] (Figure 4.52).

As you become familiar with thinking this way, you'll realize that your chance of getting nothing is $.40 \times .70 = .28$, so the mean award is $(1 - .28) \times 5 = \$3.6$ million.

In graphical form, the separate probability spaces (dividing up the total probability) for each of the two independent arguments for recovering the same damages is shown below (Figures 4.53 and 4.54).

[6] Just to make sure it's completely clear, simply adding up the first two trees double-counted the win amount when you won A and B, for an extra $.3 \times .6 = .18$ chance at the \$5 million, contributing an extra $.18 \times 5 = \$.9$ million to the mean value. And, of course, $3.6 + .9 = \$4.5$ million. You'd get the same .18 over-counting probability by subtracting the .12 lose-A/win-B probability from the .3 win-B to arrive at the .18 win-A/win-B.

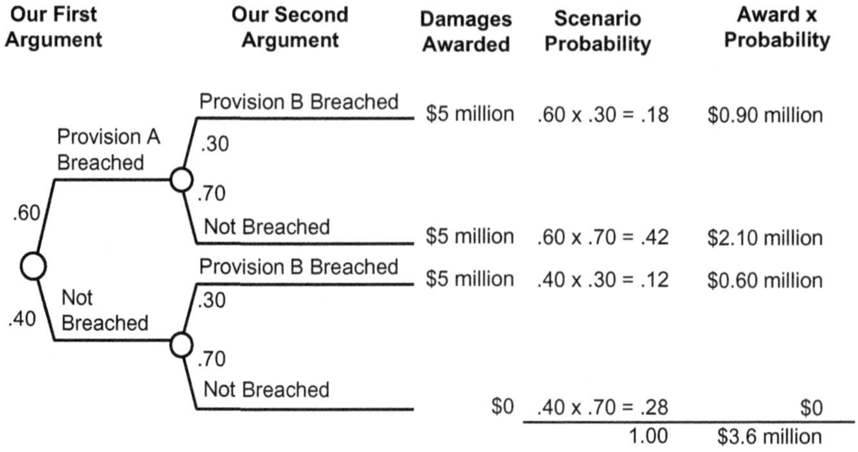

Figure 4.52 Correct way to calculate the mean damages awarded with two independent arguments for the same damages

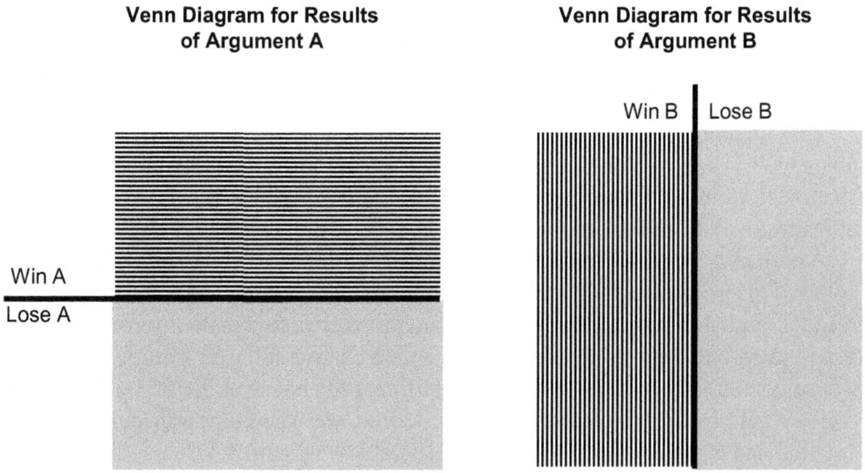

Figure 4.53 Separate Venn diagrams for probability spaces to win each separate argument

As you can see, probably account for the overlaps is precisely the need to do the calculation carefully.

The following pointers will help you avoid these problems of unclear nodes and double-counting,

- Make sure the branches of a chance node comprise a set of mutually exclusive and collectively exhaustive events.
- Make sure the roll-forward probabilities tell a sensible story (that is, they add up to 100%).

Figure 4.54 Venn diagram for joint probability space for two independent arguments to recover the same damages

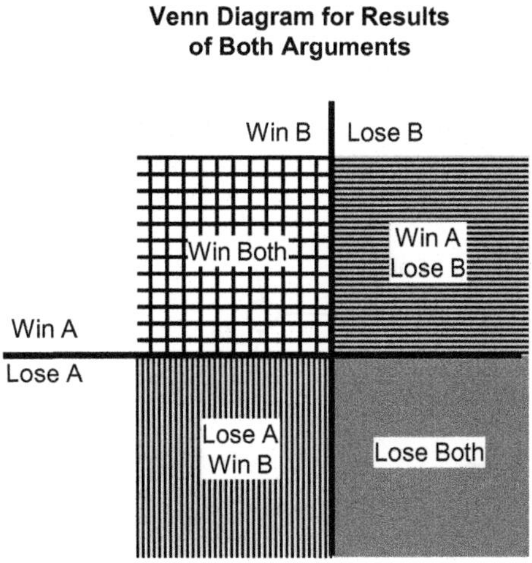

Figure 4.55 Choosing between two mutually exclusive arguments

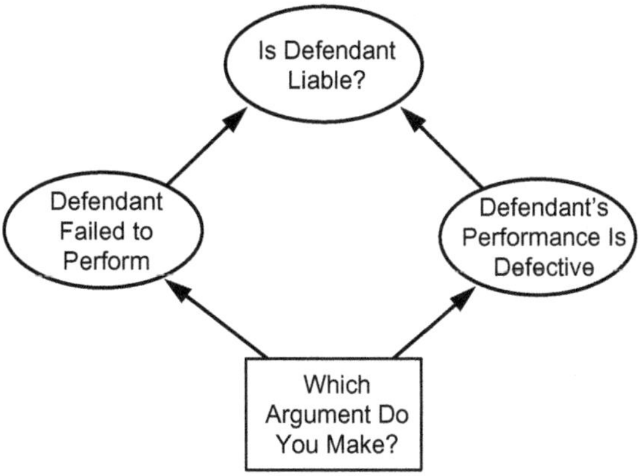

When You Can Only Make One Argument

Occasionally, you may have a situation where you must choose between trying two different theories, rather than just presenting both. It would, of course, be rare that you were prevented from trying both arguments. However, you might feel that, because of spill-over, presenting two contradictory arguments to the jury would destroy your credibility and cause you to lose on both (such as, for instance, arguing that the defendant did not complete the work and that the defendant did complete the work late). You might therefore think it advisable to choose the better argument and present only it. The relevance diagram for this situation appears below (Figure 4.55).

A possible decision tree for this situation appears below. Note that the spill-over effect is expressed in the different probabilities of finding the defendant liable, depending on which argument you make. These probabilities would need very careful assessment (Chapter 4) and, indeed, the usefulness of your analysis depends on your thoughtful consideration of how spill-over would operate (Figure 4.56).

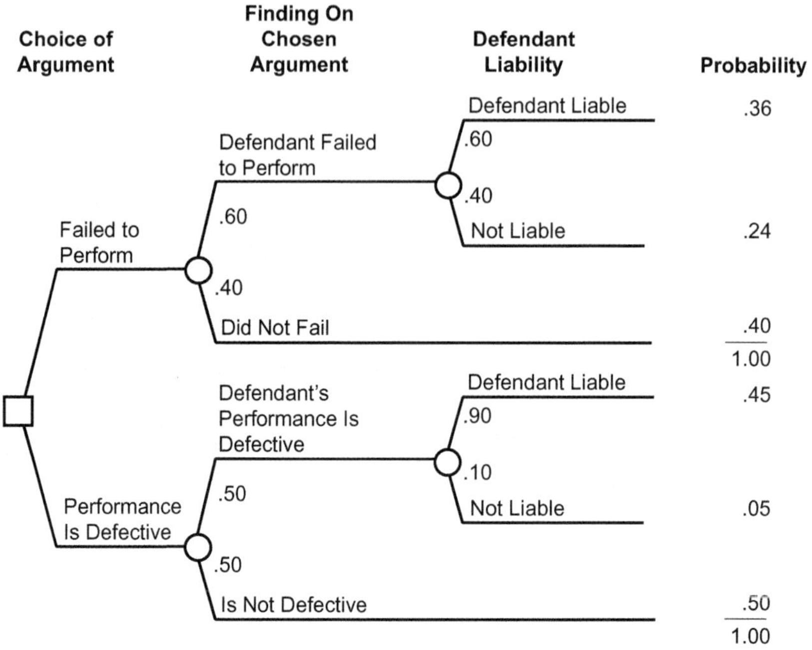

Figure 4.56 Decision tree for choice between two mutually exclusive arguments

With the probabilities of winning on each argument, probabilities of winning the case with a finding of defendant liability can be calculated. For this tree, arguing the defendant's performance is defective offers a .45 of winning the case, while arguing the defendant failed to perform offers only a .36 chance of winning. Thus, according to this analysis, assuming the potential damages are the same, plaintiff would be better off arguing that defendant's performance is defective rather than arguing defendant failed to perform. Of course, in many actual cases, potential damages would vary as well and they would likewise need to be considered. Clarifying trade-offs between alternative legal strategies (chances of liability versus potential damages) is one of the main ways applying Decision Analysis allows one to craft winning legal strategies.

4.2.2.6 When Can I Draw One Node *or* Two? (Structure Equivalency)

In this section, you've seen a number of examples where it was incorrect to draw one chance node instead of two, or vice versa. One final wrinkle is that, as you already know, there are many equivalent ways to draw a single decision tree—that are *all* correct. You've likely played with a number of variations in drawing trees already, but, as a special example, any uncertainty that has three possible outcomes can be drawn as a single three-branch node, or as two two-branch nodes. An example will help make the point clearer.

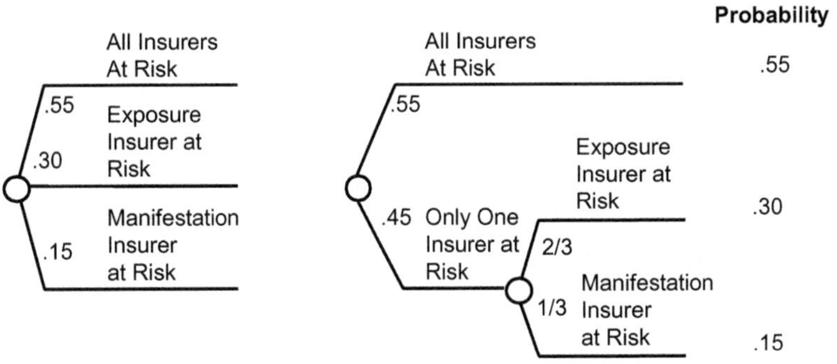

Figure 4.57 Equivalent trees for which insurer is at risk

Suppose that an insured has a claim for a health condition, and one insurance company provided health coverage when the person was exposed 20 years ago to the alleged cause of the condition, while another company provided coverage when the condition manifested itself more recently. A court or jury could find the insurer at risk when the person was exposed, the insurer at risk when the condition manifested, or all (both in this case) insurers at risk from the date of exposure to the date of manifestation. The two ways of structuring the tree for this uncertainty are illustrated below (Figure 4.57).

How would you draw the equivalent trees if you include the possibility that no insurer be held at risk?

Note that the probabilities in each tree could be different if you thought the jury will be affected by which way you argue your case (that is, a one- or two-stage determination of which insurer should be held at risk), or if you thought the jury was going to decide the issue one way or the other, with differing chances of various results. This kind of a spill-over effect would mean that the two trees were no longer equivalent because of different probabilities—even though the number of nodes and branches hadn't changed.

If you don't think there's a spill-over effect (making the two possible structures equivalent), then the best one is whichever corresponds to your thinking about the problem. Ask yourself:

- How do you think the judge or jury will decide the issue?—or—
- How do you find it easiest to assess probabilities?

These questions are, of course, just specific applications of the Trier-of-Fact Rule: structure your problem the way you think the trier will think about it.

4.3 Determine the Value Formulas

The final step in structuring the problem is determining how to value each scenario in the decision tree. In most cases, this determination will be very simple for your first pass analysis.

- The value of a settlement scenario is the amount of the settlement.
- The value for a winning defendant or losing plaintiff is zero.
- The value for a losing defendant or winning plaintiff is the amount that the court awards.

Important: all numbers should be present value numbers. That is, all future payments should be discounted back to the present so that the present value goes at the end of the tree. Besides handling the time value of money (a big industry in itself), this step also keeps your value calculations simple by making all dollar figures in the tree equivalent present value numbers.

For cases which will be resolved in a year or two, the effect of discounting is small enough you can ignore. Discounting only becomes important when numbers extend years into the future, as in future lifetime lost wages or medical costs.

The judgement should be based on what you expect the court to award in your case, and not on past losses. Thus, because the court award will normally be one of the questions you assess probabilities for, you won't be able to fill in the actual numbers until after the probability assessments. That's why this stage only asks you to determine *how* you will value each scenario.

It is perhaps easiest to keep all the numbers straight by using positive numbers for amounts you receive and negative numbers for amounts you pay out (thus, an award would be positive for the winning party and negative for the losing). If you're only looking at different possible amounts you could pay out, you can use positive numbers, but you'll need to remember that you want to minimize your payouts, rather than just choosing the maximum positive amount.

Your value formula could be more complicated if you consider one of the more advanced topics covered in Part III, such as legal fees, prejudgment interest, and complicated damages. The discussion in Chapter 6 on analyzing the results will help you decide if you need to include these other costs.

If you feel very uncomfortable leaving all consideration of legal fees out of your first-pass analysis, a simple way to handle them is to make a best guess of the future legal fees to litigate the case and subtract that estimate from the mean value of litigating the case. The section of sunk costs under Legal Fees and Costs in Part III discusses at greater length why you should only consider future legal fees and expenses.

You look over your decision tree, and see that there will be four formulas for the values at the end of the tree: the settlement amount if you settle, zero if you litigate and win; the amounts of the comps if you are held liable but avoid punitives, and comps + punitives if you're liable and hit with both (Recall that you're omitting avoidable legal fees, possible codefendant apportionment, the effect on precedence, etc., from your first-pass analysis.). You added those formulas to your tree drawing (Figure *4.58*).

Note also that you changed the tree drawing to show the expected punitive and compensatory damages when you're hit with both. The summary mean value of damages from your subtree will go there. You decided you'd rather see the node for the comps with intentional misrepresentation and no punitives, and it only makes the tree a little bigger.

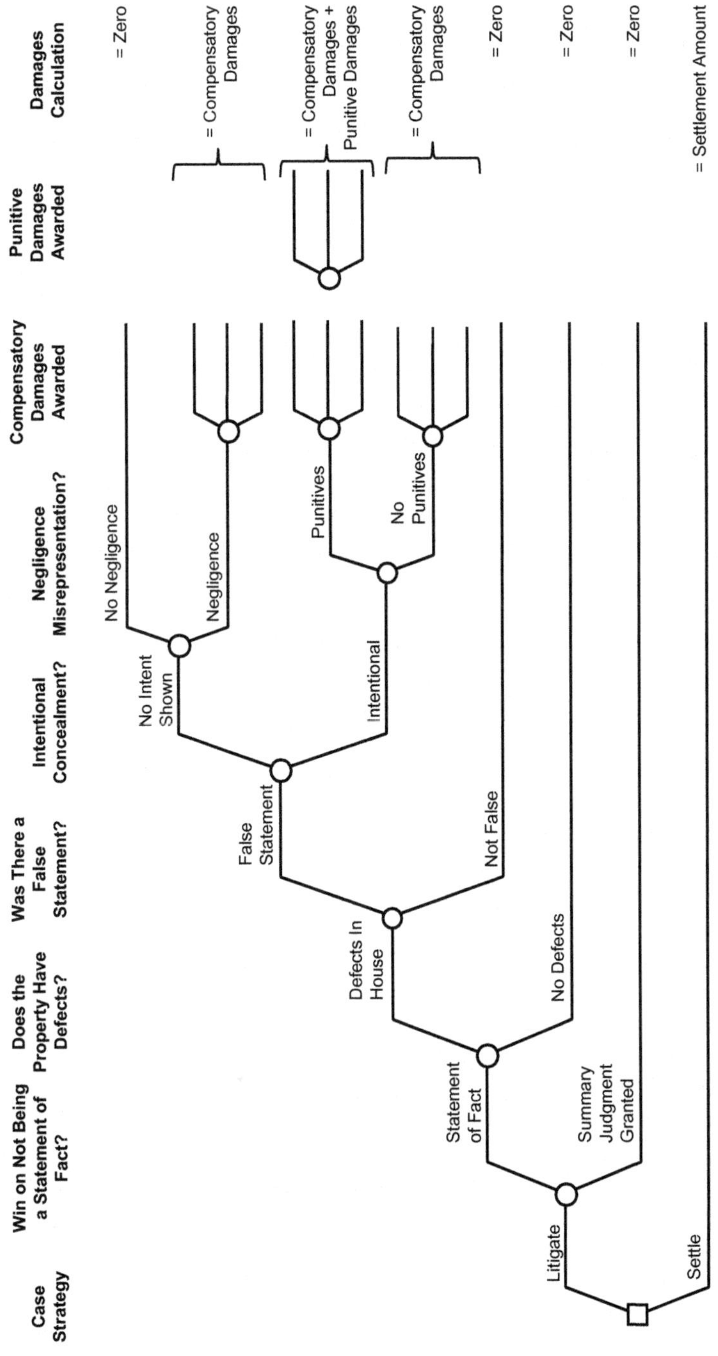

Figure 4.58 Howitzer v. Crabtree decision tree with method of calculating damages awarded

4.4 Problems

4.4.1 Draw the Relevance Diagram

1. Relevance diagrams can be described as a kind of flow diagram. What is "flowing" from one circle to another via the arrows, and what happens when the "flow" arrives? (In general terms).
2. List and explain two kinds of relationships that an arrow in a relevance diagram can represent (There are at least four.). Do you have to keep in mind what kind of relationship the arrow represents when you're drawing the tree? Why or why not?
3. For each of the following situations, try drawing a relevance diagram (Note that the answers will vary with the person responding.). Did it help you to think about the problem? What further questions were raised?

   3.1 Your client received a severe electric shock while using a portable automobile vacuum. The accident happened at the client's workplace in state A; the client lives just across the border in state B, and the company which makes the product and distributes it nationwide is located 1000 miles away in state C. You need to decide where to file the case.

   3.2 Your client accuses the contractor who built his home of shoddy workmanship. The client is withholding final payment and wants to sue for additional damages. The client's complaint is that the floor of the intentionally unfinished attic above the garage collapsed when the client nailed heavy, 1-inch thick sheets of plywood across the $2'' \times 4''$ ceiling joists and had a grand piano placed there. The contract disclaims any liability for such an intended use. The construction contract has a general warrant of habitability and fitness for residential use.

   3.3 The defendant's predecessor in title owned a six-unit apartment building in a very desirable section of the city. When she had the building converted to condominiums, she had the garden area deeded personally to her ground-floor unit, rather than following the usual practice of designating it as common area. She soon thereafter sold her unit to the defendant, who, upon learning of the arrangement, told the other owners he did not want it for himself and would deed it to the condominium association so that it could become a common area. The plaintiff relied on these representations in purchasing his condominium.

   For a variety of reasons (the association needed to concentrate on painting the building etc.), the defendant never executed a deed as promised. He has now reconsidered the value of the land and wants to keep it, or at least sell it to the association. The plaintiff is considering bringing suit to force the conveyance as promised.

   3.4 You're an attorney in the Commerce Department charged with producing an opinion on whether action is required in a particular case under a statute requiring retaliation against the Japanese whenever an American product has been "unfairly excluded" from the Japanese market. The article in question is American-made Japanese baths.

The market in Japan for such baths is enormous. Every dwelling unit has one or has access to one, and very few have western-style baths or showers. A number of California manufacturers of hot tubs hungrily eyed the market and produced special models for Japan. They've had great difficulty getting them approved for sale in Japan, and, even when approved, have had difficulty importing and selling them. The manufacturers contend that the Japanese have been systematically excluding them to protect local industry.

The Japanese Ministry of Trade contends that the tubs were unsafe because the water heaters had insufficient electrical insulation (posing a danger of electric shocks), and because they drew too much electricity for most Japanese homes' wiring (posing a danger of short-circuits or fires). They further contend that, even when the tubs were modified to address these concerns, the Japanese shied away from them because they required unusual and difficult installation procedures, and because the average Japanese found them ugly.

4. Give at least one reason why dependent probabilities are helpful, and one clue that would help you know when they're needed. Are there any drawbacks to using dependent probabilities? How would you decide which uncertainty to assess first?
5. Draw the relevance diagram for each of the following situations. For each issue, indicate whether the dependency will be expressed in the structure of the tree (the "if yes" or "if no" labels), dependent probabilities, or dependent outcomes. Why might your preliminary answers on this question change as you draw the tree and assess probabilities?

   5.1. A number of years ago, your client invested a great deal of time and money in developing a distinctively shaped rear spoiler for its sports cars. Since then, the spoiler has become a big hit and client now has a large business in selling spoilers for cars not originally equipped with them (both for its own make and others). Competitors now sell identical spoilers, and you are seeking to stop them under trademark and unfair competition theories.

   The spoiler has gained wide recognition, and thus may qualify for trademark protection as identifying the maker. And, although the spoiler is functional in that it reduces aerodynamic lift in the rear of the car to maintain better traction at high speeds, there are numerous other shapes that do the job. Thus, the spoiler may or may not be disqualified for trademark protection by being functional.

   Under unfair competition, misappropriation seems the most suitable theory. To establish a right, you need to show that client has invested, labor, skill, and money in developing a salable product (that is, there is a property right in the product). You further need to show that, by duplicating or imitating the original, competitors have the name, reputation, or business good will of client.

5.2. You represent Canada in a dispute before the International Court of Justice (ICJ) in the Hague regarding the Alaska/Vancouver boundary delimiting the offshore continental shelf between Canada and the USA. Because there is no governing treaty, the case is being tried under the principles of customary international law.

The ICJ has most recently stated the relevant principles in a similar dispute between the USA and Canada regarding the Gulf of Maine. Those principles are proportionality (that is, allocating shelf territory according to the respective length of the local coastlines), and non-encroachment (each state has an inherent right to those portions of the continental shelf that form a natural prolongation of its land territory into and under the sea, without encroaching onto the prolongations of others.)

Applying those principles is especially tricky in this case. Regarding proportionality, each country has many offshore islands in the area. If the court counted the coastline of all islands, Alaska would be awarded approximately 3/4 of the disputed area, though the coastline measurements are subject to dispute and the results of USA and Canada surveys vary by 20%.

Applying the non-encroachment principle would seem to point towards dividing the territory equally because both countries have islands that extend equally far out into the Pacific. The court could also reach this result by applying the method used by Norway in The Fisheries Case, where it marked baselines along its offshore islands and asserted territorial claims from there. However, unlike Canada, Norway had previously asserted this method for some 30 years. The court may or may not find this factor important.

Finally, the ICJ may apply the general principles of equity that underlie the aforementioned rules to find that, since Alaska has a very long coastline otherwise, Canada should get the continental shelf in this area right up to the US territorial waters (11 miles out), or perhaps even farther in. The 11-mile line would give Canada all of the disputed area, and a line farther in would cede even more.

5.3. You represent a group of minority shareholders in a corporation that's being merged. Under the laws of the state of incorporation, the merger gives rise to appraisal rights for the shareholders; that is, they can elect to have their shares purchased by the corporation rather than taking stock in the new corporation in exchange. Accordingly, you are advising the shareholders on the best course of action.

The stock of the present company was trading at $25 per share before trading in it was halted. The stock of the acquiring corporation has not previously been publicly traded, but it has historically paid dividends of $2.10 per share each year. Capitalized at a rate of 6% (3% real rate of return plus 3% inflation), this yields a stock value of $35. However, those dividends may decline as a result of the merger.

Alternatively, the shareholders could have the court determine a fair price for their shares. The traditional method of determining this value is a weighted average of the earnings, asset, and dividend value of the stock.

The accounting earnings over the past 5 years have averaged \$5 per share and the dividends have only been \$1 per share (yielding a capitalized value of \$16.67 per share), but the asset value is \$40 per share. However, the fact that accrual accounting changes the timing of revenues and expenses without discounting them makes you very suspicious of accounting earnings, and you know that the company has many assets that have been completely depreciated and are thus carried at zero book value, though they continue to be productive and have significant resale value. Further, the company has expanded significantly over the past few years and expects to generate large amounts of cash over the next few years. Thus, you expect greatly increased dividends either now or later in the future, depending on how long the company continues expanding.

Accordingly, you would argue that capitalizing the increased dividends is the proper way to value the stock. This precedent for this in that Delaware, which historically used only the weighted average method of valuing stock, has now recognized the dividend discount model as a legitimate alternative. You estimate that dividends could increase to as much as \$4 per share, but you would have to persuade this state's courts to accept the dividend discount measure, and would have to use a probabilistic measure of future dividends to establish the higher stock value.

4.4.2 Draw the Decision Tree

6. For each situation below, draw the relevance diagram and then the decision tree. Be especially alert for where nodes should or should not be repeated. Where a node is repeated, indicate whether or not there is a dependency.

   6.1. You represent the unhappy purchasers of a home, Loretta and George Mutton. Loretta and George toured the premises prior to purchase and noted nothing wrong, but, upon closing escrow and occupying, they found the pipes so clogged with debris that the bathtub filled up as you took a shower and the toilet clogged frequently. They first engaged a rooter man to clean out the pipes, but he found that prior occupants had tried lots of crystalline drain cleaner—producing deposits of cement-hardness in the pipes. As a result, large sections of pipe had to be ripped out of the walls and basement and replaced at a cost of thousands of dollars.

   You've concluded that was no contractual misrepresentation, and your jurisdiction has not yet recognized an implied warranty of habitability in the sale of single-family homes, though it is close to doing so. You're not sure if bad drainage would violate the warranty even if a court recognized it. Nor are you sure what damages the court would assess, since the Muttons have admitted also using crystalline drain cleaner before calling the rooter man. The seller has offered \$4,000 to drop all proceedings.

6.2. You are the general counsel for The Neutron Glow Boys, an electric power utility that's into nuclear. The Glow Boys have a contract to supply all the electric needs of Bronze Age, Inc., for a fixed price for 10 years more years. Unfortunately, Bronze Age's business of bronze electroplating baby memorabilia has taken off with the common decision of aging yuppie-dom to have babies and their tendency to bronze everything from the first pair of Reeboks to the first Play-Do pasta salad. This boom is unfortunate for the Glow Boys because their spent fuel storage costs have skyrocketed and no federal storage facility looks to be available until after the contract has expired.

You have been asked, therefore, to advise the Glow Boys on whether or not to breach the contract. If the Glow Boys honor the contract, then, unless there's a drastic reversal in their fuel storage situation, the cost to store and Bronze Age's use will determine how big a loss they take. On the other hand, if they breach the contract, they might get away with paying a moderate amount specified by the liquidated damages clause in the contract. You suspect, however, that Bronze Age might contend that the clause does not apply and ask for a variety of remedies, ranging from the benefit-of-the-bargain to specific performance, and that damages in each case could vary widely.

6.3. You represent the Greater Arcata Truffle Pickers and Packers, a union that includes almost all the licensed truffle pickers in Arcata (one of only two areas in the whole USA where truffles grow). The spring picking for the brief, semiannual truffle harvest is fast approaching and the union contract is up for renewal. Negotiations have been extremely frustrating and the membership is considering calling a strike.

You're unsure how much of a wage increase the pickers will be able to obtain if they just keep negotiating because, so far, the pickers are at 9 percent and the growers have only offered four. However, going on strike presents a number of problems.

The contract has a special clause specifying no strikes during the critical 2-week ripeness period that comes twice a year (A strike would not be barred, then, during the subsequent flash-freezing, pickling, and repacking.). If a strike is called now, it might or might not extend into the truffle picking season. If it doesn't, then the pickers are presented with the same question of how much of a raise they'll get (perhaps with more bargaining leverage as a result of the strike).

However, if the truffles do ripen (a function of the weather), then the union will be faced with the choice of whether or not to go back to work. If they don't go back to work, they can be fired. If they do go back to work, then a special binding arbitration clause comes into play. You, for one, have a dismal view of the possible results from the arbitrators. And, of course, defying the arbitration award could lead to injunctions, fines, contempt citations, etc.

6.4. You are the general counsel for the Morning Crisp Co., a cereal manufacturer. The FDA has just announced new labeling requirements for products containing the so-called "tropical oils" (palm, palm kernel, and coconut) under which any mention of the oils in the product must be followed by the legend "A Saturated Fat." This includes such large-print banners as "Contains Only Vegetable Oil" and the actual names of the oils in the ingredients. Unfortunately, all of Morning Crisp's products contain tropical oils, and the brand-new Oat Bran Multi-Form Facility has been specifically tooled for their range of saturation and solidity temperatures. Further, all the consumer avoidance of saturated fats likely spoil much of the market share gains from Morning Crisp's "Health & Bran Campaign."

Your task, then, is to advise the company on whether to oppose the regulation. If Morning Crisp decides not to oppose the regulation, then it has to decide whether or not to change its products to non-saturated fats (like soybean oil), with a cost to reformulate the recipes and re-tool and an impact on sales either way. If the company decides to oppose the regulation, then they'll be in the same boat if they lose, they'll probably not contemplate a product change if they win, and the publicity generated will likely impact sales either way.

6.5. You are the general counsel for Lotus Motorcars, Ltd., of the UK. The director of research, Dr. Octane, has come to you with his suspicions that Ferrari has gotten hold of the basic design for Lotus' experimental Variable Density Swirl-charge System (VADSS), which allows an engine to be lower compression, higher efficiency at small throttle openings and higher compression, higher power at large throttle openings—in effect giving you an economy engine and sports car engine all in one. The technology is still under development, and, if successfully developed, could be introduced in the 2018 model year. The question now is whether to sue Ferrari for unauthorized acquisition of the proprietary technology.

A suit would probably be expensive and drawn-out. And even if Lotus won the suit, they might only obtain damages from Ferrari for use of the technology, rather than an injunction against Ferrari's use. Lotus would still have to complete its own development on VADSS and introduce it, although, presumably, any VADSS model would have greater sales success if Ferrari could not also use the technology. Dr. Octane has also posed the intriguing question of whether any Ferrari progress with the technology could be obtained in the course of the lawsuit.

Finally, in response to your questions about whether there was any question of the VADSS technology being too expensive, even if successful, Dr. Octane replied in the affirmative. The tooling would be horrendously expensive because the engine block would be a hollow centrifugal aluminum casting with nickel-titanium cylinder liners—possibly making the engine just too expensive to introduce into a production model.

7. Roger von Schwartzwasser owned and farmed the northeast corner of the township of Delaney, his holdings comprising 80 acres. A public road was put through the center of the property in a north–south line, running straight into the center

of the property from the north border before veering off to the west before leaving the property. There was thus a triangle of land that lay between the southern portion of the road and the true centerline of the property.

Roger occupied a house on the western half of the property until his son Carl become of age, at which time Roger deeded him the western half and moved into a new house on the eastern half. Roger continued to farm all the land east of the road until he died 15 years later. The property then passed to his wife Frieda. Frieda, in turn, deeded the property to her younger son Alfred, though she continued to live in the house. Seven years later, Carl himself died and his holdings devised to Carl, Jr.

You represent Carl, Jr. in an action for ejectment to recover possession of the triangle from his uncle Alfred, who has also continued to farm it. Carl, Jr.'s pleadings note that all the conveyances in question uniformly refer to the eastern or western halves of the tract. Alfred raises two defenses.

First, Alfred contends that he acquired title to the land by adverse possession. You note that the requirements for adverse possession are that it be: (1) hostile and under claim of right; (2) actual; (3) open and notorious; (4) exclusive; (5) continuous; and (6) that all these elements are present for a period longer than the relevant statute of limitations, which is 20 years in this case. You plan a two-pronged attack on this defense. First, that the necessary elements are not currently extent, and, second, that even if they are, they were not established until a time more recent than the statute of limitations.

Alfred's second defense is that the Roger and Carl, Sr. made an oral agreement that the road should divide the tracts, and the road should therefore be regarded as the legal boundary. He has introduced Frieda's testimony to this effect. Such an oral agreement would normally be invalid by either the parole evidence rule for written agreements or by the statute of frauds as applied to transfers of land, but Alfred contends that the agreement is relevant by the Parole Agreement Doctrine. The elements necessary to establish this doctrine are: (1) that a dispute or uncertainty regarding the boundary existed; (2) that there was an oral agreement to fix the boundary; and (3) that there were physical acts on the property referring to the agreed boundary. Regarding the first element, you note that an actual dispute is required, or that the boundary be uncertain in the sense that it is not calculable, as opposed to merely unmarked. Regarding the third element, you note that the disputed triangle has never been fenced in separately from the remainder of the eastern half.

7.1. Construct the relevance diagram for the value of the case. Be careful to include only significant uncertainties, and not points of evidence (though you may want to include key influencing facts).
7.2. Explain why or why not you think there are any dependencies between Alfred's two defenses.
7.3. Suppose that Alfred offers to settle the case. What value needs to be weighed against the settlement offer?
7.4. Draw the decision tree for whether or not to settle the case. You need not fill in numerical information like probabilities.

7.5. Suppose the award if you win the case has been complicated by Alfred investing thousands of his own money to install sophisticated irrigation equipment in the triangle, including a small catfish pond. Alfred has advanced the theory that he made the improvements in the good-faith belief that he owned the property and before he knew there was a dispute, and that he therefore should received offsetting damages for the value of the improvements. You suspect that, if the court agrees with him, it will give you the choice between paying for the improvements or allowing Alfred to destroy them (Because the materials are not reusable, removing the system is equivalent to destroying it.). You're not sure if Alfred would rather destroy his improvements than hand them over to his nephew. Draw the relevance diagram for the value of the award if Carl, Jr. wins.
7.6. Draw the decision tree for the value of the award if Carl, Jr. wins.

8. You are the chief counsel for Western Dry Goods, Inc., a distributor of dry (that is, non-refrigerated) food goods to grocery stores. About a year ago, Western was having trouble collecting bills from Russian Hill Grocers, a family-owned and incorporated business with three stores. When Western threatened to cut off credit, Frank Gianini, the son of the recently deceased founder, responded with an ambitious program to compete against the big grocery chains by expanding and changing its product mix. Russian Hill cater to the increasingly eclectic young professionals by offering upscale foods, ranging from a variety oat bran and whole-grain cereals to complete selections of Chinese and Thai foods to assorted oriental and European fresh mushrooms to fresh pastas. He predicted that the dollar volumes on these foods would be the same or higher than the current business, but with higher profit margins. His offer to Western was that they forego their current payment due-on-delivery policy and ship him goods on consignment (no payment due until sale). In return, he would guarantee them a fixed monthly payment (renegotiated biannually) plus half of all operating receipts after direct operating costs, up to a maximum of 125% of the standard wholesale cost of the goods. With some trepidation, Western signed accepted and signed a contract to that effect. Your predecessor as the counsel approved the offer.

The arrangement worked well for about 9 months, with Russian Hill enjoying initial success and Western receiving an average of 110% of standard wholesale prices. However, the large chains in the area responding by adding the same goods to their offerings, using their volume discounts on acquisition to undercut Russian Hill's prices. Frank had anticipated this move and planned that superior service would retain his customers, but his business has been steadily declining nonetheless. Other suppliers are now suing Frank for amounts due them, and are trying to hold Western liable.

The plaintiffs advance two theories. First, they contend that Western is in partnership with Russian Hill and thus jointly liable for its debts. They base this on the contentions that the consignment arrangement constitutes co-ownership of the assets of the business, and that the payment arrangement is a share in the profits of the business. You note that sale on consignment is not usually deemed co-ownership, but, if the court does find co-ownership of business assets, it will be a factor in determining whether a partnership exists. Participation in the profits of a business creates a presumption of partnership, but that presumption may be rebutted if the profits are received in payment of a debt (such as amounts owed for merchandise). You therefore believe that you can effectively contest these two grounds. Further, you would contend that the contract is not a partnership agreement, although, in general, courts generally interpret oral partnership agreements as terminable at will, rather than void because of the statute of frauds.

The plaintiffs' second theory is that, even if there is no partnership, Western should be equitably estopped from denying it. Establishing this theory requires (1) a false representation, and (2) detrimental reliance on it. The plaintiffs assert false representation to them in Russian Hill representing Western as co-venturer in the expansion, and in Western knowing of the representation and failing to take any steps to disclaim. You dispute the second assertion, and don't know if the plaintiffs can establish the first. You further don't know if the plaintiffs can show they relied on this representation (if any) to change their contracts with Russian Hill to their detriment.

Your conversations with Frank have revealed one more wrinkle. Frank was very sloppy about separating his personal finances from those of the business. There is therefore a question of how much of his debts will be classified as those of Russian Hill, of Frank personally, or of a grocery delivery business he started in conjunction with the expansion.

8.1. Draw the relevance diagram for the amount of Western's liability.
8.2. Consider the possible evidentiary sources relevant to the uncertainties (such as deposing plaintiffs' contract signatories). List at least four of them. Are they additional uncertainties that could be added to the relevance diagram? Why or why not would you want to do so?
8.3. Prepare an outline discovery plan on the issue of whether a partnership exists. How can the relevance diagram help you manage the discovery project? How would you update the diagram and discovery plan as you received information?
8.4. Would you find it useful to assess the information for a decision tree and evaluate it at this point? Why or why not? What would indicate to you that a decision tree was appropriate?
8.5. Draw the tree for the amount of Western's liability. Can you draw it as two separate trees, or must it be one larger tree? How can you tell from the relevance diagram? If two trees, what information (if any) is needed by both trees? Is this the same as a dependency?

4.4.3 Determine the Value Formulas

9. For each situation in problem 6, add the value formulas to the decision tree (You will first need to draw the relevance diagram and decision tree if you have not already done so.). Indicate any additional information you require or assumptions you need to make in order to do so.

10. For each of the following situations, structure the problem and then go through the following steps to complete and appraise your structure. You need not include all the possible uncertainties and decisions in the problem; only those that seem most important to include in a preliminary analysis. Feel free to flesh out the problem where the information given seems inadequate.

10.(a) Structure the problem by drawing the relevance diagram and decision tree.

10.(b) For each tree you've drawn, read the tree from left to right and examine the sequence of events. Are you satisfied that the tree is a reasonable representation of the problem? What omitted decisions or uncertainties would you like to include—even if they weren't mentioned in the original problem statement?

10.(c) For each tree you've drawn, describe any dependencies you see in the tree (That is, where the probabilities or outcomes of an uncertainty or the alternatives of a decision depend on something that comes before in the tree.). Is it clear which of each pair (or group) of dependent variables should be assessed first? Does this sequence of assessment correspond to the time sequence that the events will be resolved in?

10.(d) For each tree you've drawn, describe what value measure you'll use to value each scenario and to distinguish among the alternatives, and fill in the value formulas. What information do you have to calculate the values with, and what information do you still need?

10.(e) For each tree you've drawn, list all the information you still need to complete a preliminary analysis and a reasonable source or two for it. Which information must come from the decision-maker? Which information can you, as counsel, supply? For which information would you recommend recourse to a third-party expert (who may be in-house)?

10.(f) Finally, for each tree, comment on the level of complexity of your problem description and whether that level is needed to accurately represent the problem. Is it already unwieldy and looking computationally difficult? Can a simpler structure capture the essence of the problem?

10.1. A California resident was killed while flying on an Washington State-built airplane that crashed in Germany. The plane was owned and operated by a German airline, but the jet engines were built in England. You represent the estate of the deceased in an action filed in California against the airline and the two manufacturers. All parties have filed for dismissal on the ground of forum non conveniens, and the German airline also contends lack of personal jurisdiction. The Washington State manufacturer has agreed to drop its motion if you agree to dismiss the case in California and refile it in Germany. Note that Germany is a civil law country, as opposed to England and the USA being common law countries.

10.2. You represent a criminal defendant who is currently charged with an offense under the laws of the District of Columbia. You could, however, contend for dismissal on the ground that your client should be charged under the applicable federal laws, rather than under the D.C.

laws. Further, your client is addicted to crack and contends that he has no control over his own actions—that he acts only to serve his habit. D.C. has adopted the Model Penal Code version of the insanity defense, under which a "person is not responsible for criminal conduct if at the time of such conduct or as a result of mental disease or defect he lacks substantial capacity either to appreciate the criminality [wrongfulness] of his conduct or to conform his conduct to the requirements of law." The federal insanity standard, under 18 U.S.C. § 20 is that it "is an affirmative defense to a prosecution under any Federal statute that, at the time of the commission of the acts constituting the offense, the defendant, as a result of a severe mental disease or defect, was unable to appreciate the nature or quality or wrongfulness of his acts. Mental disease or defect does not otherwise constitute a defense." The applicable D.C. penalty is 10 years without probation, versus only 5 years under the federal statute.

10.3. Two years ago you made a loan to a friend's failing business. The loan was payable in 5 years and, in the meanwhile, paid periodic interest. Further, the loan was convertible to common stock and conferred common stock voting privileges if the interest was not paid for 6 months.

Your friend never actually paid any interest, but your tax situation required you to report it as income as it accrued. The business has now gone bankrupt and you're seeking to collect your loan and interest in bankruptcy court. Given the meager remaining assets, you only hope to collect 5–10¢ on the dollar. Further, other creditors have moved to reclassify your loan as equity, in which case you would only be paid any remainder along with other shareholders after all debts were paid (that is, probably nothing).

Alternatively, you could drop your claim in bankruptcy court and try to deduct the loan as a business loss. You're unsure of whether your loan would then still be subject to reclassification, and what effect reclassification would have on your eligibility to deduct the loss. There is also a question of whether you should just try to deduct the lost interest on this year's tax return, or whether you should try under I.R.C. § 1341 to recover the tax paid on the interest in prior years when they were actually claimed and the tax rates were higher.

10.4. Your client is the assistant treasurer for a company and has been subpoenaed by a group of the shareholders who are suing the directors for unfair practices in a recent tender offer. Those practices may have also violated S.E.C. regulations, but the S.E.C. has thus far been satisfied with allowing the shareholders to pursue a civil remedy, rather than also seeking criminal sanctions.

Aside from the tender offer, your client was particularly involved with a bond issue used to finance the offer. Attention has so far not focused on the bonds, but your client is worried that, if it does, the many stock-like features of the bonds could lead the S.E.C. to contend that they should have been registered as securities under the 1933 Securities Act. If so, then your client could personally be subject to criminal sanctions.

Your client could plead hardship or travel and perhaps avoid testifying altogether, or he could show up and take the Fifth if the questions turn towards the bond issue (they might not). He's not sure which would create the least suspicion. Finally, your client is a lawyer and is worried that any manifest irregularities could lead the state bar association to investigate him and perhaps impose discipline, including disbarment. He noted grimly that the bar association can impose discipline when no violation of the law has been shown, or even if he's acquitted of any wrongdoing.

10.5. You represent a defense contractor that's suing its cloud computing company (The Cray Crew) for shutting down the system for maintenance in the middle of a complex and critical product simulation, causing the company to miss delivering an interim progress report the next day and suffer the imposition of a $50,000 fine. The cloud computing company has a policy of delaying scheduled outages on request from clients at least 12 hours in advance, and Dr. Paula Circuit, the director of research simulation, claims to have left such a message for the Cray Crew scheduling manager on his voice mail machine. However, in a preliminary deposition the manager denied remembering receiving any such message that day. To avoid tipping him off, you've not yet directly challenged the messenger on the voice mail issue.

You anticipate calling the manager to testify at trial, but are unsure of how to proceed if he persists in claiming not receiving or not remembering the message. You could just call Dr. Circuit to impeach his testimony, but you're hoping to present something with more punch.

The voice mail system used by The Cray Crew records a number of statistics, such as when messages arrived and when they were listened to, and, if you could obtain the statistics, you might be able show circumstantially that the manager listened to all his messages that day. However, you're unsure whether the statistics still exist at that level of detail or have been summarized into some sort of a monthly report, or, if they still exist, whether they're in a written form you could obtain via discovery. If they're not in written form, you wonder if you could compel the manager with a subpoena duces tecum to produce a written report and bring to a deposition. Of course, if the mail system does routinely delete or summarize older statistics, your requests could tip off the manager and give him the opportunity to destroy the evidence before you knew if it existed or not.

Alternatively, a hired expert tells you that you could just try for a court order for a backup of the entire memory of the voice mail system. This could produce a mass of data that would be difficult to decipher and pick through, or the Cray Crew might contest such a request as overly broad and requiring the divulgence of other classified information, and ask the court to allow them to prepare an abstract—leaving you with the same problems as before.

You're also considering asking for an injunction to prevent the Crew from tampering with the mail system in any way until your expert has had a chance to examine it, but, given that such a course would effectively shut down the whole mail system while your expert examined it, you doubt that a judge would issue it without an adversarial hearing, which would again tip off the Crew as to what you were after (And now your head is really beginning to hurt.).

Chapter 5
Elicit Probabilities

5.1 Background on Probability Theory

Eliciting probabilities is a crucial step to capture judgment on the important uncertainties in your case and allows them to be applied to drawing conclusions. The first part of this chapter discusses the basis, meaning, and use of those probabilities more thoroughly than in Part I, while the second gives the procedures for assessing probabilities for different kinds of uncertainties.

The subjective probabilities employed in Decision Analysis are fundamentally different than statistically calculated probabilities, though they look the same (e.g., 0.05 or 0.45). We first discuss what subjective probabilities are, then why they are needed.

5.1.1 Subjective Probabilities Represent a Person's State of Knowledge

Subjective probabilities represent a person or persons' belief and judgment as to the chances an uncertain event will (or has) occurred. They use numbers (.50) to avoid the ambiguity in verbal descriptions of uncertainty (e.g., "very likely").

Because subjective probabilities are based on a person's state of knowledge and because no one knows and sees all (except the mythical clairvoyant), it would be rare to include a probability of 0 or 1.0 in a Decision Analysis. Things which are certain (either 1.0 or 0 probability) are instead treated as key influencing facts (part of the data) rather than as uncertainties.

J. Celona, *Winning at Litigation through Decision Analysis*,
Springer Series in Operations Research and Financial Engineering,
DOI 10.1007/978-3-319-30040-5_5

Because subjective probabilities are based on a person or persons' state of information, they can be based on sources as rich and varied as that person's experience, such as the following.

- History of similar situations;
- Evaluations of how relevant past situations are to the present one;
- Judgment developed in analyzing past situations;
- Intuitive relation together of all knowledge and judgment—quantified or not—to arrive at a probability for the instant, unique uncertainty.

Indeed, the purpose and unique value of subjective probability lies in its expression of all this information and judgment in a simple form: a number between 0 and 100 percent.

Furthermore, because subjective probabilities are only rated by how well they reflect the giver's state of knowledge, there are no "right" or "wrong" probabilities; consistency with that state of knowledge is the test. Thus, different people can have different probabilities, and there is nothing inherently objectionable about that.

However, some people may have *better* probabilities than others because they have greater knowledge or experience with the question under consideration or more experience thinking in terms of subjective probability. Thus, you can give better probabilities of the jury finding for or against you on a specific legal issue than the average lay person can, and a neurologist can give better probabilities on future neurological effects than you. You are always better off with the best probability you can obtain—though it may not be worth the cost to obtain it (an issue taken up in discovery strategy). Chapter 9 in Part III also discusses working with experts in general, and when you need to resolve differences of opinion on probabilities.

Assessing high-quality subjective probabilities is a learned skill. Practice is highly recommended. For this reason, assessment exercises are included in the problems at the end of this chapter.

5.1.2 *Why Do We Need Subjective Probabilities?*

People schooled in statistics often object that subjective probabilities aren't "objective"; they don't result from calculations on an underlying data set. That is true and exactly the point.

Experts in statistics are quite familiar with the many choices in a statistical analysis which drive the calculated results. How do you define the system being measured, how was the data obtained, and how representative is it? How was it determined which "outlier" data should be excluded? Which mathematical form was chosen to model the data, and which analysis results were presented?

Indeed, it is quite possible to structure a statistical analysis to reach the conclusion you desire. However, there is a more fundamental problem with applying

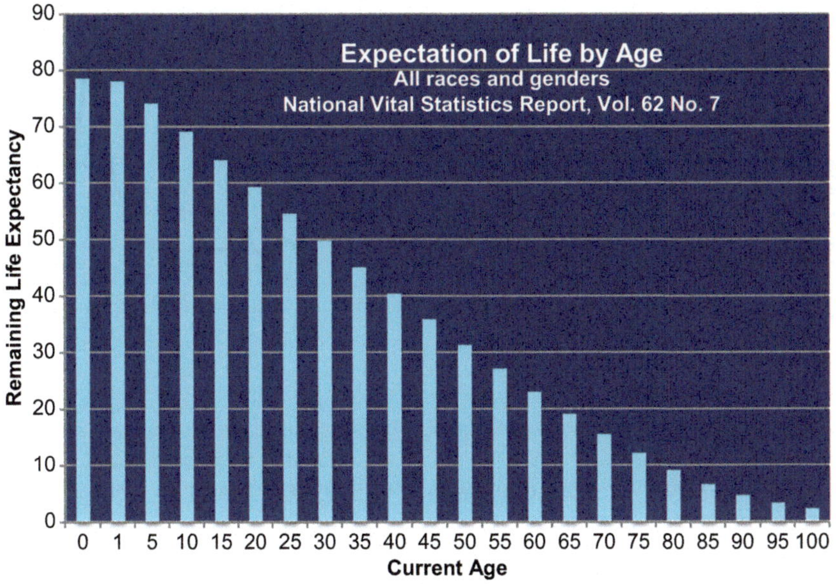

Figure 5.1 Remaining life expectation by current age group

statistical analysis of historical data to make predictions which drives the use of subjective probability in Decision Analysis.

The problem is this: statistical analysis of historical data can only be reliably used to make predictions when the following two conditions are met:

- There is a statistically significant data set.
- A stable system.

Most people are either unaware of these restrictions or ignore them for want of a better means: feeling more comfortable pounding away with the hammer they know.

Consider, for example, the following chart (Figure 5.1).

As it states, this is from a statistical analysis. How well do you think statistical analysis of data works for predicting remaining life expectancy by age group for tens of millions of people?

I would guess it works pretty well because there is a lot of data and a stable system: barring a major pandemic, large numbers of people aren't suddenly dying a lot sooner or living a lot longer. Of course, this data is useless for predicting life expectancies 50 or 100 years ago, and would look utterly ludicrous 500 years ago.

In contrast, let's look at housing prices in the USA. Below is data from the Case-Shiller Index of housing prices for the entire USA expressed in 2014 dollars (Figure 5.2).

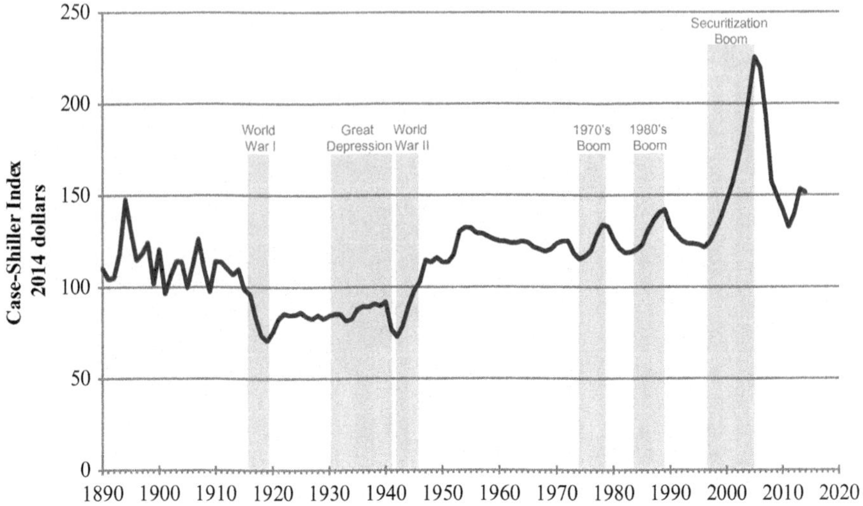

Figure 5.2 Case-Shiller Index of real (inflation-adjusted) housing prices in the USA

How well do you think statistical analysis of historical data works for predicting housing prices? I would say: lousy.[1] Certainly, there is a lot of data available, and you can fit any variety of model you like to the data (linear, nonlinear, dynamic, etc.) A simple linear regression fitted to this data would like something like this (Figure 5.3):

Would you rely on this data analysis to decide whether or not to purchase a house? I certainly hope not.

Let's look at one last example. Here is a weather forecast for St. Louis, Missouri in June of 2014 (Figure 5.4).

Note that probabilities of precipitation appear, as they do in virtually ever other weather forecast. Do you think they're calculated from statistical analysis of historical data? Certainly, there is a lot of data on precipitation, some going back many decades. Those probabilities could have been produced by looking at the percentage of June 10ths it rained going back to the beginning of the data.

In fact, any forecaster who worked this way would soon be out of a job. There is a lot of data, but the weather is a famously unstable system.

Those probabilities are subjective, Bayesian probability estimates. They express the forecaster's subjective belief as to the probability it will rain at some point in the

[1] Although it is beyond the scope of this text, I have taken an in-depth look at where the securitization boom went bust. For our purposes here, I will simply note that I have attended presentations by senior personnel at one of the major rating agencies describing in detail their analyses by PhD statisticians and economists which failed utterly at predicting the prospects for the housing market and the securities based on them. In the end, the key questions for any uncertainty quantification and risk management methods are: Is it working? How do you know? See Hubbard, Douglas W., The Failure of Risk Management. Wiley 2009: Hoboken, NJ.

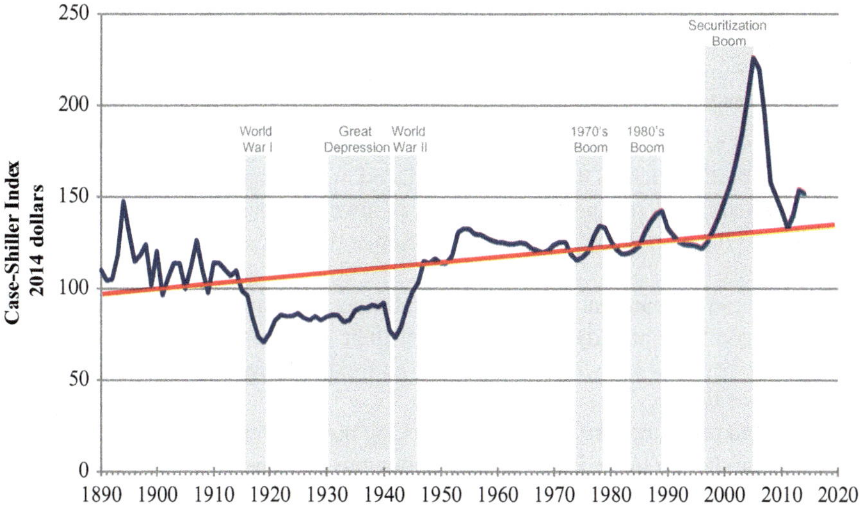

Figure 5.3 Linear regression fit to Case-Shiller Index

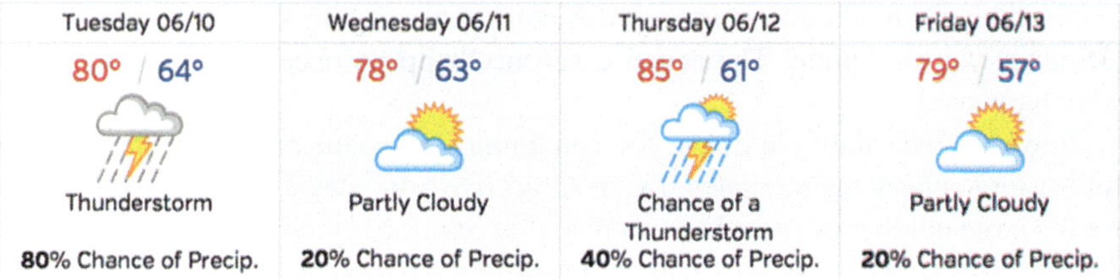

Figure 5.4 Weather forecast for St. Louis, Missouri

place and on the date in question. They incorporate his or her integration of all the available information: radar data, satellite data, maps of pressure systems and where they are moving. The forecaster carefully evaluates all the data and expresses the chance it will rain at some point in the form of a subjective, Bayesian probability. People see them all the time. They just don't realize what they are.

Certainly, forecasters do use quantitative models to help predict the weather. But, in the end, the forecasts themselves are Bayesian estimates based on all the available data and model results. For example, in predicting an "El Niño" condition, the National Oceanic and Atmospheric Administration (NOAA) provides a "Consensus Probabilistic ENSO [El Niño-Southern Oscillation] Forecast"[2] in addition to the results of individual quantitative models.

[2] http://www.elnino.noaa.gov.

To sum up, whenever we have:

- An unstable system; OR
- Not "enough" data

we need to use subjective, Bayesian probabilities to make predictions. That applies to pretty much all the uncertainties faced in decision-making. Unfortunately, decision-making is prospective, and data is always retrospective. Most inconveniently, we have no data from the future. Subjective, Bayesian probabilities fill in the gap.

This doesn't mean that subjective probabilities are entirely arbitrary. Quite the contrary, you have already seen one test of their quality: whether they accurately reflect the person's knowledge, intuition, and experience without falling prey to heuristics or biases.

A second test of quality comes from tracking how well the guesses did over time. With the prediction in the form of an unambiguous number, we can analyze the predictions to see how well the estimates did. For example, forecasters can track the days on which they estimated an 80% chance of rain. Did it rain on around 80% of those days? If not, the forecaster is likely still looking for another line of employment.

Note that the forecasts are in round numbers. There is seldom a solid basis for making a subjective probability estimate of, for example, a 78% chance of rain.

For a single prediction it is very difficult to see if your subjective probability estimates were accurate. The actual case only happens once and you only know what happened.

However, over multiple cases, you can employ the same process as the weather forecaster to check the accuracy of your subjective estimates. I have in fact done this for the large number of cases I've evaluated for Stanford University Medical Center. All the cases which closed out with no indemnity payment were ones for which we estimated a very low (<30%) chance of liability if it went to trial. For the cases in which there was an indemnity payment, the total costs to resolve all the cases added up to around 80% of the predicted indemnity exposure if the case went to trial—about as good of proof as one can get that their strategy of resolving cases for less than the exposure at trial was working.

And now on to how subjective, Bayesian probability actually works.

5.1.3 *Probability Arithmetic*

The rules for probability arithmetic are so simple that you likely haven't even noticed them being applied. They are as follows.

- Probabilities must be between zero and one ($0 \leq p \leq 1.0$).
- Probabilities for all possible answers to a question must add up to one.
- Probabilities for "and" events like "Lose on A" and "Lose on B" multiply together, and probabilities for "or" events like "Lose on A" or "Lose on B" add together.

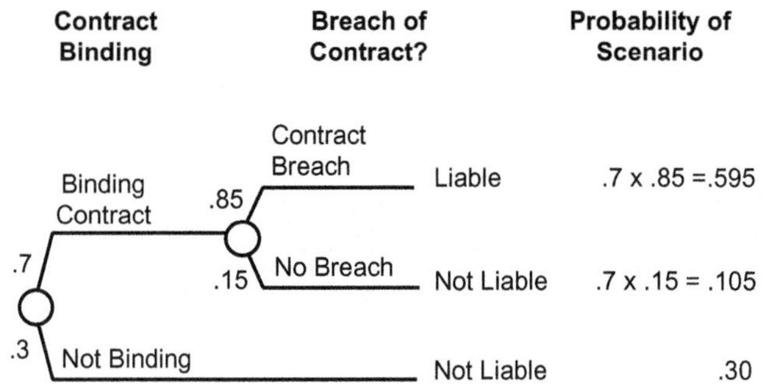

Figure 5.5 Tree for two events required to lose the case

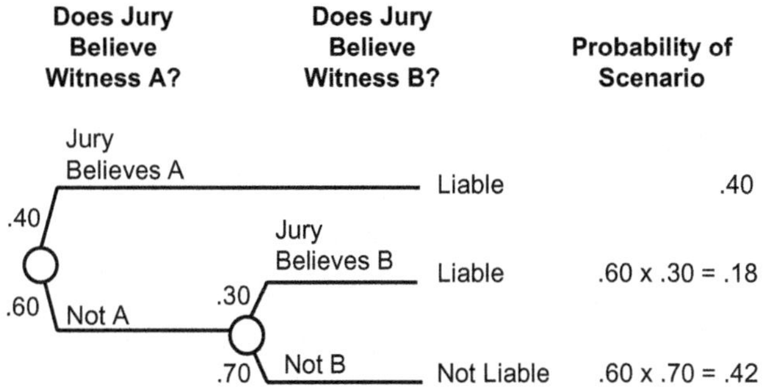

Figure 5.6 Tree for either event leads to a loss in the case

The last rule is the one most people have trouble on—but only if they don't draw the tree! The rule is simple enough in tree form that you've already seen it with needing long explanation.

For instance, take "Lose on A" and "Lose on B." This is nothing more than the tree for "Only Lose If Both" presented back in the beginning of the relevance diagram discussion. This tree is repeated below, with probabilities added (Figure 5.5).

The chances of losing on binding contract and losing on contract breach multiply to give you your chance of losing on both. Accordingly the probability of losing this case (if you're the defendant) is about 60%.

Thus, the "and" events are when you need several things to happen together to lose or win your case. But, since you don't even have to think about this when you draw the tree, we recommend you draw the tree to help make the right answer obvious.

Similarly, the "or" events are when you have more than one way that you can win or lose your case all by itself. "Lose on A" or "Lose on B" is the "Lose Either Way" tree presented in the relevance diagram discussion and repeated below, with probabilities (Figure 5.6).

Your overall chances of losing the case, then, are .40 + .18 = .58. Notice that this is not simply adding up .40 + .30 = .70! That's because you have a .40 chance of the jury believing witness A and never even getting to the issue of B's credibility. These kinds of over- and under-counting problems are frequent when people don't draw the tree and easy to avoid when they do (You may recall that the discussion of Two Arguments to Recover a Single Loss involved the same kind of over-counting problem.).

In summary, then, just remember the ***Probability Arithmetic Rule***: multiply out the probabilities in the tree and work with that column and you won't run into problems on when to add and when to multiply.

5.1.4 Mutually Exclusive and Collectively Exhaustive Events

All the possible answers to a question and all the possible scenarios shown in a tree should be mutually exclusive (it's crystal clear when you are in when scenario and not another) and collectively exhaustive (they describe all the possible things that could happen). Assuring this condition is met for all your assessments of individual probabilities and for all scenarios described by the tree is important for two reasons.

- **Intuitive Reason**. You can't think about the relative likelihood of different possible answers to arrive at your probabilities when the answers don't span and separate all possible answers. For instance, how would you think about probabilities for the following chance node? (Figure 5.7).
- **Formal Reason**. Your probabilities have to add up to 1.0 to make sense, and they won't if your answers overlap and thereby double-count some of the probability (mutually exclusive), or if they leave out part of the probability (collectively exhaustive). For instance, in the above example, you could lose on your motion for summary judgement and still win on a jury verdict (overlap problem). A correct tree would look like the one below (Figure 5.8).

You've already seen the solution to the problem: the ***Clairvoyant Test***. To review, the Clairvoyance Tests asks if the clairvoyant could answer your question with yes, no, or a number. For the first tree above, the answers given on the branches are so

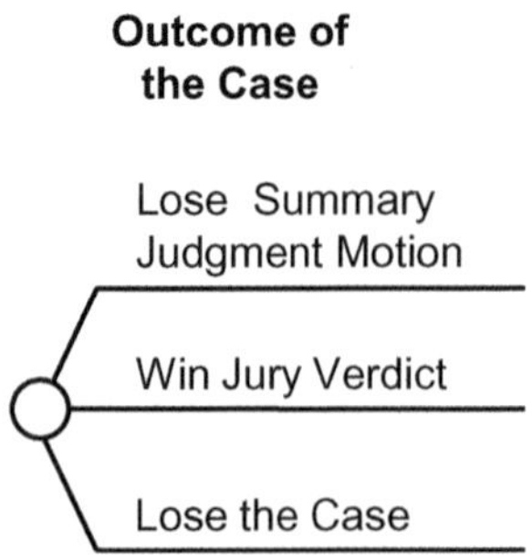

Figure 5.7 A chance node that flunks

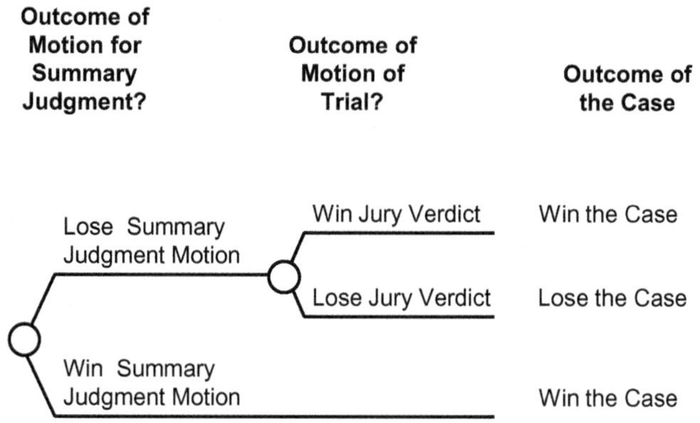

Figure 5.8 Correct tree for outcome of the case

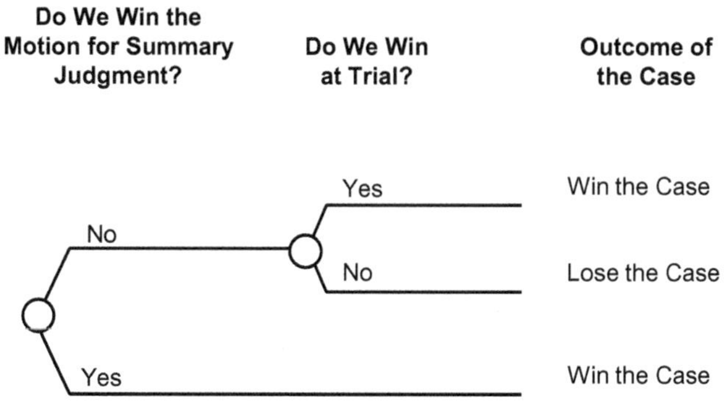

Figure 5.9 Correct tree with questions restated to pass the Clairvoyant Test

vague that the question would have to be something like "What happens?" You can almost see the clairvoyant scratching his head.

That the second tree passes (although, of course, there are lots of different and valid ways to set up the questions), becomes clear when you put the questions above the chance nodes and try to put just the answers on the branches (Figure 5.9).

The Venn Diagram below illustrates graphically how you're trying for answers that divide up the whole possible world without overlap between them. Note that there is overlap between *questions* because you only get to the question of a jury verdict if you've already lost on your motion for summary judgment. You can see this overlap in the diagram because both jury verdict hatchings are also shaded by "Lose S.J. Motion" (Figure 5.10).

Generally, people's questions or answers will fail the clairvoyant test for one of two reasons.

- They are too vague, so they don't make the distinction you're trying to get at.
- They mix up different questions, like the example above.

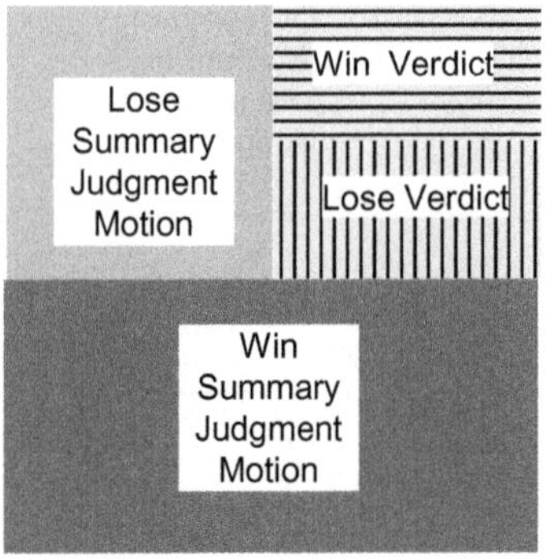

Figure 5.10 Venn diagram showing mutually exclusive and collectively exhaustive events

The solution, of course, is to think about and to redefine your question and answers until they pass the clairvoyant test. This does not mean simply breaking things down into a further level of detail, like you do when you add another level to a relevance diagram. The question, rather, is whether at *this* level of detail you have a clear idea what you're talking about.

5.2 Procedures for Assessing Probabilities

There are several methods for assessing subjective probabilities. Which one you choose depends on the kind of uncertainty you're considering (yes/no versus a continuous amount like damages), and what it takes to get a meaningful answer. These choices are discussed in more detail throughout this section.

It's hard to over-emphasize the importance of the probability assessments. You've already seen how the probabilities lead directly to the mean value of litigating your case, which determines how much you should be willing to settle for. Thus, probabilities are crucial to combining your judgment on particular issues with judgments on all other issues of liability and damages.

However, the whole process of assessing probabilities has other benefits because of the way it forces you to think hard about your case and its important uncertainties.

- Probability assessments sharpen your understanding of the role of the particular question.
- Probability assessments force a more thorough evaluation of the chance of success on a particular issue.
- Probability assessments often identify the best tactical approach to arguing an issue.

In short, the real value of probability assessments they force you to think more clearly about your case.

As already hinted at in the discussion of the basis of subjective probabilities, you may not necessarily provide all the probabilities used in the analysis of your case. For example, a potential increase in market share as a result of a favorable patent infringement judgment would be provided by client's marketing manager (a judgment easily integrated with your own because all the judgments are expressed in numbers). Chapter 9 on Special Probability Approaches discusses working with experts, and working with a group of attorneys.

5.2.1 *Procedure for Assessing Discrete or "Yes/No" Variables*

Many legal questions can be framed for a yes or no answer by the trier-of-fact (and, indeed, this is the way you argue them). Is defendant liable? Was the conduct intentional? Was the harm reasonably foreseeable? This section gives you full details on the procedure for assessing the probabilities for those kinds of questions.

The next section gives the procedures for continuous variables—that is, questions like damages whose answers are numbers and could turn out to be any number within what you think the possible range is.

5.2.1.1 Apply the Clairvoyant Test

Could the clairvoyant could give you a yes-or-no answer to the question as-stated, or would he have to ask for further clarification? The following questions will help to focus your inquiry.

- Is the issue unambiguously described?
- Is it clear exactly what a good resolution would be?
- Is it clear how the issue will arise? When it will arise?
- Is it clear how opposing counsel will argue the issue?
- What's the jury thinking about? What other evidence is already in?

5.2.1.2 Examine and List the Reasons for a "No"

Imagine a "no" outcome. Take a few minutes to consider how and why it might have happened. Write down all the reasons you can think of why it might have happened. ***The quality of your probability assessment depends on how much effort you put into the reasons.***

- Are there any more plausible reasons?
- Are you taking something for granted (certain) that is really uncertain and could have a significant impact on the uncertainty you're assessing?

Listing the reasons is the single most important step in the process because writing them down forces you to think clearly about what could happen, why, and how likely it is. Further, writing them down helps avoid biases because it forces you to think more broadly about how and why things could turn out "good" or "not good".

Listing the reasons also helps to make your case evaluation realistic and complete by identifying all possible reasons for each possible outcome. Thus, you can use this step to think about all the factors not individually important enough to include in the relevance diagram. Possible factors include:

- Undisputed facts;
- Possible eyewitness testimony and the quality of witnesses;
- Characteristics of the plaintiff, such as age, family situation, and sympathetic nature;
- Quality of opposing counsel;
- Judge and jury biases in the jurisdiction;
- Legal precedent;
- Broad policy arguments;
- Any "spill-over" effect from other parts of the case.

Finally, for maximum effectiveness, you should have present both a content expert (someone who has deep knowledge of what happened) and a legal expert (who can think about the legal implications of what happened). Although doing so requires more professional time, your probabilities are critical to your case analysis and they will only be as good as the people and time you spend on them.

5.2.1.3 Examine and List the Reasons for a "Yes"

For the possible answer of "yes," go through the same process you just went through for the "no" answer.

5.2.1.4 Pick Numbers or Use the Wheel to Pick Probabilities

Only after arguing and listing the reasons for "no" and for "yes" answers should pick numbers. Over time, you will become comfortable with using this method.

Initially, you may want to use a probability wheel to pick numerical probabilities. Use of the helps people to pick probabilities by visualizing how big or little they are.[3]

A probability wheel is like a roulette wheel: with one spin of the wheel, where will the pointer land? Colored areas help to visual the chances of landing in the blue area versus the orange area. The colored areas can be adjusted so you can change the probability assigned to each. Below is the front of a probability wheel with the blue area set to a probability of 65% (Figure 5.11).

[3] I am not presently aware of a source to purchase these wheels, and have needed to have them specially made when I run out. If you know of a source, please drop me an email.

Figure 5.11 Front of a probability assessment wheel with the probability of the spinner landing in blue set to 65%

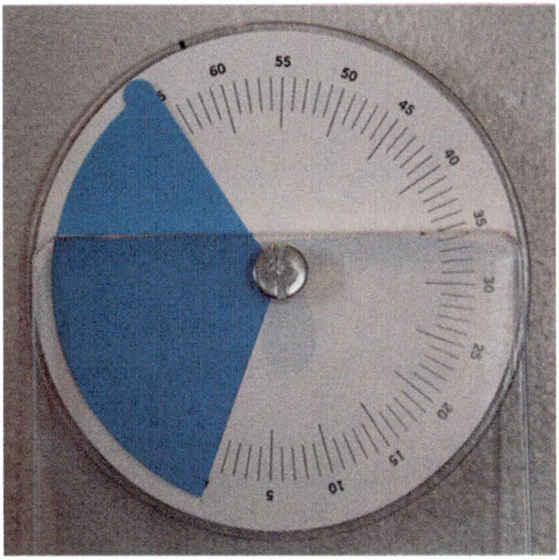

Figure 5.12 Probability assessment wheel with probability of blue on the front set to 65%

Calibrations on the back of the wheel take the guesswork out of setting the probabilities, as shown below (Figure 5.12).

Most people have a very imperfect idea of probabilities, and will think 75 and 90 percent probabilities very similar until they *see* on the wheel the difference between 3-to-1 odds at 75% probability and 9-to-1 odds at 90% probability. The wheel thus helps make differences in probability apparent.

The wheel also improves your judgment by asking you to think about probabilities visually and as a choice between uncertain prospects (the spinner landing in blue or the question being answered "no", for example). The wheel thus forces you to be more honest about your chances. Probabilities of .98 and .02 look very extreme on the wheel. When you see these kinds of probabilities in a tree, it almost always indicates the wheel wasn't used.

For most people, simply showing them the wheel is enough to help visualize the probability. Vary the area of blue until it's equal to their estimate of the probability for the event in question.

If people are having difficulty, follow is a procedure to take them through.

11. **Set almost All Blue**. If you could win $100,000, would you rather take your chances on the spinner landing in blue, or on a "no" answer to the question? With almost all blue showing, it should be hard to turn down the spin of the wheel—almost no matter how strongly you felt about a "no"—because your odds on the wheel are so good.
12. **Set almost No Blue**. Now would you rather take your chances on the spinner landing in blue or on a "no" answer? Similarly, with almost no blue showing, it should be hard not to go with the "no" answer—almost no matter how strongly you felt about a "yes"—because your chances of winning with the wheel are so bad.
13. **Set Varying Amounts of Blue**. Always ask yourself: would you rather put your chances of winning $100,000 on a spin of the wheel, or on a "no" answer? Look at your lists of reasons, think hard, then pick the wheel or the "no" answer.
14. **Try Another Setting of Blue**. Keep picking the wheel or the "no" answer.
15. **Stop When You Are Indifferent Between the Wheel and the "No" Answer**. The percentage showing on the back of the wheel is the probability to use in the decision tree for the "no" branch. One minus that probability is the number to use for the "yes" branch. Don't round off your answer: .70 is not more precise or a better answer than .71.

5.2.1.5 Use the Lists of Reasons to Refine Your Strategy

Listing the reasons for "yes" and for "no" answers should have pointed out the strengths and weaknesses in your case and case strategy. Use these reasons to help think about and refine your case.

You sit down with your Howitzer v. Crabgrass tree and begin to go through the procedure for the yes/no questions. Because you've been rethinking and reformulating the questions as you went along, it's not too much trouble going through the clairvoyance test. You set down with your lists of reasons for the various possible answers (from when you made up the tree) and think hard about whether there are any new reasons as you go through the answers one-by-one. After about an hour of pondering the reasons and adjusting the wheel, you've filled all the yes/no probabilities in on your tree, as below (Figure *5.13*).

Note that the attorney talked to on the negligent misrepresentation question assessed probabilities (without use of the wheel) at 1/3 and 2/3. That's perfectly fine. For analysis, they simply translate to .33 and .67—probably the only time we see probabilities that aren't round numbers.

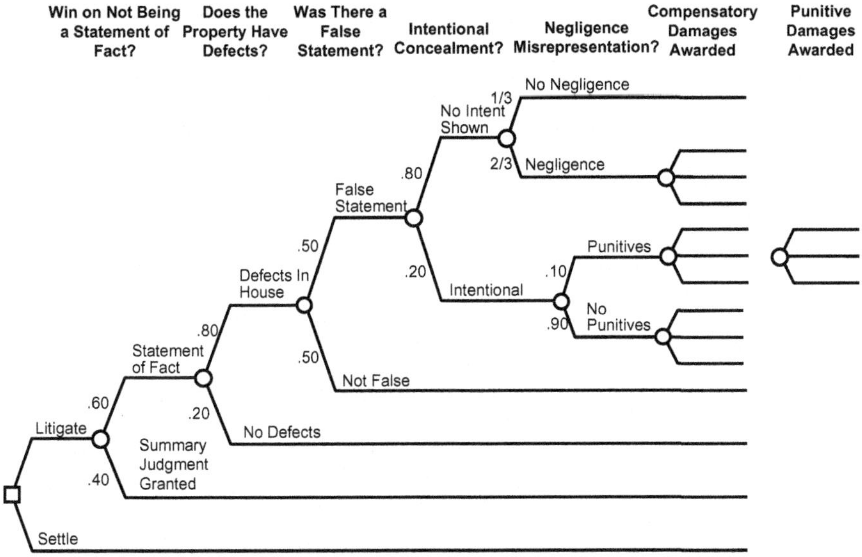

Figure 5.13 Probabilities for discrete events assessed for Howitzer v. Crabtree

5.2.2 *Introduction to Continuous Variables*

5.2.2.1 Why Do Continuous Variables Need a Special Method?

Unlike questions for which there is a discrete set of answers (liable, not liable, etc.), some questions have a large or even infinite number of possible answers (amount of damages awarded, future medical costs, etc.). We refer to these as continuous variables.

For decision tree calculations, we need to break these continuous variables down into discrete pieces which can go on branches of a chance node (See the note at the end of this chapter of a discussion of Monte Carlo simulation methods using continuous distribution definitions.).

Many uncertainties in litigation are continuous. In particular, all types of damages are continuous. The broad categories are compensatory damages and punitive damages. Compensatory damages further divide into general damages (pain and suffering) and special damages. Special damages include:

- Any actual costs paid (repairs, other damages, etc.)
- Past or future medical bills
- Past or future lost income
- Loss of consortium (loss of the value of the services provided by a spouse)

Basically, most uncertainties where the answer is a number (e.g., share of verdict allotted to a codefendant) rather than a distinct state (e.g., the jury does or does not

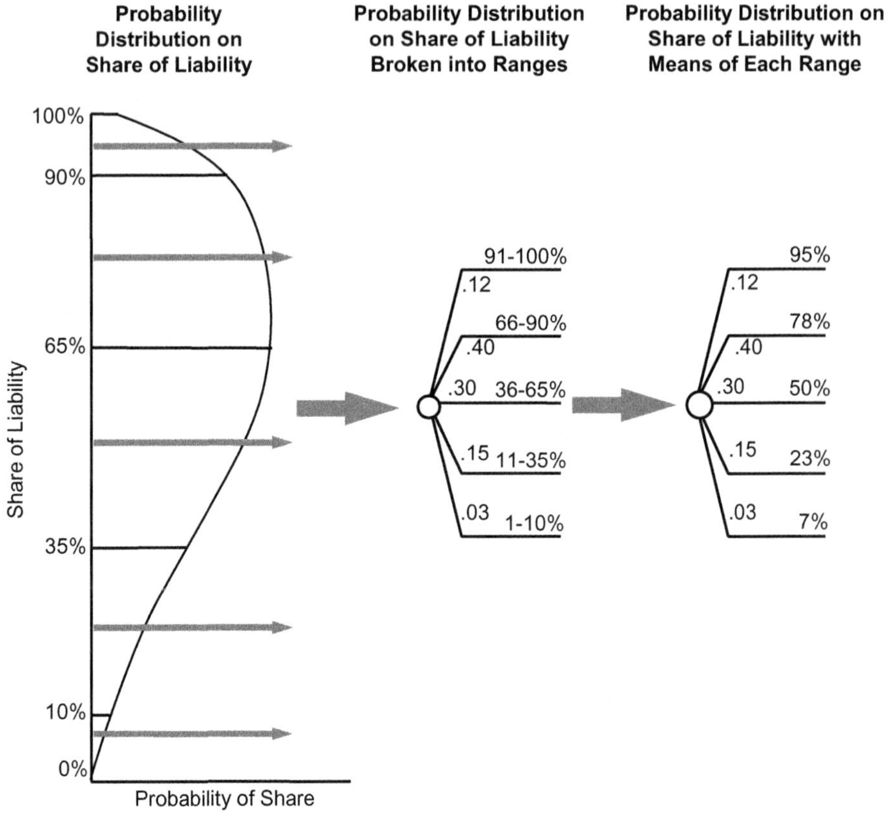

Figure 5.14 Visual depiction of discretizing a continuous probability distribution

believe witness A), is a continuous variable rather than a discrete one. There are an infinite number of possible answers to continuous uncertainties. For decision tree calculations, we need to:

- Break the continuous possible answers into discrete ranges to go on individual branches of a chance node; and
- Find a value and probability that properly represent each range.

This process is visually depicted in the graph below (Figure 5.14).

The probability distribution for share of liability is shown as a probability density function (turned sideways relative to the usual display). The area under each section of the curve is the probability of being in that region of the curve, which is shown as the probabilities on the branches of the chance nodes. The grey arrows show the mean of each region, which appear as outcomes on the branches in the right-most chance node.

There are two methods for assessing probabilities for continuous variables. The first method is to assess a simple three-point range, and the second is to assess the full distribution on the variable.

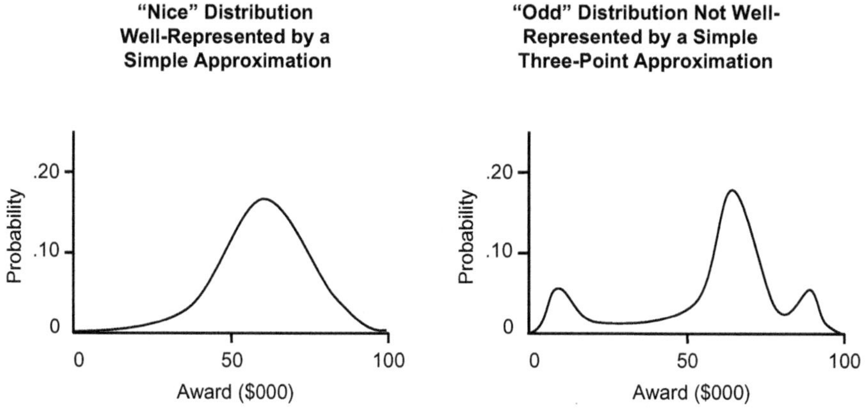

Figure 5.15 Graphic illustration of whether a distribution is well-represented by a simple three-point approximation

5.2.2.2 How Do I Pick Which Method to Use?

Use the simple, three-point approximation:

- Most of the time and especially for a first-pass analysis.
- When assessing probabilities with a group.
- When answers to the question are not very skewed or bimodal—that is, values in the middle of the potential range of outcomes are much more likely than values in the extremes of the range. The graph below on the left illustrates just such a "nice" distribution, while the graph on the right illustrates a skewed distribution that would not be well-represented by picking the a simple range (Figure 5.15).

Use the second method to assess the full probability distribution when:

- You have a hard time representing it with a simple three-point approximation.
- The uncertainty is very important or your subject doesn't have a good sense of how it might turn out.
- You want to challenge your thinking more on how well you've assessed the high and low ranges of the variable.
- You want to pick up irregular variation in how the uncertainty might turn out.
- You want to gain a better representation of the uncertainty.

Although it is possible to pick any number of points to assess rather than three (4, 5, or 6, for example), if a 3-point approximation doesn't seem adequate I've found it best to assess the full probability distribution and then look at how you might like to break it up to show the various possible scenarios (e.g., a low-probability high-value scenario).

In many cases, when an assessment of an uncertainty starts to look really "funny," this is an indication that multiple factors are at work which are best separated out as

Figure 5.16 Bimodal probability distribution

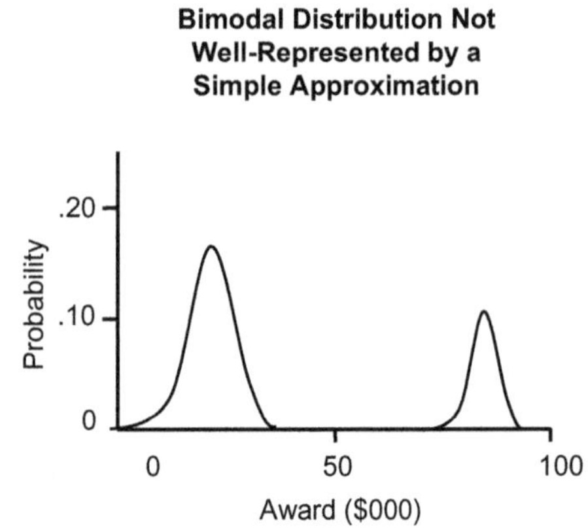

Figure 5.17 Bimodal distribution disaggregated into multiple uncertainties

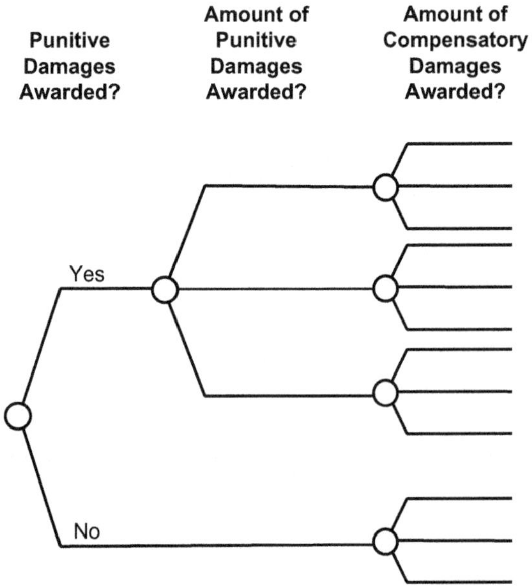

separate factors. For example, a bimodal distribution above might indicate that there are two elements of damages which could be awarded (Figure 5.16).

A sample tree for disaggregation of these two factors is shown below (Figure 5.17).

You would then separately assess the yes/no uncertainty on punitive damages being awarded and the range on each category of damages.

Although you know that our objective is to pick probabilities and points to represent the ranges the total distribution is broken into, we don't recommend trying to think in terms of ranges. Few people can think meaningfully about the probabilities of ranges or of the values bounding a range. Ranges are helpful for explaining the process (as illustrated above), but generally not for assessments.

5.2.3 Procedures for Assessing Continuous Variables

The three-branch approximation works for almost all uncertainties, and especially for a first-pass analysis.

5.2.3.1 Three-Branch Approximation

When using this method, it is very important to apply the clairvoyant test and to list out the high and low reasons before starting to pick numbers. The most common bias in assessment is to make the ranges too narrow, so be sure to push people's thinking on the potential variation in how the uncertainty could turn out. With subjective probability assessments (as with all analysis), garbage in means garbage out. With a new group, it is highly recommended to take them through an assessment exercise (as shown in the problems in this chapter) to improve their thinking before starting on real assessments.

1. **Apply the Clairvoyant Test**. Can the clairvoyant answer your question with a number, or does he need more clarification?
2. **List High Reasons**. Imagine an amazingly high outcome that really surprises you. Push yourself for the possible reasons why it might have happened and write them down, as with the yes/no questions.
3. **List Low Reasons**. Imagine an amazingly low outcome that similarly surprises you. Again, think hard about all the possible reasons for it and write them down.
4. **Pick A High Number**. Use the wheel to help visualize what 90% odds look like. Pick a number for which you think there's a 90% chance the actual answer will turn out lower (the blue area), and thus also a 10% chance the actual answer will turn out higher (the orange area).
5. **Pick A Low Number**. Use the wheel to help visualize what 10% odds look like. Pick a number for which you think there's a 10% chance the actual answer will turn out lower (the blue area), and thus also a 90% chance the actual answer will turn out higher (the orange area).
6. **Pick A Mean Number**. Now consider the mean. Again, use the wheel to help visualize 50% odds, and pick a number such that you think the actual answer is equally likely to come out above or below it. This number need not be (and usually will not be) exactly midway between your 10% high and 10% low numbers.
7. **Add Probabilities**. Put the numbers into your tree with a .25 probability on the low and high numbers, and a .50 probability on the mean number.

To explain how this method works, we first need to introduce the cumulative probability distribution, which is the integral of the probability density function (Figure 5.18):

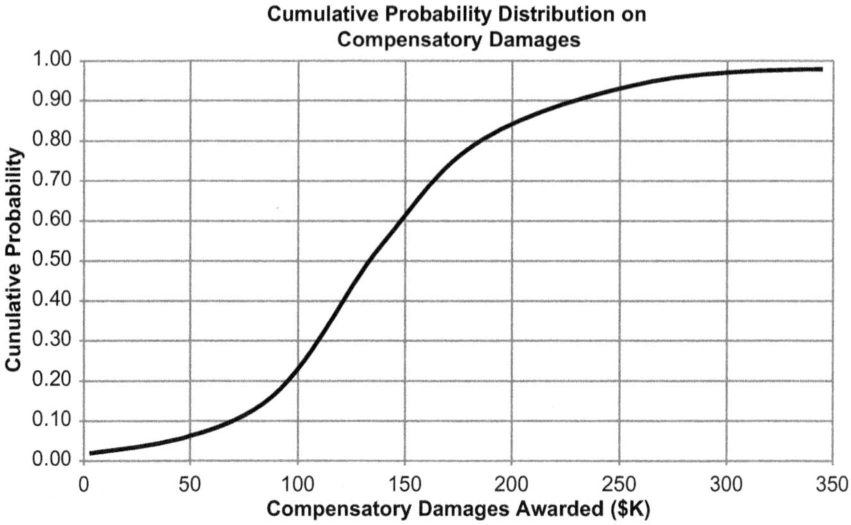

Figure 5.18 The cumulative probability distribution is the integral of the probability density function

We use this form for two reasons:

- It's easy to read numbers off of it. For example, given that compensatory damages are awarded, there's a 10% chance they're less than about $75K, and a 90% chance they will be greater. Likewise, there is about a 10% chance they will be greater than about $220K, and a 90% chance they will be less.
- They are easier to work with than probability density functions. To correctly draw a probability density function, you need to ensure that the total area under the curve equals 100% (rather hard to do). For a cumulative probability distribution, you just need to make sure it starts at zero probability and goes up (at a possibly varying slope) to 100%. The curve just can't go down.

The three-point approximation works by breaking the distribution down into three regions (Figure 5.19).

- There is a 25% chance of ending up somewhere in the lower region of the distribution.
- There is likewise a 25% chance of being in the upper portion of the distribution.
- There is a 50% chance of being in the middle portion of the distribution.

The three assessment questions then work like this:

- A number of which there's a 10% chance the actual is lower is the mean of the lower portion of the distribution, of which there's a 25% chance of ending up in this region.

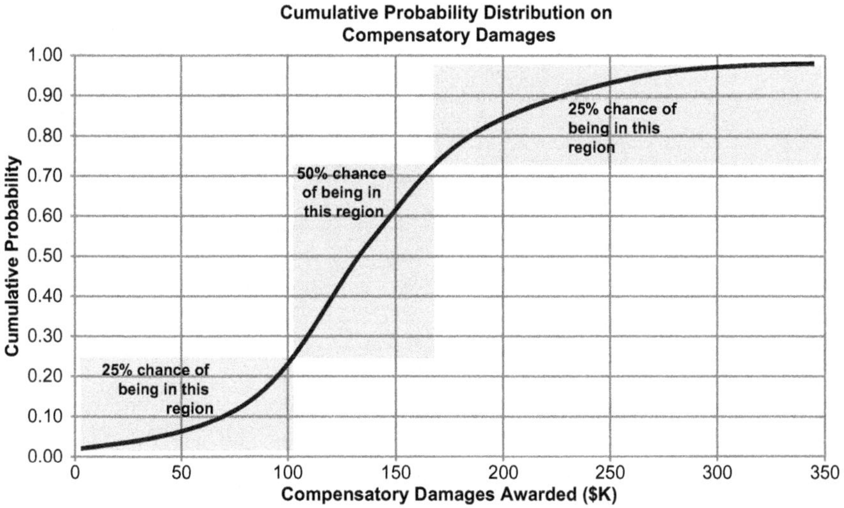

Figure 5.19 Cumulative probability distribution divided into three regions

- A number there's a 10% chance the actual is higher is the mean of the upper portion of the distribution, with a 25% chance of ending up in this region.
- A number where it's equally likely the actual is higher or lower is the mean of the middle part of the distribution, with a 50% chance of ending up here.

This relationships arc illustrated in the graph below (Figure 5.20).

Visually, you can see that, for each region, the area above the mean of that region is equal to the area below, indicating that the mean divides that portion of the curve into sections of equal probability.

The resulting relationship between the assessment questions and the chance node for the tree are illustrated below (Figure 5.21).

5.2.3.2 Assessing the Full Cumulative Probability Distribution

Use the procedure when the uncertainty is very important (large potential impact on case value), or very uncertain so it's really hard to get a rough sense of how it might turn out. Most of the time, the three-point approximation works well enough. Whether you choose the three-point method or assessing the full distribution, that assessment may be updated as new information emerges as discussed in the next section.

To assess the full cumulative probability distribution:

1. **Apply the Clairvoyant Test**. Can the clairvoyant answer your question with a number, or does he need more clarification?
2. **List High Reasons**. Imagine an amazingly high outcome that really surprises you. Push yourself for the possible reasons why it might have happened and write them down, as with the yes/no questions.

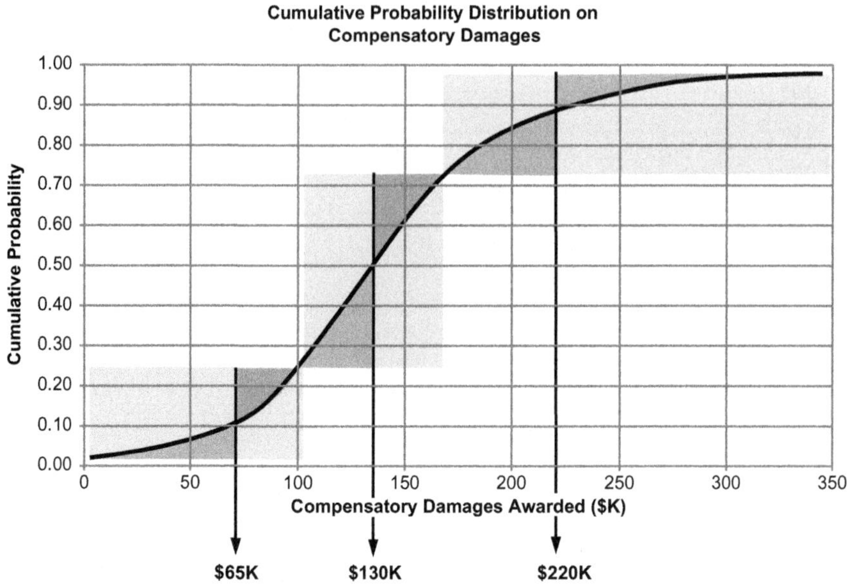

Figure 5.20 Assessing the means of the three regions of the distributions

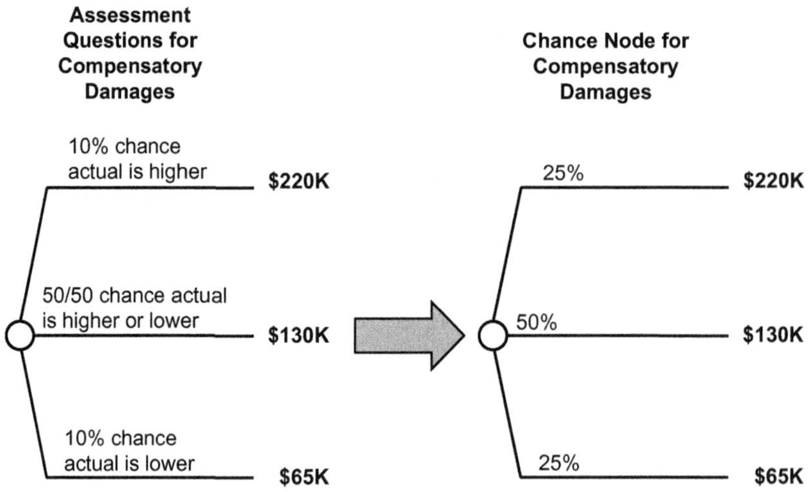

Figure 5.21 Relationship between assessment questions and node for the decision tree

3. **List Low Reasons**. Imagine an amazingly low outcome that similarly surprises you. Again, think hard about all the possible reasons for it and write them down.
4. **Pick A High Extreme**. Use the wheel to help visualize what 95% odds look like. Pick a number for which you think there's a 95% chance the actual answer

will turn out lower (the blue area), and thus also a 5% chance the actual answer will turn out higher (the orange area). Plot that number on your cumulative probability paper (This step and the next one allow you to lay out the bottom axis on your graph.).

5. **Pick A Low Extreme**. Use the wheel to help visualize what 5% odds look like. Pick a number for which you think there's a 5% chance the actual answer will turn out lower (the blue area), and thus also a 95% chance the actual answer will turn out higher (the orange area). Plot that number on your cumulative probability paper.
6. **Develop the Low Region**. Pick a number not too far from the low extreme. Ask yourself whether you would rather bet on the wheel in blue, or on the answer coming up more than the number. Adjust the wheel as described in the yes/no procedure until you're indifferent between betting on blue and coming up higher than the number. Plot that number and its probability as point number one on your cumulative probability paper.
7. **Develop the High Region**. Pick a number not too far from the high extreme. Again, ask yourself whether you would rather bet on the wheel in blue, or on the answer coming up more than the number. Adjust the wheel as described in the yes/no procedure until you're indifferent between betting on blue and coming up higher than the number. Plot that number and its probability as point number two on your cumulative probability paper.
8. **Fill in The Range**. Go through the same process of picking a point and finding a probability until you have enough points plotted that you feel comfortable with your representation of the range (usually an addition 2–4 points). Number each point as you plot it.
9. **Draw the Curve**. Draw a smooth curve through or near as many of the plotted points as possible. If it is not possible to draw a smooth curve without being too far away from any of the points, remember that the later points are likely more accurate than the early ones. Rethink your assessments and assess additional points if necessary. Y
10. **Verify the Curve**. Pick numbers on the curve that you didn't specifically assess and ask yourself if the probability given by the curve for those numbers agrees with your judgment. If not, do additional assessments.
11. **Discretize the Curve**.
    11.1. Pick three or four probability ranges to divide the curve into. Make the two ranges at the extremes somewhat narrower than the others (this will help capture the full range of possible answers). The ranges can be wider in the center of the curve, where the curve usually tends to be steeper than at the extremes. Where to divide the curve is largely a matter of judgment, dividing it at the inflection points usually gives you a good range of values (Figure 5.22).
    11.2. Draw horizontal lines separating the curve into those cumulative probability ranges. Then draw vertical lines through the places where the vertical lines intersect the curve.

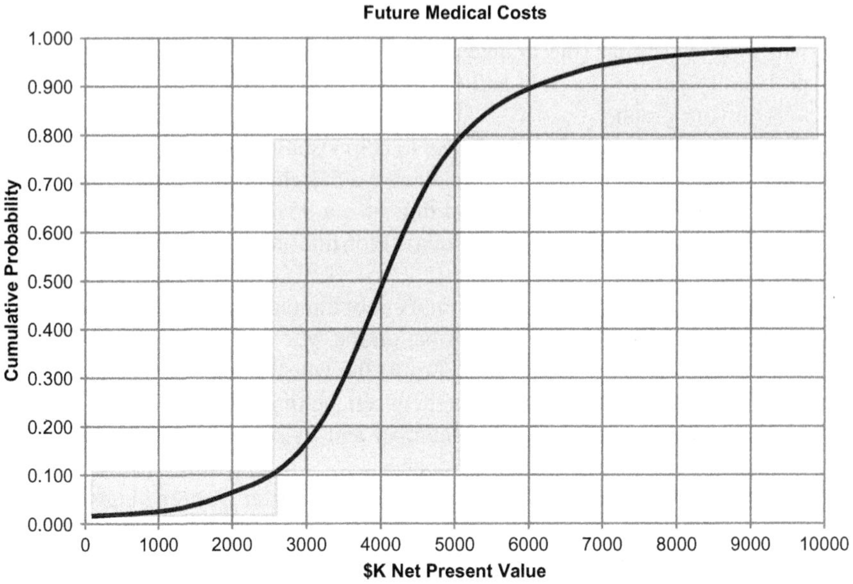

Figure 5.22 Illustration of places to discretize a cumulative probability distribution

11.3. In each little block of the graph, draw another vertical line through the curve so that you create a below-left region with the same area as the above-right region.

11.4. In each block, that last vertical line gives the number that represents the block, and the height of the block is the probability that goes with the number. These numbers and probabilities will go on the branches of your chance node.

11.5. Verify your discretization by computing the mean value of the chance node and comparing it with the vertical line that divides the whole curve into equal below-left and above-right regions.

12. Congratulate yourself on having completed the best and most demanding procedure, and reflect on what kinds of questions you would really want to use this procedure for.

Here are some further tips to help you as you go through this process:

- Jump around from high to low numbers, and leave the middle of the range for one of the last assessments. Do not start at one extreme and gradually move to the other because you will impose a false consistency.
- Randomly switch your questions from betting on greater than the number to betting on less than the number (still against the choice of betting in blue).
- Be alert for how the assessment develops. Later points may be inconsistent with earlier ones, but are usually a better representation of the probability.
- This process goes much better with one person asking the questions and the other answering, and even a third person to plot the answers is very helpful (your secretary can do it).

This method gives you the best possible representation of the potential answers to a question. It does the best job of avoiding cognitive and motivational problems, and it produces the best valuation of your case. It is also the most time-consuming and complicated, and, as previously discussed, the three-branch shortcut does a reasonably good job most of the time. Accordingly, you will probably not use this method frequently—but do practice it on some less important uncertainties. You will then be confident of the method when major imponderables come along that you want to understand as well as possible.

You begin to think about the probabilities for the continuous variables, the punitive damages and the various compensatory damages scenarios. As you study the tree, you first of all decide that, when the jury awards punitives, the amount of compensatory damages isn't going to vary very much according to the amount of punitives awarded. Thus, the compensatory damages will just be a function of what the plaintiff is able to demonstrate and what the trier-of-fact (judge or jury) decides to award.

Because punitive damages, if awarded, are a function of what the court decides are appropriate to "punish" the defendant, you're worried that they could much larger than the compensatory damages. You decide to definitely assess the full probability distribution for those.

For the compensatory damages, the assessment process yielded the curve shown above, resulting in the chance node shown below (Figure *5.23*).

By breaking the curve into 25%, 50%, and 25% probability regions, you arrive at a node that you could have assessed directly using the shortcut method (if you or your expert felt comfortable doing that).

You proceed to the process for punitive damages.

You start with reviewing your lists of high and low punitives. Lots of reasons that could send the punitives shooting up or down. Thinking first about the extreme values, you feel that, with the need to "punish," a court would award at least $500,000 in punitive damages (the 5% low number). On the other hand, unless some major and

Figure 5.23 Assessment for compensatory damages node

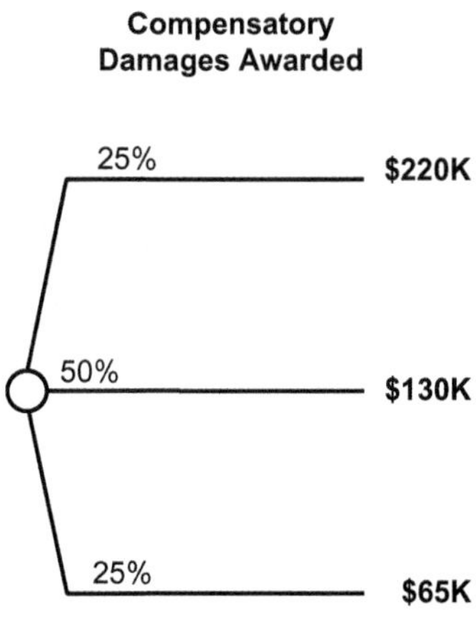

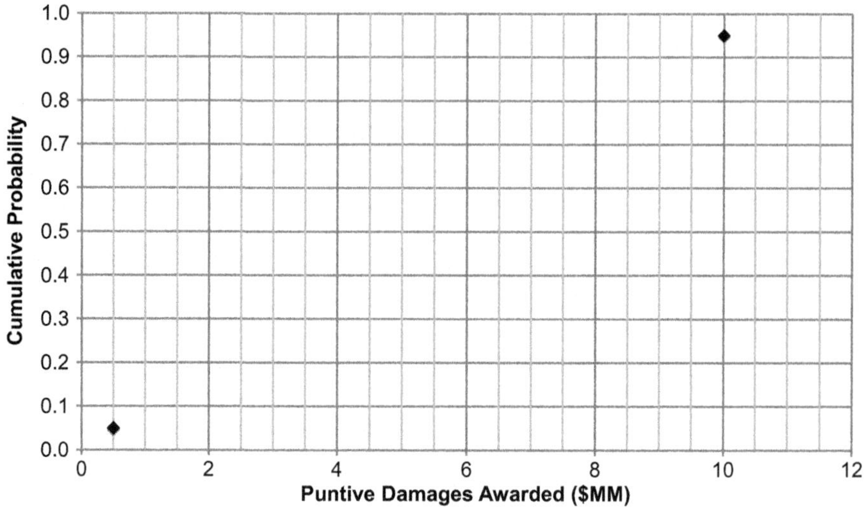

Figure 5.24 Start of probability assessment for punitive damages

systematic misconduct were shown on the part of your client, you'd be very surprised if they awarded more than $10 million (the 5% high number). You start with those values (Figure 5.24).

You then begin going through questions and using the wheel to develop the high, low, and middle regions of the curve. Would you rather bet on blue, or punitive damages being more than $6 million? That yielded squeezing the blue down to about 7%. Greater-than questions are can either be converted to less-than questions by calculating 1 – 7% = 93%, or can be plotted directly on graph paper also having an Excess Probability scale. This scale is the inverse of the cumulative probability scale (Figure 5.25).

Alternating between less-than and greater-than questions is a great way to avoid bias in the estimates, especially anchoring on someone's initial guess as to what the number should be. You work back and forth until you have the chart populated, then draw in a smooth line that comes close to all the points. The idea is to end up with a "fitted" line that minimizes the variation between it and individual points (with plus or minus variations roughly canceling out) (Figure 5.26).

Finally, you break the curve into regions at where the inflection points in the curve are and estimate the conditional mean in each region (the mean given you are in that portion of the curve) (Figure 5.27).

This yields the chance node you need for the decision tree (Figure 5.28).

Looking over it all and considering it, you feel this is a reasonably good picture of the low and high possibilities and where most of the probability lies. There's a 70% chance that, even if awarded, punitive damages will be fairly low (still very large compared to compensatory damages!), and shows the possibilities of an extreme award. This discretization is very different than what would result from the

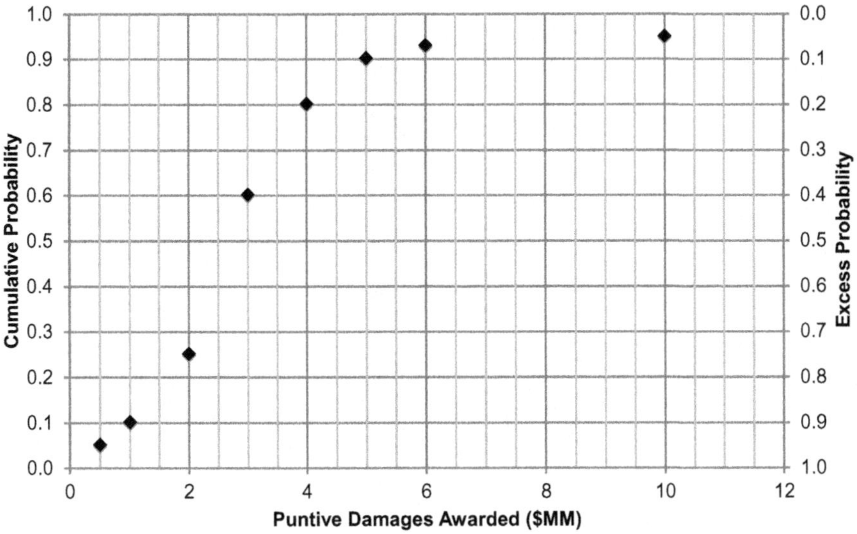

Figure 5.25 Graph with cumulative probability scale for plotting answers to less-than questions and excess probability scale for plotting answers to greater-than questions

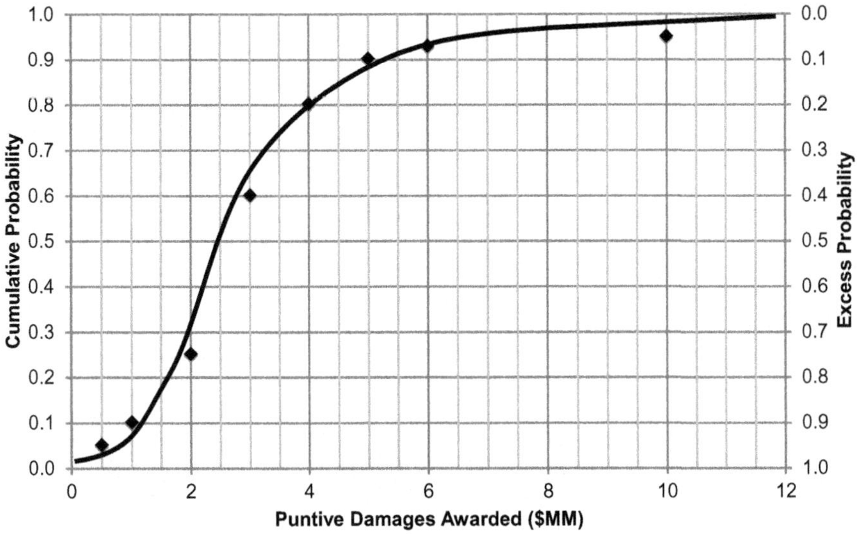

Figure 5.26 Complete assessment with smooth "least squares" line connecting points

three-point approximation method but, for very important uncertainties, that's exactly the point!

With the probability assessments completed, you're ready to go on to integrating all information, judgment, and consequences.

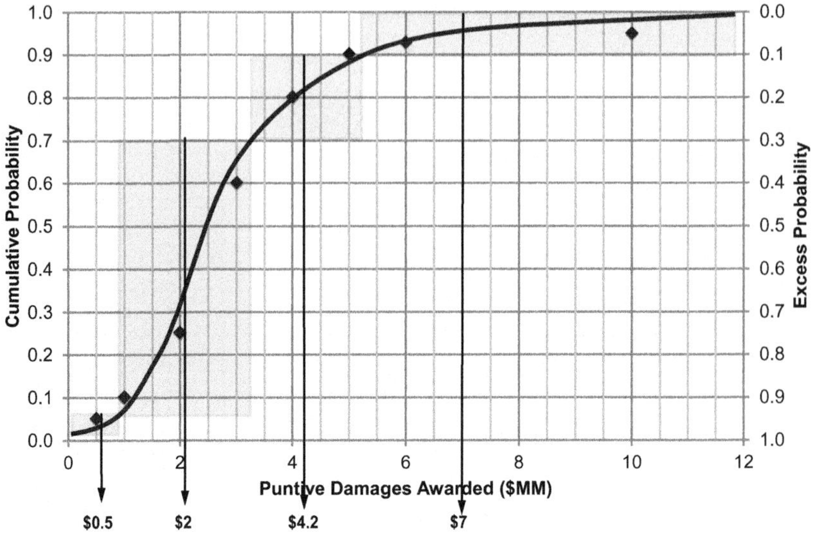

Figure 5.27 Curve broken into regions with conditional means

Figure 5.28 Chance node for punitive damages awarded

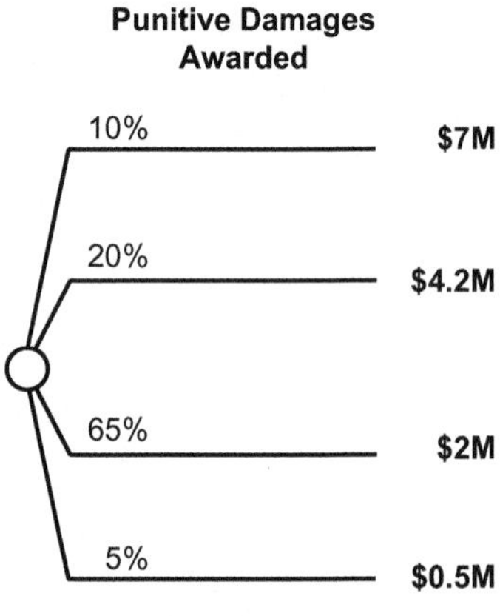

5.2.4 *When to Revise Your Assessments*

A probability assessment is a representation of a person or persons' information and judgement at a particular point in time. That picture may change as new information emerges, or as people reevaluate their initial thoughts. Assessments are most often revised when:

- Sensitivity analysis reveals that a particular uncertainty has a large impact on case value, or that it changes a preferred decision. Then, knowing how important it is, people often want to take a second look at their assessments.

- New information emerges which changes the basis on which the assessment was made. This information could be additional data (from discovery or depositions), or a legal ruling relevant to one or more of the issues in the case (such as whether statutory damages for breach of privacy can be supplemented with punitive damages).
- A new expert is brought in to opine on an important uncertainty.

The list of reasons helps to document the state of information at the time an assessment is made. When changing (updating) an assessment, it is highly recommended that you write down and date the new information so you can track the changes in an assessment over time and the reasons for them.

5.3 A Note on Monte Carlo Simulation Methods

Much of this chapter dealt with assessing and discretizing continuous variables for decision tree calculations. One might observe that, with Monte Carlo simulations methods, no such discretization is required because they work by sampling from any number of continuous variables to generate an overall probability distribution.

Done well, Monte Carlo simulation methods can be equivalent to decision trees: they are just alternative calculation methods for creating probability distributions on the values of interest. However, the way Monte Carlo methods are commonly applied don't prompt the hard thinking we are looking for from a decision analysis. They therefore don't create the key insights as to which course of action is best and why.

First off, when using Monte Carlo tools it is all too easy to simply pick one of the many preprogrammed probability distributions to model a particular variable (normal, lognormal, geometric, Weibull, etc.). This would only be appropriate if the conditions for statistical analysis were present (stable system and "enough" data). That is seldom the case for the variable of interest in decision problems. We typically don't know what the distribution is; we're just making our best guesses about its range and shape.

Second, Monte Carlo analysis can accommodate as many distributions on input variables as you like. It doesn't care if there are 10 or 10,000. It simply samples from all of them until criteria are met for generating a "stable" distribution (one that doesn't change much with further iterations). There's no discipline about which variables are most important (generated by sensitivity analysis in complex models), which forces you to think harder about what to include and about the assessments for the most critical variables.

Also, because of the input and sampling methods, distributions generated by Monte Carlo all tend to look like suspiciously well-behaved S-curves. For litigation decision analysis in particular, we in fact look for "corners" (discontinuous jumps) in the distribution indicative of switches from one scenario to another (e.g., not liable to liable, no punitive damages to punitive damages awarded).

Further, because a "seed" drives the pseudo-random generator employed in Monte Carlo methods, the answer will change if you change the seed. This is a spurious source of variation.

Lastly, some things are extremely difficult to do with Monte Carlo methods. Complex conditional probability distributions or distributions dependent on decisions are very difficult to include. Trees with "downstream" decisions (uncertainties followed by a decision followed by more uncertainties) are simple with good tree software (like Supertree), but I only know of a handful of people in the world who can finagle Monte Carlo software into handling these.

Consequently, Monte Carlo methods don't force you to think as hard about the uncertainty, about which uncertainties to include in the decision tree, or about the resulting overall distributions. They do force you into unrealistic simplifications for complex conditional distributions or downstream decisions. They therefore don't generate the clarity of thinking and insight as to which alternative is best and why.

There's no magic to how decision trees tend to work better than Monte Carlo simulations; they just force you to think harder about the problem.

For these reasons, we employ decision tree methods rather than Monte Carlo methods. In particular, as presented in this book, we urge people to start at a paper-and-pencil level rather than immediately diving into complex computer models and software. The latter are a very necessary next step, but one to take after having mastered the basics.

5.4 Problems

5.4.1 Background on Probability Theory

1. You've been asked to justify why the probability assessments you used in a case analysis are correct. How would you respond?
2. List at least one or two techniques for improving a case other than the listing and elicitation that go along with a probability assessment. What advantages and disadvantages does each technique have relative to probability assessments? Does it apply differently, and, if so, how?
3. What formal or informal procedure can you use to combine judgments on different issues other than probability assessments and arithmetic? (One example is having a meeting and arguing until you reach a consensus or a majority position.) In which situations would this procedure be more helpful (or work better) than probability assessments and arithmetic? In which situations less helpful (or worse)?
4. For each of the following, indicate whether it passes the clairvoyance test, and rewrite it as a more precise event if it does not.

    4.1. The court found the absent party necessary and indispensable under F.R.C.P. 19.

4.2. A "judgment rendered in the person's absence might be prejudicial to the person or those already parties…."
4.3. The court ruled on the motion.
4.4. The court found causation on the second and third counts, but not on the first and fourth.
4.5. Counsel performed to the utmost of her ability.
4.6. The jury found counsel innocent of any violation of his professional responsibility.
4.7. The court might find "a violation of law which reasonably might be imputed to the organization, and is likely to result in substantial injury to the organization." (M.R. 1.13b)

5. State whether each pair of dependent events below passes the clairvoyance test. If not, then redefine one or both of the events so that they both pass.

5.1. A document was in existence when the will was written. A court gives the document testamentary effect via incorporation by reference.
5.2. The victim died of a gunshot wound. The victim bled to death.
5.3. The court found a general charitable intent in the trust instrument. The court applied the doctrine of cy pres and allowed the charitable trust to continue after its specific purpose failed.
5.4. The court declared the matter a political question. The court declared the matter non-justiciable.
5.5. The witness was belligerent and uncooperative on the stand. The court allowed leading questions as if on cross.
5.6. Under the parole evidence rule, the court excluded a prior draft of the contract that supported defendant. The court granted a motion for summary judgment.
5.7. The court found reckless disregard for the safety of others. The court found defendant liable.
5.8. The doctor made an error of judgment in choosing which of the standard treatments to apply. The patient charged the doctor with malpractice.
5.9. Lawyers are subject to discipline. Lawyers have a codified professional responsibility.

6. For each of the following events or list of events, complete the list to make it mutually exclusive and collectively exhaustive.

6.1. The damages awarded exceeds $10 million.
6.2. The court found intentional or reckless defamation in the newspaper article, but found no damages.
6.3. The court ruled against his motion for summary judgment and submitted the matter to the jury.
6.4. The additional discovery cost at least $30,000 and less than $50,000.
6.5. You and one of your two codefendants are found liable.
6.6. The court found that the codicil was validly executed but the will was not, and thus they both failed.

7. The Porcelain Bus Corporation is being sued in a class-action suit for producing a semi-translucent alabaster toilet bowl with an excessively cold seat, startling unsuspecting people. They could be found liable or not, and, even if liable, the court could find no damages.
   7.1. Define a set of mutually exclusive events.
   7.2. Define a set of collectively exhaustive events.
   7.3. Define a set of mutually exclusive and collectively exhaustive events.
8. Your client is the Greasy Riders, Inc., specializing in oil pipeline and machinery repair. They've had a running dispute with MegaPetro regarding work under a maintenance contract, and MegaPetro has been delaying and withholding payments. One day after the Riders performed scheduled maintenance on a pipeline pumping station in Louisiana, the station blew up and burned. MegaPetro is now suing the Riders.

   Case law makes it clear that oil pipelines are not considered an abnormally dangerous activity, so there's no question of strict liability. Accordingly, the court could find no liability, normal negligence, or even that the Riders' conduct was reckless or intentional. You estimate that possible damages in case of normal negligence range from \$2–5 million, and from \$8–15 million in case of reckless or intentional conduct.
   8.1. Define a set of mutually exclusive events.
   8.2. Define a set of collectively exhaustive events.
   8.3. Define a set of mutually exclusive and collectively exhaustive events.
9. Consider a weather forecast for today (You will probably need it either written down or recorded so you can examine it closely.).
   9.1. Was the forecast given in terms of probabilities? Does the forecaster give any hint of a subjective approach to probabilities?
   9.2. Assume that the forecaster is using subjective probabilities. Does this change your interpretation of the forecast?
   9.3. Look outside. What is your probability of rain today? Does it agree with the forecaster's? List reasons why your probability might be the same or different than the forecaster's.
10. Consider the standard mortality table below (Table 5.1).
    10.1. How does this information differ from a subjective probability?
    10.2. Locate your age in the table. According to the table, what is the likelihood that a white US male born alive will die at that age?
    10.3. According to the table, what is the likelihood that a white US male born alive will die in that age interval, given that he has already lived to that age? (Hint: ignore the table up to that age.)
    10.4. Describe how this information should affect your own subjective estimates of your life expectancy, and what other factors should enter in (Hint: this table is constructed from a statistically average sample of white US males 30 years ago.).

Table 5.1 Life table for white males, US of 100,000 born alive, number dying during age interval

Age	Number dying during age interval	Age	Number dying during age interval	Age	Number dying during age interval
0	2592	37	229	73	2775
1	149	38	251	74	2815
2	99	39	278	75	2841
3	78	40	306	76	2853
4	67	41	339	77	2855
5	60	42	376	78	2844
6	55	43	415	79	2821
7	52	44	458	80	2789
8	47	45	505	81	2738
9	43	46	556	82	2639
10	40	47	613	83	2482
11	40	48	681	84	2280
12	46	49	754	85	2096
13	56	50	835	86	1898
14	73	51	916	87	1693
15	90	52	995	88	1490
16	107	53	1071	89	1288
17	121	54	1144	90	1086
18	134	55	1216	91	888
19	143	56	1295	92	709
20	153	57	1383	93	548
21	163	58	1486	94	413
22	167	59	1598	95	300
23	163	60	1714	96	216
24	157	61	1827	97	152
25	149	62	1935	98	103
26	141	63	2039	99	70
27	137	64	2136	100	45
28	137	65	2231	101	29
29	141	66	2323	102	17
30	147	67	2409	103	11
31	154	68	2487	104	6
32	161	69	2559	105	3
33	170	70	2621	106	2
34	180	71	2678	107	1
35	194	72	2729	108	1
36	210				

11. Carmen Cardozo had the misfortune of trying a case in a town in Tennessee where all the jurors worked nights sorting packages for Federal Express. As a result, all 12 jurors have slept soundly through the whole trial, defying all attempts to wake them. Accordingly, Carmen thinks it will be a coin toss (50/50

each way) as to how each juror votes. She further doubts that any jurors will be awake long enough to sway another's initial vote.

11.1. Draw the relevance diagram and decision tree for Carmen's problem.
11.2. According to Carmen's estimate, what are the odds that she will obtain an acquittal?
11.3. What are the odds that Carmen will obtain a conviction?
11.4. What are the odds of a hung jury?
11.5. What are the odds that at least one juror will vote for acquittal?
11.6. What are the odds that the jury will split exactly down the middle?

12. Xavier Estoppel is going to trial for a client with a claim of intentional infliction of emotional distress against a husband who, before the divorce, insisted on making and burning Tuna Helper for himself each night at 2 a.m., and never cleaned up any of the pots. Xavier's case will either be assigned to Judge Prosserer, who thinks very little of the emotional distress cases, or to Judge Frisco, who heads the local Melvin Belli Fan Club. She thinks she still has a .7 chance of winning the case with Judge Prosserer (her client saved three dozen burned pots), and a .9 chance with Judge Frisco. She thinks there's a .6 chance of getting Judge Prosserer because his calendar is less crowded.

12.1. Draw the relevance diagram and decision tree for Xavier's problem.
12.2. What should Xavier think her overall chances of winning the case are, given that she doesn't yet know which judge she'll get?
12.3. Similarly, what are her overall chances of losing the case?
12.4. What are Xavier's chances of winning and getting Judge Prosserer? Of losing and getting Judge Frisco?
12.5. Plot Xavier's chances of winning the case as a function of the probability of getting Judge Frisco (Hint: the probability of getting Frisco should be on the bottom axis.).

13. Sid Liberty is defending a student charged with criminal misdemeanor gaming on school premises for organizing a pool on which teacher the principal would take to the school dance. The key witness against the student is Karen Kiwi, a near-sighted teacher who caught a glimpse of him taking money and distributing slips, and then picked him out a few moments later when the principal lined everybody up against the cafeteria wall. No evidence was found on the student at the time.

Sid is going to contend that the student had a Wade–Gilbert right to have counsel present during that line-up, and that Karen should be excluded as fruit of the violation. He figures that he has a .90 chance of excluding Karen and winning the case if the court agrees with him, but only a .05 chance of exclusion and winning if the court says no Wade–Gilbert violation.

13.1. Draw the relevance diagram and decision tree for Sid's case.
13.2. If Sid's chance of losing on Wade–Gilbert and losing the case is .57, what is his chance of winning on Wade–Gilbert?
13.3. What are Sid's overall chances of winning the case?

14. Sean Twilliwit was a sales clerk at the Banana Republic safari clothing store. During the frequent times when no one was in the store, he worked on sketches for "Khaki for Freedom," an exhibition on the role of clothing in various political reform movements that he was putting on at the Museum of Fashion Art in a few months. Sean's manager saw him at work one day, and told him to "Get that crap out here." Sean replied "I don't have to take this!" and left in a huff. When he returned to work at his usual time the next day, he was told he no longer worked there.

    Sean has retained you to sue, and you've developed a number of theories. The first obstacle, however, is whether Sean was fired, or whether he quit, as his employer contends. If the court finds he quit, you lose the case.

    If, however, the court finds that he was fired, you think the Republic's personnel manual give good grounds for finding an implied clause to only fire for good cause. You give a .8 chance the court will agree. You would then have to convince the court that the clause was violated.

    Further, even if the court finds no implied clause, or if they find that the implied clause was not violated, you'll still contend that the firing violated Sean's constitutional rights of freedom of political expression. You think that if the court finds the implied contract clause, there's about a .7 chance they'll also find the constitutional violation (an issue they'd only come to if they found no violation of the implied clause). However, you believe that if the court rejects the implied clause, there's only a .3 chance of them finding the constitutional violation.

    14.1. Draw the relevance diagram for Sean's problem.
    14.2. Draw the probability tree and fill in the information you have. Remember not to repeat chance nodes where they're not relevant.
    14.3. If there is a 0.288 chance of "yes" on Sean being fired, "yes" on the implied clause, and "yes" on violation of the clause; and a 0.084 chance of the court finding "yes" on firing but "no" on implied clause and "no" on the constitutional violation, then what is the chance of the court "yes" on the implied violation after it found that the implied clause exists? (Hint: this is not the compound probability at the end of the tree.)
    14.4. What's the chance the court will find that Sean was not fired?
    14.5. What is the overall chance of winning the case?

5.4.2 Assessment Calibration Exercises

15. For each of the questions in the table below:

    15.1. Fill out your answers for numbers low enough there's 10% chance the actual is lower and high enough there's a 10% chance the actual is higher (Do this for all the questions before you start Googling the answers.).

15.2. Google the answers.
15.3. Give yourself one point for each time the answer is outside of your range. Add up your total points.
15.4. What is a perfect score on the test?
15.5. What does it indicate when a score is higher than a perfect score?
15.6. What does it mean when a score is lower than a perfect score?

	Question	10% chance actual is lower than:	10% chance actual is higher than:
1	Distance from Honolulu to Cape Town, South Africa in miles		
2	Mean temperature at Murdoch Station in Antarctica in August (°C or °F)		
3	World diamond production in 2013 in millions of carats		
4	Population of China in 1800		
5	Percentage of silver in pewter		
6	Cost of the Louisiana Purchase		
7	Number of full time employees of Google as of September 30, 2005		
8	Number of full time employees of Google as of September 30, 2014 (including Motorola Mobile)		
9	US total stock market capitalization, 2012		
10	China total stock market capitalization, 2012		
11	Average per person daily calorie consumption in the USA, 2011		
12	2014 percent of population over 60 in Japan		
13	2014 percent of population over 60 in Zimbabwe		
14	2011 cinema attendance in India		
15	2012 percent of population in the labor force in Iran		
16	Worldwide number of McDonald's restaurants		
17	Number of words in the Bible (King James authorized edition)		
18	Maximum takeoff weight of a Boeing 787-9 Dreamliner in kilograms		
19	Number of lawyers in the USA in 2014		
20	Total number of cattle and calves in California in 2015		

5.4.3 *Procedures for Assessing Probabilities*

16. Obtain an ordinary, old-fashioned-style thumbtack (the newer push-pins won't work). As you realize, the tack can sit on a flat surface in two ways:

    16.1. Assess your probability that the thumbtack will land pointing up. Recall that this must be 1 minus the probability of landing on its side. In assessing your probability, you may want to consider such factors as:

    - where you think the center of gravity of the tack is;
    - what effect you think the shape will have on air resistance as the tack falls;
    - how the tack is likely to bounce on the surface before landing.

    16.2. Flip the tack once and observe how it lands. Now reassess your probability.
    16.3. Flip the tack ten more times, keeping track of how it lands each time. Reassess your probability once more.
    16.4. Discuss why your probability assessment did or did not change. What was your initial state of knowledge? What was your final state of knowledge? How did any additional information affect your probability assessments?

17. Use the procedure described in the text to assess probabilities for the following discrete events (as opposed to continuous events). Be sure to write down the results of each step. You may need to revise some of the questions in order to pass the clairvoyance test (Note: pairs of people with alternating roles work best for this exercise.).

    17.1. What is the probability that you will gross more than $40,000 in the calendar year following this one? (Use a higher or lower figure if doing so will make the question more interesting.)
    17.2. What is the probability that the next case you take on will go to trial?
    17.3. What is the probability that you will change jobs within the next 2 years? (Consider being a student a job.)
    17.4. What is the probability that you will be sued in connection with damaging another's property (including in automobile accidents) within the next 10 years?
    17.5. What is the probability that you will read a federal case this week?
    17.6. What is the probability that you will be paying legal fees to another lawyer within the next 6 months?
    17.7. What is the probability that you will be admitted to the bar of an additional state within the next 5 years?

18. For each of the following problems, assess the required probabilities and answer the question. Again, check for the clairvoyance test (Your instructor may want

to indicate which assessment procedure to use, or pick the method that seems most appropriate to you in each case and justify your choice.).

18.1. You are being sued for $15 million in a malpractice case involving the Rule Against Perpetuities (there was a big tax bill when the trust failed). You will lose the case if the court finds that your law school property professor spent less than 30 minutes on the subject. What is your expected liability?

18.2. You're a plaintiff in a telephone long-distance class-action suit that has a .60 chance of winning. If you win, you'll recover 1 percent of your total long distance charges over the past 12 months. What is your expected recovery? (No checking all your old phone bills.)

18.3. You're a defendant in a gender discrimination case. You will lose if the female percent of your firm's clients (or the female percent of your law school class) is less than the female percent of the population as a whole. What is the chance you'll win the case?

18.4. You're one of the plaintiffs in a civil suit against the Chinese restaurants in your state for using MSG in violation of FDA guidelines. You will recover 200 percent of all the money you spent on Chinese meals that contained MSG. What is your expected recovery?

18.5. You're offered a malpractice policy that provides lifetime coverage for a fee equal to twice your gross earnings last year (or twice your expected earnings your first year out of law school). Assume that you could arrange to pay the premium all at once. Should you buy the policy? On an mean value basis, how much would you come out ahead or behind? (Hint: be sure to consider the timing of the various cash flows.)

18.6. Assume that you have a 90 percent chance of winning a case if you can accomplish a valid service of process on your mother (or sister, etc.) within the next two business days, and a 60 percent chance of winning if you can't. The judgment if you win will be twice the odometer on your father's (or brother's, etc.) car. What is the least amount you should be willing to settle the case for?

19. Contested damages in a personal injury case consist of immediate property damages and medical bills, and medical bills for a condition that did not manifest itself until 6 months later. Assume there's a 50/50 chance the court will find causation for the later condition.

19.1. Choose an immediate injury and later condition that you would feel comfortable assessing the special and general compensatory damages on (If you're stuck, assume that plaintiff suffered neck whiplash in a car accident, immediately sustaining $1,500 in property damage and $4,000 in medical bills, and that 6 months later he'd spent $1,500 in traction and chiropractor bills for lower-back pain. Compensatory damages will, of course, be extra.).

19.2. Assess the uncertainty in the total value of the case, depending on whether or not the court finds causation for the later injury, using the same numerical outcomes (the dollar values on the branches) with dependent probabilities. A three-branch estimate will be sufficient.

19.3. Asses the uncertainty in the award again, but using dependent outcomes and the same probabilities on each branch (perhaps .25, .50, .25).

19.4. Compare the mean values by each method. Which method do you feel gave you the better estimate? What does this tell you about how you thought about the uncertainty in this particular case?

Chapter 6
Integrate All Information, Judgment, and Consequences

In this step, we put together all the pieces of the decision basis (alternatives, information, and preferences), calculate the results, and start the process of asking whether they make sense and what they tell us. In this chapter, we focus on how to put the calculations together. In the next chapter, we focus on what they tell us.

In calculating scenario values and overall mean values for alternatives, there are three types of errors to look for:

- A math error in the calculation
- A logic error in that the calculation is correct, but it's not really what you meant
- What we might call a "greater understanding" error, where the calculation is correct and rcflccts what you meant but, on further consideration, you have a better idea.

For math errors in the calculation, software is extremely helpful and eliminates worrying about them. However, as illustrated in this book, it is necessary to start with a paper-and-pencil level of analysis to make sure you understand how to structure the problem before trying to enter it into software. Diving right into the software typically makes matters worse: it is all too easy to enter a complex problem formulation that neither you nor anyone else can make sense of. If you don't understand the problem structure, you won't understand the results it produces. After more than 3 decades as a professional decision analyst, I still draw the tree on a piece of paper before entering it into software.

For the remainder of this book, we'll continue with paper-and-pencil level of analysis supplemented with software where especially helpful.

There are many programs on the market for decision tree analysis. Some are quite good; some have significant conceptual shortcomings. We recommend Supertree, a free student version of which is available at Supertree.net.[1]

[1] Full disclosure: the author worked on the development of Supertree in the 1980s and 1990s, but has no ownership or financial interest in it.

J. Celona, *Winning at Litigation through Decision Analysis*,
Springer Series in Operations Research and Financial Engineering,
DOI 10.1007/978-3-319-30040-5_6

A logic error is indicated when you can't understand why the results are what they are. The value in each scenario should be transparently traceable to the inputs that produced it, and likewise the overall mean values should be a clear result of the scenario values.

If you don't understand why a number turned out they way it did, you need to figure that out. It probably indicates an error you'll want to find before going forward with a recommendation. This tends to be more of an issue when you need to use a computer model for complex damages calculations.

"Greater understanding" errors are possibly the most valuable: they point out where a different structure or assessment would make more sense. These are a key part of how decision analysis forces you to think harder to make you litigation strategy a winning one.

For example, an earlier draft of this manuscript had an assessment in the Howitzer case for punitive damages about equal to the potential compensatory damages. On reflection, I realized this didn't make sense: since they are meant to "punish" the defendant, punitive damages typically dwarf the compensatory damages (which is what makes them so scary!). Hence the assessments you see in this text.

Best to keep these three types of errors in mind as you calculate scenario values and mean values and correct them along the way.

6.1 Calculate Scenario Values

To calculate the scenario values, you simply take the formulas you wrote when you were structuring the decision tree and plug in any missing numbers that were waiting for probability assessments. Remember to discount all numbers for present value. You should end up with a numerical value for each scenario in the decision tree. This value will just be a simple number describing what it will cost you if everything described by that scenario (path through the tree) happens. The probabilities don't get used until the next step.

For your Howitzer v. Crabgrass tree, calculating the scenario was values was just matter of writing in the numbers from the damages probability assessments at the end of the tree and adding them up. You started with the subtree for damages, as below (Figure *6.1*).

Doing the values for the main tree was even easier—not even any addition involved (Figure *6.2*).

You then remembered about only including present value numbers in the tree. However, given the potential time to judgement and current interest rates, you decided to put aside time issues for now. Looking more closely at discounting and pre- and post-judgment would be something to consider after your first-pass analysis.

After seeing all the numbers written out in the trees, you were getting very curious to see how the mean values worked out. The $100,000 settlement offer didn't seem attractive at all at the time, but now you're really wondering given how large the punitive damages exposure could be.

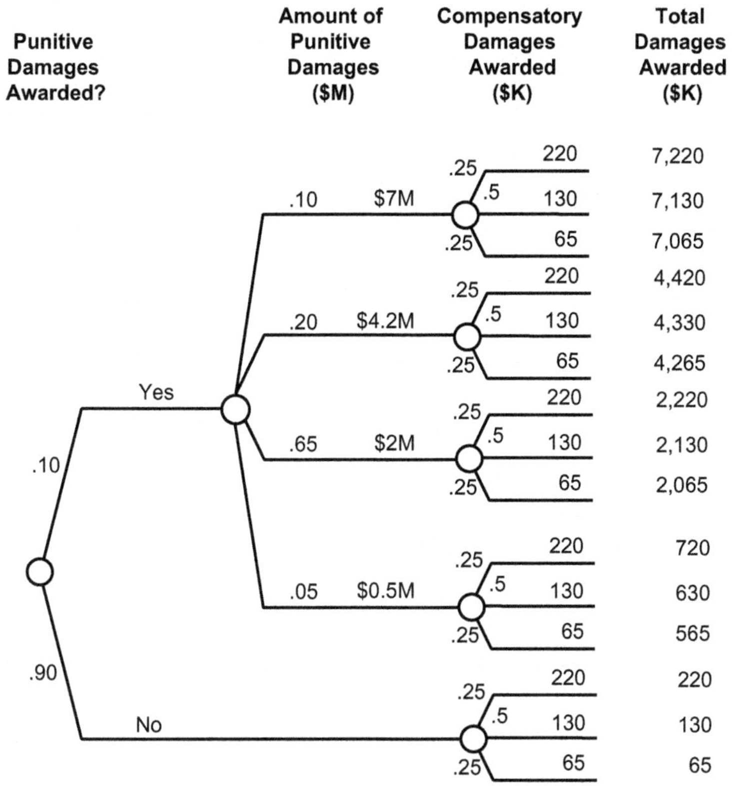

Figure 6.1 Subtree for damages calculation

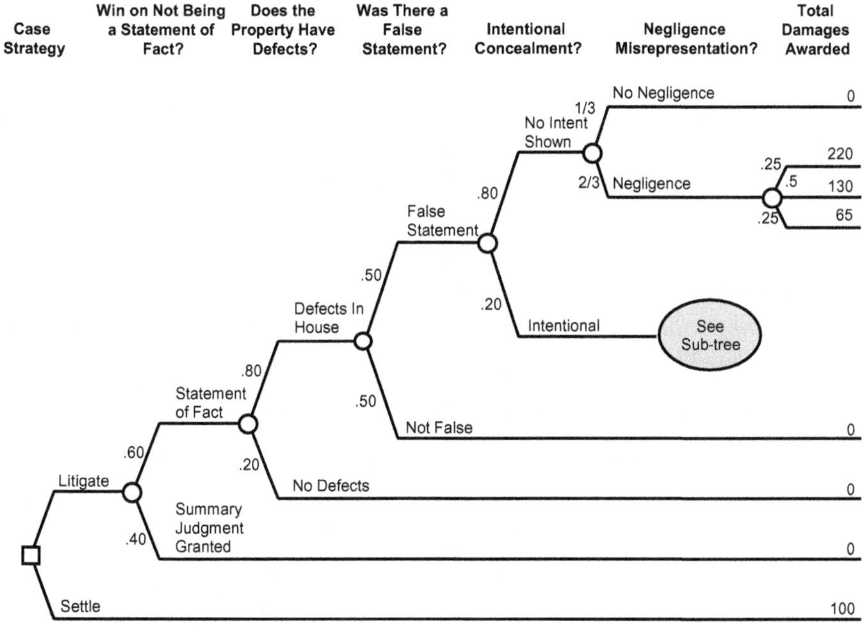

Figure 6.2 Total damages for all scenarios

6.2 Calculate the Mean Value

As you've seen, the mean value is the probability-weighted average value of all the possible outcomes of the ultimate issue (such as the chances of being held liable or the value of the case). This single number is what you'll use to represent the value of litigating the case, and to compare against settlement offers. The mean value can be computed by the roll-forward/compounded probability method, or by the roll-back method. The two methods are equivalent.

6.2.1 Calculating the Mean Value with the Roll-Forward

- **Multiply Probabilities**. Start at the beginning (left) of the tree, and, for each path leading through to the end (right side), multiply together all the probabilities on all the branches you go through to reach each end point of the tree. Thus, each end point of the tree represents a scenario, and the probability of that scenario is the "and" probability of everything that happens in it. Check to make sure all your multiplied-out probabilities add up to 1.0.
- **Multiply Probabilities and Values**. Multiply together the probability and value for each scenario. This gives you the probability-weighted value of that scenario, or that scenario's contribution to the overall mean value.
- **Add the Resulting Products**. Compute the overall mean value by adding up all those probability/value products.

The only wrinkle to this whole procedure is that you treat each alternative for a decision node separately. Thus,

- The multiplied-out probabilities *for each alternative* (litigate or settle) should add up to 1.0.
- You add up the probability/value products for each alternative separately. Thus, each alternative will have its own mean value.

You probably intuitively assumed these two points, since you get to pick which alternative you want (instead of there being an uncertain answer), and because decision nodes don't get any probabilities.

Continuing with your tree drawings for Howitzer v. Crabgrass, you multiplied out the probabilities and calculated the roll-forward mean value for the punitive damages subtree (Figure *6.3*).

Of course, for even this level of analysis something more than a cell phone is helpful. Here's the simple Excel spreadsheet you used to figure it (Figure *6.4*).

Once you had the $422,750 figure to plug back into the main tree, you were ready to do the roll-forward calculation for the main tree. Here's the Excel spreadsheet to do the calculation, then values plugged into the three (Figures *6.5* and *6.6*).

Here is at least a first crack at your answer: even though the potential punitive damages exposure is quite large, so many things need to happen to get there that the overall exposure of litigating the case to judgment is only $38,000—versus the $100,000 settlement offer on the table.

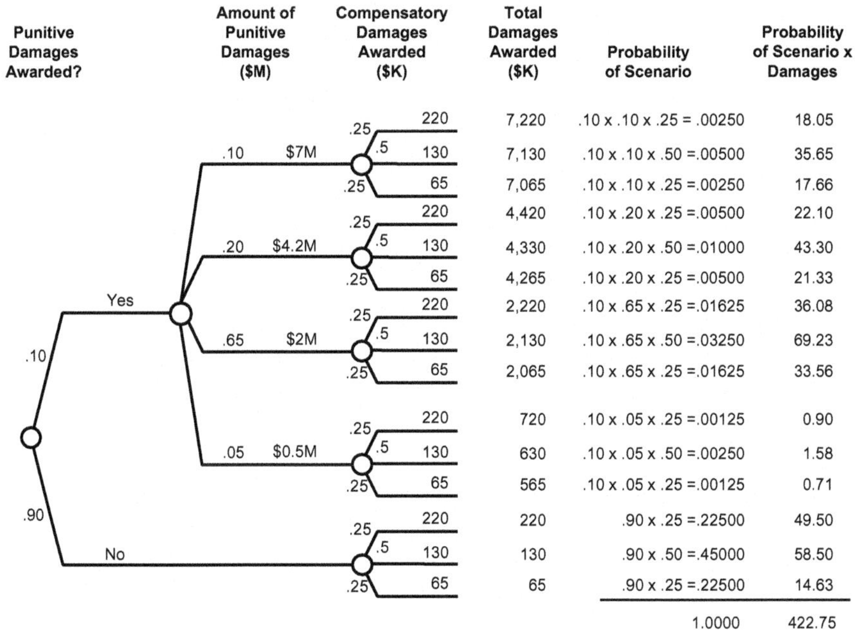

Figure 6.3 Roll-forward mean value for damages subtree

	A	B	C	D	E	F	G	H	I
2	Punitive Damages Awarded?	Amount of Punitive Damages	Compensatory Damages	Product of Probabilities		Amount of Punitive Damages	Compensatory Damages	Total Damages	Damages x Probability
3	0.1	0.1	0.25	0.00250		7000	220	7220	18.05
4	0.1	0.1	0.5	0.00500		7000	130	7130	35.65
5	0.1	0.1	0.25	0.00250		7000	65	7065	17.66
6	0.1	0.2	0.25	0.00500		4200	220	4420	22.10
7	0.1	0.2	0.5	0.01000		4200	130	4330	43.30
8	0.1	0.2	0.25	0.00500		4200	65	4265	21.33
9	0.1	0.65	0.25	0.01625		2000	220	2220	36.08
10	0.1	0.65	0.5	0.03250		2000	130	2130	69.23
11	0.1	0.65	0.25	0.01625		2000	65	2065	33.56
12	0.1	0.05	0.25	0.00125		500	220	720	0.90
13	0.1	0.05	0.5	0.00250		500	130	630	1.58
14	0.1	0.05	0.25	0.00125		500	65	565	0.71
15	0.9		0.25	0.22500			220	220	49.50
16	0.9		0.5	0.45000			130	130	58.50
17	0.9		0.25	0.22500			65	65	14.63
18									
19	Totals			1.00000					422.75000

Figure 6.4 Excel spreadsheet for roll-forward mean value calculation for damages subtree

	A	B	C	D	E	F	G	H	I
21	Statement of Fact?	Defects In House?	False Statement?	Intentional Concealment?	Negligence?	Compensatory Damages	Scenario Probability	Value	Probability x Value
22	0.6	0.8	0.5	0.8	0.33333333		0.064	0	0.00
23	0.6	0.8	0.5	0.8	0.66666667	0.25	0.032	220	7.04
24	0.6	0.8	0.5	0.8	0.66666667	0.5	0.064	130	8.32
25	0.6	0.8	0.5	0.8	0.66666667	0.25	0.032	65	2.08
26	0.6	0.8	0.5	0.2			0.048	422.75	20.29
27	0.6	0.8	0.5				0.24	0	0.00
28	0.6	0.2					0.12	0	0.00
29	0.4						0.4	0	0.00
30									
31	Totals						1.00		38

Figure 6.5 Excel spreadsheet for roll-forward mean value calculation for main tree

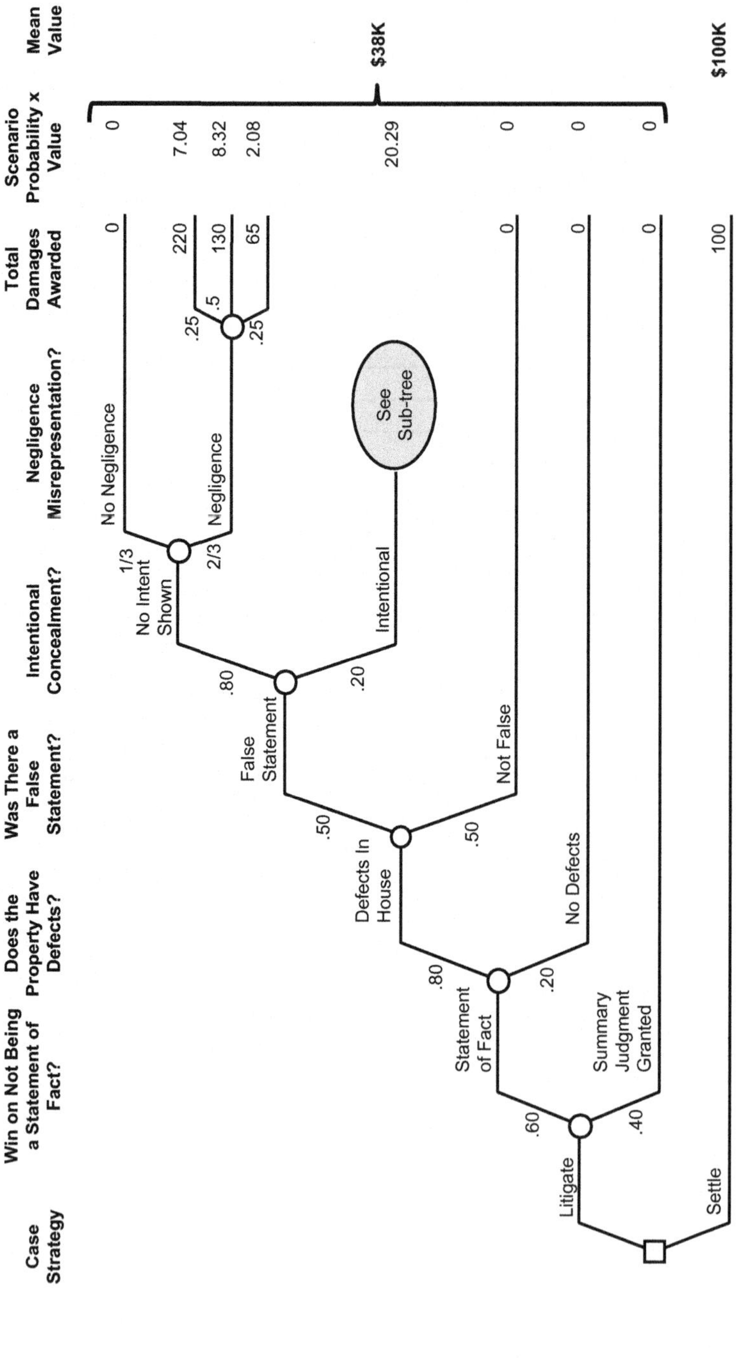

Figure 6.6 Roll-forward mean values plugged into the main tree

However, you note that this analysis doesn't yet include the cost to try the case. If the costs to go to trial looked to be more than 100 – 38 = \$62,000, the settlement offer might still be a good deal—unless your client is worried about getting a reputation for easy settlements which prompts additional lawsuits.

As with all good analyses, answers to your first set of questions prompt additional questions.

For practice, you decide to try the rollback calculations in addition to the roll-forwards already completed.

6.2.2 Rollback Mean Value Calculations

Rollback mean value calculations start at the end of the tree (the rightmost side) and work backwards to left. The mean values for each alternative are the same as the ones calculated by the roll-forward method.

1. **Start at the End of the Tree**. Just as roll-forward starts at the beginning of the tree and rolls forward, multiplying probabilities along the way, roll-back starts at the end of the tree and works back. So go to the last, right-most column in your decision tree (If you are following the hint in the tree drawing section, each question in the tree should have its own column.).
2. **Compute a Mini-Expected Value for Each Node**. For each node in the column (each uncertainty or decision), compute a mini-mean value. For chance nodes in the last column, this means multiplying thc probabilitics on those branches by the scenario values at the end of the tree, then adding up the results. For a decision, of course, just pick the branch with the best value (which will be the highest number if you followed the hint about making payments to you positive and amounts you pay out negative).
3. **Go Back to the Next Column in the Tree**. Now move one column to the left in the decision tree and repeat the process. The only difference is that the numbers at the end of the branches will be the mean values you just calculated, rather than the scenario values at the end of the tree.
4. **Repeat Till You Hit the Start**. Repeat step 1 until you get to the root/beginning of the tree (typically, the litigate/settle decision). At that point, you should have an mean value for litigating the case, while the settlement amount will be a simple number.

You started the rollback calculation with the damages awarded if your client were found negligent. That is $(.25 \times 200) + (.5 \times 130) + (.25 \times 65) = 136.25$ *and filled that in the tree. Two-thirds of that was 91. Then you filled in the 422 from the damages subtree* (Figure 6.7).

You continued working back through the tree $(.8 \times 91) + (.2 \times 422.75) = 158$, *and so on until you completed the tree (All numbers rounded off)* (Figure 6.8).

As you'd hoped, the mean values for each alternative calculated by the two methods agreed.

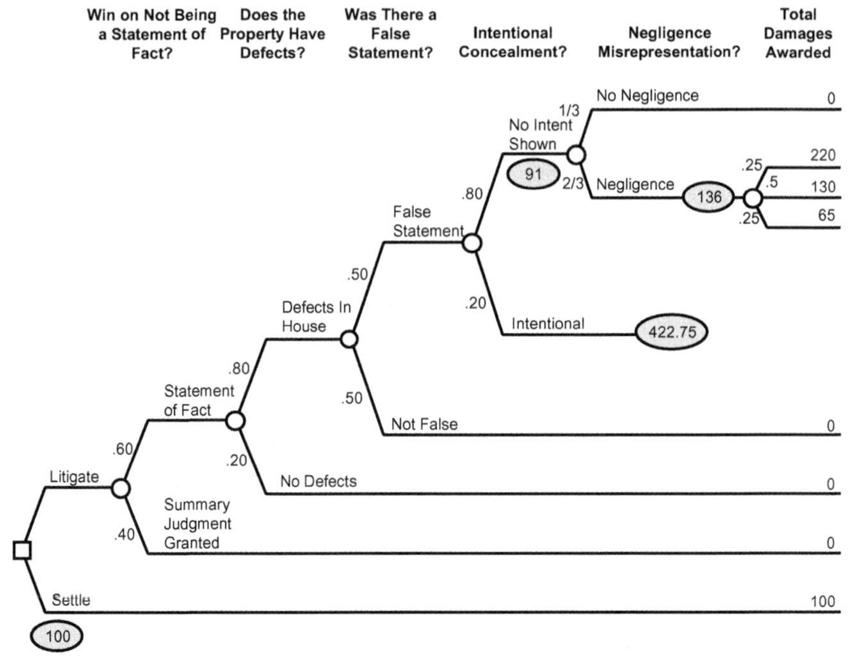

Figure 6.7 Start of rollback mean value calculation

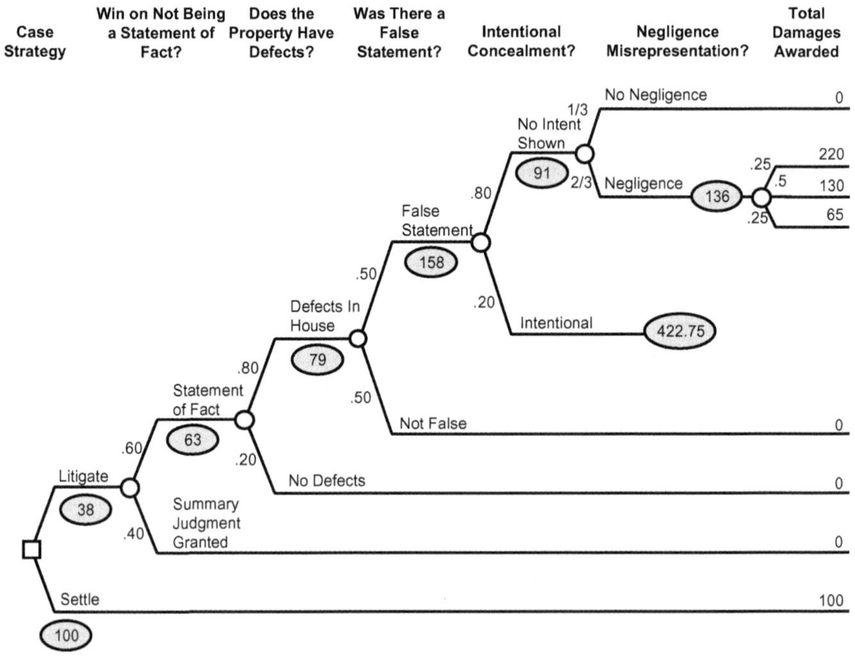

Figure 6.8 Completed rollback calculation

6.3 Cumulative Probability Distribution and Histogram

You've already seen how to assess directly a cumulative probability distribution for the answer to a question. The decision tree can be used to produce a similar cumulative probability distribution for the value of litigating your case, for example. Thus, the cumulative probability distribution can provide a complete description of the uncertainty in the value of litigating your case; first, by showing the range of all possible values (along the bottom axis) and, second, by showing what the odds are of ending up above or below a particular number (by reading from the vertical axis). The steepest parts of the cumulative also show you what regions you have the greatest chances of ending up with.

The cumulative distribution is also needed if you want to handle your client's risk aversion. However, constructing a cumulative distribution is a little tricky, so that subject is deferred until Part III. All the same, the cumulative will be referred to in the next chapter in the discussion of how risk aversion affects your settlement strategy.

Being the incredibly curious and adventurous sort, you looked ahead to Part III and read how to construct a cumulative probability distribution (briefly, you arrange all the scenario values from highest to lowest, and then plot them in a "stair step" pattern: the probability of each value is the height of its step, and then you draw a horizontal line over to the next value, and so on).

Then you decide to download the free student copy of Supertree and use it to draw the cumulative probability distributions for both alternatives (much easier than drawing it yourself!) (Figure 6.9).

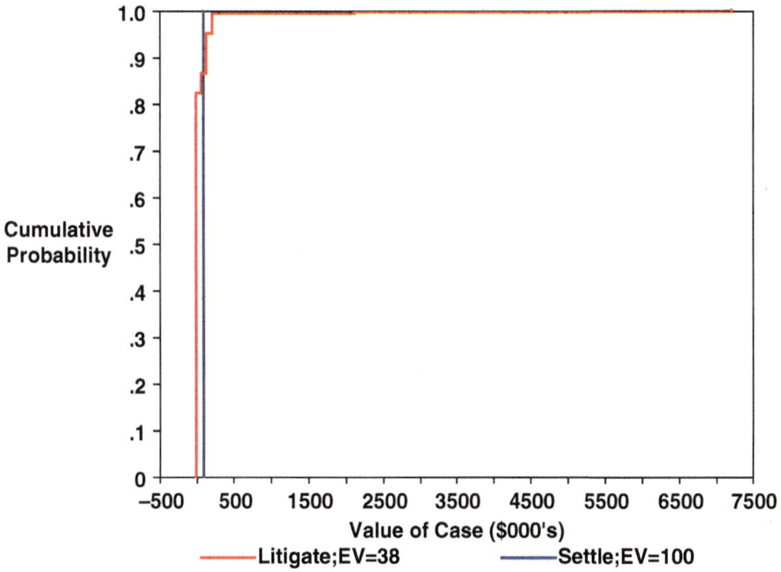

Figure 6.9 Cumulative probability distributions for litigate and settle alternatives

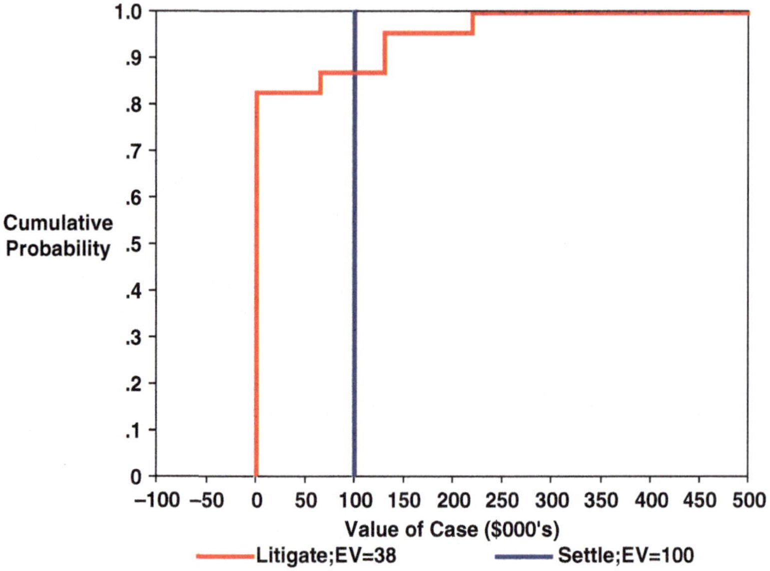

Figure 6.10 Cumulative probability distributions for values between $−100K and $500K

You note the "long tail" on the litigate distribution for the tiny-probability scenarios in which your client ends up paying millions of dollars in punitive damages. Because it's difficult to see what's going on at values closer to zero, you plot the distribution again, this time only showing part of it for more detail (Figure *6.10*).

Here you can begin to see some of the detail. Overall, there is a better than 80% chance of escaping with no liability if you go to trial. Even then, your client is most likely to be held to something in the range of the $65K–$130K–$220K scenarios for compensatory damages (the next three steps in the stairs). Overall, chance of having to pay punitive damages are tiny. You refer back to the tree for a quick calculation. The chances of being hit with punitive damages are $.6 \times .8 \times .5 \times .2 \times .1 = .005$ *– less than 1%.*

You can also look at a distribution in the form of a histogram rather than a cumulative probability distribution. A histogram is rather like a discrete form of a probability density function. It's created by:

- Dividing the x-axis into intervals with each interval defining a bin.
- Sorting the scenario values (the numbers at the end of the tree) into the bins. Each value number *v* goes into the bin where it is greater than or equal to the lower limit and less than the upper limit (e.g., $x_i \le v_i < x_{i+1}$).
- Plot a graph where the height of each bin is equal to the sum of the probabilities of all the scenario values in the bin (using the product of probabilities associated with each scenario at the end of the tree).

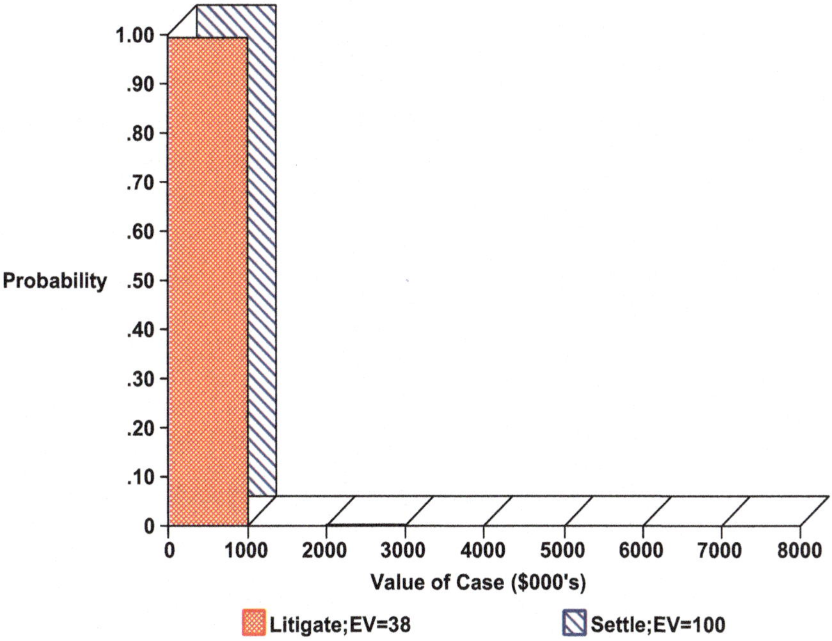

Figure 6.11 Overall histogram for Howitzer v. Crabtree

Below is the histogram for the litigate and settlement strategies in our example (Figure 6.11).

As with looking at the complete cumulative probability distribution, it is hard to see the detail with what is going on because the chances of punitive damages are so small. We can likewise see more detail by looking only at the value between zero and $300K (Figure 6.12).

With this graph, it is likewise easy to see that, with litigating the case, there is better than an 80% chance of winning the case and paying no indemnity, versus settling brings a 100% chance of costing $100K if you pay the offer on the table.

Of course, in an actual case, you would be trying to negotiate down the settlement amount, keeping in mind your mean exposure if litigating is $38K. Possibly a settlement amount of $40K or $50K would be a good deal for the client.

Alternatively, you could roll the dice on litigation, with an 80% chance of no indemnity at all, a 20% of paying more, and a less than 1% chance of paying a lot more. A final recommendation would need to take into account the costs (legal fees, experts, discovery) of litigating, which are analyzed the same way as damages and covered in more detail in Part III.

Either way, even though we have only begun to dive into analyzing the results, you're well-informed to put together the best possible legal strategy for you client given what you have to work with. This is how you build winning litigation results.

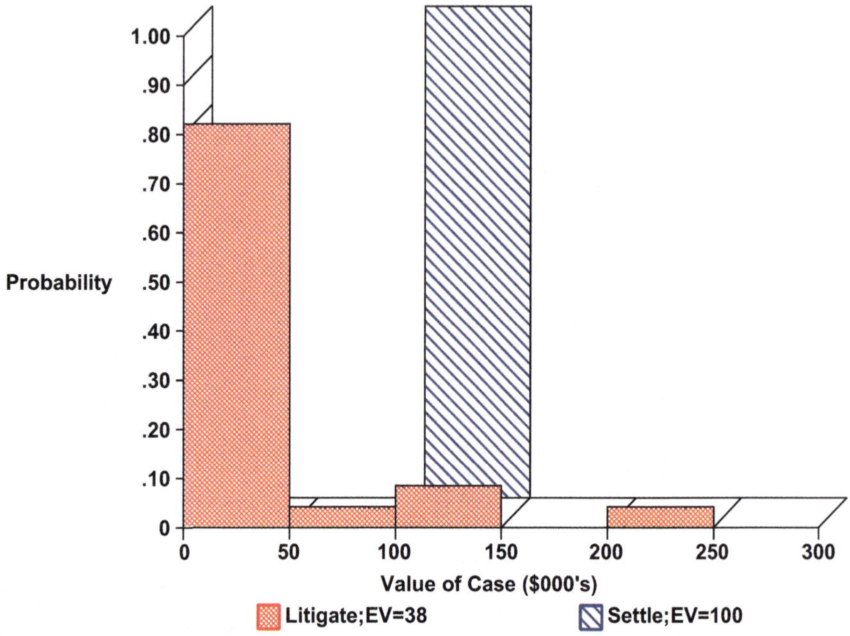

Figure 6.12 Histogram for values between zero and $300K

6.4 More on Subtrees

Subtrees are a way of breaking the analysis into tractable pieces and not making the main tree too complicated while you're becoming familiar with decision analysis. In the Howitzer example, we used a subtree to handle the punitive damages piece. We recommend everyone learning decision analysis begin at the paper-and-pencil level before starting to use software. With tree software, it's all to easy to get lost in the forest and put an analysis together neither you nor anyone else can make sense of.

When you have graduated to using software, one of the most necessary and valuable steps is using deterministic sensitivity analysis to identify which uncertainties have the most impact on overall value. See *Decision Analysis for the Professional* for details on how to do this. Those most critical uncertainties are candidates for inclusion in the decision tree.

With this important step of using deterministic sensitivity analysis, using software for calculations is the preferred route. Then, it's not unusual to evaluate trees with thousands of scenarios.

While you are still in the learning and paper-and-pencil mode (the focus of this book), here are further suggestions on the use of subtrees.

6.4.1 Definition and Purposes of Subtrees

A subtree is simply a portion of the main tree that you separately analyze. You can use a separate subtree for any questions whose answers aren't impacted by factors (decision or uncertainties) in the main tree.

For instance, suppose you had a case for injuries in an auto accident, and damages depended on if the jury found plaintiff's back was injured in the accident (and not a preexisting condition), and whether they found him disabled from working. First of all, you might have your main tree for the overall value of the case, as shown below, with damages analyzed in subtrees (Figure 6.13).

There might then even be two subtrees to analyze the potential damages, as shown below (Figures 6.14 and 6.15).

Figure 6.13 Main tree for auto liability

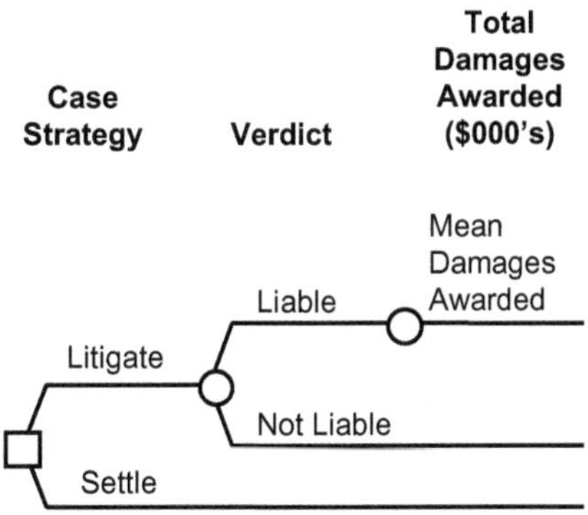

Figure 6.14 Subtree for medical costs

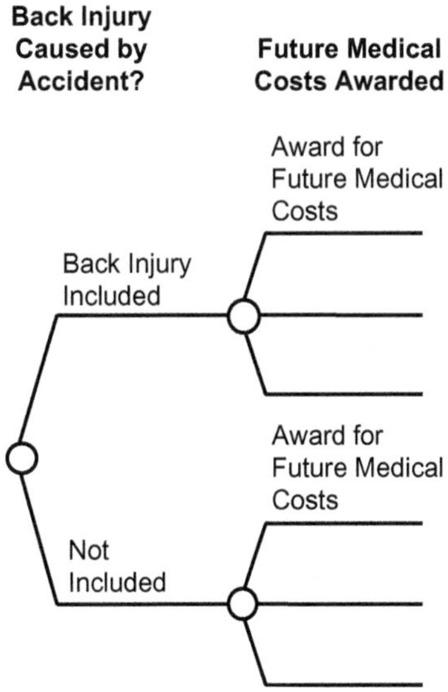

Figure 6.15 Subtree for lost income

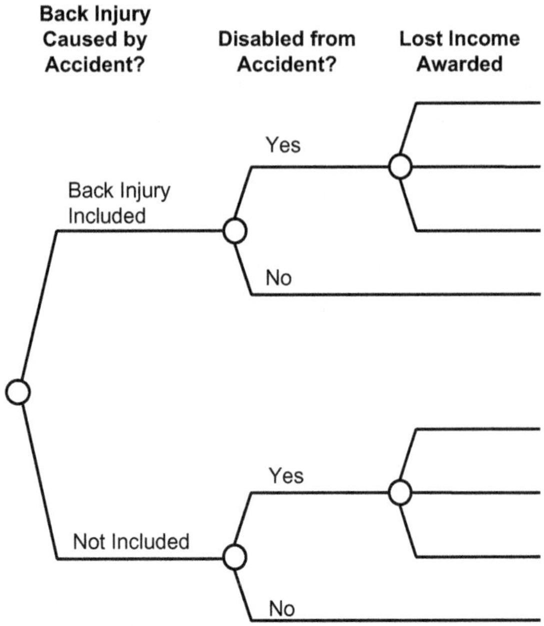

The total mean damages award in the main tree, then, would be the sum of the mean award for future medical costs and the mean award for lost income. In contrast, calculating everything in one main decision tree would produce quite a large tree (below). And this tree doesn't even include any questions affecting liability! (Figure 6.16).

Using subtrees, then, serves two important purposes.

- Subtrees allow consideration of important influencing factors …
- … while keeping the analysis in small, tractable pieces.

6.4.2 *Requirements for Subtrees*

There is one main reason for wanting to use a subtree, and one requirement that determines whether you can do so.

- **Reason for Using a Subtree**: there is a factor in the main tree that you can't think meaningfully about without first knowing how other uncertain factors turned out.
- **Requirement for Using a Subtree**: those other uncertain factors are otherwise independent from the main tree. That is, they don't influence any factor in the tree other than the one the subtree will explore in more detail.

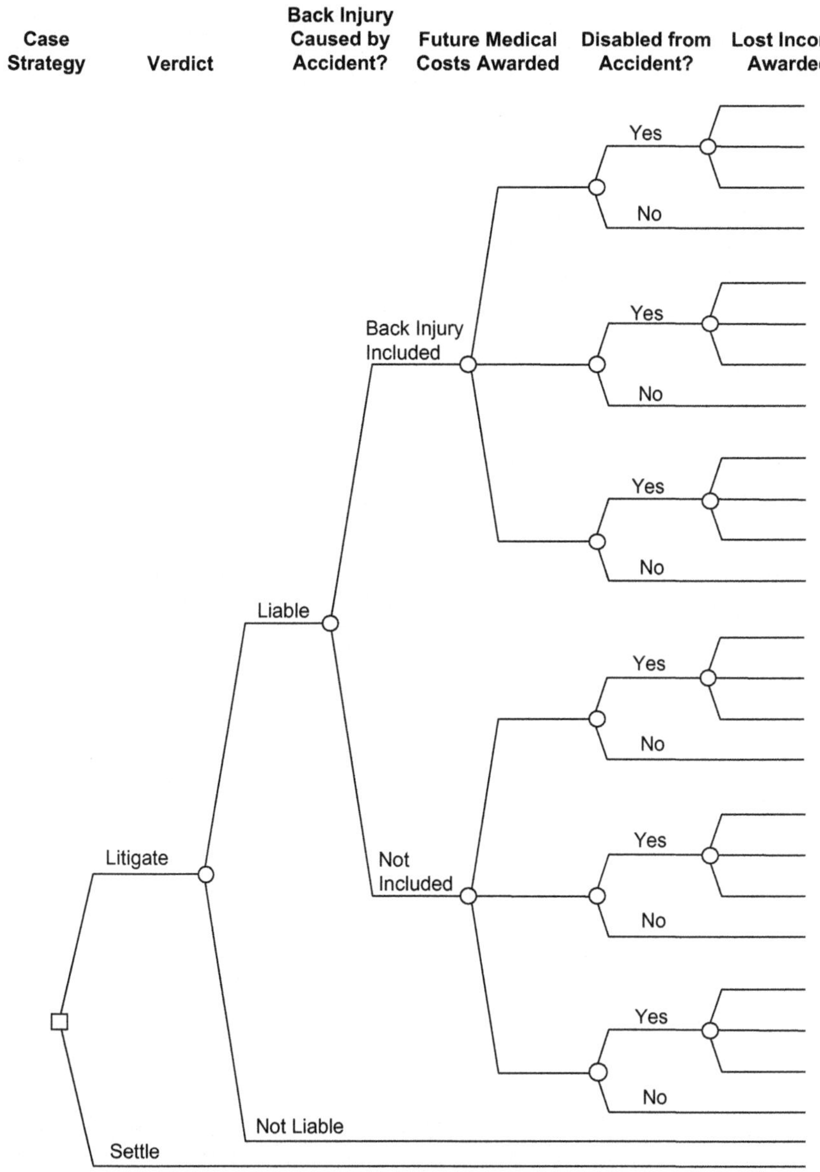

Figure 6.16 The tree gets large fast

Identifying whether you want to use a subtree for a particular question in the main tree is a matter of applying the clairvoyance test: can you answer the question with yes, no, or a number, or do you need to know something more? In other words, do you need to know the answers to other factors (uncertainties or decisions) before you can think about the potential answers to that question? If the answer is yes, you may want to consider doing a subtree for that question.

Whether you can use a subtree for that question depends on whether the subtree is independent from everything else in the main tree. Because dependencies can run two ways (it depends on …, or something else depends on it), the following two conditions must be met.

- The questions in the subtree must not depend on any other questions in the main tree. For instance, in the above example you can't use a subtree for damages if damages depend on the legal theory you're found liable under.
- Nothing else in the main tree must depend on a question in the subtree. Likewise, you couldn't use a damages subtree in the above example if there was also an uncertainty on obtaining indemnification that depended on the kinds of damages awarded, such as your only being indemnified for awards for future medical costs.

Both of these questions can be answered by applying the following steps to your relevance diagram (illustrated with the diagram following these steps).

- Draw a circle around the questions you want to do a subtree for.
- There must be no arrows going into the circle (the first condition above).
- Any arrow coming out of the circle must only mean a number, or the odds of a yes/no answer to a question (the second condition above) (Figure 6.17).

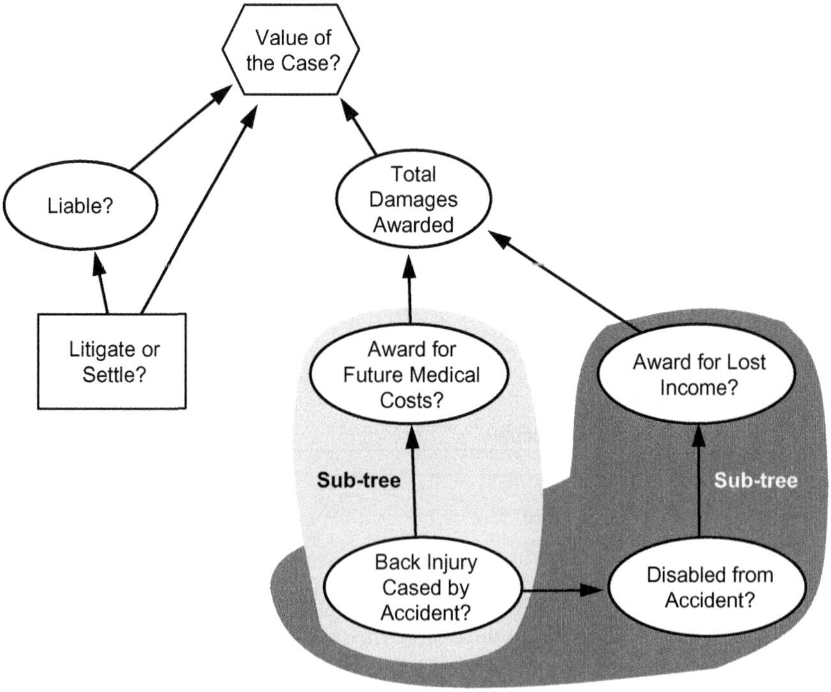

Figure 6.17 Specification of subtrees

Thus, in this relevance diagram, you couldn't use the two subtrees indicated if Total Damages Awarded was not a simple sum, but, rather required an assessment of probabilities depending the back injury and disability.

6.4.3 Procedure for Subtrees

The procedure for using a subtree, then, is as follows.

- Draw the relevance diagram for your problem if you haven't already done so.
- Use the test described in the preceding section to identify the questions you want to do subtrees for.
- Use the procedures for assessing probabilities for the questions in each subtree, as for the two shown below. The lists of reasons for these probabilities aren't shown, but, as you now, the reasons are crucial for good probabilities. Also, all these values should be net present values because both future medical costs and lost income are for damages accruing over time (Figures 6.18 and 6.19).
- Compute the mean value for each subtree, as shown below (Figures 6.20 and 6.21).
- Plug the mean value into your main tree and compute the mean value (Figure 6.22).

For this case, a maximum settlement value for the defendant (minimum for the plaintiff) would be $176,000.

Alternatively, you can use the subtree to calculate the probabilities for a yes or no answer to the question you're interested in, and then plug those probabilities into your main tree, as illustrated below (repeated from the discussion in Chapter 4 on which influencing factors should go into the tree) (Figure 6.23).

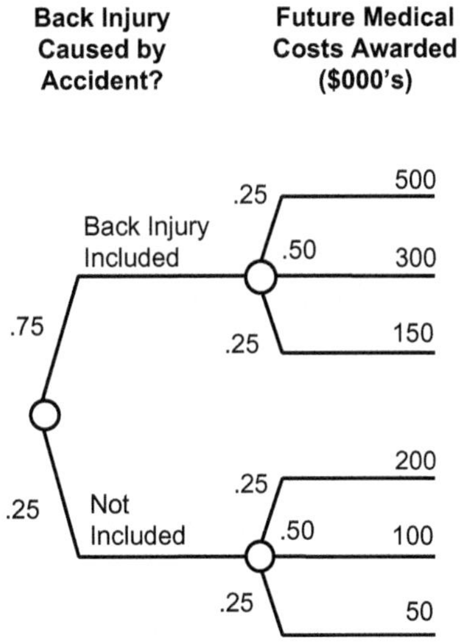

Figure 6.18 Assessed probabilities for future medical costs

Figure 6.19 Assessed probabilities for lost income

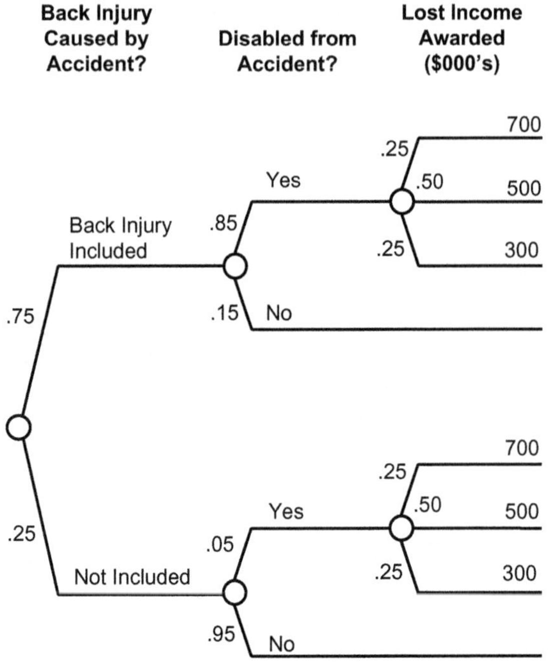

Figure 6.20 Rollback mean values of future medical costs awarded

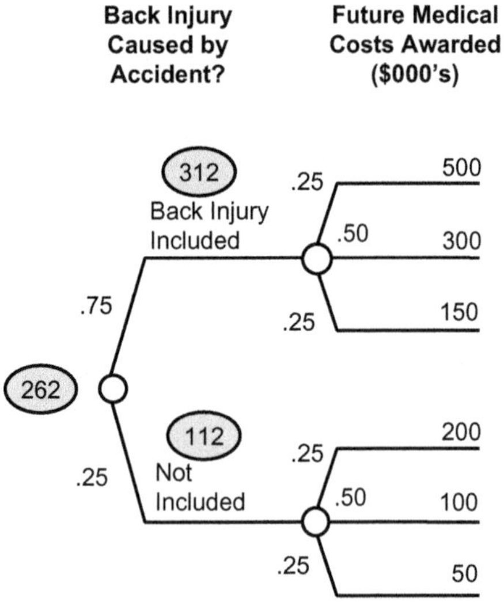

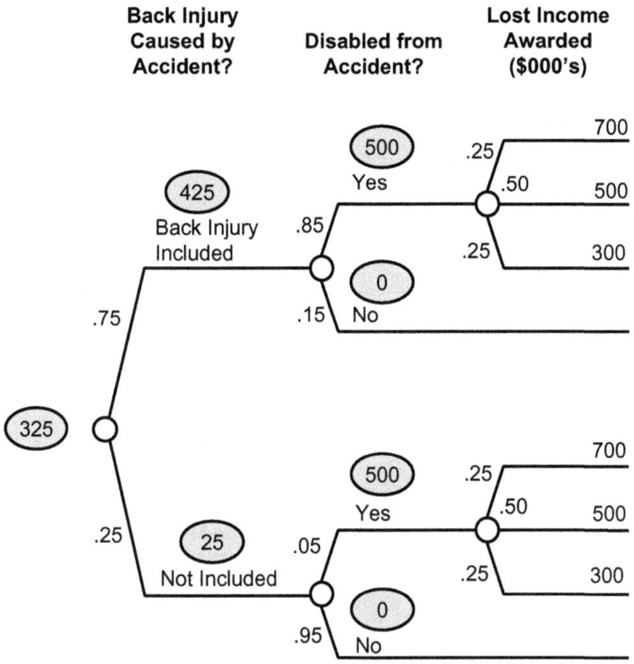

Figure 6.21 Rollback mean values of future lost income awarded

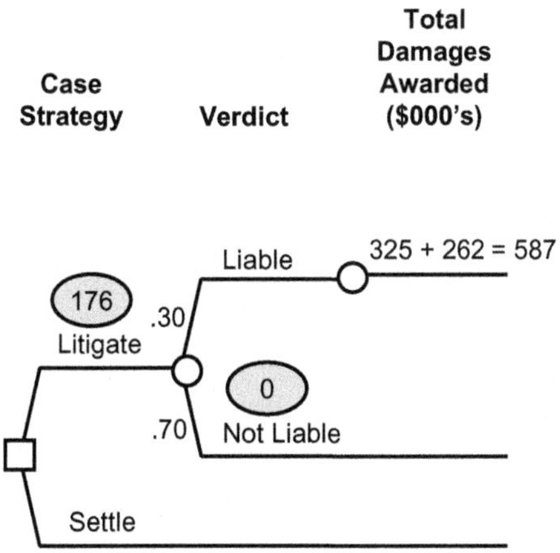

Figure 6.22 Subtree values plugged into main tree

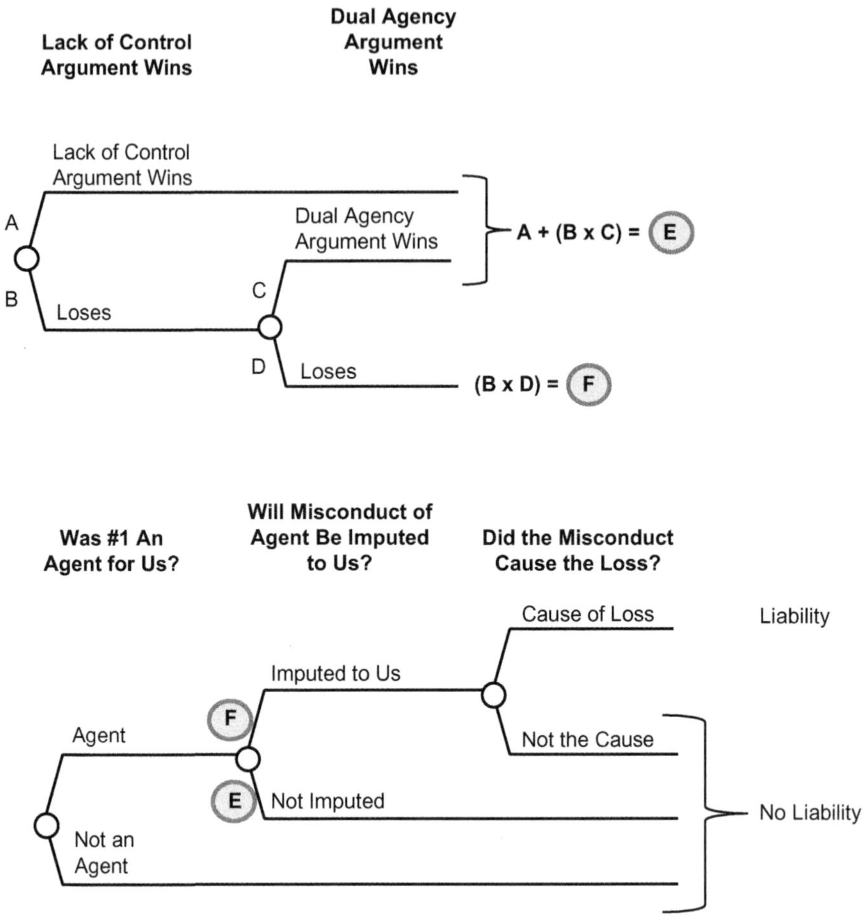

Figure 6.23 Subtree to calculate probabilities for main tree rather than values

The disadvantage of using subtrees is that the main tree doesn't show the full range of uncertainty in the value of the case. If you would like to show the full range of uncertainty, structure and evaluate the full tree. Software is helpful for this.

6.5 Problems

6.5.1 Calculating Scenario and Mean Values

1. Consider a method of quantitative analysis that you use (One common method is calculating and comparing the per-unit costs of items in a grocery store to see which is cheapest on that basis.). What qualitative insights do you obtain from the analysis? Can the level of analysis vary, and how do you know when more analysis is required?

2. Considering the emphasis on qualitative insights in decision analysis, do you think quantitative analysis is required at all, and why or why not? What kinds of insights would you lose by omitting quantitative analysis? What are the drawbacks of quantitative analysis (and not just in the Decision Analysis context)?
3. What reasons might a person give for refusing to consider the implications from quantitative analysis in a decision analysis? What counter-arguments could you respond with?
4. Explain how an mean value/roll-back calculation is different from a compound probability/roll-forward calculation, and how the two are equivalent.
5. When you moved out of your apartment, you cleaned it as if your grandmother was coming to visit, but the landlord still withheld your entire cleaning deposit, contending "The apartment smelled of sushi and basil." If you go to small claims court to get it back, you figure there's a 0.8 chance of the court finding in your favor, and then even odds of it awarding you half the deposit ($1,000) or the whole amount ($2,000).

   5.1. Draw the relevance diagram and tree.
   5.2. If the filing fee for small claims court is $75, what is the mean value of going to court? (Omit the value of your time because you'd enjoy going.)

6. You're representing a dental corporation in suing a dentist who left and set up shop 8 miles away in violation of a no-competition contract specifying a 10-mile radius within which a departing dentist could not practice for 5 years. In the process, the dentist also took with him a new technique for cold-spraying the gums with clove oil which allows for painless novocaine injections and leaves the delighted patients smelling gingerbread for the rest of the day. Accordingly, you've included a claim for conversion of the proprietary technique.

   It looks like a battle of the experts is shaping up over whether the simple but new technique is new enough to be covered by a technological innovation clause in the contract, so you'd assign a 50/50 on the court finding either way. However, if the court does find that the technique is proprietary, you think there's a .8 chance the court will also find the dentist competing in violation of the contract, versus only a .6 chance of violation if the technique is not proprietary. Similarly, you've assessed .25/.50/.25 chances of the court awarding $200K, $500K, or $1 million in damages if they find proprietary and competition, but only $50K, $100K, and $200K in damages if they find no proprietary and competition (same probabilities). Given the fact that the defendant makes his living as a dentist, the access rights of patients, and the nearness to the contract boundary, you don't think the court will award specific performance instead of damages.

   6.1. Draw the relevance diagram and decision tree. Include the value formulas and calculate the scenario values.
   6.2. What is the mean value of the case?
   6.3. What is the overall probability of winning the case?
   6.4. What are the compound probabilities for the various damage awards, ordered from lowest to highest?
   6.5. What is the probability that the damages will be at least $100K? What is the probability that they'll be less than $500K?

6.6. What is the expected damage award given that you've won on the proprietary technology issue? Given that you've lost on that issue?
6.7. What are the compound probabilities for damages given that you've lost on the proprietary technology issue?

6.5.2 Cumulative Probability Distribution and Histogram

7. Your client was riding her bicycle through an intersection with a green light when she was hit from the right by a car entering the intersection and making a right turn on the red light. The motorist denies any negligence and contends that your client was traveling at an excessive rate of speed.

   Your jurisdiction has the comparative negligence rule whereby the plaintiff cannot recover unless she was less than 50% at fault, in which case defendant is fully liable. In considering how the jury will make their decision, you've decided that they'll effectively decide first whether your client is less than 50% responsible (whether she gets anything at all), and then decide how much to award. You think that at this second stage, your client's award will be affected by whether or not the jury feels she was negligent at all, even though this issue is not formally before them.

   Accordingly, you've assessed a probability of .85 that the jury will find your client less than 50% liable (After all, your client will be in court in a cast, and you plan to introduce the mangled bicycle into evidence.). However, given that your client previously stated to defendant's insurance company that she was traveling about 25 miles per hour, and that defendant has a spotless prior driving record, you feel there's about a .45 chance that the jury will think your client somewhat at fault. You've assessed a distribution in each case, and found .25/.50/.25 chances of awards of $75K/$200K/$350K thousand is your client is found somewhat at fault, and awards of $150K/$300K/$500K if she's not (same probabilities).

   7.1. Draw the relevance diagram tree and fill in the probabilities. Calculate the value for each scenario.
   7.2. What is the mean value of the case?
   7.3. Calculate the compound probabilities and plot the cumulative probability distribution. What is the probability that the case is worth more than $20 thousand? Less than $10 thousand?
   7.4. Plot the histogram. Does the pattern of the histogram tell you anything?
   7.5. What is the mean value given that your client has been found less than 50% negligent? Would it be worthwhile to stipulate to being 25% at fault and avoid the whole issue altogether?
   7.6. Plot the mean value of the case as a function of the probability of your client being found less than 50% negligent? Does this plot have any implications for your case strategy?

Chapter 7
Analyze the Results

The objective of a decision analysis is a clear, cogent, and defensible understanding of which alternative is best and why.

Sometimes the answer becomes clear immediately on calculating the results. Sometimes, only more questions are prompted. Do the results make sense? Do they lead you to question your framing of the alternatives, structure of the problem (as set down in the relevance diagram and tree), your probability assessments, or all of these?

For these reasons, we describe decision analysis as a dialogue: both an internal dialogue between your own two thinking modes and external dialogue with you and other people involved in the case (the clients, counsel, analysts, and experts).

As we have discussed in Chapter 1, decision analysis works by framing the both the internal dialogue between your Affective/Intuitive thinking and Deliberative/Logical thinking and the external one you have with others. Your team should be going back and forth between the two thinking modes repeatedly as you analyze your case. Here are some examples of the questions one bounces back and forth with (Figure 7.1).

We refer to this iterative process as the Decision Analysis Cycle. It breaks finding the best course of action down into stages (Figure 7.2).

We have focused so far on basis development, structuring the problem, and assessing probabilities. We now move on to analysis and evaluation. All along the way, we continually appraise the basis: do we understand the recommendation, the rationale for it, and all the pieces leading to that conclusion?

7.1 Do the Results Make Sense?

You've already seen a number of results from a Decision Analysis, including the tree itself with all the probabilities and answers filled in, the roll-forward and roll-back mean values the cumulative probability distribution and the histogram.

J. Celona, *Winning at Litigation through Decision Analysis*,
Springer Series in Operations Research and Financial Engineering,
DOI 10.1007/978-3-319-30040-5_7

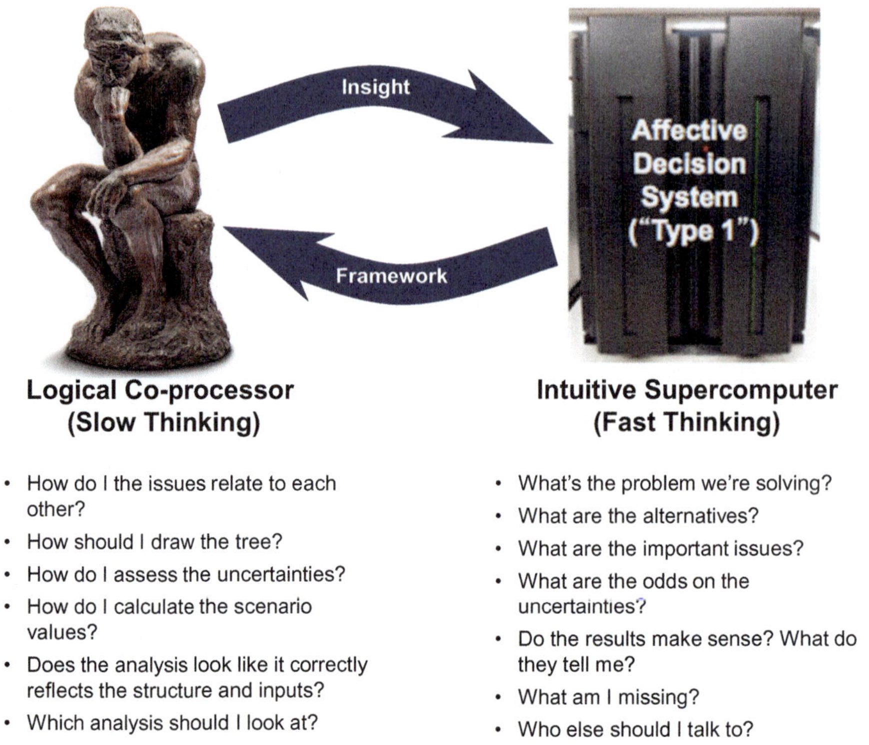

Figure 7.1 The internal and external dialogue of Decision Analysis

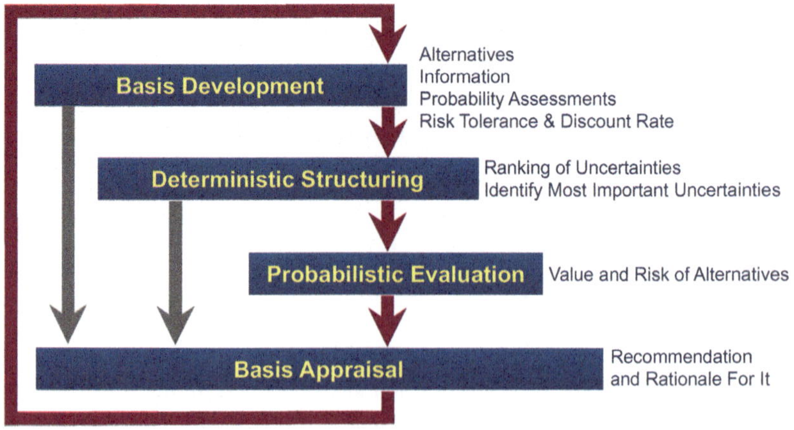

Figure 7.2 Steps and outputs of the Decision Analysis Cycle

You've seen how the process of arriving at these results should present questions to you about what the important issues in your case are, and how you should structure them in the analysis. In other words, does the way you're analyzing your case make sense to you?

Likewise, the first step in analyzing the results is asking yourself whether they make intuitive sense. Does it seem intuitively reasonable that the numbers are coming out the way they are, and can you explain them? (Your lists of reasons should help you decide.) As discussed in Part I, Decision Analysis employs both your Qualitative/Affective and Logical/Deliberative thinking modes. The analysis and results should make sense to both of them.

Your analysis might not make sense to you because of things you've done intentionally, or unintentionally. The intentional omissions are discussed in the next section. Following are two unintentional reasons why the results may not make intuitive sense to you.

- Your intuition might be wrong.
- Your analysis might be wrong (for unintentional reasons).

Regarding your intuition your intuition being wrong (and the analysis correct), remember that intuition is not good at handling uncertainty and complexity (the cognitive and motivational problems again). Also remember that the mean value is not what you think you can bet the other side to settle for; rather, it's the probability-discounted value of litigating the case—which should affect what you are willing to settle for. Thus, the mean value may be a correct number even though it's not a number you would have assessed directly for valuing the case (indeed, you do the analysis *because* it's not a number you would have assessed directly!)

After all these considerations, if you still feel the results are intuitively wrong, then the analysis may very well be wrong for unintentional reasons (the next section deals with inadequacies from issues intentionally left out). There are a number of common "goofs" in doing a Decision Analysis, especially when you're just familiarizing yourself with how to do one.

- **Wrong Ordering in the Tree**. Are the questions in the tree correctly ordering according to time sequence and/or how the trier-of-fact will consider them? Do all the scenarios make sense? Would changing the order change your probabilities, and thus the mean value? (This last point is a kind of spill-over.)
- **Issues Unintentionally Omitted**. Besides wrongly ordering the issues (according to your intuition), might you also have unintentionally left out issues necessary to think about the case? Are all the scenarios completely described, or, in assessing liability at the end of the tree, would the clairvoyant need to ask how something else turned out?
- **Damages Not Correctly Calculated**. Did you correctly add up all the damages and types of damages in computing the values at the end of the tree? Did you discount future damages and use present values at the end of the tree? Does figuring in pre- and/or post-judgment interest (usually left out of a first-pass analysis) so change the numbers as to make the current mean value seem wrong?

- **Probabilities are Not the "Best."** You say in the chapter on eliciting probabilities that there are no "right" or "wrong" probabilities, only probabilities that are better or worse because they do or don't correctly reflect all your information, or because they come from someone with more or less information than you. Thus, your probabilities in the tree might not be the "best" because you overlooked possible reasons for yes/no or high/low outcomes, or because you omitted considering subtle dependencies between the questions (spill-over). Both of these instances are, of course, information not reflected in your probability assessments. And, since the probability assessments are half the mean value equation, bad probabilities would produce a mean value that didn't make sense to you.

People also sometimes obtain results they don't agree with for reasons having more to do with their becoming familiar with Decision Analysis and the demands it places for explicit, unambiguous estimates. For instance, perhaps in the past you shied away from assigning too high a probability of winning (like 80%) for fear the client would interpret it as a sure win (100%). Putting in an accurate estimate would make the Decision Analysis value of your case seem too high by comparison.

Alternatively, perhaps you're being limited in the past to giving one number and one conclusion influenced you to reduce your estimated value of the case for client's risk aversion (the fact that the client is willing to take less for sure than risking an uncertain higher judgment if he wins). A Decision Analysis would then also produce a value for the case that seemed too high. Note also that if the client was also intuitively reducing his valuation of the case for risk aversion, you'd have the risk-aversion effect double counted—resulting in a much lower case value compared to the analysis result.

These last phenomena thus cause unrealistically low case valuations and settlements that won't make economic sense in the long run. The solution is better communication about what the numbers are supposed to mean (mean value versus risk-adjusted value, for example), and frank assessments. Chapter 12 (Communication) provides useful tips on discussing and verifying Decision Analysis results with clients.

Pondering the tree and mean value for Howitzer v. Crabgrass et al. (repeated below), you asked yourself if you really believed what the numbers said. Did you really think: that their claim against you was only worth $38,000; that there was a better than 80% chance of no finding of liability against you at all (This number was from the cumulative probability distribution.); that, even if you were found liable, your mean damages for negligence were $136,000 versus $422,000 for intentional misrepresentation? (These conditional or "what if" values can be read from the rollback mean value decision tree.) (Figure 7.3).

Carefully, you reviewed your relevance diagram and list of reasons why the different questions might come out one way or another. Certainly, a lot of elements had to be established for a finding of liability, and all of them had strong reasons to go both ways. Further, question of damages was especially problematic because you could argue that there were no damages; that Howitzer had a twenty-year-old house built up to the standards of the time and that was likely to be there for another twenty years.

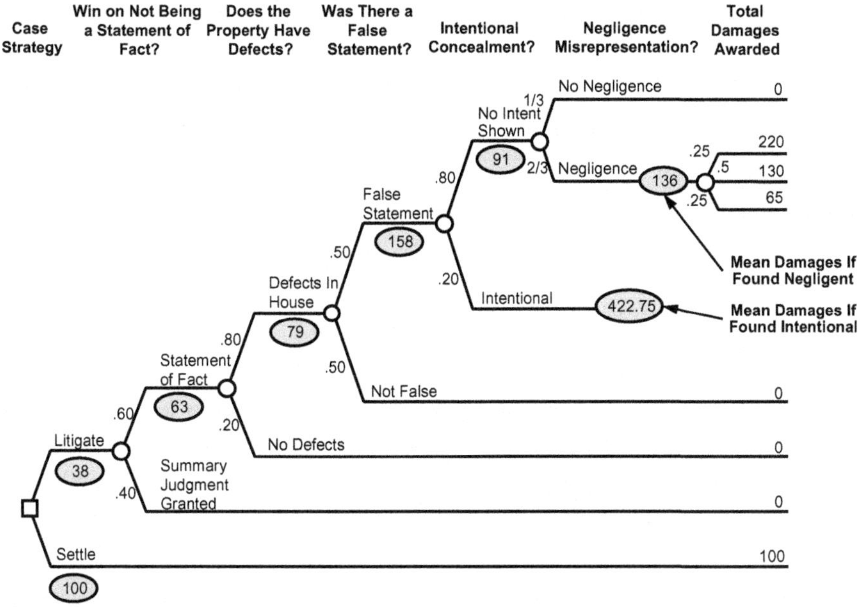

Figure 7.3 Rollback tree also shows conditional "what if" values

Was the value of the case too high then? Was it all as groundless as the Consumer Fraud Act claim? Again you went back to the decision tree. Maybe the probabilities on the jury finding defects in the house and on finding Zelda's statement false could be a little lower. However, on the defects issue, you had a strong feeling that photos of those tilted foundation pillars would make a big impression on the jury, especially with Howitzer's home at stake. As for the falsity of Zelda's statement, you suspected that if you argued the point too strongly and in too technical a fashion, the jury would think Zelda couldn't have known what she was talking about—so your arguments might boomerang right into a finding of intentional misrepresentation. You decided that the probabilities that the probabilities were a pretty good (if uncomfortably exacting) reflection of your judgment, and left further consideration of them for probability sensitivity.

Finally, you considered how you structured the two causes of action. You had effectively set them up as an "only win if both" situation for the first three questions, before the negligence and intentional counts split into a kind of "lose either way situation." Were the two counts really that interrelated, or were they more independent, so that you should be adding up your chances of losing at an earlier stage in the tree?

You reviewed the substantive law for your jurisdiction. Regardless of the scienter element, misrepresentation plain needed a false statement of fact that was relied on and that caused actual damages. Then you argued about negligent or intentional misrepresentation.

You concluded that, at least at this stage of the analysis, the $38,000 mean value of litigating the case was a necessary conclusion of your judgments on all the issues that affected the value—even though the number was perhaps $50,000 lower than what you might have guessed directly as a reasonable settlement value for the case.

7.2 Have You Left Out Anything Important?

If you're comfortable that your analysis isn't skewed by the unintentional reasons discussed above (or if you've already fixed those problems), then the question you should ask yourself is whether, given which questions have been included and the current level of detail, you would feel comfortable or not settling the case now.

Pause on this question for a moment. In my experience that a good first-pass analysis usually provides most of the insight into settlement strategy, pretrial strategy, etc., that a more sophisticated one would. Chances are that yours does too, and that additional time spent refining an analysis would be better spent doing a first-pass analysis on *another* case.

However, as always, your (enlightened) intuition is the final arbiter, and if you feel that understanding the value of your case requires further refinement in the analysis, you have two choices.

- Include questions left out of your analysis.
- Handle more carefully questions already included in your analysis.

These questions will be discussed in turn. Part III also gives details on how to handle specific questions more explicitly.

7.2.1 Issues Omitted from the Analysis

Issues omitted from the analysis includes questions particular to your case, such as discovering a new legal theory you should include, or realizing that you could dramatically reshape the case with a successful motion for partial summary judgment. These kinds of additional questions can be handled by the problem structuring techniques you've already studied.

Omitted issues may include more complex questions like attitude towards risk or the effect on precedence (and, thereby, on future cases). Risk aversion is discussed under settlement strategy, while precedence is included in Part III.

However, before complicating your analysis, you should still carefully consider whether the complication is merited. Is the settlement offer a good choice in spite of the omission? Would explicitly including the issue make the settlement offer even better? For example, including more legal fees and risk aversion to the prospect of losing the case will lower the value of the case, and make a settlement offer even more attractive. Thus, a checklist for deciding whether to include a new issue might run like the following.

- Are you uncomfortable settling the case with your current understanding and analysis?
- Could the proposed issue significantly affect the value of the case?
- Would the affect on the value of your case call your current settlement strategy (or trial strategy, etc.) into question, or only make it look better?

As a final caveat, keep in mind that it's very easy to complicate an analysis and difficult to simplify later.

7.2.2 Issues Handled Grossly

Beside issues omitted from the first-pass analysis, issues may also be handled in gross fashion. Thus, for instance, the time value of money is normally handled by keeping all numerical answers in present value dollars. Legal fees may be omitted from the first-pass analysis altogether, although, if you're really uncomfortable with doing so, you can make a best-guess and subtract it from the mean value of litigating the case. Another possibility is damages, which you probably did a direct assessment on and could have handled more finely (handling the codefendant's share of the verdict etc.) Part III is a rogues' gallery of more sophisticated ways to handle issues.

However, the question before doing so is the same as before: is a more refined analysis merited? The checklist is the same as given above, except your considering refining an existing issue rather than including a new one. Again, it's likely not worthwhile to do so unless you're uncomfortable with your current settlement or trial strategy, and the change will have a big dollar effect on your case (or a smaller dollar effect on a large number of cases).

If you decide to refine your handling of an issue, it's often appropriate to do so with a subtree for that issue. Then, the mean value for the issue will go into the main tree where you previously had a chance node, or even just a best guess. As always, watch for spill-over between questions in your subtree and questions in the main tree.

You reviewed the possible questions you left out of the Howitzer case, and considered where you might add more detail. Should you break out individually all the elements of the negligence determination, or of the intentional one? Sure, you might be able to come up with some sort of probabilities on the individual elements, but you didn't think the jury would break them down that way. Negligence especially tended to be the kind of determination where the jury chewed on everything, swallowed, and then you waited to see if they belched.

You then thought about including legal fees, the effect on precedence, and so on. All that really seemed like gilding the lily in a case like this one, especially since it wasn't breaking any new ground, didn't promise to be discovery-intensive (like an anti-trust case might be), and since you'd decided that discounting wouldn't even be necessary because it would be resolved within a year.

However, what really made your mind up not to elaborate at this point was your feeling that $38,000 was a pretty good, defensible settlement value for the case. You'd been over in your own mind why the settlement value shouldn't be a lot lower or higher, and felt you could similarly convince others. You just didn't think that adding more detail would help you think better about the value of the case.

7.2.3 *Other Strategy Considerations*

A few other strategy considerations sometimes confuse the counsel's thinking about the mean value of the case and make it seem intuitively wrong when it isn't really. For instance, your opponent might be willing to pay much more than the mean value to settle the case, or to take much less. This doesn't mean that your valuation of the case is wrong. It simply means that your opponent hasn't thought about the value of the case as clearly and carefully as you have (the usual case), or that he has very different probabilities than you for some of the important questions (the exceptional case). In either case, the difference doesn't necessarily call your valuation into question, and, indeed, may be the basis for expeditiously settling the case.

However, suppose the reverse is true: your opponent won't pay anything near the mean value to settle the case, or won't take anything less than much over the mean value. Your first approach is to litigate the case (or, more likely, take it as close to trial as necessary to get a settlement). A second approach is to discuss your analysis with your opponent and thereby try to bring your respective valuations closer together. Show him your tree (perhaps without probabilities) and invite him to fill in his own numbers. If his case valuation was less careful than your's (either less experienced or not a Decision Analysis), then you'd be facilitating his revaluation and a settlement.

A third possibility is that you've missed an issue or dependency that figures importantly from your opponent's perspective, but not from your's. You might reveal this difference by analyzing the case from your opponent's perspective, a topic discussed in Chapter 11.

7.3 Settlement Strategy

Performing a Decision Analysis has many benefits for better understanding and communicating about your case, and these benefits have been brought out throughout the preceding text (Communication, in particular, will be taken up again in Chapter 12.). However, the analysis can also be brought directly to bear on such specific problems as how much you should be willing to settle your case for, on how you should plan your pretrial discovery, and on how you should plan your trial strategy. These three applications will be taken up in the remainder of this chapter. Other possible applications, such as how you should think about allocating a settlement between yourself and a codefendant, will be taken up in Part III.

7.3.1 *Risk-Neutral Decision-Maker*

A risk-neutral decision-maker is one who is satisfied with using the mean or mean value to base decisions on—that is, he doesn't require an additional discounting for the riskiness of uncertain losses or payoffs.

For example, if you had a chance of buying either of the two bets below for the same price, a true expected-value decision-maker wouldn't care which one she got, while a risk-averse decision maker would pick the .80 chance at $6,250. A risk-seeker, by the way would pick the .50 at $10,000—like the way people buy lottery tickets (Figure 7.4).

The next section describes risk-aversion in more detail and tells when you might want to consider it explicitly. However, as a general rule, it's best to keep things on a risk-neutral expected-value basis at least for the first-pass analysis. When you do so, the settlement rules are quite simple.

- A risk-neutral defendant should be willing to pay anything less than the mean value of litigating the case to settle it.
- A risk-neutral plaintiff should be willing to take anything more than the mean value of litigating the case to settle it.

Your settlement strategy for Howitzer v. Crabgrass et al. was a simple product of the decision tree: settle the case for anything less than $38,000. However, regardless of how reasonable you felt that value was, good negotiation skills are still required. How committed does the other side seem? How strong do you believe they think their case is? How angry is Howitzer?

Your analysis helps to inform your negotiation strategy. You feel their case is weakest (lowest probability of prevailing) on the claims of intentional concealment and on the possibility of getting punitive damages. These may be good points to hammer on. If the punitive damages pot at the end of the rainbow looks unlikely, even their intuitive sense of the case value should come down.

Also, given the low mean exposure at trial, it might make sense to come in with a low settlement offer (say $10,000) to clearly communicate you think the case value is low. That also give you some room to come up to give them the impression of having "defeated" you initial offer—though you analysis tells you that $30K or $40K would very much be a winning settlement. You decide to talk this offer over with the counsel and the client.

Figure 7.4 Different risks but same mean value

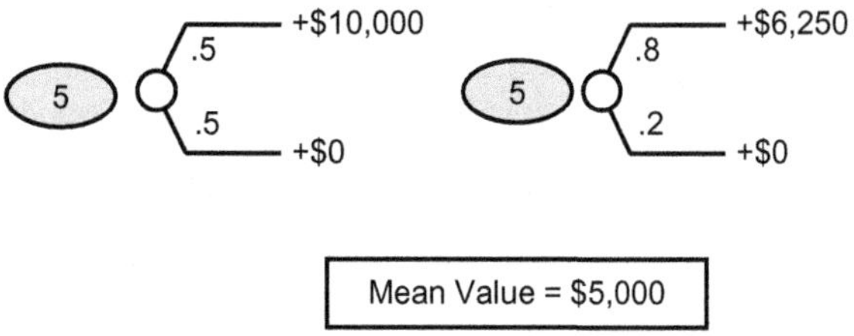

7.3.2 Risk-Averse Decision-Maker

And now this lustful lord leapt from his bed,
Throwing his mantle rudely o'er his arm,
Is madly toss'd between desire and dread;
Th' one sweetly flatters, th' other feareth harm...

—Shakespeare, The Rape of Lucrece

As mentioned above, we don't recommend considering attitude towards risk in a first-pass analysis. For completeness, we will illustrate a simple way to consider decision-maker (client) risk aversion when you make a recommendation.

7.3.2.1 Why You May Need to Consider Explicitly Attitudes Toward Risk-Taking

Risk aversion may need to be considered whenever the client is unwilling to "play the odds" by going for the long-term maximization of wealth embodied in the mean value. Usually, this happens when the amounts at stake are so large that the client is very nervous about riding out a possible large judgment against them (or, for the plaintiffs, failing to obtain any judgment at all). Thus, for instance, any corporate client might be very nervous about potential judgments large enough to put it into bankruptcy. The client might therefore be willing to pay more than the mean value to settle a case just to avoid the possibility of even worse outcomes. In other words, the client is averse to the risk of bad outcomes.

Similarly, a client may be willing to pay more than the mean value to settle a case to avoid future costs (especially of litigation will be lengthy and costly), and to avoid setting a bad precedent. For both of these possibilities, you might be advised to do a mini-analysis (considering settlement disclosure, collateral estoppel, etc.) to demonstrate to the client just how big these risks are on a mean value basis, and thus focus on whether the client is really willing to pay *that* much to avoid the risk. This usually comes up with reputation risk or the risk of setting an unfavorable precedent. Analysis can reveal just what the monetary consequences could be.

7.3.2.2 Insights and Terminology

If, after the preceding considerations and alternatives, you still feel you need to consider risk aversion, this section lays out some of the basic concepts.

The basic concept of risk-aversion (the definition has already been mentioned) is that people are sometimes willing to pay a premium for certainty and to avoid uncertain consequences. For example, consider the following simple tree (from the plaintiff's perspective) (Figure 7.5).

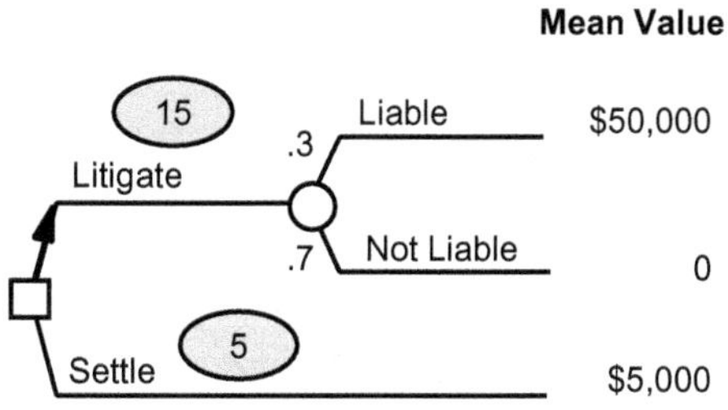

Figure 7.5 Better for the plaintiff to litigate?

The mean value of litigating this case is $15,000 (which you can get by roll-back or roll-forward). Thus, as indicated by the arrow, litigating the case is preferred on a mean value basis over a $5,000 settlement offer.

However, there is some amount the plaintiff is willing to accept for sure in settlement in exchange for giving up his .30 chance at $50,000. As you know, this amount is considerably less that the $50,000 (unless the plaintiff upset enough to be irrational—but that's a different story). For a risk-neutral (mean value) plaintiff, that amount is the mean value. However, for a risk-averse plaintiff, that amount is less than the mean value. The plaintiff could probably tell directly what that number is, like a direct probability assessment. Suppose that number is $12,000. The plaintiff, then, would be willing to accept $12,000 for certain as the "sale price" of his .30 chance at $50,000. Hence, this amount is known as the ***certain equivalent***. It is the certain amount that, to the plaintiff, is equally valuable as the uncertain lawsuit.

The difference between the certain equivalent and the mean value is known as the ***risk penalty*** or ***risk premium***. In this case, the risk penalty is 15 – 12 = $3,000. Thus, the plaintiff is, in effect, willing to pay a penalty be giving up $3,000 in mean value for the sake of his risk aversion. This is an important message for clients: ***risk aversion is not free***. Being risk-averse means giving up mean value for the risk penalty—which means you won't be maximizing your wealth over the long run.[1] Over the long run, you will be poorer as a risk-averse decision-maker than as an expected-value decision maker.

7.3.2.3 How to Calculate a Certain Equivalent

For simple situations like this one, the certain equivalent can be assessed directly. However, for more complicated situations, or to assure a consistent risk attitude (that is, the same *degree* of risk aversion), more elaborate measures are needed.

[1] You will still be maximizing your certain equivalent—if you do the math to make sure you're considering your risk attitude consistently. However, you will giving up the risk penalty; in effect, paying a premium for your risk aversion just like the way you pay a premium to insurance companies over your possible losses.

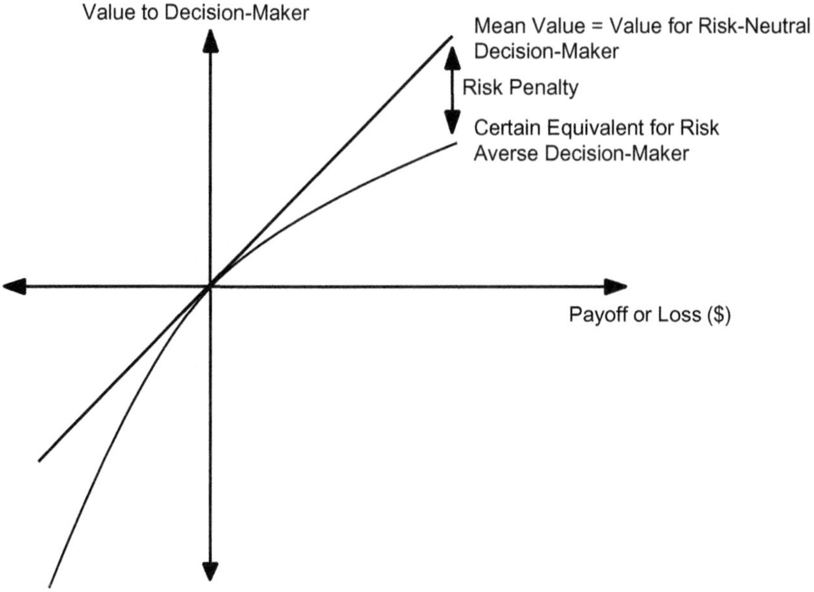

Figure 7.6 Relationship between value to risk-neutral versus risk-averse decision-maker

The key to applying risk attitude consistently is realizing that there is usually a definite relation between the amounts at stake and their certain-equivalent value to the client. This relationship is shown below for a risk-averse client. You can see that the value is the same as the mean value for an expected-value client, but a risk-averse client over-values negative amounts (a bigger negative number) and under-values positive amounts (a smaller positive number). The difference is the risk penalty: as shown in the graph (Figure 7.6).

You can see, then, how these relationships will be reflected in the behavior of a risk-averse client: he will worry much more about the possible losses than the mean value would indicate—and thus will be willing to *pay more* than the mean value to settle a case; and will discount possible awards more than the mean value would indicate—and thus be willing to *accept less* than the mean value to settle a case.

This relationship can be described by a mathematical function called a utility function. The utility function simply measures how valuable or useful a given amount is to you (that is, how your certain equivalent compares to your expected). The utility function thus works as a two-way translator: changing numbers into utility and back again.

The procedure for handling risk aversion, then, is conceptually not so very different from doing a roll-back to compute the mean value.

- **Compute Utilities**. Start with your scenario values at the end of the tree like you usually do for a roll-back or roll-forward, but use the utility function to convert those values to utilities.
- **Roll-back to Find Expected or Mean Utility**. Use the utilities just like you would damages to roll-back and find the expected utility of litigating the case.

- **Convert Expected Utility to Certain Equivalent Value**. Use the utility function in reverse this time to convert the expected utility to a certain equivalent value.

This certain equivalent, then, is the minimum amount that a risk-averse client should be willing to settle the case for. In this respect, the certain equivalent is used for settlement strategy the same way the mean value would be used for a risk-neutral client.

7.3.2.4 The Importance of the Full Probability Distribution

From the procedure outlined above, you already know that the full probability distribution (all the numbers and multiplied-out probabilities at the end of the tree) is needed to consider risk aversion—not just the mean value. You need all those numbers and probabilities to calculate a certain equivalent value.

A secondary consequence that wouldn't be obvious you actually did the math is that risk aversion has a proportionally much greater effect on low-probability numbers (positive or negative) than on higher probability numbers. Thus, the greater the risk is—*both in terms of absolute numbers and their probabilities*—the greater the effect of risk aversion.

This relationship can be seen in the graph below, which plots the mean value against the certain equivalent as a function of the overall probability of winning the case (Figure 7.7).

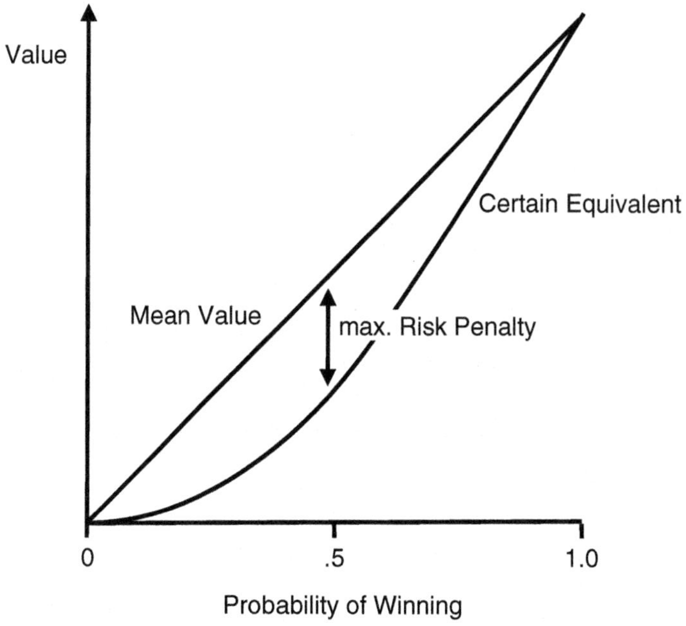

Figure 7.7 Relationship between riskiness (uncertainty) and mean or certain equivalent values

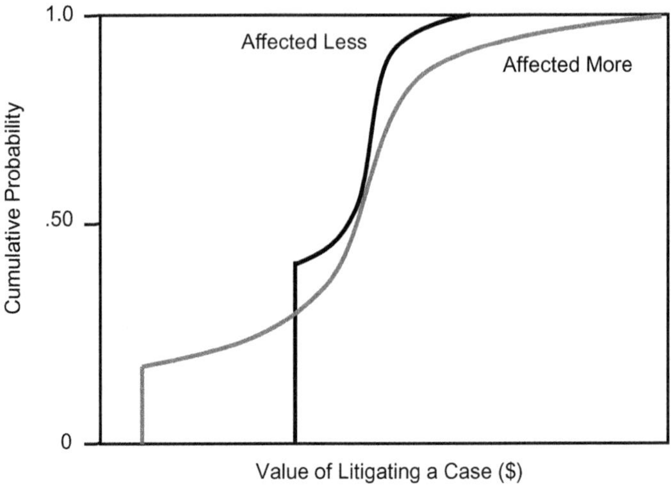

Figure 7.8 Effect of risk aversion varies with the uncertainty (riskiness) of the case

This relationship means that a case whose probability distribution has more low-probability possibilities in the high and low extremes will be more affected by risk aversion than one whose distribution is not so spread out. This comparison is more evident when you compare distributions in graphical form, such as with the two cumulative probability distributions below on the value of two cases (Figure 7.8).

Note that this effect still occurs even though these cases have about the same mean value! Risk aversion depends on the *distribution* of damages—*not* on the mean value.

7.3.2.5 A Simple Way to Look at the Effect of Possible Risk Aversion

Whether you do a risk-aversion calculation or not, you should take two ideas away from this discussion.

- Risk aversion means the client is willing to pay more (or take less) to settle—and thus risk aversion will end up costing him more money in the long run.
- Risk aversion depends on the full distribution of the value of the case, rather than on the mean value. The more widespread those possible values are, the greater the difference between the certain equivalent value and the mean value (that is, the greater the risk penalty).

These insights are useful even if you don't assess a client's attitude towards risk because clients often change their attitudes towards risk taking when the consequences are made clear.

A common example of changing attitudes with cognizance of the effects of risk aversion is insurance buying habits. Most people don't realize how much of a premium they're paying over their expected losses for the security of insurance. As people become more aware of how much risk penalty they're paying (such as through learning Decision Analysis), they typically limit their insurance coverage, or, at least, raise their deductibles. This same phenomenon of reducing insurance also happens almost automatically as people become better able to afford the risk. Affluent people often don't buy more than the legal minimum auto insurance. For some potential losses, they may even self-insure: not purchase any insurance at all!

Companies typically self-insure for many risks: employee medical costs, workers' compensation, employee motor vehicle liability, etc. Further, many health care organization will likewise self-insure for medical malpractice with captive insurance companies. They have realized the cost of risk averse behavior when they pay to transfer risks. These companies would especially benefit from the application of decision analysis (which, to date, very few avail themselves of).

Thankfully, software provides a simple way to look at the possible effects of risk aversion with the sensitivity to risk attitude calculation. This shows how the certain equivalent value of the alternatives (typically, litigate or settle) vary with differing levels of risk aversion.

The graph of certain equivalent value shown in Figure 7.6 captures people having increasing aversion to increasing large losses, and lower utility to increasingly large gains. This model does a decent job of capturing how people actually value simple risky situations, and extends it to more complicated ones. It is actually an exponential function of the form

$$u(x) = a - be^{-x/R}$$

where

- $u(x)$ is the utility of a particular x-value.
- R is the Risk Tolerance, or the maximum loss the decision-maker could tolerate.
- a and b are parameters set by the choice of two points to calibrate the utility curve (e.g., u(0) = 0 and u(100) = 10).[2]

To convert back from the mean utility of a probability distribution A to a certain equivalent value, the formula is

$$CE(A) = -R\ln(mean\,utility)$$

[2] A utility scale, like a temperature scale, can be calibrated to any two desired points. For example, Celsius is calibrated to water freezing at 0 °C and boiling at 100 °C. Fahrenheit is calibrated to water freezing at 32 °F and boiling at 212 °F.

The formula showing the individual scenarios x and respective probabilities is

$$CE(A) = -R\ln\left[\sum_{i=1}^{n} p(x_i \mid \&) e^{-x_i/R}\right]$$

where *CE*(*A*) is independent of the choice of *a* and *b* provided that *b* is not equal to zero. & represents that state of information of the person assessing the probabilities at the time they were assessed.

The sensitivity to risk attitude calculation shows what the certain equivalent values of the respective distributions are for vary values of the risk tolerance. If the preferred decision switches when the risk tolerance drops below a particular value, you can look at whether that risk tolerance might in fact be an unsustainable loss for the client.

For a corporation with tens of billions in annual revenue, their risk tolerance is certainly greater than, say, $10 million: they make bets of that size all the time in business decisions. For an individual, this may very well be a ruinous loss.

If the preferred decision switches at around a risk tolerance that might be what the client can tolerate, that is a matter for discussion. If the preferred decision never switches at all, or switches at a risk tolerance clearly much smaller than what the client typically risks, the risk sensitivity graph can be just filed away in case the question of risk aversion ever comes up.

Since you already had the tree entered into software, you went ahead and ran the sensitivity to risk chart (Figure 7.9).

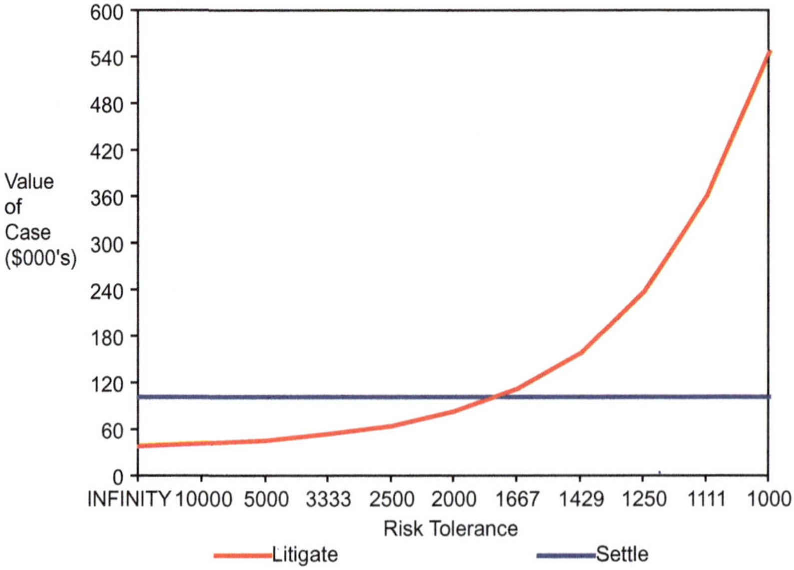

Figure 7.9 Sensitivity to risk tolerance for Howitzer v. Crabtree

The graph showed that the value to the client of the settlement offer stayed flat at $100K (because it wasn't uncertain). However, the client's certain equivalent value of litigating the case began to rise dramatically as the risk tolerance dropped below around $2 million—at which point the settlement began to look preferable. At a risk tolerance of $1 million, the client would value the exposure in the case at around $500K—much more than the settlement offer. This made sense considering that, even though probability of being hit with punitive damages was less than 1%, the amounts were in the millions of dollars.

This indicated it was worth having the discussion with the client regarding how they felt about the potential punitive damages exposure. Would it be an insured loss? The $100K settlement offer might not be such a bad deal after all.

7.4 Discovery Strategy

From the mean value, you've gotten the amount you should be willing to settle your case for. From the cumulative probability distribution, you can see the overall probability of escaping liability and what the chances of various outcomes are. The risk sensitivity graph shows whether aversion to losses would affect case strategy. With a few more calculations, you can extract results from your tree to help you plan discovery strategy.

7.4.1 Values of Clairvoyance (Information) and Wizardry (Control)

Two key questions present themselves in planning discovery strategy.

- What information should you try to obtain, and at what cost?
- How can you affect the result, and is it cost-effective to do so?

Decision analysis can provide tremendous value in addition to settlement amounts by answering these questions. Instead of simply launching into discovery, retaining experts, and deposing on all the issues (as professional responsibility considerations often dictate), you can instead determine which issues are most worthy of focusing additional resources on and which ones are not very important at all.

We do this by explicitly determining what the value is of knowing how an uncertainty turns out (value of perfect information), and by determining what the value is of controlling how an uncertainty turns out (value of perfect control).

Although in practice you can seldom perfectly predict or perfectly control an uncertainty, these values are the maximum possible values. It is not worth paying more than these values for information or influence on an uncertainty.

Most of the time, knowing the maximum possible value of information or control is by itself enough to guide your efforts. We therefore start with these values, which are fairly easy to calculate.

It is also possible (but more complicated) to calculate the values of imperfect information and imperfect control. For those, we refer you to *Decision Analysis for the Professional.*

The questions of value of information and value of control are thought of in terms of the uncertainties described in your decision tree.

- How can you improve your probability assessments for the questions in your decision tree, and how much is it worth to you to do so?
- How can you assure that a particular answer for one of the questions in your tree happens, and how much is it worth to you to do so?

7.4.1.1 How the Value of Information Works

The value of information is the value of improving the probability assessments for the questions in your tree. Finding this value is computationally fairly simple, but requires a little careful thinking about probabilities and your decision tree.

The best possible probability assessment would be perfect information: if you knew exactly what was going to happen, and could assign a 1.0 probability to that answer and 0 to all the rest. In decision analysis, this perfect information is called clairvoyance. We have been using the mythical clairvoyant in the Clairvoyant Test to clarify definitions of uncertainties. We look at what if he or she were actually going to answer the question.

We start with a simple settle/litigate tree (Figure 7.10).

The clairvoyant is going to give you an answer for the only question in your tree: whether you're held liable or not. However, you don't yet know what the answer is—even though the clairvoyant does—because you're figuring how much you should be willing to pay for that answer (the value of the information). Thus, there's a new uncertainty of what answer the clairvoyant will give (because you don't know it yet), and you have a new decision on whether to buy the answer or not. The tree below illustrates these additions (Figure 7.11).

Note that the probabilities for liability are different when you purchase the clairvoyant's prediction; the clairvoyant, by definition, is perfect, so his prediction of "Liable" should give you a 1.0 probability of being held liable, and likewise for a prediction of "Not Liable." And, of course, if you don't buy the prediction, you face the same situation as before.

Figure 7.10 Start of value of perfect information calculation

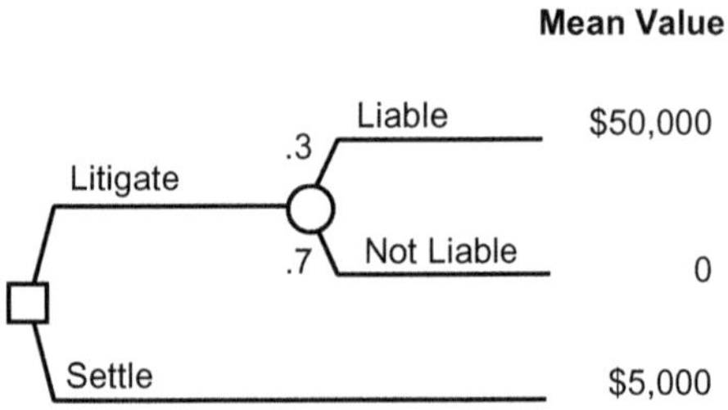

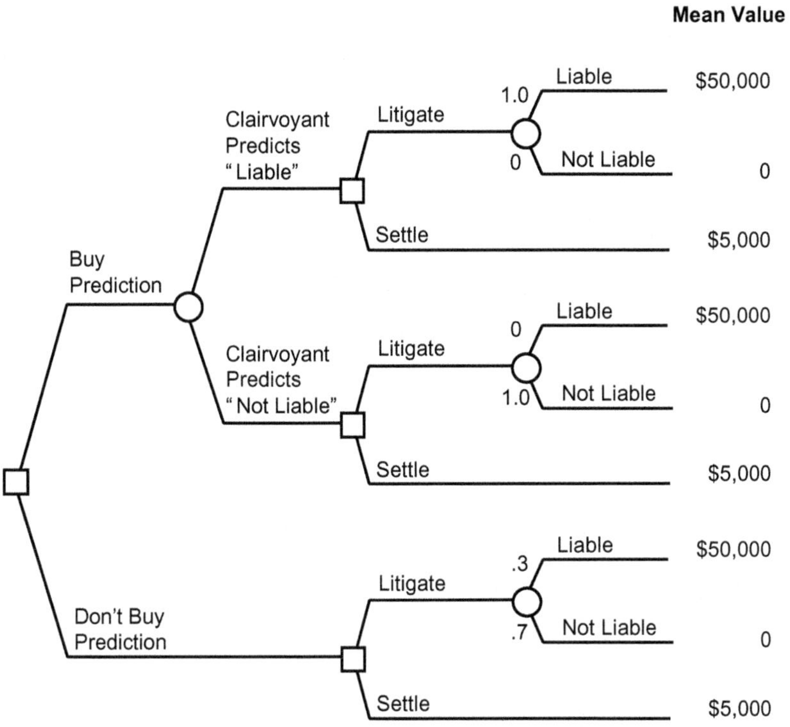

Figure 7.11 Decision on whether or not to purchase clairvoyance

The next step is to recognize that, because they don't make any difference in the mean value (or cumulative distribution, for that matter), you can prune the zero-probability branches, as below (Figure 7.12).

The only remaining question, then, is what probabilities you should put on the clairvoyant predicting "Liable" or "Not Liable." Perhaps the answer presents itself to you already. If you really believe there's a .30 chance of your being held liable (a "best" probability issue), then you should also believe there's a .30 chance of the clairvoyant predicting liable—and likewise for "Not Liable." These are illustrated in the tree below (Figure 7.13).

Now you're ready to do a roll-back calculation (or roll-forward) to compute the mean values for buying or not buying the information. Starting in the bottom of the tree, you face your original situation, where litigating for a mean value of $15,000 is preferred over a $5,000 settlement value (assuming you are the plaintiff; the reverse would be true for the defendant).

However, your decision varies in the top part of the three where you've bought the prediction. With that prediction, you can choose to litigate if you know you'll win, and to settle if you know you lose. Thus, the mean value of buying the prediction is $18,500 because you can use the information to change your decision and avoid the zero outcomes. This is a fundamental characteristic of the value of

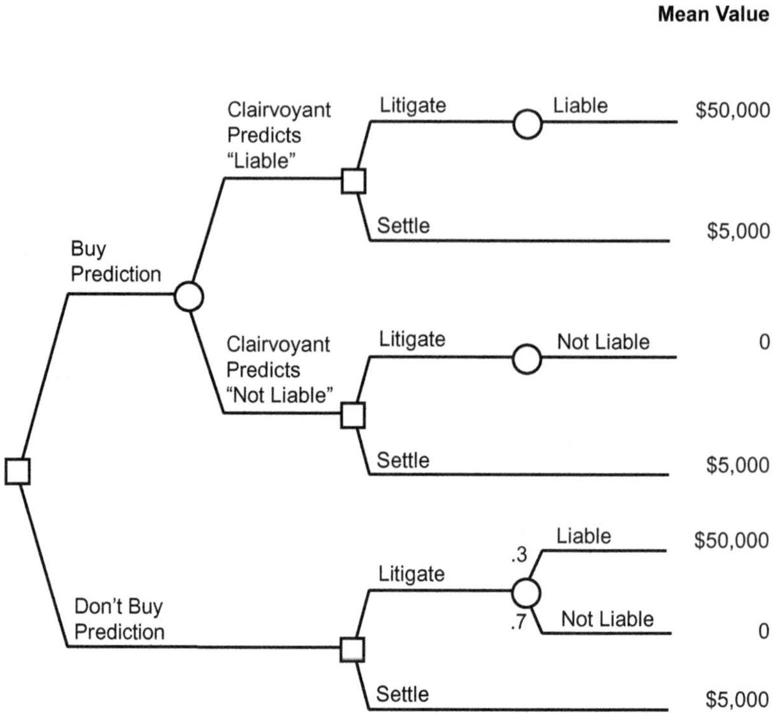

Figure 7.12 Clairvoyance tree with zero probability branches removed

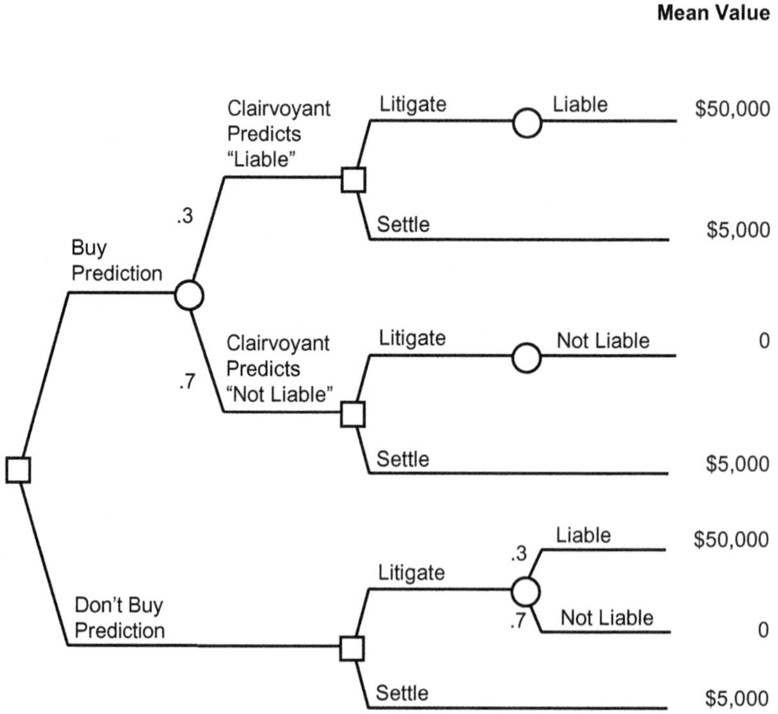

Figure 7.13 Adding the probabilities of what the clairvoyant predicts

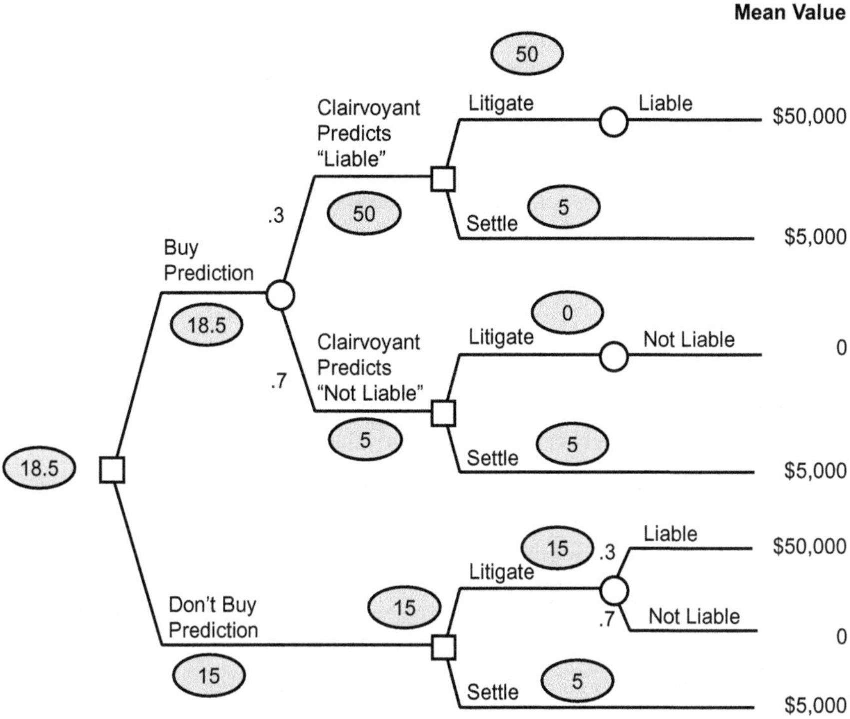

Figure 7.14 Rollback values for clairvoyance from the plaintiff's perspective

information: ***information only has value if it changes your decision***. In the context of the Decision Analysis focus on what happens to you, there is no such thing as information being valuable for its own sake. It has to change what you would do to affect the value of your case (Note also that you can't even calculate a value of information if you don't have a decision in your original tree. Do you understand why?).

Of course, it's almost always worthwhile to do a little discovery and research to help you structure your problem and get a clearer idea of the important issues. However, when you've already determined the most important questions and are considering which ones are most worth undertaking discovery on, the value of information is the way to think about it (Figure 7.14).

Now that you have the value *with* clairvoyance (the $18,500), the only remaining step is figuring the value *of* clairvoyance. The value *of* clairvoyance, then is simply the difference between your original tree, and the tree where you buy the prediction, or 18.5 – 15 = $3,500. Thus, for this case, you should not be willing to pay more than $3,500 to find out if you will be held liable or not. In other words, $3,500 is the maximum you should be willing to spend on discovery on the liability issue.

7.4.1.2 Shortcut for Calculating the Value of Information

Once you understand how the value of information calculation works, there's a shortcut method to calculate it (without going through all the steps given above).

The really tricky part of the shortcut is realizing that the value of information is just a comparison between the mean value of your original tree and the mean value with the node for the question being answered brought to the front of the tree, as below (And, of course, you have to write down the value numbers in the right places.) (Figure 7.15).

Thus, the shortcut procedure for calculating the value of information goes as follows.

- Compute the mean value of the tree with nodes in the original order.
- Redraw the tree with the question in front and compute the mean value again.
- Take the difference between the two mean values.

To illustrate this procedure, let's do one more value of information calculation. We'll look at a slightly different tree from the defendant's perspective (Figure 7.16).

The preferred decision for the defendant would be to settle for $5,000 rather than litigating with a mean exposure of $12,750. But is there any value of information for damages? That is, is it worth anything to undertake discovery or further research to

Figure 7.15 Shortcut tree for calculating value with clairvoyance

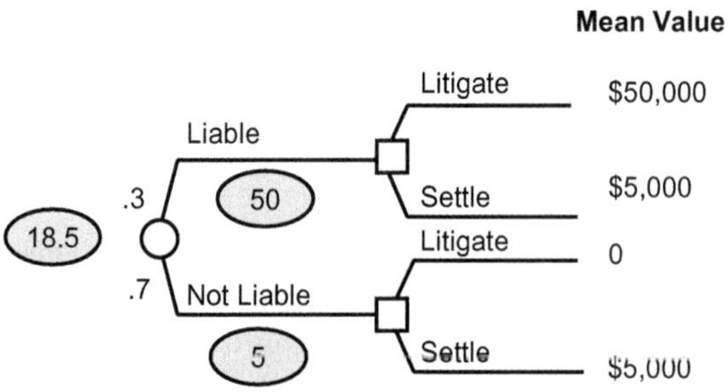

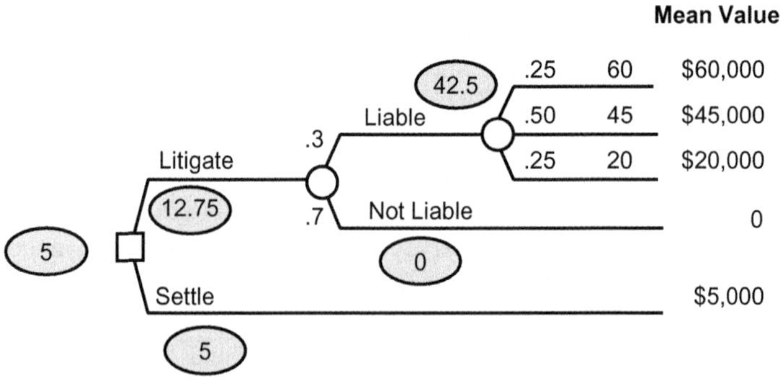

Figure 7.16 Settle versus litigation from the defendant's perspective

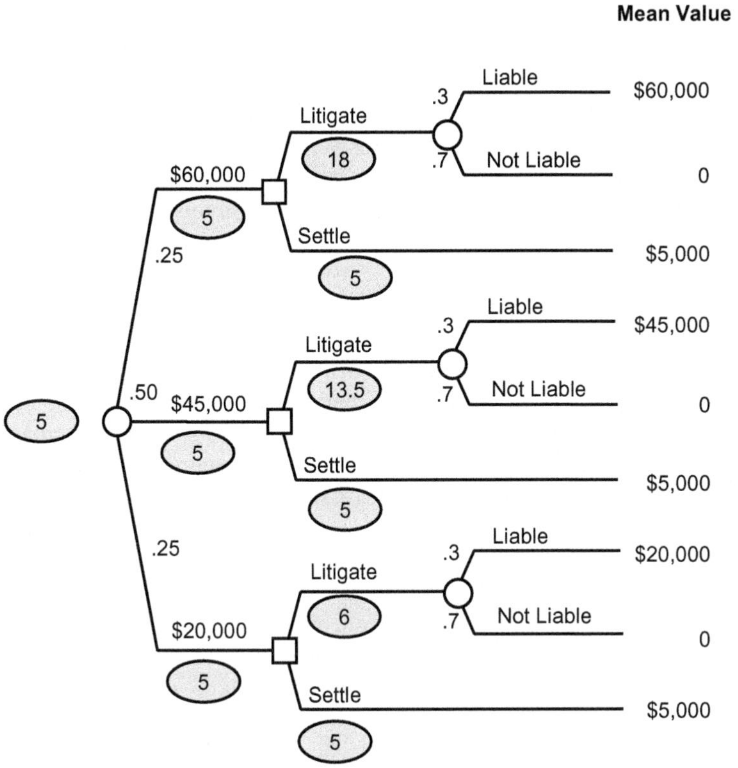

Figure 7.17 Clairvoyance on damages from the defendant's perspective

find if the plaintiff would be awarded $60,000, $45,000, or $20,000? To find out, you need to solve the tree with the damages node in front, as below (Figure 7.17).

Even though damages range from $20,000 to $60,000, there is absolutely no value to learning more about the damage which may be awarded because the best decision is always to settle for $5,000—the same as in your original tree. The worst you can do litigating is a .30 chance at $20,000, is $1,000 more than mean value than the $5,000 settlement offer. Accordingly, it's not worth it to spend any time or money at all on the improving your damages assessment because the settlement offer is so low that you're always better off taking it rather than litigating this case. In this way, the value of information should also make intuitive sense when you get it set up properly, although it is nearly impossible to intuit the value of information directly.

Of course, a cost of $6,000 is not a specific scenario in the tree. The possible scenarios described in the tree are zero indemnity cost if not found liable, and indemnity costs ranging from $20,000 to $60,000 if found liable. Because that range represents a continuous uncertainty, outcomes of $6,000 or $65,000 are still possible even if not a scenario in the tree.

The main point, though, is that making a good decision does not guarantee a good outcome: it just maximizes your chances of a good outcome. Over multiple decisions, you will come out ahead.

Consider one additional question. Is there still any value of information on the liability question? You may be able to answer yes or no merely by examining the damages uncertainty tree with the nodes in the original order. Can you calculate the value of information (if any) on the liability question in this tree?

7.4.1.3 Calculating the Value of Control

Just as the value of information focuses on whether it's worth anything to improve your probability assessment, the value of control focuses on the worth obtaining one particular answer to a question. You may, for instance, to assure a particular answer with a motion for partial summary judgment, or by obtaining a stipulation (perhaps in return for dropping a particular damages claim). What is it worth for you to do so?

To control an uncertainty, we need someone more powerful than the clairvoyant. In decision analysis tradition, that person is the wizard. He can control how an uncertainty comes out, not just predict it.

As with the value of perfect information, even though the perfect control is seldom available, it furnishes the maximum you should be willing to pay to try to effect the outcome of an uncertainty.

The pleasant surprise is that you've already done the hard work by calculating the value of information. In the same way that you find the value of information by bringing the chance node to the front of the tree (because the prediction controls what actually happens later in the tree), the value of control requires bringing the chance node to the front of the tree—but this time you get to pick (control) which answer happens instead of facing the prospect of an uncertain prediction.

Accordingly, the procedure for calculating the value of control is as follows.

- Redraw the tree with the uncertainty brought to the front of the tree (if you haven't already done so for value of information calculations).
- For that uncertainty, pick whichever answer has the best mean value.
- Subtract the original mean value from that number.

For example, going back to the earlier litigate/settle tree with clairvoyance on liability from the plaintiff's perspective, control on liability is like making that chance node into a decision node, so you get to pick which answer you like best, as below (Figure 7.18).

The value *of* control then, is 50 − 15 = $35,000—much greater than the $3,500 value of information. As you would expect, controlling the answer to a question is much more valuable than simply knowing what it will be. Thus, it would be worth a maximum of $35,000 to you to control the answer on liability. This result should also make sense to you because if, for instance, you obtain a stipulation on liability in return for reducing your damages demand by $35,000, you would be asking for a settlement amount of 50 − 35 = $15,000—the mean value of the case.

Figure 7.18 Value of control on liability from the plaintiff's perspective

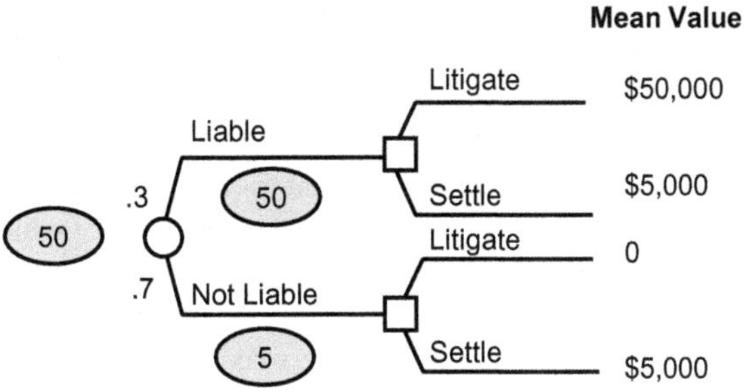

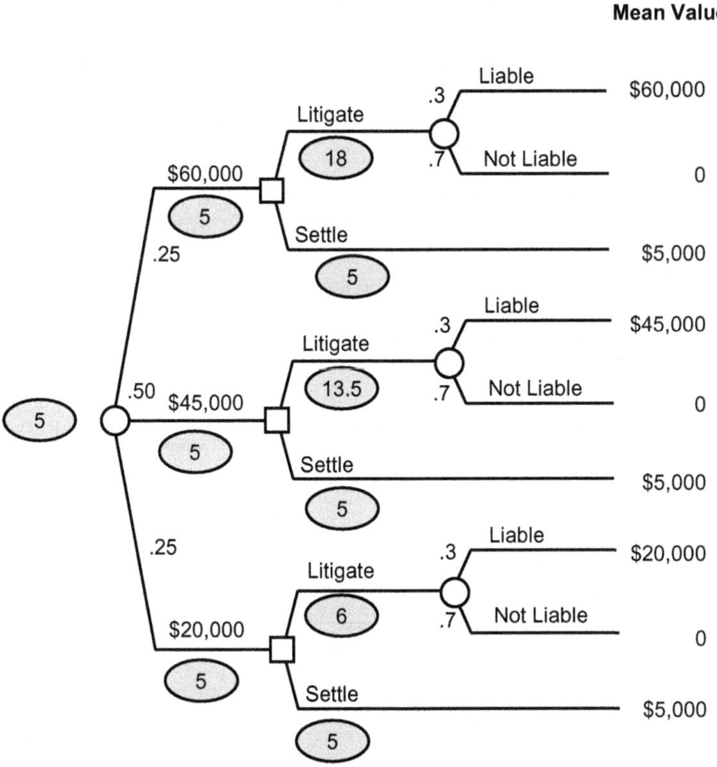

Figure 7.19 Value of control on damages from the defendant's perspective

But what about the clairvoyance tree with the damages uncertainty from the defendant's perspective? Is there value in controlling damages, even if there's no value in knowing in advance what they'll be? You should be able to answer intuitively, but the tree gives the numerical answer (Figure 7.19).

Even though you could elect to have the wizard control the damages awarded if you are found liable, there is no value in doing so—assuming the plaintiff is willing to accept the $5,000 settlement offer. If not, we know from the tree in the original order (Figure 7.16) that you mean exposure at trial is about $13,000, so it would be worth offering up to this amount to settle the case.

As illustrated, the answers to the questions of what perfect information or control is worth depends: it depends on the choices and uncertainties you lay out in your analysis. Change those and the answers may also change. With each updating, they furnish an updated guide to the maximum it is worth to learn more about or influence the outcome of an uncertainty.

Of course, there has to be a decision to calculate a value of uncertainty: information has no value if it wouldn't change the preferred alternative. However, you can calculate a value of perfect control even if there is no decision. Do you see why?

You looked over your Howitzer tree and counted eight uncertainties you could do value of information and value of control calculations for. You decide to start with values of information and control on whether the judge granted your motion for summary judgment on the ground that Zelda's statement was not a statement of fact (Figure 7.20).

The mean value via roll-back came out the same at $38K. Was that right? You looked at the tree, and saw that the preferred decision was still to litigate, so that made sense: information only had value if it changed your decision. However, value of control was 38 – 0 = $38K because this would assure you of no liability. You did the defects question next (Figure 7.21).

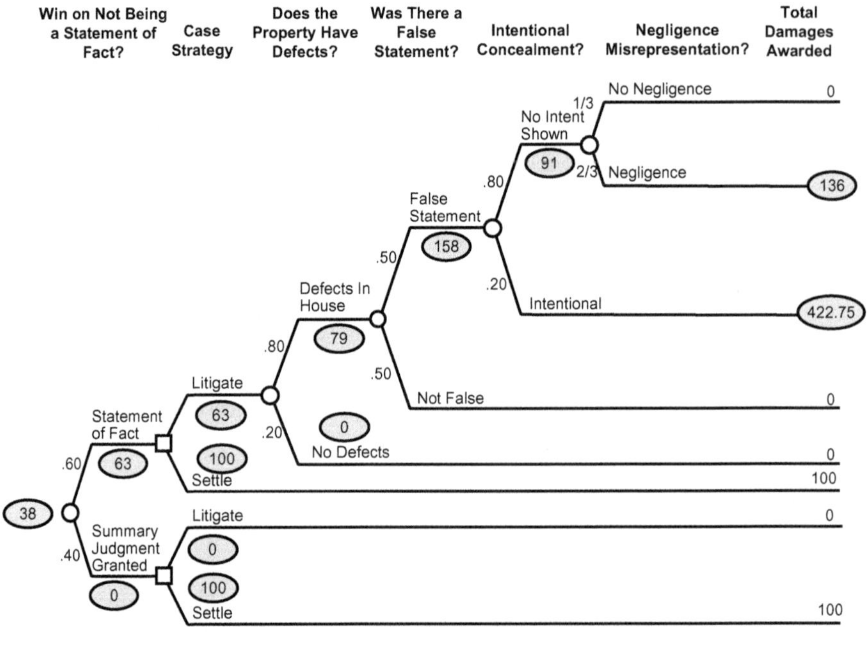

Figure 7.20 Tree for values of information and control on motion for summary judgment

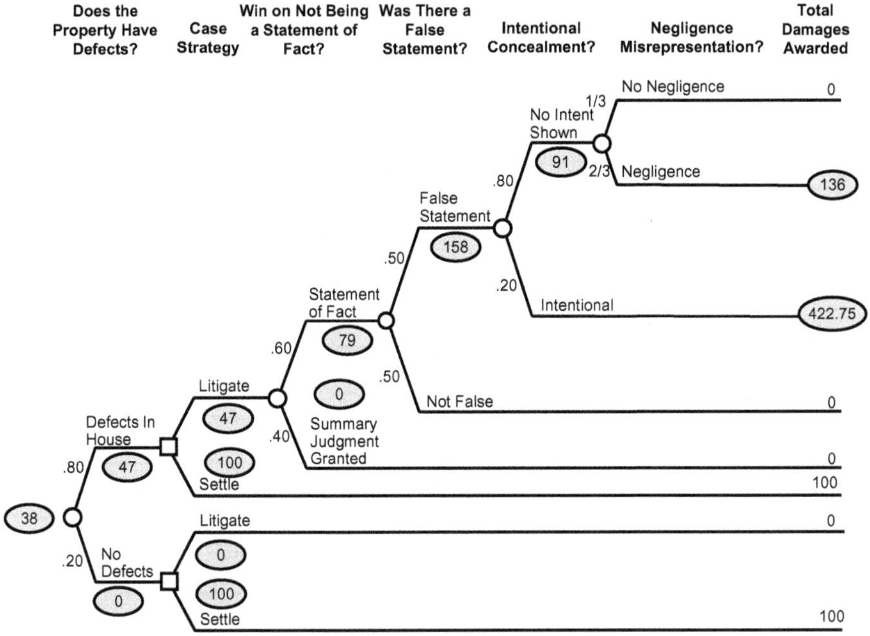

Figure 7.21 Tree for values of information and control on defects in house

Again, the value of information zero and the value of control was 38 – 0 = \$38K. Again this made sense because the question determined whether you would be liable or not but didn't affect your damages exposure if found liable.

You were beginning to see a pattern here: there was no value of information on the questions on which liability could turn because they would either result in the mean exposure at trial of \$38K or no indemnity at all, and both were less than the \$100K settlement offer on the table. You suspected that the results would be the same for whether there was a false statement or not, and so it turned out (We leave this as an exercise for the reader.).

You turn to the question of intentional concealment (Figure 7.22).

This result was surprising to you so you rechecked your numbers. In fact, there was a value of information on intentional concealment, but it was lost in rounding off the numbers. Not rounded off, the value of perfect information on intentional concealment is \$292 (dollars, not thousands).

You felt there must be a significant value of information of whether punitive damages are awarded, but that was a factor included in your subtree. You would have to combine your subtree into your main tree to do the calculation. This called for software to ease the calculation.

Indeed, it turned out that the value of perfect information on whether punitive damages would be awarded was \$12,000—the only uncertainty in the tree with a significant value of information. The value of control was only slightly higher at \$14,000. The amount of compensatory or punitive damages awarded likewise had no value of information at all (We leave this as an exercise for the reader.).

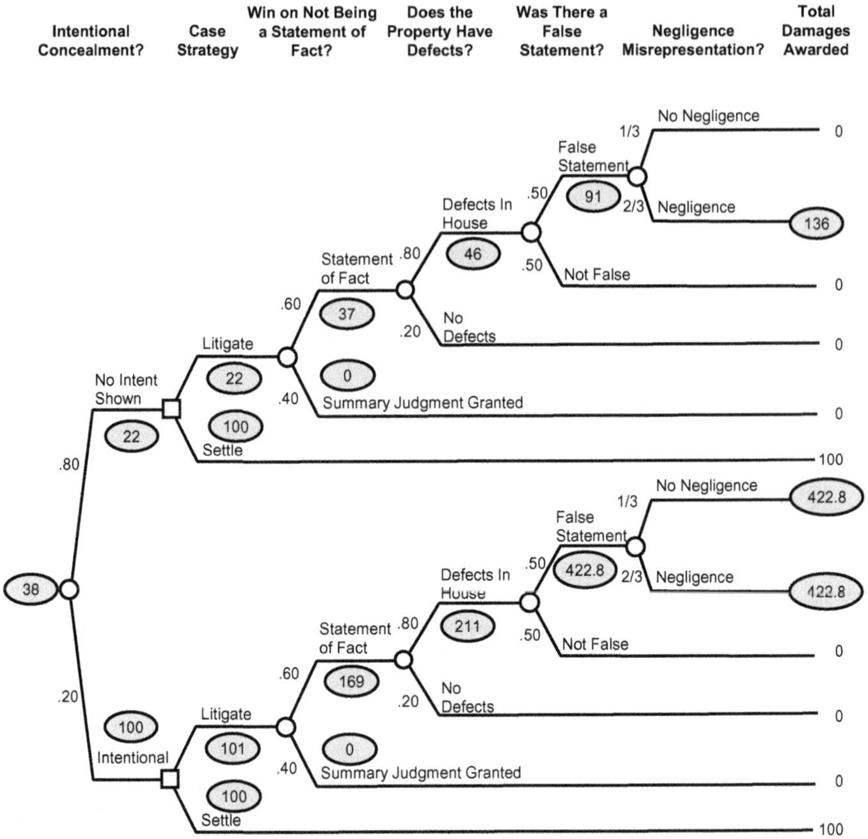

Figure 7.22 Tree for values of information and control on intentional concealment

At this point, you decided to stop drawing trees and consider for a moment. Only the uncertainties on intentional concealment and award of punitive damages had any value of information at all, and only information on awarding punitive damages was significant. Thus, knowing that any one of these questions would be found against you was not enough to make settling a better deal than chancing the mean value of litigating.

You studied the tree and realized that, because there were so many findings required for liability, any one of them just didn't make enough difference by itself to sway you towards the settlement. This would indicate that a perhaps typical approach of researching and trying to nail down every significant question wasn't a good use of your time or the client's money. You needed to go over prospects with the client and alert them that, although overall mean exposure was far less than the settlement offer, there was a less than 1% chance of being hit with punitive damages on the order of $500K to $7 million, with a mean award of about $3 million. Was the client willing to take that risk?

At a meeting with the client and the cocounsel, you laid out your thoughts and solicited their feedback. They were shocked to see that, even though the potential compensatory and punitive damages were much higher than the settlement demand, the mean exposure at trial was much less. You explained that, given the little or no value in extensive research or discovery on the other issues in the case, resources were better spent exploring the punitive damages possibilities (what was the record in this jurisdiction for these kinds of cases?) and on crafting an aggressive strategy to dispose of the case.

First off, given the low overall probability of liability, was there a way to strengthen the motion for summary judgment? Your initial probability of success on this motion was only 40%. Was it possible to craft stronger arguments that the law was clear and there was no material issue of fact? The team brainstormed for a bit, then outlined an amended motion for filing the next day.

Lastly, given the overall probability of liability and the improved prospects for summary judgment, the counsel and the client were in agreement to push hard in settlement negotiations with numbers that made sense in view of the case.

The settlement conference was a week later. You went in with a polite but firm position. You were sorry that Mr. Howitzer was having issues with the house, but that was not the fault of your client. You saw a lot of issues with the plaintiff being able to recover and were quite willing to contest them all but, to save time and expense on everybody's parts while it was still early in the case, your client was willing to make a reasonable counter offer. You offered $15,000.

The opposing counsel looked apoplectic. You calmly asked them to convey the offer to the client.

The next day, they came back with a counteroffer of $70,000. You countered with $20,000 and advised them this was a final offer. It was declined.

Two days later, the court granted the motion for summary judgement and dismissed the case without prejudice. The client was thrilled. You were similarly pleased, but advised them Howitzer could appeal the decision. You would need to wait out the 30 day filing period to see what they would do.

Thankfully, the period passed without an appeal filing, making the judgment final. Then it was time for a celebration.

As you sipped a bit of single malt scotch, you reflected on the course of the case. Certainly, given the complexity of the case, it was not apparent at all at the outset what its merits were. Using decision analysis had helped tremendously in clarifying the uncertainties and issues in the case, and in figuring out how best to craft a winning legal strategy. It was a challenging exercise, but very much worth the effort. You looked forward to the applying it to the next case.

7.4.1.4 The Values of Imperfect Information and Imperfect Control

Of course, perfect information or perfect control is seldom actually available. We can the value of imperfect information or imperfect control, though the calculation is more involved. Most of the time, it is enough to know the value of perfect information or control because it furnishes the maximum you should pay for additional information or to affect the outcome of a particular uncertainty. This value is often surprisingly low, or even zero when it doesn't change the preferred alternative. For those wishing to tackle this calculation for a particular opportunity to gather more information or influence the outcome of an uncertainty, we refer you to *Decision Analysis for the Professional* or *Foundations of Decision Analysis*.

However, even without explicitly doing the calculation, we can bound the values of information and control.

To do this, we need to consider the accuracy of the prediction or the effectiveness of the control. A prediction that is either 100% or 0% accurate is perfect information. We know how to calculate that (A prediction that you know is 0% accurate perfectly predicts what will *not* happen—which allows you to know what will happen.).

In contrast, a prediction that is 50% accurate is completely worthless: instead of purchasing the information, you could simply flip a coin.

This relationship between the accuracy of information and its value is illustrated below (Figure 7.23).

In contrast, the relationship between the values of perfect and imperfect control is illustrated below (Figure 7.24).

Unlike the value of perfect information, the value of perfect control should always be greater than zero.

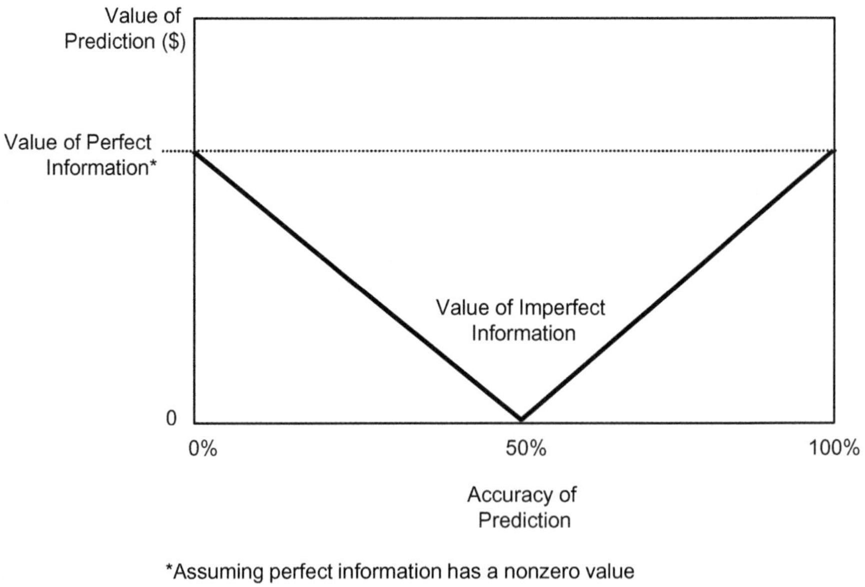

Figure 7.23 Relationship between perfect and imperfect information

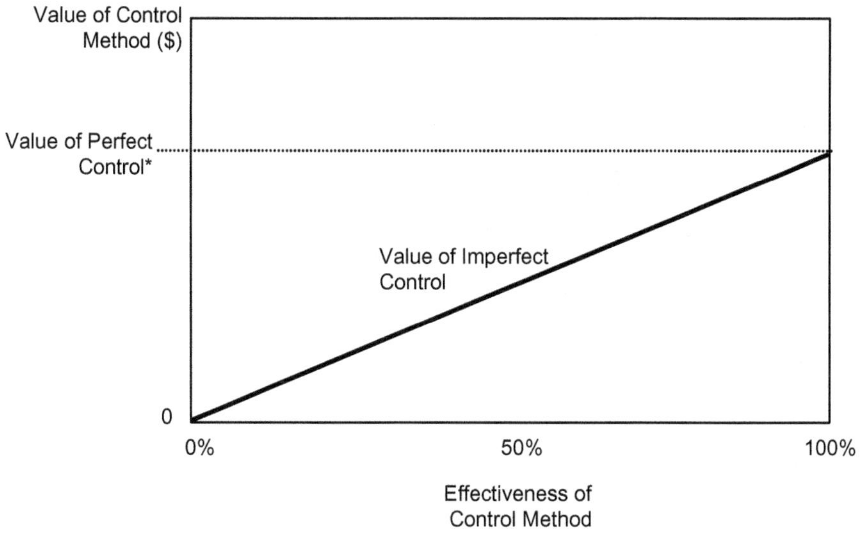

Figure 7.24 Relationship between values of perfect and imperfect control

7.4.2 Probability Sensitivity

From values of information and control you can see which questions justify further discovery and how much, and how worthwhile it is to try to affect how those questions turn out.

Another way to help plan discovery strategy is sensitivity to probability. This looks at how the value of the case and each alternative changes as specific probabilities change. This helps you plan discovery strategy in two ways:

- By showing you when differences of opinion on probability assessments need to be resolved (such as by further information/discovery or retaining additional experts);
- By showing you which questions have the biggest impacts on the value of your case.

Probability sensitivity gives you this information by showing how the mean value of your case varies with your probability assessment. An example will show how, and the litigation tree with damages uncertainty will do (repeated below) (Figure 7.25).

The procedure for constructing a probability sensitivity graph is slightly different for a two-branch chance node than for a three-branch one. Here is the procedure for a two-branch node (like the liability node).

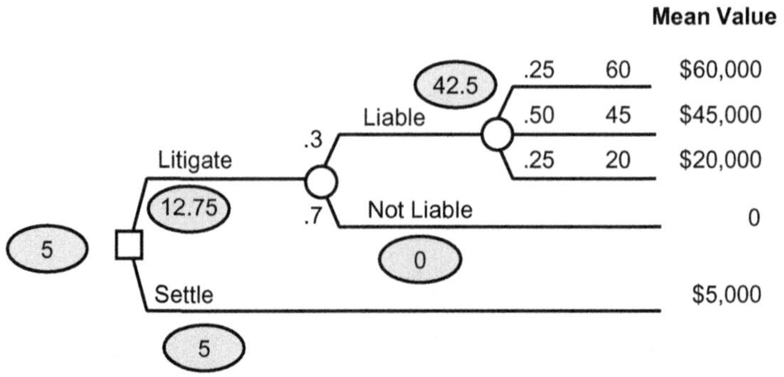

Figure 7.25 Litigate or settle tree with damages uncertainty

1. Set the probability for the top answer to 0, and the bottom to 1.0. Compute the mean value of litigating the case, and plot the resulting value on a graph like the one below (vertical axis is the mean value; horizontal axis is the probability of either the top or bottom answer).
2. Set the probability for the top answer to 1.0, and the bottom to 0. Again compute the mean value of litigating the case and plot the resulting value.
3. Draw a straight line connecting the two points.
4. Plot the mean value of litigating the case with your original probabilities. If this point does not fall on the line you've drawn, check your calculations.
5. Plot the value of settling the case (This will be a horizontal line because the settlement offer is unaffected by the odds of being found liable.).

For the liability uncertainty, this procedure produces the plot below (Figure 7.26).

From the plot, you can see that the mean value of litigating the case varies from 0 to $42,500, and that litigating has a higher mean value than settling unless you think the probability of winning the case (finding the defendant liable) is less than about .10. Thus, in answer to the first question raised above, differences of opinion on the chances of winning the case need not be resolved unless someone thought there was less than a .10 chance of winning the case—because, otherwise, you're better off litigating whether you think the chance of winning is .30 or .60. Just as perfect information only has value if it changes your decision, differences in opinion only need be resolved if they imply different settlement decisions. If the settlement decision is the same for everybody's opinion (probability), it's not worth putting any more discovery resources into that question.

In answer to the second question (where you can greatly increase the value of your case), you can similarly see that increasing the chance of winning can increase the mean value up to a maximum of $42,500 (the mean damages multiplied by a 1.0 probability of winning the case and getting them). However, just as every uncertainty should have a value of control, every uncertainty should have *some* potential for increasing the mean value of your case. The real question is which uncertainty has the potential to increase the mean value the *most*. In other words, what is the most valuable uncertainty to undertake discovery on?

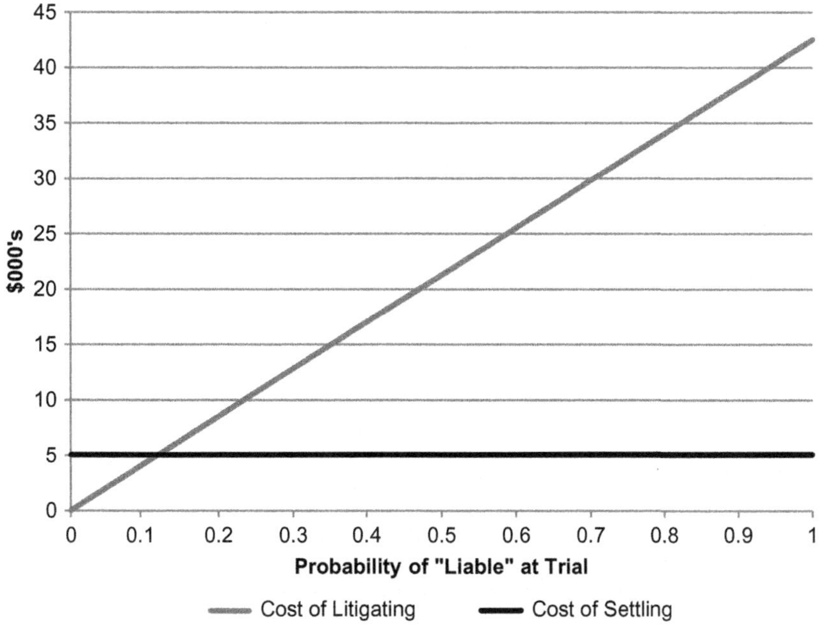

Figure 7.26 Sensitivity to probability of "liable" at trial

Answering this question requires comparing the probability sensitivities for the various uncertainties in the tree. For the damages uncertainty (or any uncertainty with more than two potential outcomes), this requires a method of allocating probability to the middle branches as you vary the probabilities on the first and last branches.

For most continuous variables discretized into branches for the tree, it works well to simply set the probabilities in the middle branches to zero and simply vary the probabilities in the first and last branches as if it was a two-branch node. Here is the procedure for this:

1. Set the probabilities for the middle branch(es) to zero.
2. Set the probability for the top answer to 0, and the bottom to 1.0. Compute the mean value of litigating the case, and plot the resulting value on a graph like the one below (vertical axis is the mean value; horizontal axis is the probability of either the top or bottom answer).
3. Set the probability for the top answer to 1.0, and the bottom to 0. Again compute the mean value of litigating the case and plot the resulting value.
4. Draw a straight line connecting the two points.
5. Plot the mean value of litigating the case with your original probabilities. If this point does not fall on the line you've drawn, check your calculations.
6. Plot the value of settling the case (This will likewise be a horizontal line because the settlement offer is also unaffected by the odds of various damages awards.) (Figure 7.27).

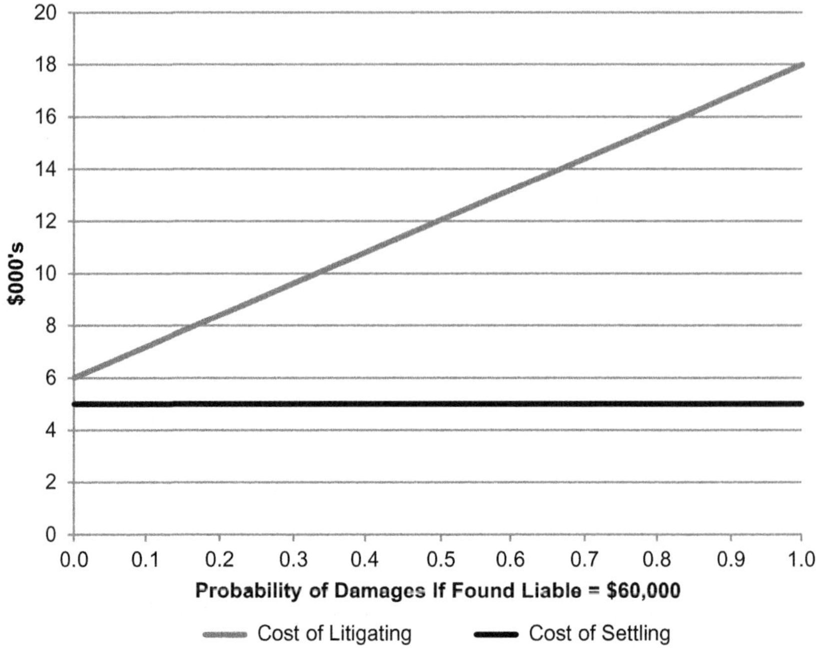

Figure 7.27 Sensitivity to probability of damages = $60,000 if found liable at trial

From this plot, you can see that damages does not offer as much potential for increasing the value of your case (yielding a maximum mean value of $18,000 versus a maximum of $47,500 with 100% chances of winning the case). Thus, liability offers better prospects for improving the mean value of your case through additional discovery (Or, from the defendant's point of view, the settlement offer looks better regardless of the damages award if found liable.).

Further, under the worst damages scenario (100% chance of damages of $20,000), the mean value of litigating is still $6,000—a thousand dollars higher than the settlement offer. However, deciding whether any differences of opinion need to be resolved is a slightly more subtle question. Recall that the $20,000 damages means "damages in the region of the continuous possible damages distribution represented by $20,000." Thus, the worst case really means that, with mean damages of $20,000, you would still litigate the case. To find the point where the decision changes, you need to find where the value of litigating is equal to the settlement offer of $5,000. Because you have a .30 chance of winning the case and obtaining the damages, litigating would have the same value at 5 ÷ .30 = $16,667.

Thus, differences of opinion would only need be resolved if someone felt that the mean damages (the "best guess") would be less than $16,667. Otherwise, you wouldn't devote any additional discovery to damages because you would be better off litigating.

On an even subtler level, a probability sensitivity plot can give you a rough idea of:

- A "safe" settlement level for your case; and
- An appropriate level of investment in discovery on a particular question.

These can be determined by interpreting the sensitivity probability plot. For instance, consider the sensitivity plot for liability repeated below (Figure 7.28).

Suppose you and a colleague disagreed about the chances of winning the case (the probability of "Liable"), and she thought it was more like 50% rather than your 30%. From the graph, you can see that the mean value of litigating at a 50% probability of winning is around \$20,000. Thus, \$20,000 would be a "safe" amount to settle for because, at this amount, you would be better off settling than litigating—regardless of who had the better probability.

Similarly, suppose you could undertake some discovery (or other case development work) that would raise your probability of winning from .30 to .70, at a cost of around \$7,000. Is this an appropriate level of investment in the question of liability? Again, the graph can give you an indication. Raising the probability of winning from .30 to .70 increases the mean value of litigating from \$12,750 to around \$30,000 (you can calculate the exact amount)—an increase of around \$17,000. Accordingly, you would conclude that it is worthwhile to invest the extra \$7,000 to improve the mean value of your case by \$17,000.

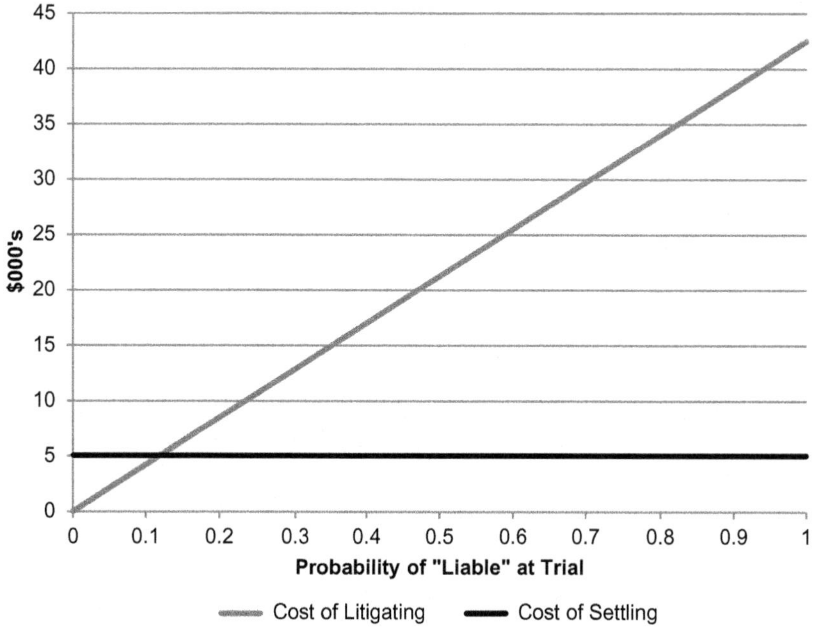

Figure 7.28 Another look at sensitivity to probability of "liable" at trial

It may seem strange at first to think of investing "real" dollars (money you're actually spending) to improve the mean value of your case—which is, after all, still an uncertain prospect and not actually winning the case. However, the entire object of trial preparation is to increase your chances of winning—like increasing the probability of "liable." The new aspect of using probability sensitivity to evaluate investments, then, is not that you can increase the value of your case by increasing your odds of winning, but, rather, that it offers you a way to think about improving the value *other* than through the chances of winning or losing (like changing the damages probabilities). It generalizes the value of the case with the mean value and lets you compare all the possible ways of improving the mean value. It helps you to think consistently and broadly about all aspects of your case's value when planning discovery strategy (as do the values of information and control).

To sum up, then, probability sensitivity can help you plan discovery strategy in four ways.

- Identify when differences of opinion on probabilities need to be resolved;
- Identify the best candidates for further discovery;
- Identify "safe" settlement levels when there are differences of opinion on probabilities;
- Evaluate the cost-effectiveness of discovery according to how it may change your probabilities.

7.5 Trial Strategy

Analyzing trial strategy decisions doesn't require any new methods. Any trial strategy decision can be analyzed using the same process of structure the problem, elicit probabilities, calculate, and analyze (For that matter, Decision Analysis can be used to help with pretty much any decision.). Rather than focusing on a litigate versus settle decision, the analysis would focus on the decisions you need to make to set trial strategy. These could include which legal theories to assert, which witnesses to call, possible stipulations, etc.

For example, in one actual case the issue of negligence looked really bad for the defense (clear failures in the standard of care), but, on the other hand, the plaintiff's alleged damages were also very squishy (no clear evidence of harm). Using decision analysis helped the counsel to develop the novel strategy of stipulating to negligence (to avoid presenting all the evidence that looked bad to the defense) while vigorously contesting damages (on which the plaintiff looked really bad). This somewhat counterintuitive strategy led to a highly favorable result.

To clarify how the process works, we'll take a look at a famously disastrous trial strategy. In hindsight, we only know there was a bad outcome for the defendant; we don't know whether they made a good or bad decision because it was never revealed whether the defendant employed decision analysis (probably not). However, we can show how a different trial strategy might have led to a very different outcome, and how a solid basis for this could have been established.

Possibly you're familiar with the epic battle between Pennzoil and Texaco in the mid-1980s. Getty Oil Company signed a memorandum of understanding agreeing to be acquired by Pennzoil Co. for a total of $1.8 billion. However, Getty continued shopping around for other buyers and shortly thereafter the board agreed to sell Getty to Texaco for $2 billion (an extra $10 per share).

Pennzoil sued Texaco in Texas state court for interfering with its acquisition of Getty. The key question at trial was whether the memorandum of understanding constituted a binding agreement or not. If so, Pennzoil would be entitled to damages for the future profits lost from the Getty acquisition. The Texas court awarded Pennzoil $7.53 billion in compensatory damages and $3 billion in punitive damages.

Following extensive further proceedings in US District Court, the Second Circuit Court of Appeals, the US Supreme Court, and the Texas Court of Appeals, the punitive damages award was reduced to $1 billion. Texaco filed for bankruptcy. This was the largest bankruptcy in US history up to 2001 (when Enron and Pacific Gas and Electric Company went bust). Pennzoil then agreed to settle the case for $3 billion. Texaco was eventually acquired by Chevron.

One key trial strategy decision Texaco made at the time was, as the defendant, not to introduce their own damages theory and experts at the state court trial. Instead, they only cross-examined Pennzoil's experts testifying as to what the asserted damages were.

Suppose instead Texaco had looked at introducing their own damages theory. What might a decision analysis of that strategy have looked like?

An actual analysis might have involved detailed spreadsheet modeling of the future potential Getty profits which Pennzoil complained of losing. Alternatively, it might have been as simple as assessing a range based on the damages Pennzoil asserted in the pleadings: the worst case scenario is Pennzoil being awarded what they were asking for. It all depends on what's needed to create confidence in the damages analysis.

For this example, we'll speculate that the counsel for Texaco felt they could reduce potential damages awarded by introducing their own theory of liability (such as Getty being far less profitable than alleged), but were worried that doing so would increase their chances of being found liable because presenting a damages theory might be interpreted by the jury as an admission of liability.

The relevance diagram for this is shown below (Figure 7.29).

Assessments on a tree show the potential for reduced damages award but increased chances of being found liable.

This assessment, then, asserts that whether or not Texaco puts in its own damages theory will affect both the chances of their being held liable and the possible amount of damages awarded if found liable. Accordingly, the assessments for these two questions should vary and depend on the strategy decision. A resulting tree might look like the following (Figure 7.30).

This tree shows the beliefs that (1) introducing a damages theory would reduce the possible damages if found liable; and (2) increase the chances of being found liable—just the kind of potentially two-edged sword that requires further analysis (Figure 7.31).

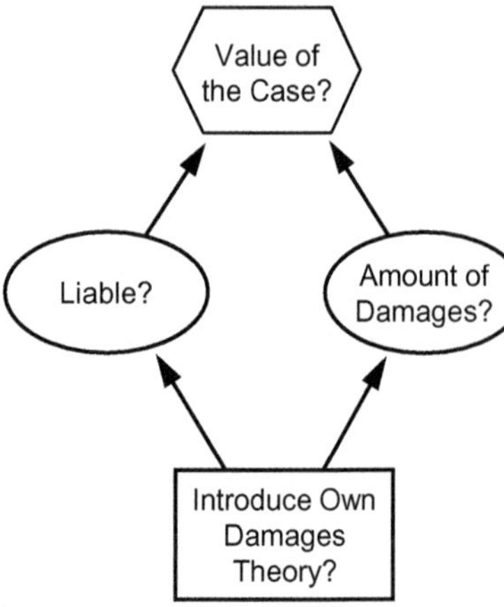

Figure 7.29 Hypothetical relevance diagram for Texaco trial strategy

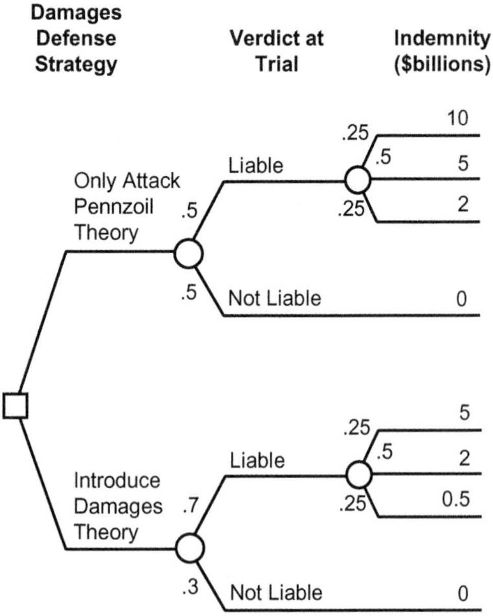

Figure 7.30 Possible assessments for Texaco damages defense strategy

Solving this tree for the rollback mean values shows that it is better for Texaco to introduce its own damages theory even if it increases the chances of being found liable from 50% to 70%. The mean exposure is reduced by $1.1 billion.

However, suppose there were a disagreement among the counsel about how much putting on its own damages theory would increase the chances of being found liable. The answer to whether that matters can be found in the graph of sensitivity to probability of liability if Texaco introduced its own damages theory (Figure 7.32).

This graph shows that, even if the probability of liability went up to 100%, this option is still better than taking your chances just cross examining the Pennzoil theory and experts.

Figure 7.31 Rollback mean values for Texaco damages defense strategy

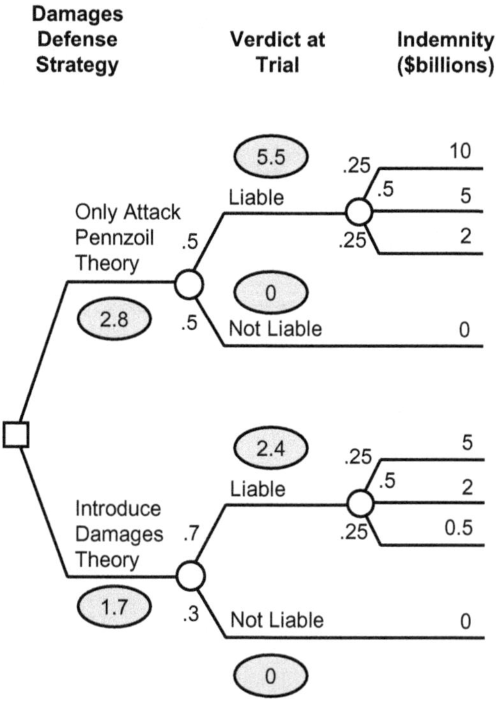

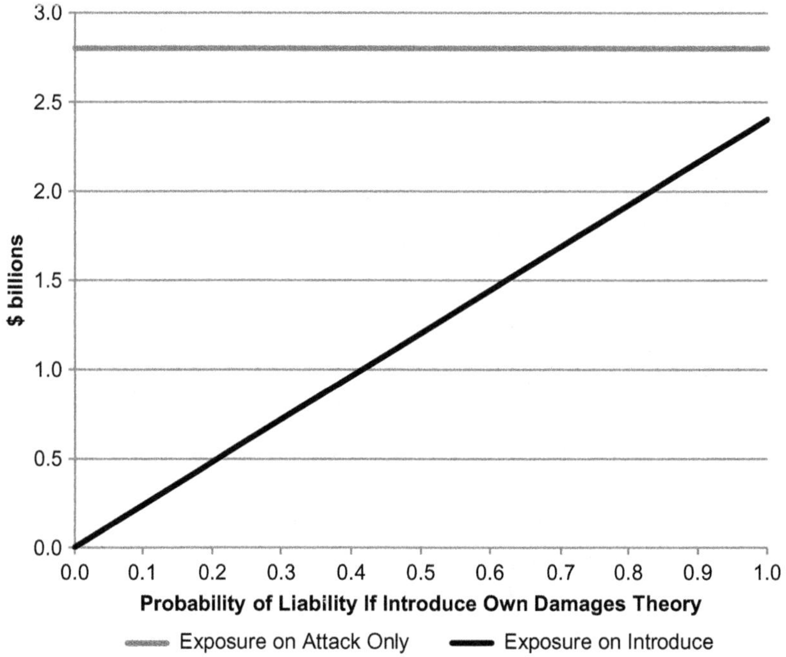

Figure 7.32 Sensitivity to probability of liability if Texaco introduces its own damages theory

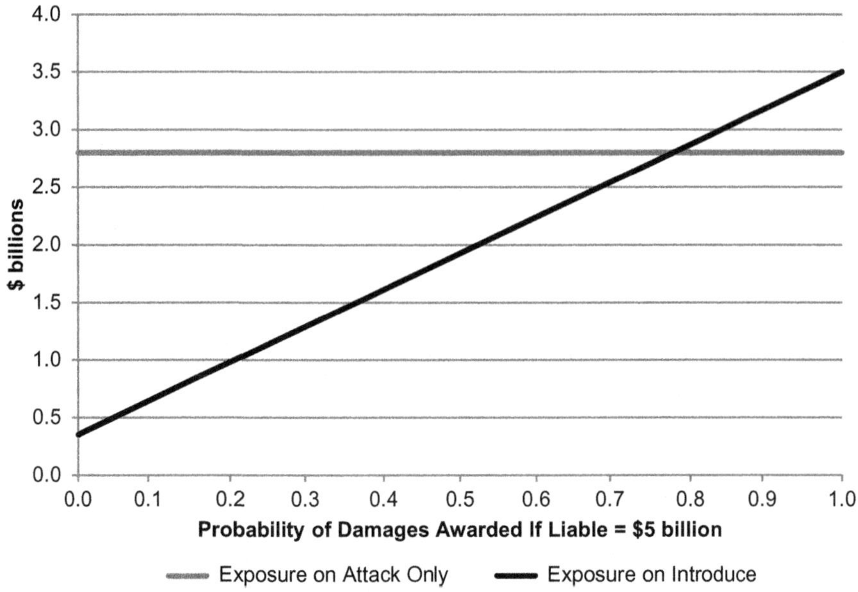

Figure 7.33 Sensitivity to probability of maximum damages awarded if Texaco introduces its own damages theory

Suppose there was also a question of how successful Texaco would be at reducing the damages awarded if liable, and how that would affect the preferred damages strategy. That question can also be answered with a sensitivity to probability (Figure 7.33).

This graph shows that the odds of being hit with the maximum damages under Texaco's theory would have to be around 80% or greater to make this strategy worse than just attacking the Pennzoil strategy. In other words, despite the increase chances of liability, as long as there is more than a 20% chance of reducing the damages the jury would award, it would still be better for Texaco to take the chance of putting on its own damages theory.

Similar means can be used to analyze any trial strategy decision the counsel can conceive and assess the chances for.

7.6 Crafting Winning Legal Strategies

We have now gone through the entire process of applying decision analysis to crafting winning legal strategies:

1. Structure the problem
2. Elicit probabilities
3. Integrate all information, judgment, and consequences
4. Analyze the results

As you've seen, the process doesn't create winning legal strategies by magic. It works by providing a structured way for the analysts, counsel, and clients to consider the alternatives and their possible results at each step of the case. By creating thought experiments in the form of quantitative models, insight is created as to what should be done and why.

This process employs both your affective/intuitive and deliberative/logical thinking modes and creates clarity of thought as to what should be done and why for the many decisions arising in the course of a dispute. By forcing you to think harder about what you might do and why at each step, decision analysis creates an advantage of superior understanding which can be employed to achieve winning results.

7.7 Problems

7.7.1 Do the Results Make Sense?

1. What means other than decision analysis might you use to evaluate possible legal strategies? How do they differ from a decision analytic perspective?
2. In considering whether an apparently better course of action makes sense, list at least three deficiencies or possible areas for improvement you might consider other than the ones listed in this book.
3. In structuring an analysis, we emphasize the Trier of Fact Rule: only making the analysis as detailed and complex as the trier of fact will consider it. How is this different from the approach one would take in preparing a pleading or writing a law journal article?
4. What is the difference between qualitative and quantitative issues in an analysis? How are the means for addressing each different?
5. List at least three indicators that you need to revise your analysis.
6. Your case is being tried before Judge Learned Eye, who is reputed to have a keen awareness for a pretty leg.[3] He is also a celebrated "hang-em high" judge with a well-known record of ruling against the defendant in 80 percent of all his cases. You therefore wonder if your male client is particularly disadvantaged against a female plaintiff, and whether it would improve your chances to join Angelina Jolie as a codefendant (a possibly complicated and time-consuming ploy in this case).

   A few days later, your team of summer interns emerged bleary-eyed from the library with the following statistics. When the plaintiffs won in Judge Eye's court, neither party was a woman 75 percent of the time, there was a woman plaintiff 13 percent of the time, a woman defendant 9 percent of the time, and both parties were women 3 percent of the time. Similarly, when the defendant won, neither

[3] The author wishes to note that he is utterly opposed to gender stereotyping, sexism, or discrimination of any sort. Regrettably, this problem was inspired by actual experiences and is included to illustrate that even issues one would wish don't exist can be analyzed. In particular, this sort of analysis would be very helpful in discrimination litigation.

party was female 70 percent of the time, there was a woman plaintiff 9 percent of the time, a woman defendant 15 percent of the time, and both parties were women 6 percent of the time.

6.1. Write down your intuitive guesses about your prospects before Judge Eye. Are you comfortable with them? Are you dissatisfied with the problem description? That is, does anything seem intuitively wrong to you, or are any important elements omitted?
6.2. Draw the relevance diagram and decision tree. Include the value formulas and calculate the scenario values.
6.3. What are your client's chances as a male defendant against a female plaintiff?
6.4. Do your client's chances improve by adding a female codefendant? If so, by how much?
6.5. Can you draw any generalizations about how a female defendant fares in Judge Eye's court as opposed to a male defendant? How about a male plaintiff opposing a male defendant as opposed to a female defendant also opposing a male? What does this tell you about Judge Eye's bias and whether you should attempt to join Lauren Hutton?

7.7.2 Settlement Strategy

7. A decision analysis view would be that, in a particular case and on a risk-neutral basis, it is worthwhile for the defendant to pay up to the mean exposure at trial to settle the case.

   7.1. How is this different than the actual negotiating strategy?
   7.2. How does this support a negotiating strategy?
   7.3. What other factors should you take into account in crafting a negotiation strategy?

8. When is it not adequate to look only at the exposure in the case at hand when determining maximum settlement value (for the defendant)?
9. If you have actually been involved in case settlement negotiations, describe how the figures offered or accepted differ from those from a decision analysis perspective. If not, speculate.
10. A risk-adjusted discount rate attempts to deal with uncertainty by calculating the present value of any investment (such as investing in legal fees now in the hope of obtaining a judgment in the future) with a discount rate that has been increased by some "amount for risk." The idea is that higher discount rate lowers the value of "risky" investments, giving them a higher standard to meet for approval. Note that this approach simply imposes an increasingly greater penalty on cash flows that occur farther in the future—regardless of when the relevant uncertainties will be resolved. For instance, a $1 million

judgment 10 years from now has a present value of $424,000 when discounted at 10%, but a present value of only $284,000 when discounted with a "risk-adjusted" rate of 15%.

10.1. Consider the timing of the various events in a typical suit (from the first filing of papers to the collection of any judgment) and discuss the disadvantages of using a "risk-adjusted" discount rate to value cases.
10.2. How does the mean value address these shortcomings, if at all?

11. Investments are commonly referred to as being more or less "risky."

11.1. What is your understanding of what's described by this "risk," and how does it compare to the explicit judgment of uncertainty in a probability calculation?
11.2. Assume that the "riskiness" of an investment is related to the variability of possible outcomes. What does the mean value tell you about the variability of outcomes?

12. Suppose that you've been offered a job by a law firm in New York (make that California if you're a member of the New York bar). The offer specified that you must pass the New York bar, and you signed a letter of intent sent you by the firm to that effect. The crucial language from the letter appears below.

Majorsky, Crumpmeyer, & Briggs offers to employ (you) as an associate at a starting salary of $93,000, plus relocation, bar examination, and bar review expenses. This offer is conditional on the applicant passing the New York bar. Expenses incurred in advance of beginning employment may be reimbursed by MC&B at the discretion of the firm.

After signing and returning the letter, you enrolled in a bar review course at a cost of $3,500, registered for the New York bar at a cost of $250, made a trip to New York in search of housing at a cost of $2,800, and packed and stored most of your possessions at a cost of $800. Since then, you've spent your nights and weekends for 4 weeks at the bar review course. It's now 2 weeks before the exam, and you've purchased a nonrefundable bargain-fare ticket to New York for $339 to take the exam. The hotel reservations are secured with your credit card.

A letter arrives from MC&B informing you with great regrets that they've found a more suitable candidate, that they're withdrawing the offer, and offering to reimburse your bar review tuition. You are greatly incensed by this turnaround and consider bringing suit.

The first question is whether a court would find the contract a bilateral, bargained-for-exchange, or a unilateral contract subject to revocation prior to your fully performing. An important consideration in this determination is whether a court would find taking the bar a performance you were required to render or a condition of satisfaction precedent to MC&B's obligation to perform. The court's interpretation is also relevant to whether, even if it found the contract unilateral, it would find that you had substantially performed and thus made the contract binding on MC&B.

Second, there is the issue of damages. If the courts finds a binding contract, whether bilateral or unilateral, you could ask for the benefit-of-the-bargain or for reliance damages. A court is unlikely to award specific performance with an employment contract. Finally, even if the court finds no binding contract, you could ask for damages on the theory of detrimental reliance on MC&B's promise.

12.1. Structure the relevance diagram for the value of the case. List any assumptions or changes you make to render the case more realistic to you. Assume your general knowledge of contract law, or specify a particular jurisdiction if you prefer.
12.2. Draw the decision tree for whether or not you should accept MC&B's reimbursement offer and ascertain all the information you need. Include the cost of prosecuting the cost, using a value of your own time if you would pursue it yourself.
12.3. Assess the necessary probabilities. Compute the values for each scenario to complete the decision tree. On the side, make an estimate of the time to judgment for any award so that you can discount it.
12.4. Compute the mean values for each alternative. Which one is better on a mean value basis? What is the probability of winning the case if you decide to pursue it? What should your settlement strategy be?
12.5. Plot the probability distribution for pursuing the case. What is the probability of obtaining more than the MC&B offer?
12.6. Compute the values of information and control for the uncertainties in the tree. What do they tell you about which issues would profit the most from further research? What should your discovery strategy be?
12.7. Assess your formulation of the problem. What are its strengths and weaknesses? Would you feel comfortable taking action on this basis, and why or why not? What would your next step be, and how would you use it to update or revise your analysis?

13. You represent Duane Hornet, a former software programming manager for the Department of Defense. Duane was fired 4 weeks ago after having worked for the Department for 6 years, three of them as a programming section manager. He came to you claiming that he was the victim of reverse discrimination. After interviewing Duane and investigating on your own, you've pieced together the following facts:

    Four years ago the Department initiated an internal study on possible discrimination within the programming section after a number of complaints were received. Most of these complaints were from Asian Americans and Asian resident aliens, who contended that they were being unfairly passed over for promotion. Because of the complications of differing security clearance allowances for the resident aliens and because the average tenure of the Asians as a group was less than the average Caucasian tenure, the Department felt a detailed study was necessary to determine if discrimination has indeed taken place.

Two years later, the study reported that there was a significant disparity between the percentages of junior- and management-level Asian programmers that could not be accounted for by differing security access or tenure. Promotions beyond the junior level were made solely by the section managers or section coordinators. The study therefore concluded that race discrimination was the probable cause. The Department then undertook a further study of how to remedy the situation.

The result of that second study was a reorganization plan that eliminated the two lowest existing levels of management: the programming section manager (Duane's position) and the section coordinator (who Duane reported to). Instead, it established a programming team leader with fewer people reporting to him or her than Duane did. Each team leader would then report to two different team coordinators: one working on the applications development side and the other on the applications support side. Previously, the functions had been mixed. Further, detailed standards were set up for each position and each appointment had to be approved by the reorganization steering committee. In this way, it was felt that two problems could be attacked at the same time: the Department could become more efficient while candidates for the new management positions were appraised on a stricter, "race-free" basis.

However, as the names of the new team leaders and team coordinators began to emerge, Duane saw what seemed to him a disproportionate number of Asians on the list. When his name appeared on a list of team leaders (in effect, a demotion to him—though with no pay cut), he was very upset. Although becoming a team coordinator would have been a promotion, he felt he was more qualified than the other former section manager, Diem Zhu, who became one of his team coordinators. Duane stormed into the office of his former boss, Bill Algol, to complain. Bill was now the next level above the team coordinators, and explained that it was a complicated process and Duane shouldn't feel slighted by not making team coordinator this round. Duane was unappeased and stated he would not assume a team leader post. Bill cautiously told him to take the rest of that day (Friday) off, and come back to see him on Monday. Duane fumed the whole weekend, then repeated his refusal. At that point, with some reluctance, Bill fired him. Duane was stunned, and brooded at home for a week before coming to see you.

You've been reviewing the Supreme Court cases on affirmative action to see whether Duane has a claim that his right to equal protection under the Fourteenth Amendment has been violated.

First of all, you note that there are precedents going back to the Slaughter House Cases (83 U.S. (16 Wall.) 36) in 1873 for subjecting any government discrimination by race to strict scrutiny (the closest and most exacting level of judicial scrutiny), rather than the more lenient and generally applied standard of a rational relation to a legitimate governmental end. Further, although the early cases dealt with discrimination against blacks, later cases applied the same analysis to discrimination against other ethnic groups, including the Chinese (Yick Wo v. Hopkins, 118 U.S. 356, 1886).

However, Korematsu v. United States, 323 U.S. 214 (1944) gives you reason to believe that a court might duck the strict scrutiny altogether because the military is involved. Korematsu upheld the internment of people of Japanese ancestry and, although severely criticized, has never been squarely overruled. One commentator interpreted Korematsu and a prior case, Hirabayashi v. United States, 320 U.S. 81 (1943) as demonstrating that "there are some circumstances in which the Court will, in effect, conclude that it is simply not in a position to reject descriptions by the Executive of the degree of military necessity."

Although this situation does not involve the exigencies of wartime, the Supreme Court has, in other cases, indicated its reluctance to interfere with decisions by the military (Goldman v. Weinberger, 475 U.S. 503 (1986), which declined to second-guess the Air Force's refusal to allow an Orthodox Jewish psychologist to wear his yarmulke while on duty in a hospital). You note that Goldman involved freedom of religion, which generally merits the same standard of strict scrutiny for close relation of the means to a compelling governmental end that's applied to racial discrimination. Your first challenge, then, will be to persuade a court that a suspect class is involved, that strict scrutiny should apply, and that the court should not defer merely because the military is involved.

If you get this far, you'll then need to show lack of a compelling governmental interest, or lack of a close relation between this interest and the means employed (the reorganization plan). The governmental interest claimed is, of course, remedying past racial discrimination. Regents of the University of California v. Bakke, 438 U.S. 265 (1978) and the Richmond School District Case, 109 S.Ct. 706 (1989), addressed when there is a compelling governmental interest that would justify an affirmative action program. Bakke required a judicial, administrative, or legislative finding of discrimination that violated a statute or constitution before any affirmative action program is put into place. Richmond further required that the acting body have the statutory or constitutional authority to discriminate in implementing the affirmative action. The Court noted that Congress has broad authority to discriminate and take remedial measures under Section 5 of the Fourteenth Amendment. Presumably, the executive branch would have similar authority within the Department of Defense. Finally, Richmond also required that the legislative finding show causation with ¨ concrete evidence of racial discrimination ¡ in this case ¬ that caused the noted disproportion. Thus, if the government fails to establish any of these requirements, it fails to establish a compelling governmental interest and you win the case.

Finally, the government must also defend the remedy employed. Richmond also addressed the requirements for a permissible remedy, holding that:

- the remedy must be narrowly tailored to the identified discrimination;
- the governmental body must first appraise race-free remedies before resorting to race-affirmative remedies; and
- if possible, the remedies must be directed against those who actually discriminated.

The Court commented that, in an extreme case, a narrowly tailored racial preference "might" be necessary.

At last, you begin to consider possible remedies for Duane if you succeed in overturning the reorganization plan. A first possibility is the court ordering back pay for Duane from the time of judgment to his discharge. You note that, with the complexity of the case and the inevitable appeals, it could be 5 or more years before getting to a judgment. The court might also order Duane reinstated, presumably as a team leader, which would restore his former salary. In thinking through these possibilities with your employment litigation expert, you assign a .95 possibility of the court awarding back pay if you win, but only a .40 chance of it then ordering reinstatement.

A more difficult issue is whether a court would also order Duane promoted. On the one hand, you would have a strong case after having shown violation of Duane's right to equal protection, but, on the other, the promotion would be a severe judicial intrusion into the operation of the Department. You assign a .50 probability either way. Finally, the benefit Duane gains from any reinstatement or promotion depends on how long he stays on the job. Deciding to focus on additional salary, you elicit from Duane and discretize probabilities of .1/.5/.4 that he stays there 1/4/7 years if he is reinstated and gets the promotion, with probabilities of .5/.4/.1 if he does not get the promotion. Duane was making $33,000 a year, and the promotion would be worth an additional $8,000 per year (Assume a discount rate of 10%.).

13.1. Structure the relevance diagram for the value of Duane's case. Remember to start with the value of the case and work your way down (Hint: focus on the court findings you need, rather than on the facts of the case.).
13.2. Decide which parts of the relevance diagram will go into the final tree, and which will go into subtrees.
13.3. For each subtree, assess the necessary probabilities, and compute the probabilities and mean values (if any) for plugging into the main tree.
13.4. Draw the main decision tree. Plug in the information from the subtrees and assess any additional probabilities needed. Write in the values for each scenario, assuming that, because it's a discrimination case against the government, the court will award you your legal fees.
13.5. What is the value of Duane's case? What is the probability you will win the case? Lose the case? What should your settlement strategy be?
13.6. Plot the cumulative probability distribution for the value of the case. What does it tell you about settlement strategy? How does it reflect the legal structure of the case?
13.7. Assume that the government is will to settle for a cash payment of $33,000 right now (1 year's salary). Should you recommend that Duane accept it? How comfortable are you with that recommendation, and why? (That is, describe where you feel the analysis so far has the greatest deficiencies, and how should it proceed from this point.)
13.8. Using the government offer as the alternative, compute the values of uncertainty and control for the uncertainties in the main tree. What is the order in which you should research the issues? For which issues shouldn't you do any research at all? What should your discovery strategy be?

7.7.3 Discovery Strategy

14. Discovery includes requesting documents, deposing parties, retaining experts, etc. How would value of information calculations inform how much should be spent on these?
15. Values of information can be calculated on legal as well as factual issues. How should the value of perfect information on legal issues be figured into case decisions?
16. How might the value of perfect control figure into case decisions regarding factual matters?
17. How might the value of perfect control figure into case decisions regarding legal matters?
18. What is the difference between the values of perfect information and control as applied to:

    18.1. Factual issues?
    18.2. Legal issues?

19. Besides factual and legal issues, what other issues in a case might values of information and control be applied to?
20. The value of perfect information is zero if it won't change any decisions.

    20.1. Why should the value of perfect control always be greater than zero?
    20.2. If you calculated the value of perfect control to be zero, what would that indicate?

21. List at least one way to handle differences of opinion as the prospects on uncertainties in a case.
22. Assume that in a simple analysis, there is a settlement offer of $50,000 is on the table, the chance of liability at trial is 60%, and the 25%/50%/25% range on damages if found liable is $50,000/$100,000/$150,000. For a mock trial which has a 50% chance of accurately predicting liability at trial, show that the value of information is zero.
23. You are the chief counsel for a group of former shareholders of Grease Ferrets, Inc., an oil exploration company. The shareholders sold their stock to TerraFuel when that company bought out and consolidated with the Ferrets. Soon after the consolidation, TerraFuel made a major strike on properties formerly owned by the Ferrets. The shareholders contend that the tender offer jointly issued by the boards of Grease Ferrets and TerraFuel significantly understated the potential value of the fuel, thereby inducing the shareholders to accept an unfairly low price for their stock. Accordingly, you are suing TerraFuel for fraud under § 10 of the 1934 Securities Exchange Act and Rule 10b-5 thereunder.

    The requirements to prevail in a 10b-5 action are: (1) that the plaintiffs be buyers or sellers of the stock; (2) that there was a violation of the Act by a purposeful or reckless misrepresentation of a matter that the company had a duty to disclose; (3) that the misrepresentation was material; and (4) that the buyers or sellers can show reliance on the misrepresentation and damages.

Although the first requirement is clearly met, the latter three are more problematic. Given how speculative the valuation of oil properties is, it will be difficult to establish that the boards purposely or recklessly misrepresented the value of the property. At the time of the sale, the Grease Ferrets had promising seismic results and two encouraging core samples, but had not actually struck oil. Accordingly, you feel there's only a .4 chance of establishing the necessary misrepresentation.

The remaining two elements are more clear-cut. Given the context of a tender offer, there was a clear duty to disclose information on the value of the Ferrets' assets. You therefore feel there's a .9 chance of the court finding a duty to disclose. Regarding the element of materiality, you need only establish that the undisclosed information was material to the market as described in S.E.C. v. Texas Gulf Sulphur (2d. Cir. 1983). Although the value of exploration properties would reasonably affect the value of the company's stock, the tender statements were optimistic about the properties, and it is a little less certain that a more optimistic statement would have increased the stock value. You therefore assign a slightly lower .85 probability to establishing this element.

As for damages, the value of TerraFuel stock increased by $50 million with the announcement of the find, and you are seeking this amount in damages. TerraFuel has offered $2 million to settle the suit, and you estimate that it would cost another $500,000 to try the case.

One more possibility worries you. If the court finds against you on all three issues, you feel the court would award costs to TerraFuel, which would cost the plaintiffs an additional $500,000.

23.1. Draw the relevance diagram for the value of the case.

23.2. Draw the decision tree for the decision on whether to settle or continue the suit. Calculate the values for each scenario.

23.3. What is the probability of winning the suit? Of losing the suit?

23.4. Calculate the mean values. On a mean value basis, should you accept the settlement?

23.5. Calculate the values of information and control on the uncertainties. Which one has the highest value of information, and the highest value of control? Can you tell why it's the highest by examining the tree?

23.6. Suppose you have the opportunity to undertake more discovery and research on the issue of misrepresentation. You estimate that this action would cost an additional $100,000, and that it has an 80 percent chance of correctly determining whether there is misrepresentation. That is, given that the court finds misrepresentation, there is a .8 chance the study will find misrepresentation, a .2 chance it will find no misrepresentation, and just the reverse if the court will find no misrepresentation. What is the probability of the court finding misrepresentation, given a prediction of misrepresentation from the discovery? Given a prediction of no misrepresentation?

23.7. Should you undertake the discovery? (Hint: consider whether you will need to flip the probabilities for the tree.)

23.8. Given that you undertake the additional discovery, what is the probability of winning the case? Of losing the case?

23.9. What is the value of perfect information on misrepresentation given that you undertake the additional discovery? Is it different from the initial value of information, and how do you explain this result?

7.7.4 *Trial Strategy*

24. List at least three possible trial strategy decisions which can be aided by applying decision analysis.
25. Are there any trial strategy decisions which cannot be aided by applying decision analysis? Why or why not?
26. What is the difference between how trial strategy decisions are typically made and a decision analysis approach? Draw on your own experience or speculate.
27. You're representing an aging plaintiff who was represented for a long time by another lawyer before that lawyer died. The plaintiff now would like to revoke an irrevocable living trust, claiming that he never intended to make it irrevocable and that he misunderstood what his lawyer was doing. You see no particular problem since a unilateral mistake is generally sufficient grounds for a settlor to revoke a trust. Unfortunately, the memory of this settlor is fading, and he remembers the circumstances surrounding the trust creation with varying degrees of clarity.

    While the plaintiff is on the stand, you could refresh his recollection with an early draft of the trust instrument on which he wrote "Go ahead Bill. I'm not going to bother reading this." However, you're worried that the defendant trustee's counsel might insist on admitting the document itself into evidence, which would allow the jury to see also how the plaintiff went on to comment on a rather technical point of the trust instrument. Accordingly, you believe that your .6 probability of winning without refreshing the plaintiff will rise to .75 if you refresh and the defendant does not admit the draft, while it will decline to only .50 if the defendant does admit it. You think there's a .5 chance the defendant counsel will admit the document, and the plaintiff will be out $10,000 in court costs if he loses, versus gaining the $30,000 trust corpus. The plaintiff has already paid $10,000 in legal fees.

    27.1. Draw the decision tree and calculate the mean value for each alternative. Should you use the draft to refresh the plaintiff?

    27.2. Plot the cumulative probability distributions for both alternatives. For each alternative, what is the plaintiff's probability of gaining money, and or losing money?

    27.3. Draw the histogram for each alternative. Do the histograms highlight the differences between the alternatives more clearly? Do they convey any additional information?

27.4. What are the plaintiff's probabilities of winning and of losing, given that you've chosen to refresh? Compare these to the probabilities given that you've chosen not to refresh. Does this explain the difference in the mean values?

27.5. Can you calculate the value of information on winning the case? Why or why not?

27.6. Can you calculate the value of information on whether the defendant counsel admits the draft? Either explain why not or calculate the value.

27.7. If your answer to (f) was no, explain why you can't calculate the value of control either. If yes, then calculate it.

28. You're the chief patent counsel for Helixir, Inc., a biotechnology company. Dr. Pauline Peptide, vice president of research, has been working with you on the patent application for Helixir's immune-booster project. Helixir has isolated a molecule that seems to boost generally immune system response without inducing allergy-like symptoms. The latest phases of the project have concentrated on binding the molecule into a lipid(fat)-complex that slows absorption and thereby mimics the natural production of the molecule by the body. The question now is whether this molecule can be patented.

You note that the prospects for patenting naturally occurring molecules are dim. However, synthetic molecules, such as this one when married to the lipid complex, stand much better chances. Further, Dr. Peptide has indicated that efforts are now concentrating on isolating the active segments of the molecule. This effort, if successful, would greatly ease production problems because only a much smaller molecule need be produced.

The question facing you, then, is whether to proceed with a patent application for the full molecule, or for the molecule fragment being developed (Simultaneously applying for both is not practical because Helixir lacks the resources to develop both.). Given your legal research, you estimate that there is a .6 chance of patenting the entire molecule, but a .8 chance of patenting the partial molecule because it is more distinct from the naturally occurring form. Dr. Peptide has assessed probabilities of .9 that the full molecule will be active in clinical trials, and .7 that the partial molecule will be active. Further, because small variations from the entire molecule may still be active, you estimate that there is only a .8 probability of enforcing the patent to exclude other similar immune boosters, while the partial molecule offers a .9 chance of excluding similar molecules because it is smaller and more specific.

The marketing department tells you that excluding similar, competing molecules should garner a 75% share of the $500 million projected immune booster market, while failing to do so will drop the market share to 25%.

Dr. Peptide briefed you on one more wrinkle. Because of its greater complexity, it will cost something like $150 million to develop the full molecule to the point where it can be tested for activity, while the shorter partial molecule would only require $100 million. Both of you agree that failure to obtain the patent or failure to find activity in clinical trials would certainly lead to abandonment of the project (getting creamed by overseas duplicators, no therapeutic value, etc.).

28.1. Draw the decision tree for which form of the immune-booster to apply for the patent on (Hint: consider whether, for instance, patent approval for the whole and partial molecules should be represented by different chance nodes, or by the same chance node with dependent probabilities. What are the events in question?)

28.2. What is the highest mean value alternative?

28.3. Plot the cumulative probability distributions for both alternatives. What do they tell you about the relative risk between the alternatives?

28.4. Calculate the values of information and of control for all the uncertainties. What do they tell you about conducting further legal research, or about obtaining a letter of opinion from the US patent office? What do they tell you about getting more information from research about which form of the molecule is likely to be active?

29. About 9 months ago, Joe Cinelli began to experience severe knee pain during his weekly outings with the Weekend Wheelers bicycle club. At first, he took aspirin and tried to ignore it, but, as it became steadily worse, he went to see Dr. Rapier, an orthopedic surgeon. Dr. Rapier diagnosed the problem as torn and inflamed cartilage and recommended arthroscopy (surgery via tiny clippers inserted through a long, thin tube) rather than traditional open-incision surgery. Arthroscopy had the advantages of only requiring a local anesthesia and generally being less traumatic to the knee, with commensurately shorter recovery times.

Unfortunately, the arthroscopy was not successful and produced unusually severe inflammation and pain. Joe was unable to walk on the leg at all, and, after 3 months of being laid up, went to see another surgeon. The second surgeon opined that arthroscopy was not appropriate in the first place, and that, as performed, it only aggravated Joe's condition. When Joe discussed this second opinion with Dr. Rapier, Rapier became very angry, and insisted that Joe's condition was an unusually severe but not unexpected reaction to the surgery.

Unsatisfied with Rapier's response, Joe went to the other surgeon and allowed him to perform traditional open-incision surgery. Although Joe was then in a cast for some weeks afterward, his knee has steadily strengthened since its removal. Joe is now walking with some difficulty and continuing with physical therapy.

Joe has come to you regarding a possible malpractice suit against Dr. Rapier. From what Joe told you, you feel there are two possible theories for seeking damages. First, there is the possibility of a traditional malpractice suit.

To establish malpractice, you would have to show: (1) that Dr. Rapier failed to exercise or to possess the degree of skill and learning of an average orthopedic surgeon; (2) that the patient suffered injuries; and (3) that the injuries were the result of failing to meet the standard of care.

From your research into the area of arthroscopy on knees and your conversations with other experts, you feel there's a .6 chance of establishing that Dr. Rapier failed to meet the standard of care. You put the probability so low because of the difficulty of establishing the standard of care for the surgery in question. However, you feel there's .8 chance of showing causation, should the jury find failure to meet the standard of care. The only remaining question, then, is damages, and your assessment of the possible awards produced discretized damages possibilities of .25/.50/.25 on $10/40/100 thousand.

Alternatively, you could try the case on an informed consent theory. To recover on this basis, you'd have to show: (1) that Dr. Rapier failed to inform Joe of a material risk before obtaining his consent to the arthroscopy; (2) that if Joe had been informed of the risk he would not have consented to the treatment; and (3) that the undisclosed adverse consequences did occur and Joe was injured as a result of submitting to the treatment.

You feel that these elements are less problematic than establishing the standard of care and, accordingly, have assigned a .8 to obtaining a finding in your favor on the first element, a .9 for a finding on the second, and a .85 for a finding on the third. However, because you feel the jury won't find Rapier's conduct as reprehensible under this theory as under malpractice, you damages distribution has only produced possibilities of .25/.50/.25 on $10/25/50 thousand.

You have considered trying both theories in the alternative, but are worried it might confuse the jury and hurt your persuasiveness if you take more than one position on what Dr. Rapier did wrong. Consequently, you've decided to evaluate the theories separately for now.

29.1. Draw the decision tree for which theory to try, filling in all the information.

29.2. Calculate the mean value for each alternative. Which theory should you choose on a mean value basis? What should your settlement strategy be, given that you have two possible theories to try?

29.3. Calculate the values of information and control for the uncertainties relevant to the malpractice theory alternative. Does the uncertainty with the highest value of information also have the highest value of control? Explain the relative sizes of the values of information and control for the malpractice uncertainties (that is, why uncertainty X has the highest value of information, but not of control, and so on). What implications does this have for your pretrial discovery planning?

29.4. Consider the values of control you just calculated. You could, for instance, control an issue by trading a stipulation on that issue for another stipulation, or for a reduction in the requested damages. Ignoring the informed consent option for now, assume you could get a stipulation on the issue with the highest value of control by halving your requested damages (reducing your probability distribution on damages by half). Should you do it? (Halving your damages could, for instance, be interpreted as one or more particular damages claims.)

29.5. Calculate the value of information and control for the consent theory alternative and explain the relationships among the values (as before), but also relative to the malpractice values of information and of control. What does this tell you about how the alternatives compare to one another?

29.6. (Optional) Plot the cumulative probability distribution for each alternative. For each theory, answer the following questions:

29.6.1. What is the probability of losing the case (of a zero outcome)?
29.6.2. What is the probability of winning the case (a greater than zero outcome)?
29.6.3. What is the probability of obtaining a judgment less than or equal to $10 thousand?
29.6.4. What is the probability of obtaining a judgment less than or equal to $25 thousand?
29.6.5. What is the probability of obtaining a judgment less than or equal to $40 thousand?
29.6.6. What is the probability of obtaining a judgment greater than $60 thousand?

29.7. How are the values of information and control reflected in the cumulative probability distributions and in your answers to part 11.6?
29.8. Plot the sensitivity to the probability of the high and low outcomes for the damages under the malpractice theory (that is, ignore the middle branch for now). Do the same for the damages under the consent theory. What minimum probability for high malpractice damages is required to make that theory preferred? What minimum probability for high consent damages makes that theory preferred? What does that tell you about which is the stronger theory as represented in this problem?

Part III
Advanced Decision Analysis Topics

1.1 Note on Use of Part III

As you probably noted from the Introduction and from the table of contents, this section differs from Parts I and II because the discussion does not closely follow the steps of a Decision Analysis. For convenience, the chapters have been ordered according to the stage of the process they focus on the most, but, typically, each discussion will apply at several stages of a Decision Analysis. This section is therefore more of a reference than a narrative. You should read (or at least skim) through it once to familiarize yourself with the topics discussed and to learn how to apply, and then refer back to the specific topics as they arise in your own analyses.

If nothing else, be sure to read Chapter 13 on Communication. It will help a lot in credibly explaining what should be done and why for the decisions you've analyzed.

Chapter 8
Difficult Questions to Structure

This chapter discusses a number of questions that are difficult to structure and are usually left out of a first-pass analysis, either for that reason alone or because they also usually don't have a large effect on settlement strategy (like legal fees). However, occasionally these questions arise where you need to include them in the first-pass analysis to help you think about the value of the case, or the first-pass analysis itself may indicate a need to include them. If so, this chapter gives ways of structuring the questions that work for most cases. As always, be alert for where a different structure would better reflect the way you think the trier-of-fact will decide your case.

8.1 Complicated Damages

The simplest type of damages, of course, is when you feel comfortable directly assessing the possible answers for total damages because the jury is simply going to decide on a single number if they've found liability. Following are a number of possible variations on how the jury may decide on damages.

8.1.1 Sum of Two Kinds of Damages

In a number of causes of action, total damages are more typically the sum of two numbers the jury decides on separately. For instance, in the hypothetical in Part II, total damages were the sum of compensatory damages and punitive damages (if liability was on the basis of intentional misrepresentation, rather than negligent). Another common example is wrongful discharge cases, where damages are the sum of back pay and future pay (unless reinstatement is ordered).

J. Celona, *Winning at Litigation through Decision Analysis*,
Springer Series in Operations Research and Financial Engineering,
DOI 10.1007/978-3-319-30040-5_8

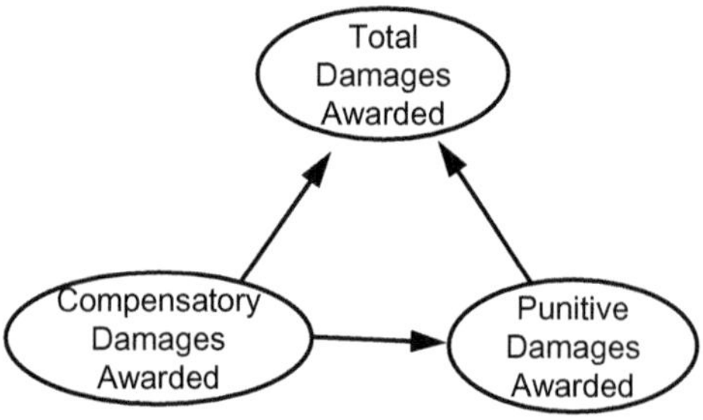

Figure 8.1 Relevance diagram for the sum of two types of damages

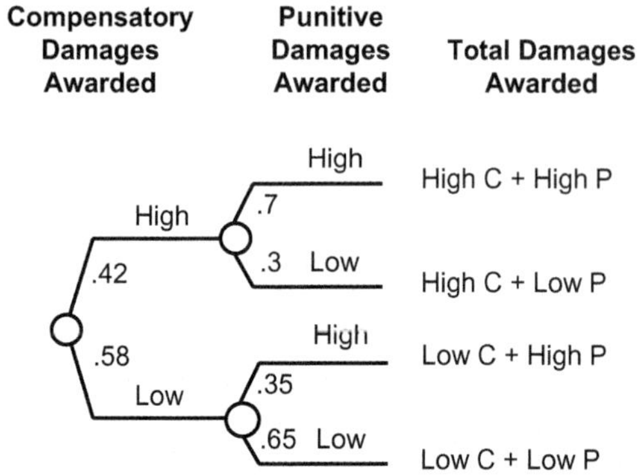

Figure 8.2 Probabilities assess for sum of two damages

For these kinds of damages, you should, as always, do your analysis the way the trier of fact will decide it: as two separate numbers. This will at least require two separate questions in your relevance diagram and two chance nodes in your tree, and be alert for interactions between the two questions (perhaps expressed as dependent probabilities).

For instance, a partial relevance diagram for compensatory and punitive damages might look like the following (Figure 8.1).

This relevance diagram expresses a belief that the amount of punitives awarded will be affected by the amount of compensatory damages. For instance, you might think the jury more likely to award punitives if they don't feel the compensatories came out high enough for the injury sustained, or you may feel that high compensatory damages mean the jury is really against the plaintiff and they're more likely to award large punitives on top of that. This latter situation is illustrated with the dependent probabilities in the tree below (Figure 8.2).

Note that there are a number of other variations for structuring damages that are the sum of two kinds of damages.

- The tree above assumes the jury will award some punitives, or the low punitives includes the probability of zero punitives. You may want to have a "no punitives" branch in your chance node.

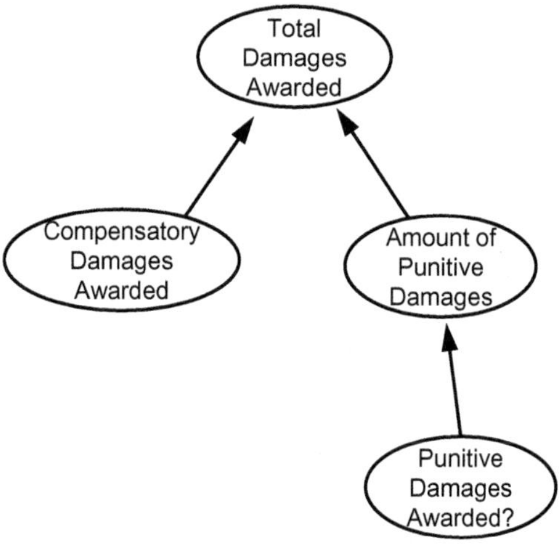

Figure 8.3 More complex relevance diagram for sum of two damages

- You may think the jury will first decide yes or no on punitives at all, and then determine an amount if the answer was yes. This would require an extra oval in your relevance diagram and an extra node in your tree.
- As always, you have a number of different methods for doing a continuous variable like damages. You might want to do the three-branch shortcut instead of the two-branch used above, for instance.
- The tree above can be done as a subtree, rather than as part of the main tree, if it isn't affected by other questions in the tree (see the section on subtrees for more details).

For instance, in the Howitzer example, there was a question of whether or not punitive damages would be awarded, but it was thought that the amounts of punitive or compensatory damages would not affect each other. Here is the relevance diagram for that portion of the analysis (Figure 8.3).

The assessed tree from the Howitzer case appears below (Figure 8.4).

Not that, for this example, we assessed and discretized the full probability distribution on punitive damages. A 25%/50%/25% assessment would also be fine for a first-pass analysis.

8.1.2 A Share of the Verdict

Another variation on damages is where, for some reason, you may only have to pay a share of the damages awarded by the jury (if any). This would happen, for instance, where you had a codefendant, or in a jurisdiction with a pure comparative negligence system.

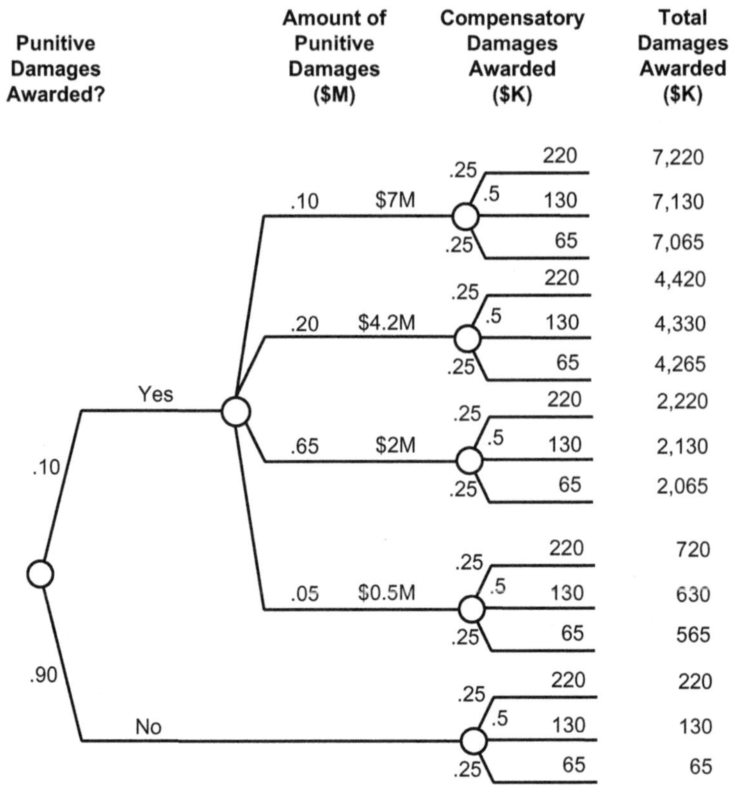

Figure 8.4 Assessed tree for sum of two damages with punitive damages question

First of all, you need to keep three questions distinct.

- Whom will the jury find liable?
- How much damages will the jury award?
- What will our share of those damages be?

The tree below explicitly handles the first and third of these questions, and assumes that, if the jury finds anybody liable, they will determine the amount of damages independently of whom they find liable (You could, on the other hand, believe that the jury is likely to award higher damages when they find two or more parties liable than when the find one party liable.).

Accordingly, the tree posits that the jury will first decide on your liability, then on your codefendant's, and only then will determine the split if it finds both of you liable. The question of the split itself assumed that the jury is either going to settle on an 80/20 either way, or will just split 50/50. It would have been equally valid to come up with the 80/20 by assessing the continuous distribution on your share of the liability and then discretizing it.

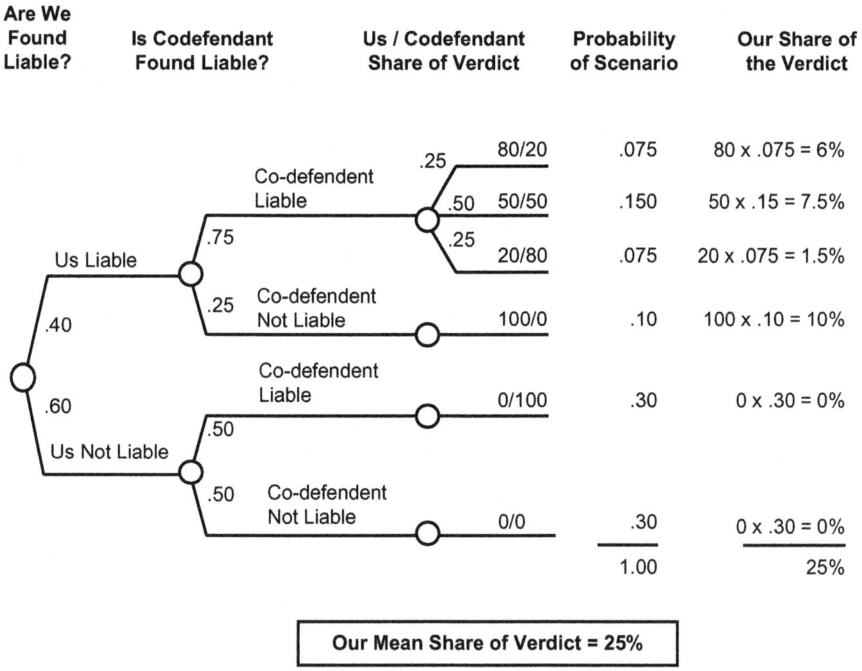

Figure 8.5 Tree for our share of a verdict with a codefendant

Determining your share of the verdict, then, is simply a matter of multiplying the probabilities of each scenario by your share in that scenario. The resulting products are added up to produce your 25% expected share of the verdict (Figure 8.5).

Thus, if your separate damages tree had produced a mean damages award of \$70,000, your expected share of that award would be $.25 \times 70 = \$17{,}500$. Accordingly, on an expected-value basis you should be willing to settle the case for anything less than \$17,500 (Note that when you do this last multiplication you don't have to worry about the scenarios in the above tree where you have no liability at all because those scenarios contribute 0% to your expected share of the verdict.).

As with all these questions, there are multiple ways to structure the problem. You could have also assessed a 25%/50%/25% range on our share of the verdict if found liable and then calculate the codefendant's share at one minus that if they are also found liable.

8.1.3 A Multiplier on Damages

For a variety of cases, the final damage award is calculated by multiplying some "actual" damages figure by a specific number. These include antitrust cases (treble damages), willful infringement in patent cases, and employee cases. Typically, the

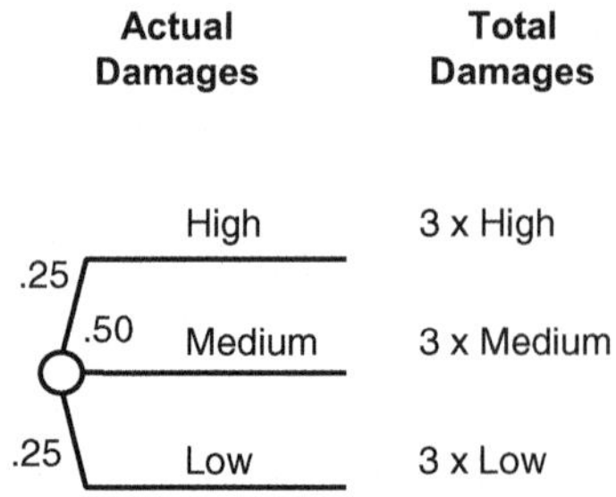

Figure 8.6 Calculation for a multiplier on damages

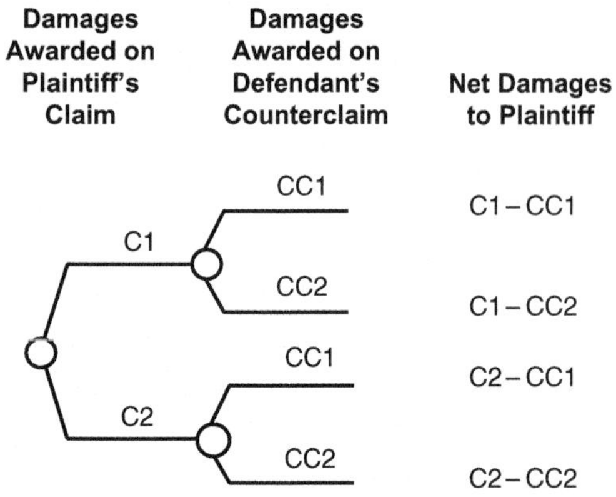

Figure 8.7 Tree and calculation for an offset to damages

way to handle these kinds of damages is simply by assessing the actual damages the jury finds and then multiplying that amount in your value formula, as below (Figure 8.6).

However, the way you handle multipliers on damages could vary according to how you think the jury will handle them. For instance, if you believed the jury would be second-guessing the statute by looking to what the trebled amount will be—rather than the "actual" amount—and by adjusting their award accordingly, then the appropriate answer for you to assess would be the final damages award (without further multiplication). As always, the way you structure the question should depend on how you think the trier-of-fact will decide it.

8.1.4 An Offset to the Award

For certain kinds of cases, the value of the case to either the plaintiff or the defendant can be affected by an offset to the award. For instance, from the plaintiff's perspective, if a counterclaim has been consolidated with her suit, then her final recovery will be the amount of damages awarded on her claim minus the offsetting counterclaim award (if any). As when total damages is a sum, this situation can be handled with two chance nodes in your tree and a value formula, as below (Figure 8.7).

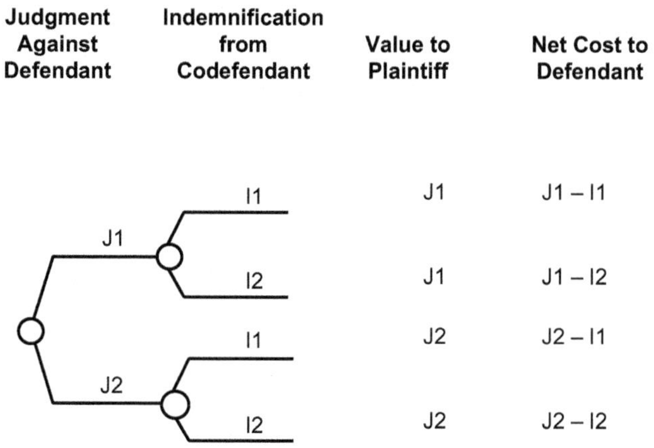

Figure 8.8 Tree for indemnification for a judgment

From the defendant's perspective, she may implead a codefendant and obtain partial or complete indemnification. Thus, the value of the case to the defendant could be represented as below (Figure 8.8).

Note that, assuming a solvent defendant, the plaintiff shouldn't care about how indemnification turns out, in contrast to the plaintiff's tree above where the value to the plaintiff and the defendant should be the same. Thus, the defendant's indemnification tree is one instance where analyzing from your opponent's point of view may produce a different case value than yours—with both case values technically correct.

As always, you also need to consider whether there will be any spill-over between the damages awarded and the offset to it. This will likely not be the case where the offset is solely a matter of allocation between the codefendants and will not affect the relative fault of the plaintiff in the jury's mind. However, where claims and counterclaims are at issue, you may feel your success on a claim will be affected by your opponent's success on a counterclaim. This might be true, for instance, where the claim and counterclaim are legally or factually related, such claim for failure to pay under a contract and a counterclaim for material failures in performance (that is, the jury may award you less money if they feel you've not rendered substantial performance). This situation might be represented by the relevance diagram below (Figure 8.9).

Note that, as mentioned before, for this particular problem it is not necessary to distinguish between the value to the plaintiff and the value to the defendant unless there are other judgment allocations (such as the insurance angle—next topic). And, of course, this relevance diagram posits that the trier-of-fact will first determine what the binding contract terms are (covering formation and interpretation issues, etc.), then whether the plaintiff rendered substantial performance, and then whether the defendant failed to make a due payment (The contract terms question also shows that the amounts due each party are separate questions from substantial performance and payment due. Each should separately pass the clairvoyance test, though the contract terms question fudges a bit.).

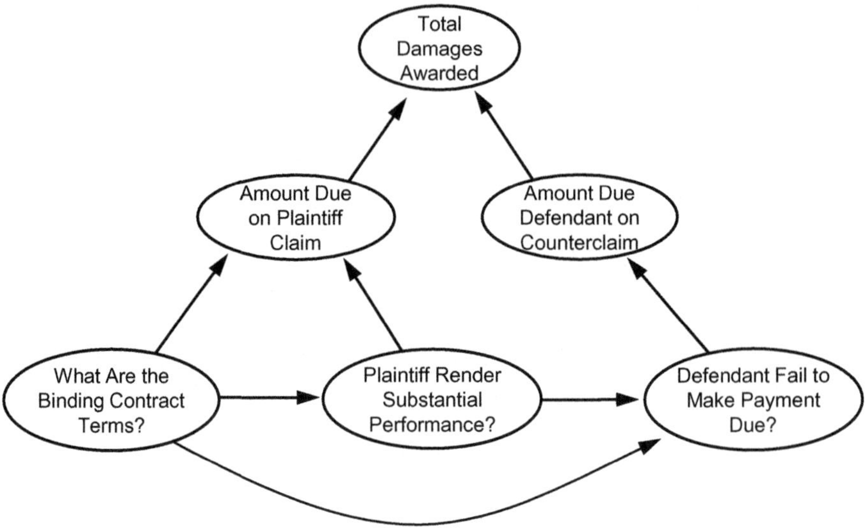

Figure 8.9 Related valuations of claim and offset counterclaim

Your relevance diagram will likely vary according to how you feel the trier-of-fact will relate the claim and counterclaim. For this diagram, the arrows, at least, should indicate dependent probabilities in the tree. Can you draw the decision tree for this relevance diagram?[1]

8.1.5 Insurance or Reimbursement

Insurance or reimbursement introduces another wrinkle on the allocation between the codefendants problem. As before, the value of the case to the plaintiff may be unaffected, or, as sometimes happens, the plaintiff's settlement demand may be framed in light of the defendant's insurance coverage. Thus, from plaintiff's perspective, the value of the case may look something like the following (Figure 8.10).

Note the double-ovals for the amounts recovered from the defendant and its insurer. As you'll recall, double-ovals indicate either significant facts (which are not uncertain) or calculations. In this case, the amount recovered from the defendant's insurer is the judgment minus the deductible up to the policy limits. The amount recovered from the defendant is the judgement minus the policy limit plus the deductible up to however much is actually collectible from the defendant (considering willingness to pay, assets susceptible to judgment liens, etc.). The value of the case to the plaintiff is still shown as uncertain in this diagram because it doesn't include the issues bearing on a finding of liability.

[1] Hint: you'll first need to rephrase the contract terms question so that it does pass the clairvoyance test. Then, as usual, start from the bottom and follow the arrows.

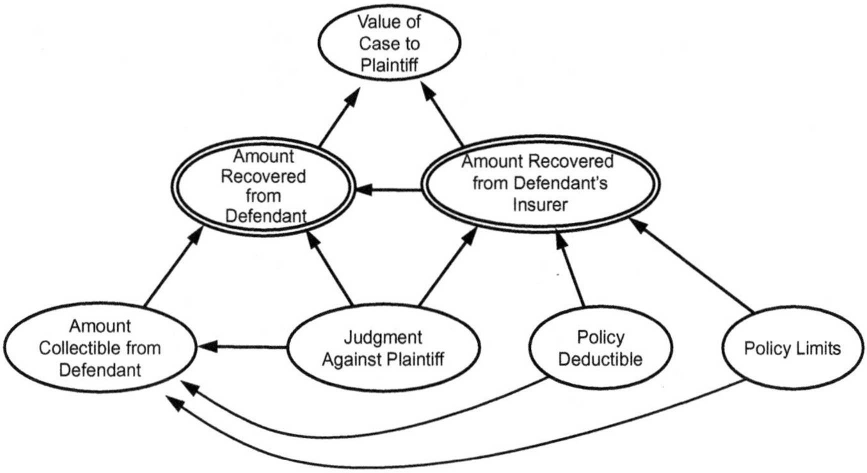

Figure 8.10 Relevance diagram for value of a case with insurance coverage

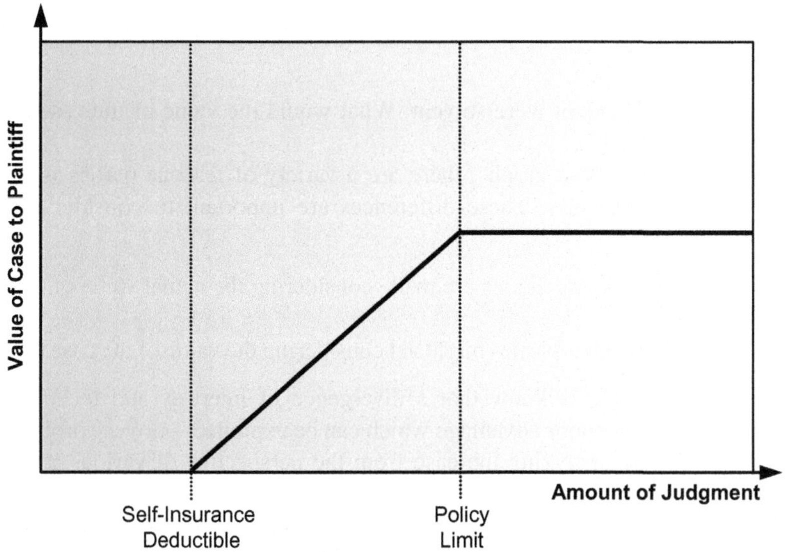

Figure 8.11 Value of case to the plaintiff with the insolvent insured defendant

In the extreme case, then, of an insolvent defendant, the value of the case to the plaintiff would look like the following graph, where the case is worth nothing to the plaintiff until the judgment has exceeded the self-insurance deductible, and gains no further value after the policy limits have been reached (Figure 8.11).

Note that this graph plots the *value* of the case—the numbers in the formulas at the end of the tree—and not the *mean value* because the uncertainties on obtaining a judgment are not yet taken into account.

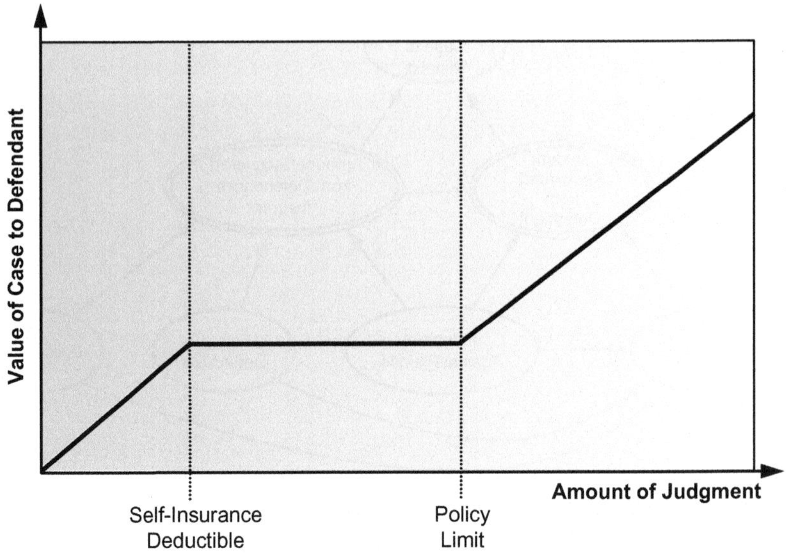

Figure 8.12 Value of case to the solvent insured defendant

Suppose the defendant were solvent. What would the value of the case to him look like? (Figure 8.12).

As illustrated in these graphs, there are a variety of reasons parties to a case would value it differently. These differences are important to consider for two reasons:

- To ensure your own decisions are made considering the actual value of the case to you.
- To examine how other parties might act considering the value of the case to them.

Experienced litigators know that a divergence of interests and incentives of opposing parties is a major advantage which can be exploited—or very problematic if on your side. By analyzing the case from the perspective of various parties (a topic we'll revisit throughout this book), you can create effective strategies to divide and conquer. These methods of looking at a situation considering divergent decisions and interests of various parties (including uncertainty in these) are among the foundations of game theory.

8.2 Legal Fees and Costs

Legal fees and costs (expert witnesses, document production, investigators, etc.) can be quite significant, in some cases larger than the potential indemnity. However, in most cases, both the amounts and uncertainty in legal fees and costs are much smaller than the potential indemnity. Accordingly, we have focused first on quantifying the uncertainty in exposure if a case proceeds to judgment and on determining case strategy. Once you're comfortable with analyzing and developing these for a matter, it absolutely makes sense to include legal fees and costs in quantifying total exposure and setting case strategy. Following are some considerations which apply to analyzing values in general and often come up when considering legal fees and costs, followed by some methods of including them.

8.2.1 Don't Include Sunk Costs

Properly handling legal fees and costs first requires understanding why sunk costs are best not included in your analysis. Sunk costs are monies that you've already spent (that is, they've already been "sunk" into the case), and, typically, they are not recoverable. A typical oil-drilling example will illustrate the dangers of including sunk costs.

Suppose you've already spent \$2 million drilling for oil and have so far come up empty. You now have the opportunity to spend an additional \$1 million and, because of extraordinary reports, are assured of getting a mean net present value of \$2.5 million worth of oil if you do so. Should you spend the money?

If you lump all the costs together, it looks like you're spending \$3 million to gain \$2.5 million, for a loss of \$.5 million. With this kind of perspective, it looks like you should quit right now.

However, if you recognize that the \$2 million is already-spent and therefore sunk cost (water under the bridge), then your prospective opportunity looks like the following (Figure 8.13).

In fact, then, looking at future expenditures and receipts means that you're facing a choice between the status quo (zero) and spending \$1 million to get \$2.5 million, for a profit of \$1.5 million. Investing the incremental \$1 million makes economic sense, and you should do it (or sell the chance to somebody else). Thus, the safest way to make sure your decisions make economic sense is to leave sunk costs out of your analysis altogether.

All the same, if you are really puzzled at simply ignoring the sunk costs altogether, there is a slightly more subtle way to include them that also makes economic

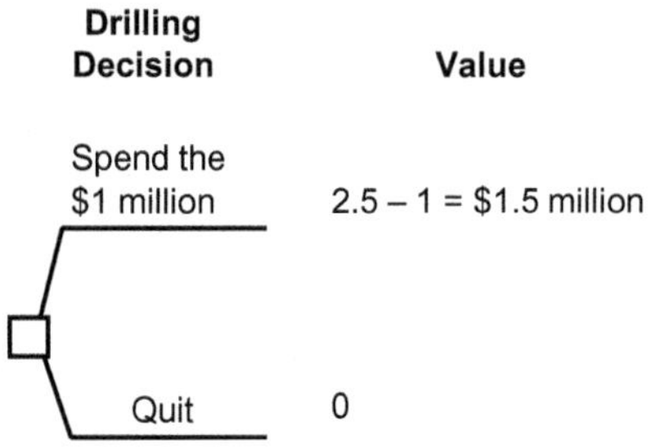

Figure 8.13 Oil well drilling decision focusing on incremental cost and value

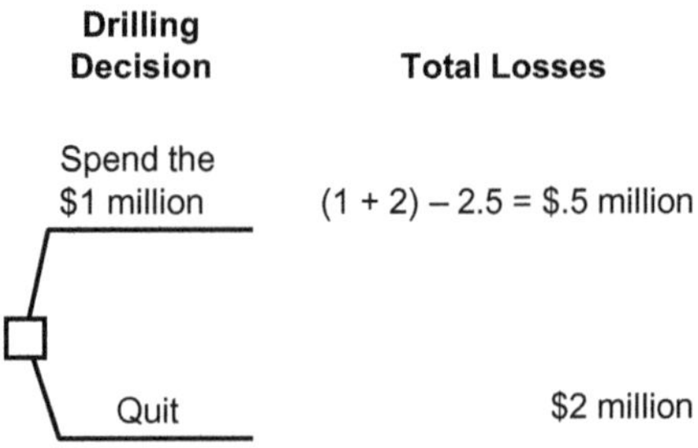

Figure 8.14 Drilling decision including sunk costs, but minimizing total losses

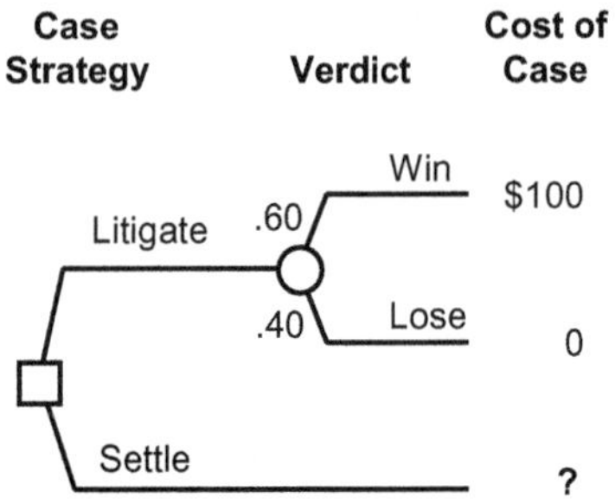

Figure 8.15 Include sunk legal costs?

sense. If you frame your tree to minimize your total losses, you see that another way to express the decision is as a choice between losing $2 million and losing $0.5 million, as shown below (Figure 8.14).

Thus, if you properly include the sunk costs, you would still choose to spend the extra $1 million because that minimizes your losses—reducing your losses from $2 million to $.5 million.

Put either way, the object is still to assure that future expenditures make sense on an *incremental basis*—future returns justify future investments. Sunk costs are irrelevant to this determination. You can see this in the tree above: sunk costs appear in the value no matter what you do, so they don't help to tell you which alternative is better. They're like a question in the tree that doesn't affect any probabilities or values elsewhere in the tree: they don't matter in thinking about what you should do *next*.

These same principles apply when you're including legal fees in your analysis. For instance, suppose you're defendant and you've already expended $50,000 in legal fees. You're facing a .60 chance of losing at trial, for a judgment against you of $100,000, as shown below. How should you think about a good settlement amount? Should you not settle for anything over $50,000 because you've already spent that much in fees and the claim is for $100,000? (Figure 8.15).

Actually, the tree gives away the answer by omitting the sunk legal costs altogether (the safest way to do it). Thus, the mean value of litigating the case is $(.6\times 100)+(.4\times 0)=\$60{,}000$, so you should be willing to settle the case for anything up to $60,000.

Figure 8.16 Sunk costs included in all scenarios with minimizing total losses

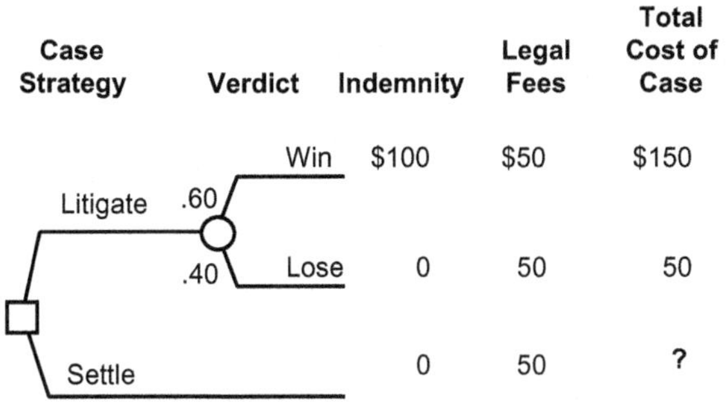

Figure 8.17 Case value from plaintiff's perspective

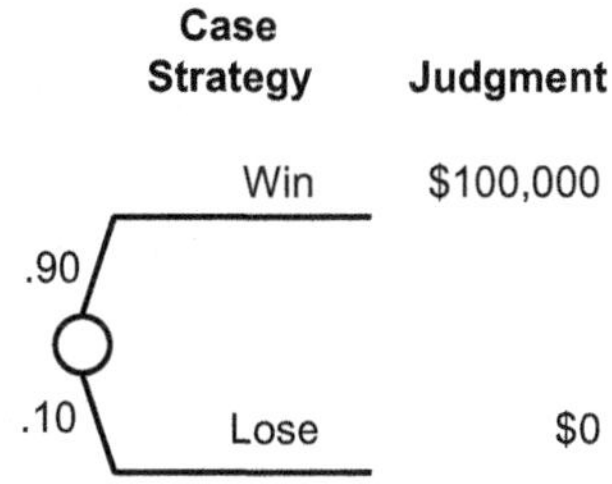

As noted above, the sunk costs can be properly included by adding them into every scenario, and doing so won't change your settlement strategy (Figure 8.16).

In this formulation, the mean value of the case is $(.60 \times 150) + (.40 \times 50) = \$110,000$. Thus, you should be willing to settle the case for anything up to $110,000, and this figure breaks down into your $50,000 sunk costs and an additional $60,000 you should be willing to pay. In other words, the answer is the same. However, note in particular that offering only $10,000 (the $60,000 mean value of litigating minus your $50,000 sunk costs) is only going guarantee you a trial.

Two illustrations from plaintiff's perspective will serve as final examples of sunk cost fallacies. Suppose you value your case as follows (Figure 8.17).

Should you accept a settlement offer of $40,000 for this case? Clearly, the answer is no: the offer is not even half your mean value of $90,000. Suppose further that your sunk legal fees on the case are $20,000. Does it make sense to accept an offer of $40,000 because it's twice your sunk legal fees? Again, the answer is no; the proper perspective is looking forward to your prospective expected case value of $90,000, not backward to the $20,000 spent. Why accept $40,000 when you can "get" $90,000?[2]

[2] If you wanted to be very detailed and consider future legal, you could make your best of guess of how much more it would cost you to litigate the case and subtract that estimate from the mean value of litigating, and then compare that figure to the $40,000. By this reasoning, you should still only accept the $40,000 settlement offer if your expected future costs of litigating are greater than $50,000—for total legal fees of $70,000 on a $100,000 claim. Not the usual case, by any means. See the discussion in the next heading for more details.

Figure 8.18 Case value from plaintiff's perspective with different odds

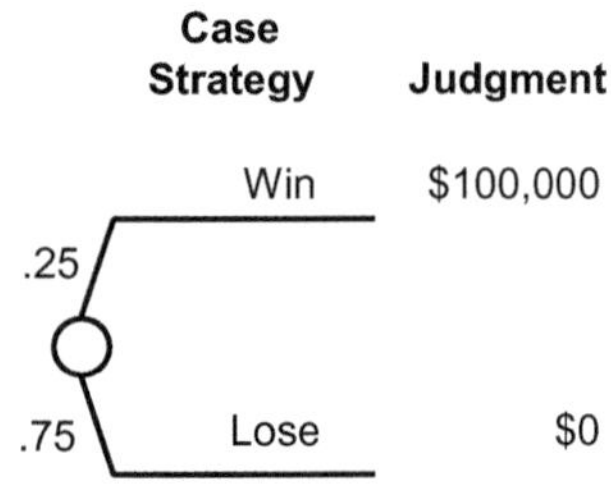

Alternatively, suppose you valued your case as follows (Figure 8.18).

Now should you accept an offer of $40,000 to settle this case? Clearly, the answer is yes: the offer is greater than the $25,000 mean value of litigating this case. Now suppose that you sunk costs are $60,000. Should you accept the offer even though it's less than your sunk costs? Again, the answer is yes: accepting the offer leaves you in a final position of 40 – 60 = –20, or with a net loss of $20,000 on the case, while refusing the offer means you're choosing a mean value of 25 – 60 = –35, for a mean loss of $35,000.[3]

Thus, whether you choose to omit sunk costs (the easy and recommended method), or to include sunk costs and seek to minimize your losses (the difficult way), the mean value clearly shows that sunk costs don't make any difference in what your settlement strategy should be.

8.2.2 *Add or Subtract Legal Fees and Costs to an Assessment from Total Case Value*

Now that you've a clearer idea of which legal fees to include in your analysis at all, you can better consider the amount and variability of possible future legal fees and decide which method for handling them suits you best.

On your first-pass analysis, the following procedure is recommended if you decide to consider legal fees at all.

1. Omit legal fees from your main decision tree.
2. After you've completed analyzing your main tree, use one of continuous variable assessment methods to assess incremental legal fees and costs.
3. For the plaintiff, subtract the mean value of incremental legal fees and costs from the value of litigating the case. For the defendant, add the mean value of legal fees and costs to total exposure.
4. Make the case strategy decisions based on total incremental costs (indemnity and legal fees).

[3] Certainly, there is a chance you could come out ahead by refusing the offer, litigating, and then winning. There is always a chance that a bad decision will be followed by a good outcome. But, by doing so, the odds and mean value are stacked against you. Over the long run, you will lose money.

As mentioned above in the discussion of sunk costs, this method lets you be more detailed in considering the future expenditures necessary to obtain the possible future revenue of a judgment in your favor.

8.2.3 Add Legal Fees and Costs to the Distribution on Case Value

If you would like to include incremental legal fees and costs but they don't vary significantly by case strategy, then you may want to include them in the pictures of total exposure in the case.

For example, suppose for a large health care organization it typically runs \$5,000–\$50,000 in legal fees and costs to handle a medical malpractice case. Assume the organization has billions of dollars a year in revenue, and the potential indemnity payments run from the hundreds of thousands to millions of dollars. For this organization, the legal fees in an individual medical malpractice case are not a major financial issue. You could include them in the pictures of uncertainty of case exposure (either a histogram or cumulative probability distribution).

1. Assess the incremental legal fees and costs using a continuous variable method.
2. Construct a histogram or cumulative probability distribution of the value of litigating the case, as described in Chapter 9 on special probability approaches.
3. Add your best guess to the horizontal scale of the histogram and use this histogram in discussing risk aversion with the client.

By using this method, the client can see the total amounts at stake, including the future legal fees, and thereby have the total scenario values for considering whether he is averse enough to the risk of the larger numbers to make a decision other than the best expected-value decision (and, you'll recall, thereby incurring the risk penalty).

8.2.4 Make Multiple Assessments If Potentially Large and Variable by Scenario

Occasionally, future legal fees may be large relative to the client and may vary greatly according to the scenario in your tree. For instance, your future costs to handle and try a case may be greatly reduced if you succeed in obtaining partial summary judgment on a critical issue in the case.

If your legal costs vary by tree scenario like this, use the following procedure.

- Make a continuous variable assessment of the legal fees and costs for each scenario in the tree where they would vary greatly. Here are the typical stages in a case where legal fees and costs vary greatly.
  - Matter arises, but no case is filed yet.
  - Case is filed and being managed with some discovery, but detailed trial preparation hasn't yet begun.

  - Detailed preparation for trial though trial judgment.
  - Appeal.

- Use a continuous variable assessment method to assess incremental legal fees and costs in each scenario.
- Subtract (for the plaintiff) or add (for the defendant) mean incremental legal fees and costs in each scenario.
- Construct a histogram or cumulative probability distribution if necessary for discussing risk aversion with the client.

8.2.5 Include in the Analysis If Potentially Large, and Variable, and Uncertain

The most detailed method for handling future legal fees is only necessary if those fees are potentially large relative to the client, variable by case strategy or outcome, and uncertain.

In such cases, you can simply add the uncertainties impacting incremental legal fees and costs into your analysis. The kinds of uncertainties you might consider for legal fees and costs include:

- How are the case progresses in the milestones described above.
- What the costs might be in each stage along the way. Duration in a particular stage an monthly total expenses in that stage are often helpful ways to think about this.

They may not particularly be variable by scenario if, for instance, the possible legal fees range from $10,000 to $90,000 in *every* scenario.

The same methods for analyzing indemnity in a case equally apply to legal fees and costs.

You might also consider whether to handle incremental legal fees and costs in one or more subtrees for different case scenarios in the main tree, or build it all into one analysis. If you use subtrees, you would add or subtract the appropriate incremental legal fees and costs for each section of the tree.

If you build everything into one analysis, software will likely be required to analyze the resulting fairly complex tree.

Obviously, this last procedure entails a lot of detailed work (you are, in effect, doing the work necessary for a first-pass analysis of the value of a case just for legal fees in particular scenarios). Before getting this elaborate, ask yourself the following questions.

- Are there really different significant influencing factors for legal fees in different scenarios?
- Are the legal fees really so large and uncertain that you need that much detailed analysis to think about potential fee amounts?

If, then, you decide that the future legal fees are really that uncertain and significant to the value of your case, then be very careful to look for other subtle interactions with potentially large effects on the value of your case.

- Other questions in the main tree may influence the amount of future legal fees—making the use of an independent subtree inappropriate. In that case, either incorporate the subtree into your main tree, or vary the subtree according to the scenarios that influence it.
- Look for correlations between different legal fee subtrees. In other words, does the same factor that makes fees high in one subtree also make them high in another? Answering this question requires careful examination of your relevance diagram(s) for fees and use of dependent probabilities.

In sum, then, you can use as much detail and care in thinking about legal fees as you feel necessary for thinking about the ultimate issue (like the value of your case). Keeping in mind all the other cases you haven't analyzed will help keep the future legal fees issue in perspective.

8.2.6 Use Discounting Where Necessary

Timing of cash flows and discounting them to present values used to be much more important when interest rates were higher. These days, the nominal risk free discount rate (which should be used to calculate net present values because we are explicitly modeling risk) runs around 1–3%, depending on the country and who you ask. Subtracting the inflation expectation to get a real risk free rate gets you pretty close to zero (as evidence in government bond yields in places like the USA and Germany).

With interest rates this low, the present value of expenditure happening in the next year or two (the time to resolve most matters) are pretty close to just adding them up. Accordingly, for most matters, it isn't necessary to project expenditures over time and calculate net present values.

The exception to this is where damages could accrue over a long period of time, as in lifetime medical costs or lifetime lost income. For these kinds of damages, it is necessary to project annual amounts and calculate net present values.

Some cases may also extend over many years. It would likewise then be necessary to project the legal fees and costs over time and discount them to present values.

8.2.7 Handling Contingent Legal Fees

If you want to explicitly consider contingent fees in your analysis of the value of your case, they'll need to be included in your main decision tree because they vary according to whether you win or lose the case (of course, no contingent fees if you lose). However, you'll note that, rather like sunk costs, contingent fees usually won't affect your client's settlement strategy. On an expected-value basis, a plaintiff will want to settle for the highest mean value alternative, and the relative ranking of the alternatives won't change because the counsel receives a percentage of any settlement or win as a contingent fee.

However, contingent fees may figure more importantly if you're analyzing the value of the case from the defendant's point of view and the opponent counsel represents the plaintiff on a contingent fee basis. In that case, your tree from your opponent's point of view may reveal the opposing counsel more likely to recommend a lower settlement (where he will receive some contingent fee) over uncertain prospects of trial (where he will receive no fee if he does not prevail).[4]

8.3 Fund Flows Over Time

8.3.1 Converting Fund Flows Over Time to a Single Value

When potential fund flows happen over a period longer than a year or two, it will be necessary to project the year-by-year fund flows and calculate net present values for the decision tree. We typically use Excel for this purpose. Many decision tree software packages (including Supertree) include the ability to link the decision tree to an Excel model for calculating scenario values.

A net present value uses an interest rate to convert payments occurring over time to an equivalent amount today. The interest rate used is also called the discount rate. The idea is that if you had that present value today as a lump sum of money, you could invest it and the principal and interest payments would be sufficient to make the periodic payments over time.

The formula for a net present value is

$$NPV(x_{1,}x_2,\ldots x_{n)} = \sum_{t=1}^{t=n} \frac{x_t}{(1+i)^{(t-1)}}$$

where x_t is the payment or receipt occurring in period t and i is the interest rate for that period. t and i need to be for the same time period (1 year, 1 month, etc.). For t = 1, the payment in the first year is divided by one plus the interest rate raised to the zero power, which equals one (any number raised to the zero power equals one).

Excel has an NPV formula for calculating net present values. However, Excel uses an end-of-period convention for calculating net present values, which means it assumes that the fund flow occurring in the first period occurs at the end of the first

[4]This issue, of course, poses difficult questions of professional responsibility and loyalty to the client. I assume that counsel will always have the client's best interests at heart, in which case all of the methods given here will contribute to pursuing the client's best interests.

period and therefore needs to be discounted. When the actual flows don't match this pattern, we calculate a net present value by just adding the payment in the first year to the net present value of the remaining payments.

Year-by-year payment projections are required for handling prejudgment interest and injuries and damages occurring and accruing over time.

8.3.2 *Time Preference: Choice of a Discount Rate*

As discussed in Chapter 3, two preferences need to be specified in order to show which decision alternative you would prefer: a risk preference and a time preference. The risk preference specifies how you value uncertain and risky prospects. As noted earlier, for decisions without ruinous prospects risk neutral decision-making usually makes sense. When the consequences are potentially ruinous, you need to consider an explicit model for attitude towards risks, such as using a utility function to calculate the certain equivalent value of uncertain prospects.

The other preference is time preference, which describes how you value money today versus money in the future (or in the past). That time preference is specified by an interest rate for translating between past and present or future sums. This interests rate is also called a discount rate. To calculate present values of sums occurring over time, you need to pick a discount rate.

Many courts have a statutory interest rate for calculating prejudgement interest. In California as of the time of this writing, the rate for tort and other non-contractual claims was 7% and the rate for contract claims was 10%, though the contract may specify a different legal rate of interest. The statutory interest rates should be used for discounting calculations if appropriate for your case.

However, these rates are far higher than the risk free rates of return we recommend using for many discounting calculations. For many case analyses looking at injury and damages over time, we employ and courts accept using a risk free rate of return. Courts in California have discretion to do this.

We also need to distinguish between the real and nominal risk free rates of return. The nominal risk free rate of return is the real (no inflation) risk free rate of return plus the inflation expectation.

As of the time of this writing, nominal risk free rates of return (as evidenced in government bond rates in countries like the USA and Germany) run 1–3%. Inflation expectations are also presently very low: zero or even negative in some countries (e.g., Japan).

For projections in real (no inflation) numbers, the real risk free rate of return is often the appropriate discount rate.

If nominal projections (including inflation) are made using the same inflation expectation as in the nominal discount rate, the net present value is exactly the same whether you do real or nominal projections. We therefore recommend make projections using real (no inflation) numbers and a real discount rate. As of this writing, that would indicate a risk free real rate of return of around 1–3%.

This rate is far lower than the "hurdle rate" used in many companies to evaluate prospective investments. Those hurdles rates attempt to raise the required rate of return to account for risk. They are not appropriate in a decision analysis because we are explicitly modeling and analyzing the risk and uncertainty rather than trying to build it into the discount rate.

In addition, there is a fundamental conceptual flaw in using risk-adjusted discount rates because they equate risk with timing, which is often not the case and especially not in legal disputes where resolutions are far quicker and riskier than could be accounted for with even the highest "hurdle rate."

Unfortunately, because "hurdle rates" are built into the financial requirements in many organizations, choice of an appropriate discount rate for a decision analysis is often a matter of some controversy.

For many organizations and especially ones with publicly traded stock and debt instruments, it is possible to determine their weighted average cost of capital ("WACC") which is weighted average of an organization's cost of debt and equity capital. The respective rates can be determined from the financial markets; the weighting is determined by how much the organization is funded from equity versus debt. The argument is that this rate is the cost of money for the organization. If you use a WACC as the discount rate, it should likewise be the real or nominal rate to match your real or nominal projections. Whether this is the appropriate rate to use for evaluating decisions remains a matter of some controversy.

Few people understand the considerations in choice of a discount rate. Our recommendation is to use a risk-free rate of return and, if necessary, rerun the calculation with a different proposed discount rate to see if the preferred alternative changes. If not, there's nothing to argue about. If the preferred decision does change, there needs to be a serious discussion about whether and why a higher discount rate would be appropriate.

8.3.3 No Prejudgment Interest and No Continuing Injury

Perhaps the most common interest and discounting situation is where there is no prejudgment interest or continuing injury to the defendant that require compounding the claimed damages. For this case, the diagram below illustrates the necessary adjustments. Payments of potential awards need to be discounted before going at the end of the tree for the time till payment, but this discounting may be canceled out after judgment by compounding for post-judgment interest. If so, you would only need to discount potential awards for the time from judgment to today.

If not, (that is, the discounting and interest aren't very nearly the same rate), then compound by the difference between the rates if the interest rate is greater, and discount by the difference if the discount rate is greater (Figure 8.19).

A second variation on the no prejudgment interest and no continuing injury situation is where the jury, uninstructed, is inflating the claimed damages to account for

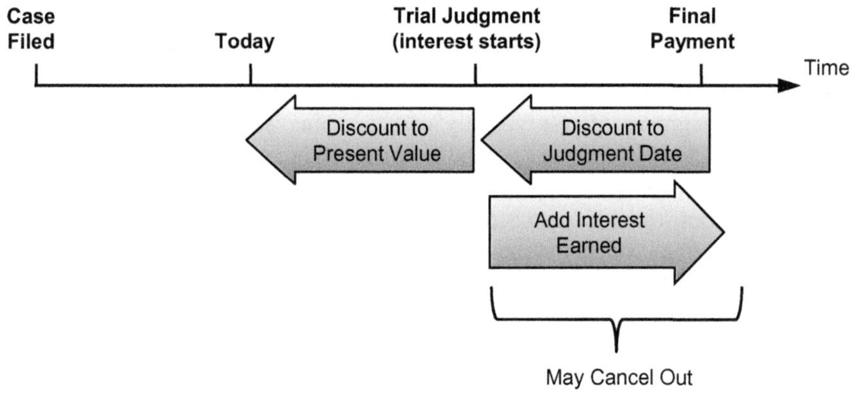

Figure 8.19 No prejudgment interest and no continuing injury

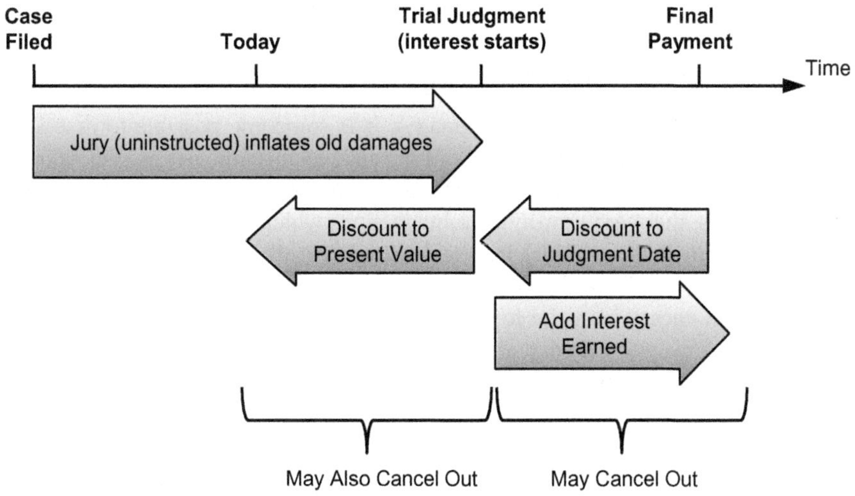

Figure 8.20 No prejudgment interest and no continuing injury, but the jury inflates damages

the plaintiff having to wait so long to recover. If so, then, as illustrated below, the different compounding and discounting factors at work may cancel out altogether, and thus your award assessment wouldn't require any discounting at all. You would then need only consider the jury's inflation of the old injury from when the case is filed till today, which may even be part of your damages assessment (Strictly speaking, of course, you should still discount or compound as appropriate for the differences between the rates, but for small differences these adjustments won't make significant differences in your settlement strategy.) (Figure 8.20).

8.3.4 *Prejudgment Interest*

Prejudgment interest presents another variation in compounding and discounting patterns. Prejudgment interest typically starts accruing when the case is filed, and, of course, is only payable as part of the judgment if the party prevails.

If prejudgment interest applies, then you'll need to compound your damages assessment from when the case is filed until today, but, as shown below, from then on discounting may cancel out the remaining prejudgment and postjudgment interest (Figure 8.21).

8.3.5 *Injuries and Damages Occurring Over Time*

Many types of injuries and therefore damages occur over time, including:

- Lost individual income or business interruption
- Continuing medical costs from an injury
- Loss of consortium
- Lost wages or loss of ability to earn an income

For these kinds of injuries and damages, it is sometime possible to assess an annual number for a particular period of time and calculate the net present value to use as part of a scenario value in a decision tree.

Many times, there are year-by-year interactions which need to be captured or it is just easier to build a model in Excel to capture all the various year-by-year amounts. For example, if the starting year of damages, ending year, and annual amount are all uncertain (as for lost income claims), a simple spreadsheet model is a more reliable way to calculate the net present values than trying to figure all the combinations and variations.

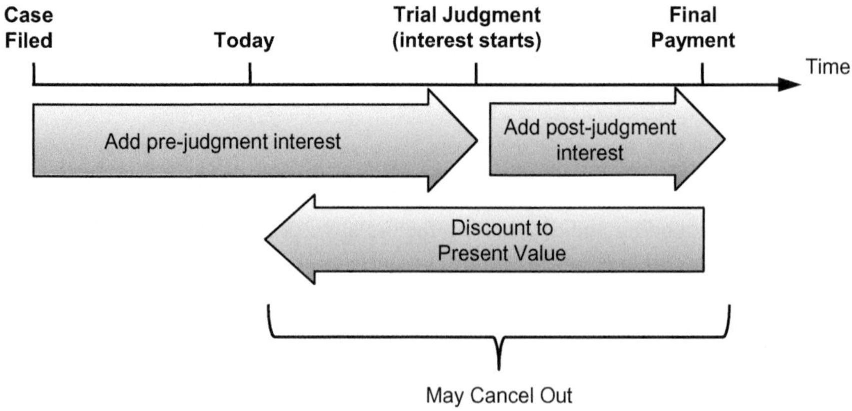

Figure 8.21 Prejudgment interest, but no continuing injury

In particular, patent cases often require, for example, a projection of future revenue and the potential license fees on that revenue for monetary damages (as opposed to injunctive relief). To illustrate this, we present a patent case towards the end of this chapter.

For a detailed discussion of spreadsheet modeling for a decision analysis, see *Decision Analysis for the Professional.*

If you are already familiar with spreadsheet modeling, we present a few basics to start you on:

- The questions defined in your tree (decision and uncertainties) need to be inputs to the spreadsheet model.
- Those inputs need to be changeable to change the spreadsheet to a different scenario.
- For each scenario (set of decision and uncertainty inputs from the tree), the spreadsheet needs to be able to calculate a scenario value.
- Software will likely be required to analyze the decision tree, send scenario definitions to the spreadsheet, and record the calculated value in each scenario.

The free software mentioned earlier (the student version of Supertree, available at supertree.net) has these abilities.

8.4 Precedence

A common concern to the counsel and clients is the resolution of the instant case will have on precedence and thereby on future cases. There are two ways this effect can come about.

First, a case could change existing case law or statutory interpretations. If a client could face future cases where those changes would come into play, there is a potential cost to those changes. In particular, insurers with many similar cases may already be facing the filed claims or suits potentially impacted by the outcome of a particular case.

Second, clients are often worried about whether the resolution of a particular case would affect their reputation as being an easy or hard target for litigation, thereby either encouraging or discouraging future lawsuits.

If either of those conditions is present, we can extend the analysis to include the possible impact on future cases in addition to the cost or value of the present case.

Accordingly, one general way to approach this problem starts simply with the usual beginning question: What would you most like to know? In this case, the client wants to know the mean values of all the cases according to how this one turns out. These future consequences will influence the settlement value of the first case.

A relevance diagram might shape up as follows, showing how the settlement decision in the first case affects the value of the first case, as well as the publicity and collateral estoppel effects on future cases (Figure 8.22).

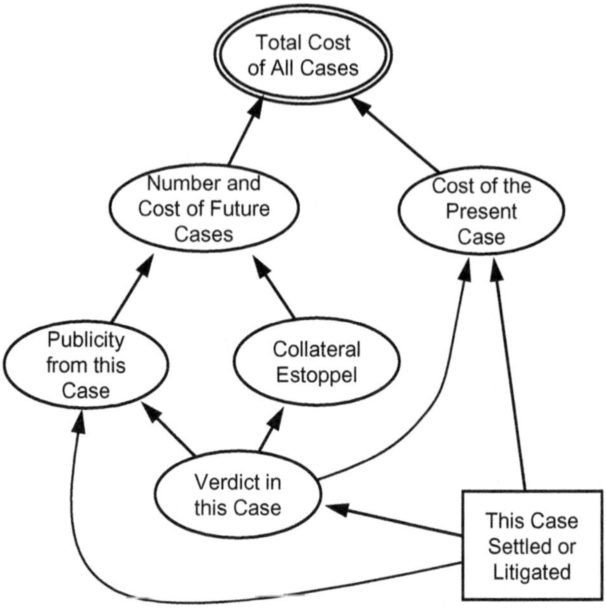

Figure 8.22 Relevance diagram for total cost of present and future cases

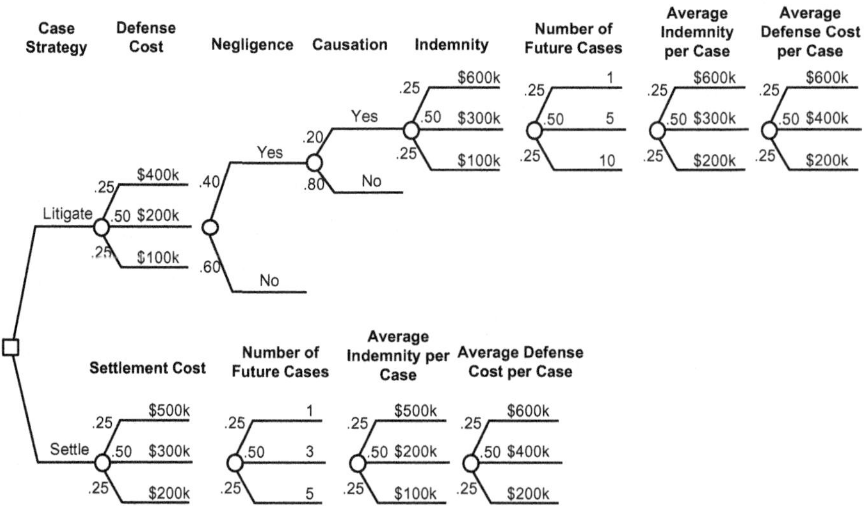

Figure 8.23 Hypothetical analysis for impact on future cases

As with any analysis, there are many possible ways to structure an analysis. One possible structure with sample assessments appears below (Figure 8.23).

Following the bottom of the tree first, there is an uncertain cost to settle the present case, plus uncertainties on the number of future cases, the average indemnity and the average defense cost per case. Note that, in this analysis, we are including defense costs to calculate the total cost to resolve cases. For this set of scenarios, the value formula would be

$$\text{Settlement Cost} + (\text{Num. Future Cases} \times (\text{Av. Indemnity} + \text{Av. Defense}))$$

On the top part of the tree, there is the cost of taking the case to trial followed by the uncertainties governing liability at trial (negligence and causation in this case). It is felt that, if defense wins at trial (no liability), there will be no future cases spurred on by the results in this one.

However, if the defendant loses at trial (only a $.40 \times .20 = 8\%$ chance of this happening), the prospects look even worse: a greater number of future potential cases and greater average indemnity costs in each case. The uncertainty on the average defense costs (a 3× spread from low to high) is the same.

For scenarios leading to no liability, the only cost would be the defense cost. For scenarios leading to liability, the value formula would be:

$$\text{Defense Cost} + \text{Indemnity} + (\text{Num. Future Cases} \times (\text{Av. Indemnity} + \text{Av. Defense}))$$

With the implied replication in the tree diagram (nodes following behind another node but not connected), this tree describes a total of 164 scenarios. Software is advised for evaluating it. We first look at the cumulative probability distributions (Figure 8.24).

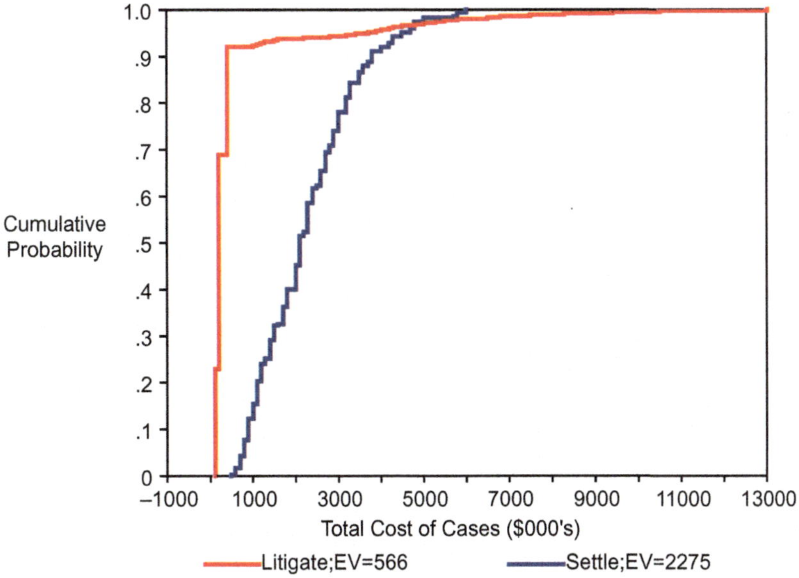

Figure 8.24 Cumulative probability distributions for litigate and settle alternatives

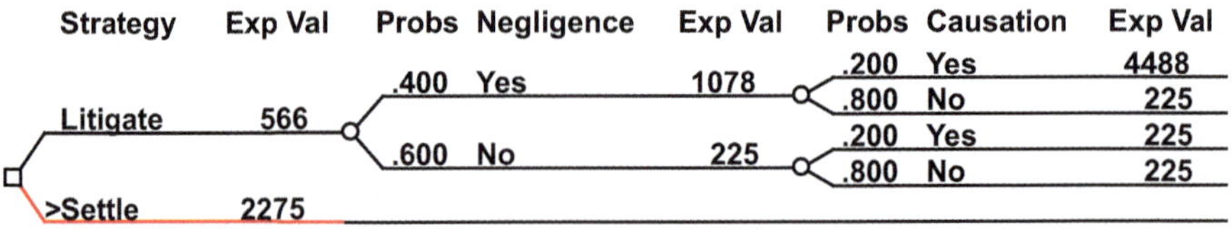

Figure 8.25 Supertree display of partial reordered tree

Figure 8.26 Summary tree for litigate and settle alternatives including total costs of future cases

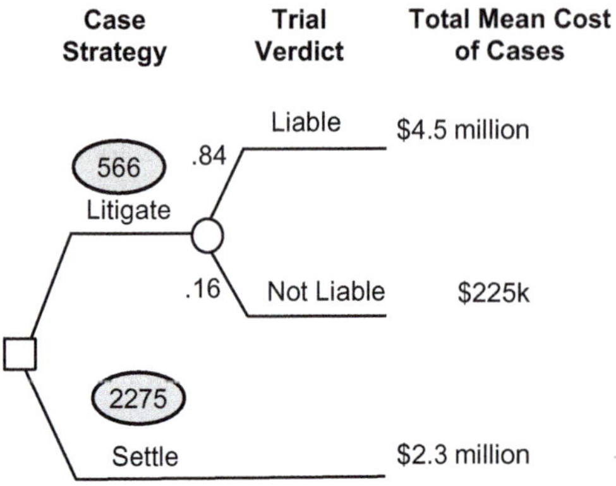

Even with the low probability potential for worse outcomes (the nasty long red tail of the litigation distribution), litigating has a total mean cost of $566K (assuming all numbers in the tree are already present values). Settling has a mean total cost of $2.275 million—4× greater. Further, looking at where the distributions cross, there is only about a 5% chance that settling the case will cost less than litigating it (and likewise a 95% chance that settling is more expensive).

It is sometimes also helpful to look at just portions of the tree when discussing the prospects with the counsel or the clients. Suppose we reorder the tree (good software can do this) to look just the decision and the liability issues. Below is the display from Supertree (Figure 8.25).

Recognizing that all no liability scenarios lead to mean exposure of $225K (92% chance of these), we can draw a summary version of the prospects in a small partial tree (Figure 8.26).

Since the probabilities of being found liable at trial must certainly switch the preferred alternative from litigate to settle, it might be helpful to see where that switch occurs. We start with the sensitivity to probability of a negligence finding (Figure 8.27).

For any probability of a finding of negligence at trial—even 100%—litigating is still a better decision because the chances of finding that the negligence caused the damages complained of are so low (only 20% in the initial estimate). Let's look at the sensitivity to probability for this uncertainty next (Figure 8.28).

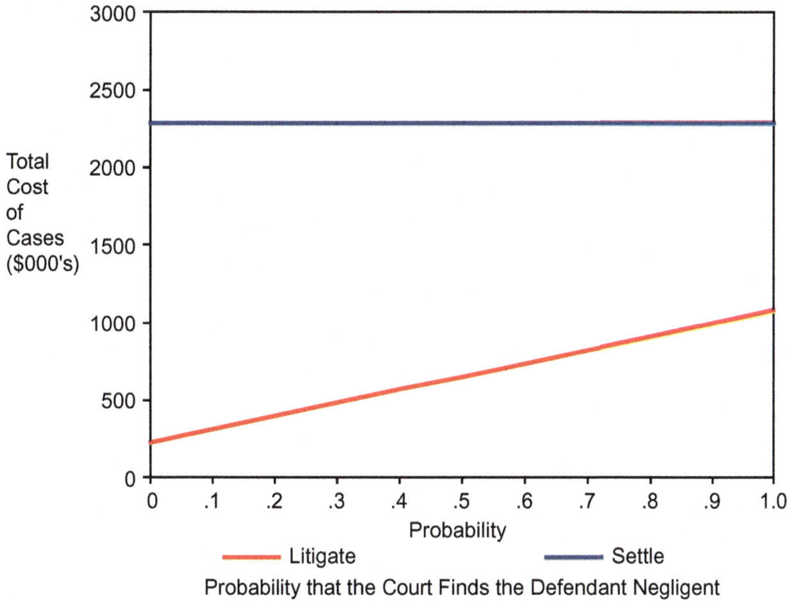

Figure 8.27 Sensitivity to probability of a finding of negligence

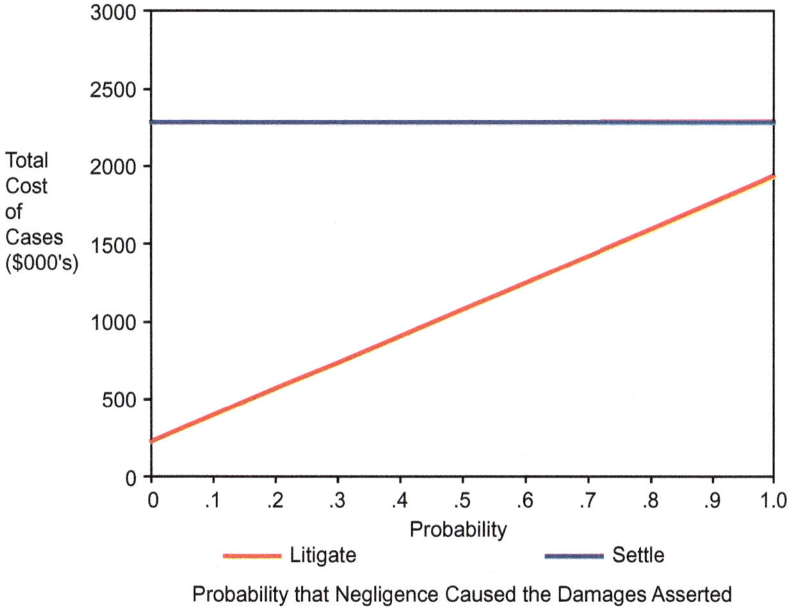

Figure 8.28 Sensitivity to probability of a finding that the defendant's negligence caused the damages asserted

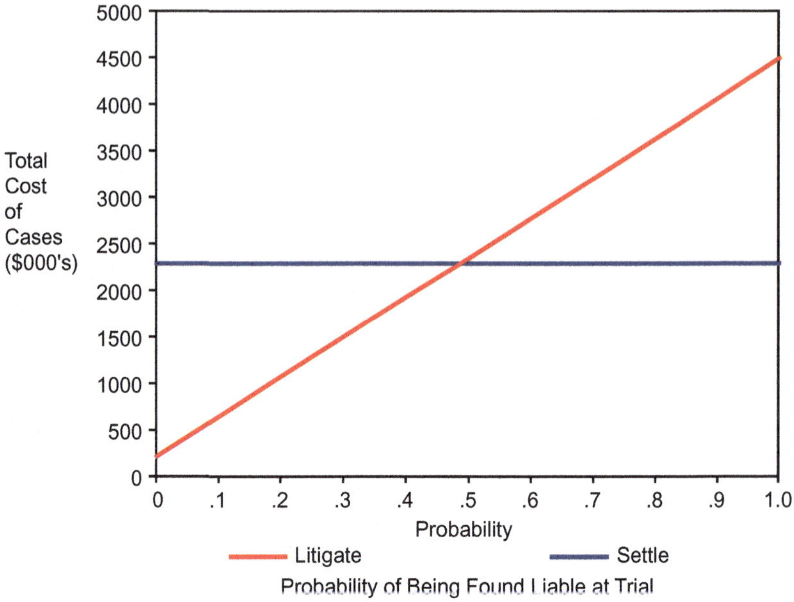

Figure 8.29 Sensitivity to overall probability of liability at trial

Accordingly, regardless of what the counsel believes the individual probabilities of negligence and causation are, litigating still looks better because it has much better prospects of heading off future cases—even though there is a small risk of ending up worse off if the defendant loses at trial.

Of course, some overall probability of liability at trial would be high enough to switch the preferred decision. This could be calculated by replacing the existing negligence and causation nodes with a single node for a finding of liability. Then we could look at sensitivity to probability for that joint uncertainty (Figure 8.29).

An overall probability of liability greater than about 50% would make it look better to settle after all. The square root of 0.50 is 0.71 ($0.71 \times 0.71 = 0.50$), so if the individual probabilities of both negligence or causation were higher than 71%, this would switch the preferred decision to settling. These chances are perhaps a long way from the initial estimates of 40% for negligence and 20% for causation.

Given the low probability, high exposure possibilities with litigating, we might also look at a sensitivity to risk attitude (Figure 8.30).

This graph shows that even a risk tolerance of only $500K doesn't switch the preferred decision, even though this amount is surely far less than the risk tolerance of any insurance company. Both alternatives have risks and settling is not safer.

These kinds of analyses should help to inform both the quantitative understand (do the results look right?) and the qualitative understanding: which alternative is better and why?

In this example, even though litigating has the low probability possibility of turning out worse, it still looks better no matter how hard we poke and prod at it.

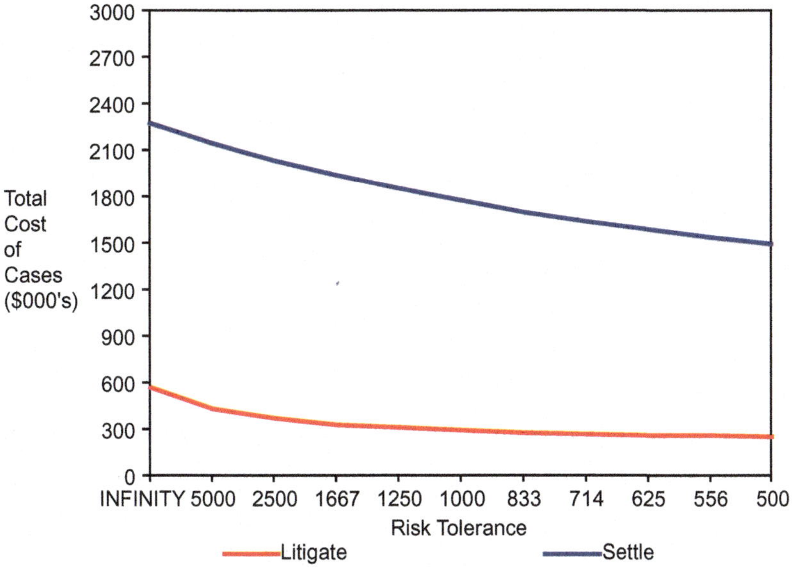

Figure 8.30 Sensitivity to risk attitude for litigate and settle decisions

This illustrates how even for complex matters where we need to consider the costs of litigation and the potential impact on future cases, we can still use decision analysis to develop a clear recommendation for which course of action is best and a clear understanding of why this is so.

8.5 Valuing Nonmonetary Consequences

For many legal disputes, the potential consequences can be described directly as monetary outcomes (such as the jury awarding $X for pain and suffering). Other times, the consequences are still monetary (e.g., loss of X income/year for Y years) but some calculation is required. These kinds of complex damages can sometimes be handled with simple calculations and other times may require spreadsheet modeling.

More difficult yet is when the potential consequences are nonmonetary, such as injunctive relief. We recommend modeling and valuing all potential consequences in dollars for a number of reasons:

- Total consequences can be added up to compare alternatives.
- Tradeoffs between consequences become explicit (what is one percent of market share worth?).
- Decision making is much clearer and better informed. Multiple nonadditive value measures only confuse the matter.
- People are quite good at economizing to make the best decision when the consequences are quantified and clear.

For business decisions, translating everything into dollars and adding them up is possibly less controversial: businesses continue to exist only so long as the dollars all work out in the end. For other decisions, we still recommend monetizing and adding all consequences. Arbitrary weightings between multiple value measures only confuse the decision. Possibly one could propose a different value measure than money, but money has the advantages of being:

- Familiar to everyone
- Fungible ($1 dollar is just as good as any other dollar)
- Easy divisible into smaller units
- Alienable in that it can easily be transferred to others
- Simple

Money is just the value measure. Like decision analysis, it can be employed to good ends or foul.

8.5.1 Nonmonetary Consequences in Business Cases

A number of kinds of nonmonetary consequences are possible in business cases. A few possibilities include:

- Injunctive relief, as in patent infringement cases;
- Bad publicity, as when a parent corporation places a subsidiary into bankruptcy;
- Frustration of a planned activity, such as when shareholders sue to block a planned merger.

In general, the same Decision Analysis techniques you've seen applied to the value of litigating a case can be applied to analyzing questions involving nonmonetary consequences. The only additional aspect is that the consequences at the end of the tree aren't already defined in dollars like a judgment is. Accordingly, the additional technique you need is a way to translate those nonmonetary consequences into dollars. Following are two techniques for doing so.

8.5.1.1 Direct Assessment of Nonmonetary Consequences

The simplest way to value nonmonetary consequences is assess a range of uncertainty on their value directly, like you would for any continuous uncertainty.

- Use the clairvoyance to refine the consequence being considered (that is, define it so the clairvoyant can answer with a yes or no whether it will occur).
  - An arbitrator orders you to finish the structural steel framing contracted for at the contract price, without any allowances for cost increases.

- Write down reasons why coping with the consequence might be more difficult or expensive than you imagined. As with other assessments, push yourself to come up with reasons, such as:
  - Construction is delayed by bad weather.
  - You need to pay higher prices than usual for materials.
  - More welds than usual turn out to be defective and must be redone.
  - The union goes on strike.
  - Structural design inadequacies are revealed as you build, requiring redesign and reconstruction of parts.
- Write down reasons why coping with the consequence might be less difficult or less expensive than you imagined. Make sure you don't confuse these reasons the arbitration might turn out differently.
  - You obtain suitable prefabricated components at a favorable price that greatly reduce on-site welding.
  - Another big job comes in as you need to do the work, so you save considerable money on materials by ordering for both in one big order at a discount.
  - Another job finishes early, so you can put extra salaried workers on this one who would have otherwise been idle until the next construction start.
  - The general contractor agrees to design changes that make the required construction faster and easier.
- Ask yourself the question: how much would you be willing to pay (or to obtain) the consequence? Think hard about how big a check you'd be willing to write.
  - Example: how much would you be willing to pay to avoid an order to complete the construction at the contract price?
  - Question: have you refined sufficiently with the clairvoyance test to know what it means to avoid the consequence? In the example, does it mean you don't have any obligation to perform further work, or that you must negotiate an adjusted price to perform the work?

Although this method is the most direct, it has all the same problems discussed in part I in relation to directly valuing a case: the cognitive and motivation difficulties the mind has in handling uncertainty and complexity. Accordingly, if you don't feel comfortable with directly assessing the value of the consequence (just as you wouldn't feel comfortable directly assessing the value of a novel or complicated case), the second method will help you to disaggregate the question and to understand it better.

8.5.1.2 Modeling Nonmonetary Consequences

If you don't feel comfortable directly assessing the monetary value of a nonmonetary consequence, your alternative is dis-aggregate and model the value of that consequence just the way you've modeled the value of litigating a case. All the same

Decision Analysis steps apply, for example, to modeling the value of injunctive relief (start with what you most want to know; add levels to the relevance diagram, etc.). The major difference is that modeling the value of the consequence will typically require much more input from the client than the value of the case typically does. You will, after all, be modeling the effect on their business.

An example will help to show how the Decision Analysis steps apply to modeling a consequence other than the results of litigation. As in Part II, the discussion of this example will appear in italic type.

Suppose you represent SoMoLo, Inc., a company that obtained a patent for a waterproof cell phone cover that enabled use of all the buttons, the headset jack, and the touch screen on the phone. SoMoLo spent years researching the combination of design, materials, and manufacturing to make the product waterproof and fully functional. The product was introduce in 2011 and has been highly successful.

A competitor of SoMoLo, Shumai Industries, introduced a similar waterproof phone case in 2012 that has only minor design changes from SoMoLo's product. On behalf of SoMoLo, you filed suit in 2013 alleging infringement of the patent and calling for damages for the past infringement.

Of course, should you win, you would also request an injunction against future infringement by Shumai (the nonmonetary part of damages). Shumai has offered to settle the case for a $300,000 payment today plus payments amounting to 1% of its 2015 and 2016 sales—and on the condition that SoMoLo issue it a license under the patent. The usual question, then, is whether you should settle on these terms.

You begin to structure the relevance diagram for the value of the case as below. The value of the case depends on your settle/litigate decision, on whether Shumai is held liable, and on the value of relief (Figure 8.31).

Next, you develop the question of whether Shumai would be held liable if you litigated. This question depends whether SoMoLo's patent is ruled valid, and whether the court finds that Shumai infringed on it. Whether the patent is held valid, in turn, depends on whether you can get the patent examiner who handled the application to testify (he is retired, ill, and in another jurisdiction), and whether the trier-of-fact concludes that the examiner was aware of other related patents when he granted SoMoLo's application. Shumai contends that these other patents render SoMoLo's patent invalid (Other evidentiary issues bearing on the patent infringement question have been omitted for brevity.) (Figure 8.32).

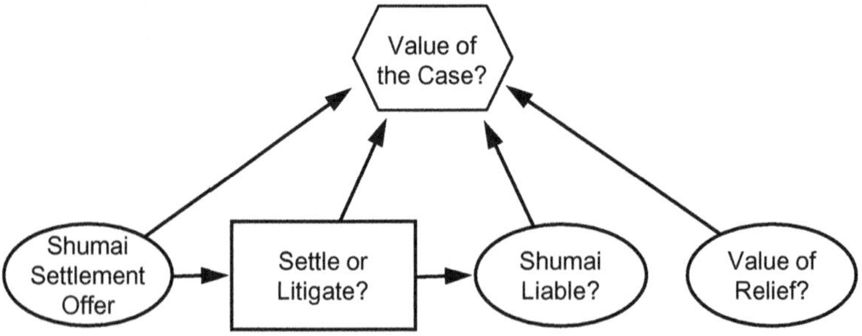

Figure 8.31 Start of structuring decision and case value

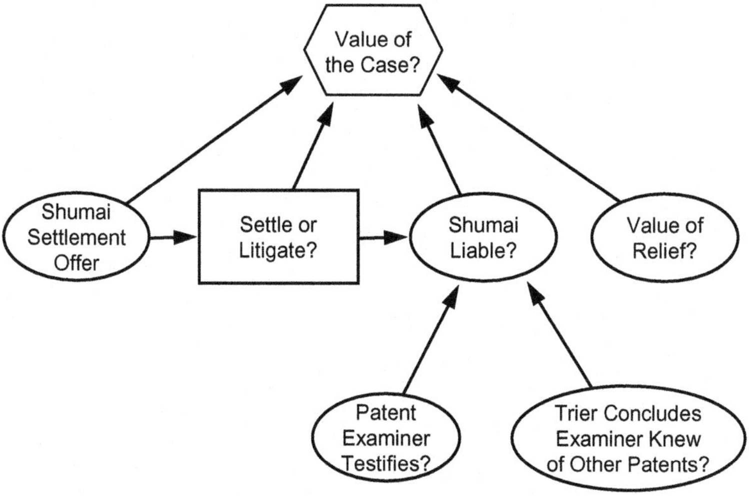

Figure 8.32 Adding evidentiary issues for patent infringement

Figure 8.33 Start of value of relief

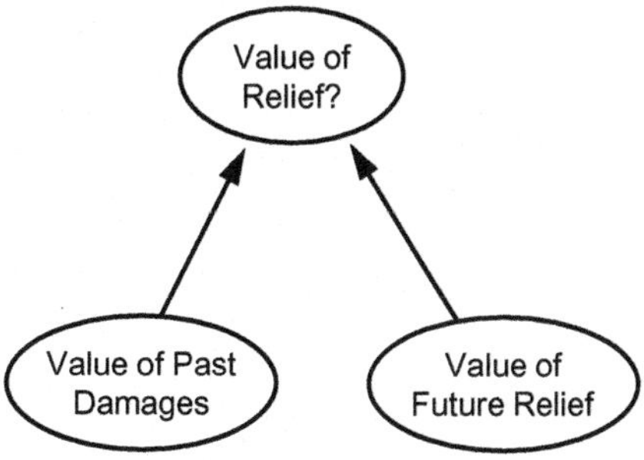

You then begin to develop the value of the relief if you win and litigate the case. This value depends on how the court assess past damages, and what the value of the future, injunctive relief against further infringement is. You start by separating out the value of the relief and developing it separately (Figure 8.33).

The value of the past damages depends on which measure of damages the court adopts. Your first theory (developed in cooperation with the marketing department at SoMoLo), is that SoMoLo should be awarded lost profits from the Shumai infringement. These lost profits would be a function of how much sales SoMoLo lost to Shumai because of the infringement, and what SoMoLo's profit margin would have been on those sales.

Alternatively, if the court finds your first damages measure too speculative, they could award a "reasonable" royalty on Shumai's sales (as if Shumai had already obtained a license under the patent). These further questions are reflected in the diagram below (Figure 8.34).

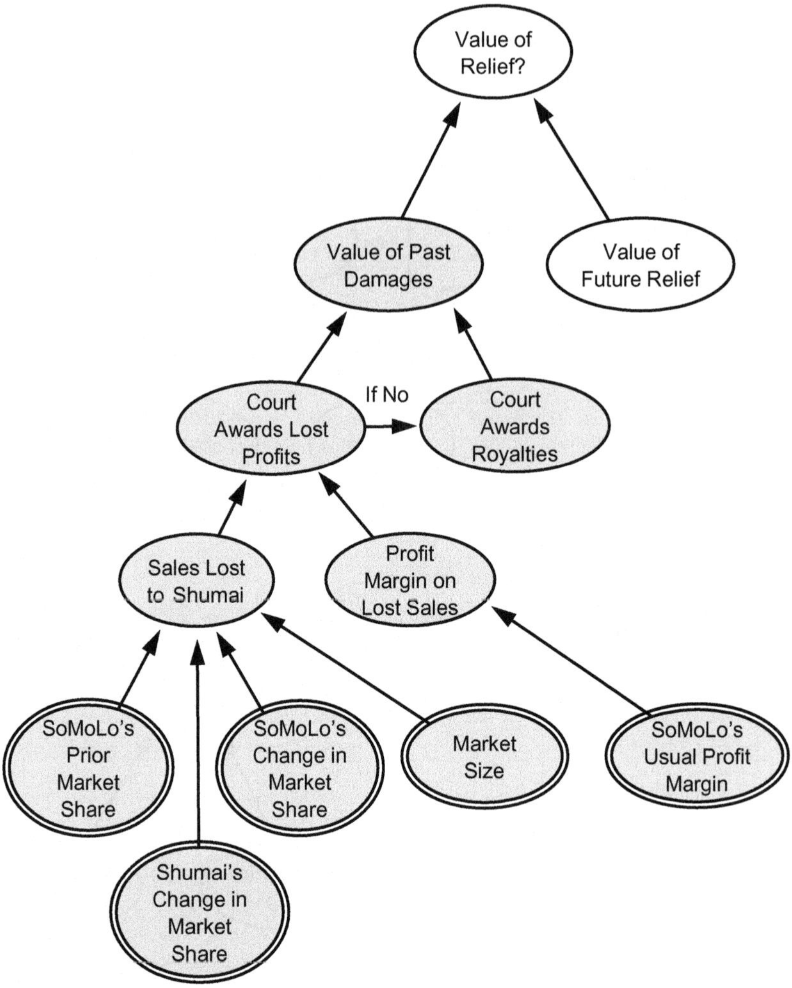

Figure 8.34 Uncertainties and facts relevant to value of past damages award

Again in conjunction with the marketing and planning departments at SoMoLo, you then begin to develop the value of the future relief. You determine that the value of Shumai being prevented from future infringement would be determined by the increase in SoMoLo's market share (if any), by the future market size (to determine sales), and by SoMoLo's profit margin on those increased sales (to determine incremental profit) (Figure 8.35).

Finally, you add one more level of questions for matters you need to consider when assessing the uncertainties. For example, how fast could Shumai redesign their product to work around any protected design elements and introduce a new competing and non-infringing product to the market? (Figure 8.36).

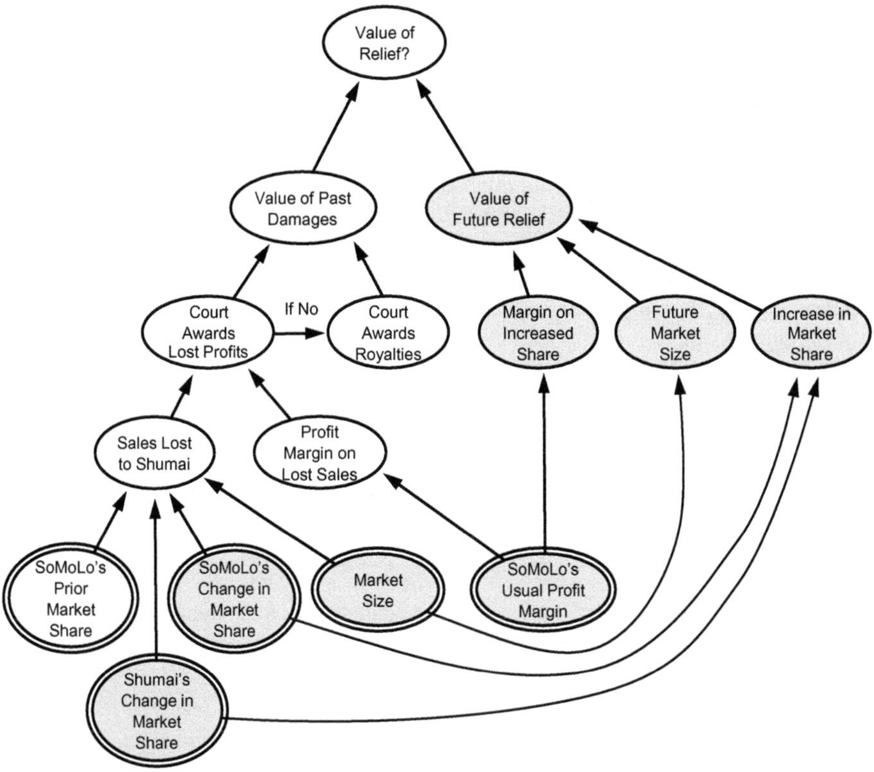

Figure 8.35 Factors relevant to the value of future relief

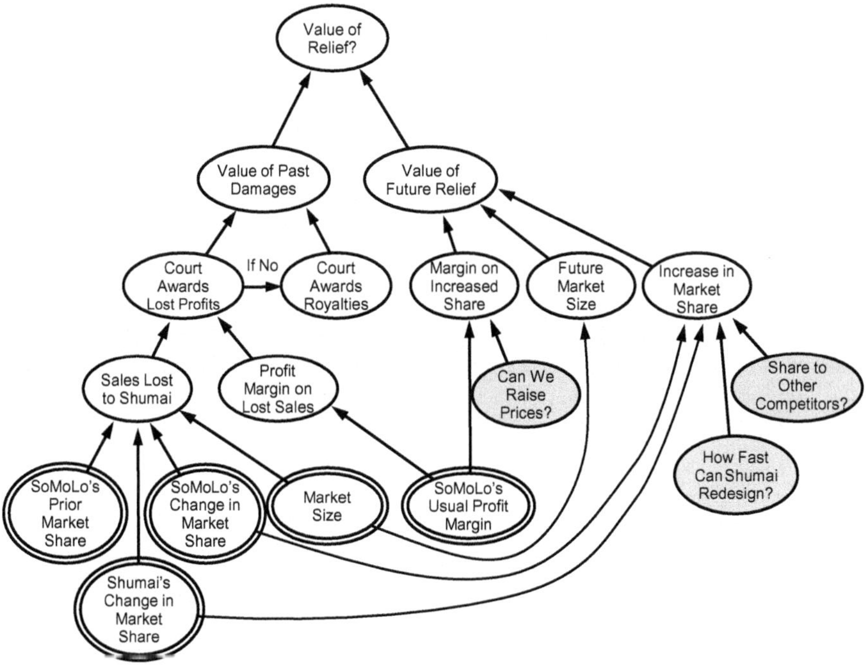

Figure 8.36 Issues to consider when assessing uncertainties

Now it's time to begin working with the marketing and planning departments to start building some simple market models to calculate potential damages of various types and in various scenarios.

First, you focus on estimating the past damages to SoMoLo according to the lost profits theory. SoMoLo's market share dropped by 10% when Shumai introduced its competing product, while Shumai's increase by the same amount—all at a time when the total market size was growing. Accordingly, you contend that the lost profits are SoMoLo's usual 20% profit margin multiplied by the lost business, as shown in the table below.

	2011 (year before other case)	2012 (year after other case)	2013 second year (suit filed)	2014 third year
Total market	$75 million	$100 million	$125 million	$150 million
SoMoLo's share	40%	35%	30%	30%
Shumai's share	10%	15%	20%	20%
Alleged lost share[5]	0%	5%	10%	10%
Market × Lost share	–0–	$5 million	$12.5 million	$15 million
Lost profits @ 20% margin	–0–	$1 million	$2.5 million	$3 million

Thus, you've expressed the value of the injunctive relief in terms of its effect on SoMoLo's business, and used this small and simple model to translate that effect into dollars.

Next, you apply these same kinds of calculations to determining the damages if the court chose the royalty measure of damages, as below.

	2011 (year before other case)	2012 (year after other case)	2013 second year (suit filed)	2014 third year
Total market	$75 million	$100 million	$125 million	$150 million
Shumai's share	10%	15%	20%	20%
Shumai's sales	$7.5 million	$15 million	$25 million	$30 million
Royalties @ 3%	–0–	$450,000	$750,000	$900,000
Royalties @ 2%	–0–	$300,000	$500,000	$600,000

Of course, properly speaking these "missed" royalty payments should be compounded to account for the use SoMoLo would have been able to make of them (like making interest at the bank), but you know the court won't go for that. Thus, the "missed" royalties would either total to 450 + 750 + 900 = $2.1 million, or to 300 + 500 + 600 = $1.4 million.

[5] Lost share due to patent infringement by Shumai, rather than differences in marketing or advertising.

You then start on the problem of valuing the future, injunctive relief by discussing the questions of Shumai's ability to redesign, how much additional market share SoMoLo could gain vis-a-vis competitors, and whether SoMoLo's prices could go up. The consensus is that prices could not go up because the market is so competitive, and there are a number of ways SoMoLo's market share could be affected. You decide that any additional market share would at most last until 2014, when redesigned competing products would be available for sure.

You use the three-branch short-cut to assess how much market share SoMoLo will gain, and obtain .25/.50/.25 probabilities of increases of 10/5/0 percent (that is, regaining all, half, or none of the lost market share). You could have gone into more detail by assessing the full continuous distribution, or by assessing the probabilities for the influencing questions and then separate market share answers for each scenario, but it didn't seem necessary to think about the possible gains in market share. You then work these estimates into the table below to produce dollar answers for the possible value of injunctive relief.

	2011 (year before other case)	2012 (year after other case)	2013 Second year (suit filed)	2014 Third year
Total market	$75 million	$100 million	$125 million	$150 million
Extra sales @ +5% share	–0–	–0–	$6.25 million	$7.5 million
Extra profit @ 20% margin	–0–	–0–	$1.25 million	$1.5 million
Extra sales @ +10% share	–0–	–0–	$12.5 million	$15 million
Extra profit @ 20% margin	–0–	–0–	$2.5 million	$3 million

Discounting the 2014 benefit at 8% gives you possible future benefits of $0, $1.25 + (1.5 \div 1.08) = \$2.64$ million, or $2.5 + (3 \div 1.08) = \$5.28$ million.

The last estimate you need to make is the total value of the settlement offer from Shumai, which you do by extrapolating the market growth and assuming Shumai's share would remain constant.

	2014 (base year)	2015	2016
Total market	$150 million	$175 million	$200 million
Shumai share	20%	20%	20%
Shumai sales	$30 million	$35 million	$40 million
Royalties @ 1%	–0–	$350,000	$400,000

Accordingly, the present value of the royalties would be $(350 \div (1.08)^2) + (400 \div (1.08)^3) = 300 + 317 = \$617{,}000$.

At this point, you draw the decision tree to show how you've structured the problem so far (below), and to get ready to work more on the probabilities (Figure 8.37).

For the question of whether the patent was valid or not, you refer to the relevance diagram and develop the following subtree to determine the probabilities of valid or invalid (Figure 8.38).

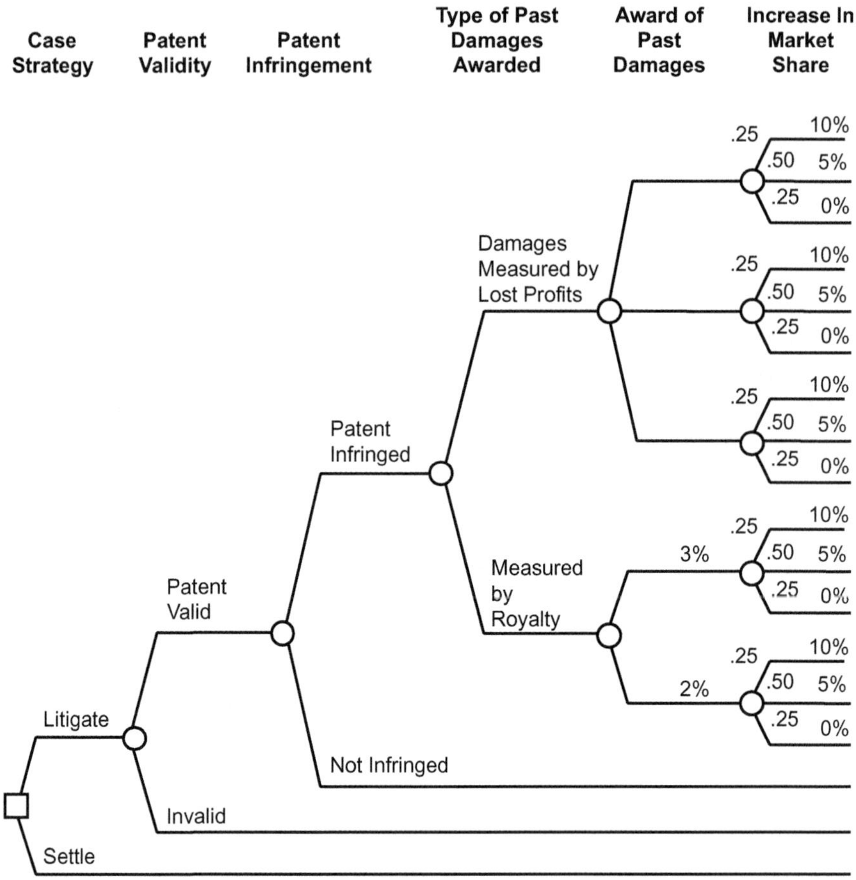

Figure 8.37 Start of the decision tree

Similarly, you assess the probabilities for the court choosing the lost profits or royalty measure of past damages, and the probabilities for the court choosing the 3% or 2% royalty levels if it chose that measure of damages. However, because of the large sums involved and the uncertainty in how the court would handle the lost profits measure of damages, you assess the full probability distribution on how much lost profits the court would actually award, as below (Figure 8.39).

Instead of assessing the distribution direction, you could have assessed ranges for the amount of lost share and what the profit margin on that lost share and done a subtree to calculate the lost damages. You make a note of that for possible additions to the analysis if the amount of past damages turns out to be one of the key uncertainties.

This assessment gives you the final information to finish filling in the tree (Figure 8.40).

Then, finally, you did a rollback to calculate the mean value of litigating the case and see how good the settlement offer really was (Figure 8.41).

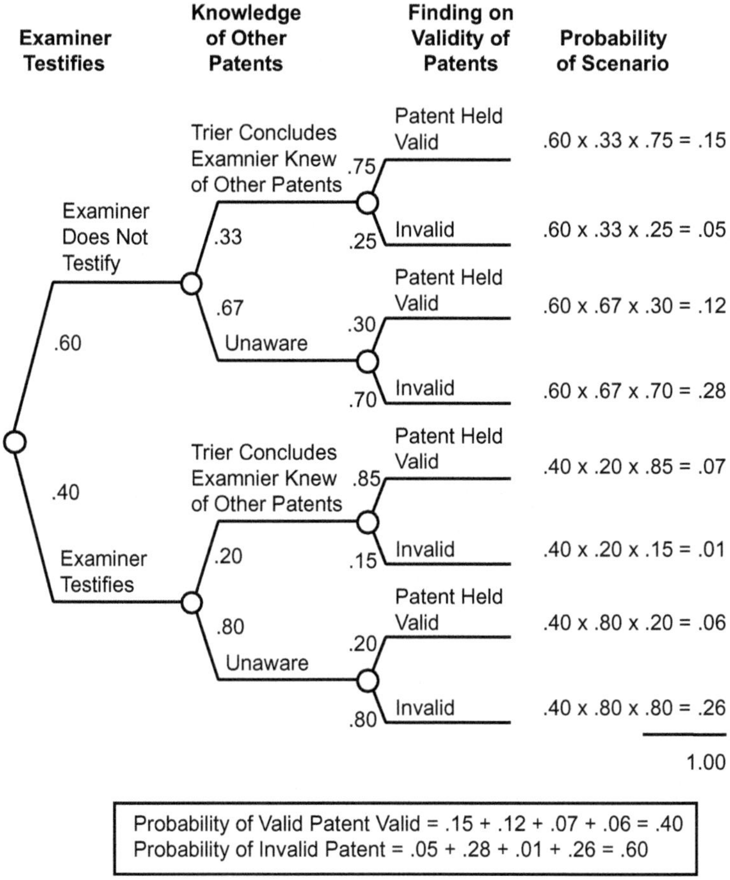

Figure 8.38 Subtree for finding on validity of patent

Thus, on a mean value basis, litigating the case was better than accepting Shumai's settlement offer by 1.5 – .92 = $580,000. Now your task was to do probability sensitivity analysis on the questions in the tree and see over what ranges the best strategy was still to litigate. You also planned on calculating values of information and control to help refine your negotiation position with Shumai. Lastly, you would need to work out the costs to litigate in view of how expensive patent trials tend to be.

To summarize, then, valuing nonmonetary consequences through modeling consists using the Decision Analysis process and tools to study and disaggregate those consequences until you can translate them into dollars. For business problems like the patent example given above, this usually means disaggregating the consequences until you can assess the effects on the business and thereby determine a change in profits. For problems where the final effect is not on the profitability of a business, you may need to disaggregate until you feel comfortable directly assessing the monetary worth of the effects, as described above under direct assessment.

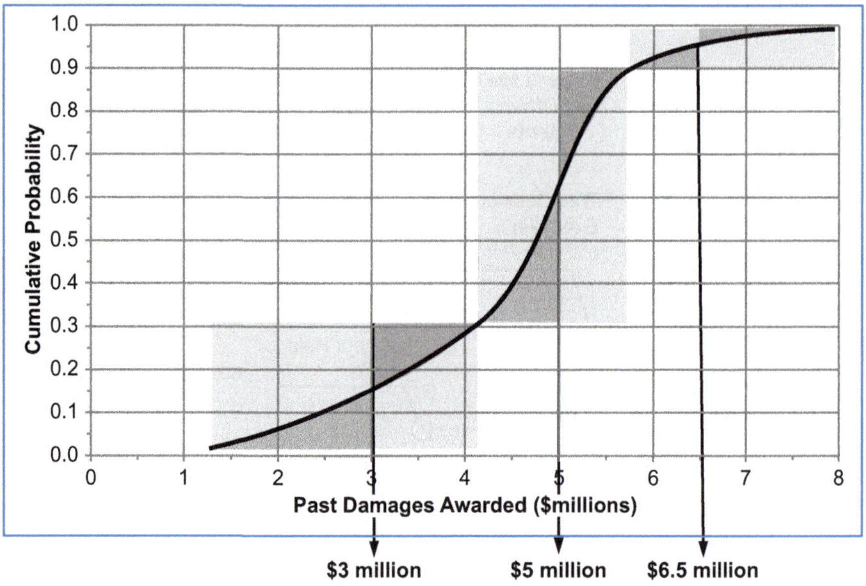

Figure 8.39 Assessed cumulative probability distribution on past damages awarded.

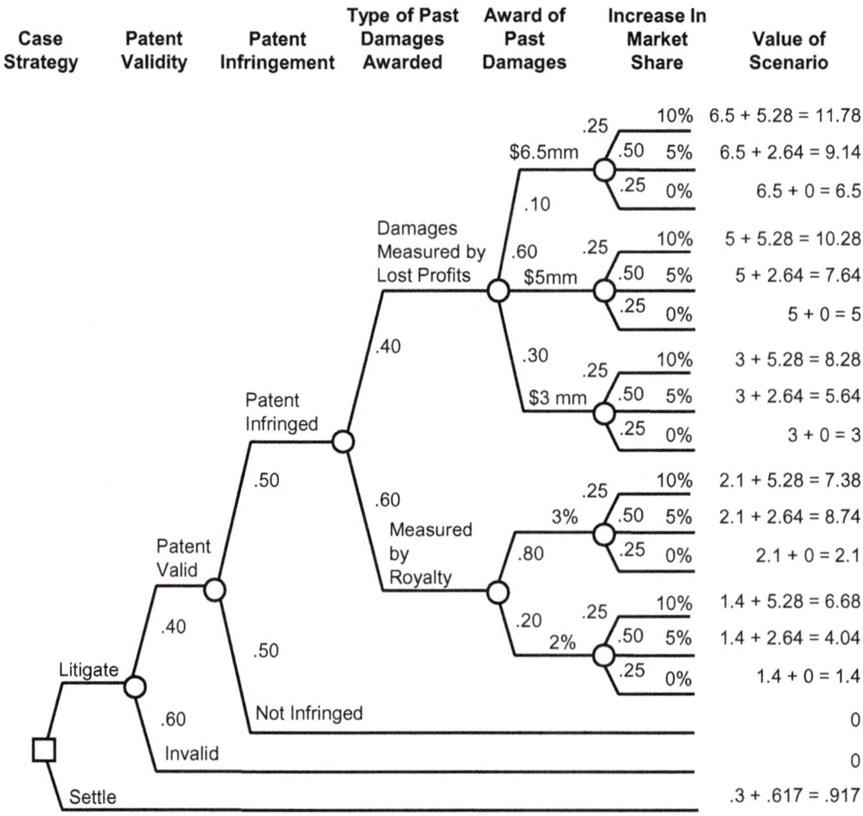

Figure 8.40 Probabilities and scenario values filled in

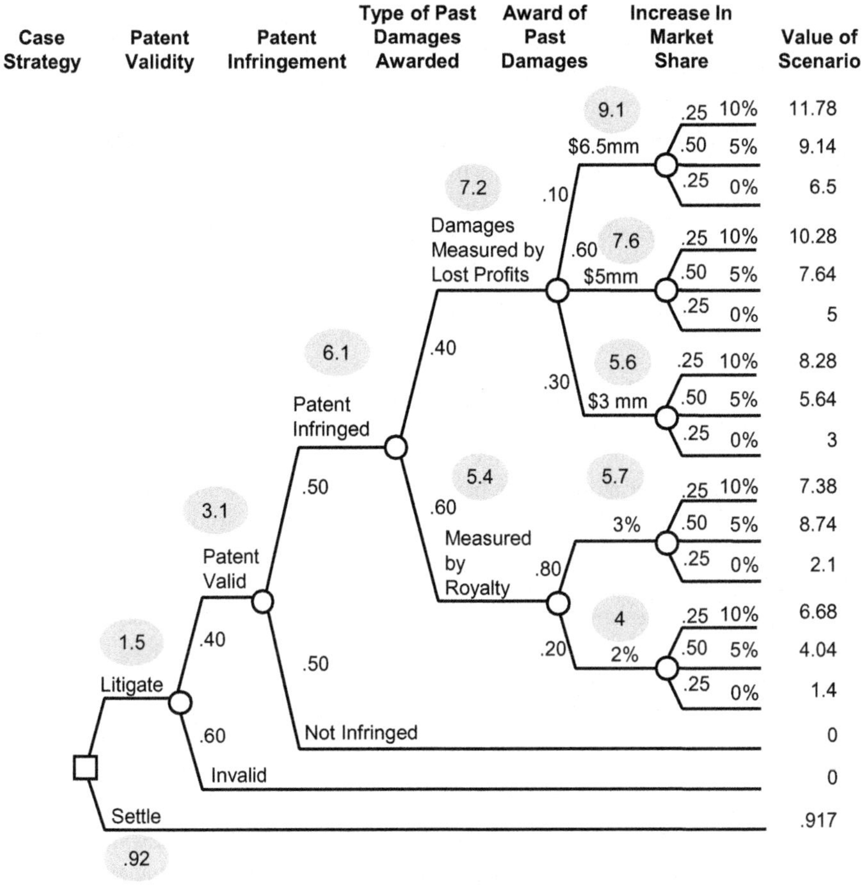

Figure 8.41 Completed tree with rollback mean values

8.5.2 Nonmonetary Consequences in Governmental Problems

Governmental decisions, whether legal or otherwise, are replete with consequences not directly expressed in dollars. A government program may improve the health of one particular group, or seek to deter a particular variety of discrimination, or determine the ease and permissible scope of union activity, or change the economics of an entire industry through taxation or regulation.

The latter situation should immediately strike you as amenable to the techniques used above in the patent example. Effects on companies or whole industries can be modeled through growth rates for gross revenues, profit margins on those revenues, level of employment or capital investment, etc.

The most difficult question, though, applies to all the situations cited, and is the first question in performing a Decision Analysis: what do *you* most want to know? The question is difficult because the "you" is no longer clearly specified—as it is when the "you" of the counsel representing a client provides a definite perspective and set of values. In contrast, government decisions typically affect a spectrum of groups, some beneficially and some detrimentally. For whom should the mean value (or certain equivalent value) be maximized?

The political solution is to let the opposing parties battle it out in adversarial fashion, with the strongest political will prevailing. However, for a lawyer working in or for the government, Decision Analysis offers a better solution.

At the least, the Decision Analysis process offers a way to clarify and to communicate clearly about the problem, using the structuring tools of relevance diagrams and decision trees to break the problem down into alternatives, information, and values (the elements of a good decision). Even this level of analysis offers great benefits.

Nor are you limited to problem structuring. Presumably, the ultimate question in a governmental problem is the net benefit to society (instead of the value of the case). This benefit can be modeled the same way the value of a case can, and a decision made that maximizes the net benefit to society as a whole. Of course, this approach makes the rather considerable presumption that someone is willing to model explicitly this benefit—perhaps a politically risky undertaking that many would blanch at.

Nonetheless, these kinds of decisions, with these kinds of consequences, are being made daily, and refusing to face up to their consequences doesn't make the consequences or the decisions any less real. Explicitly facing up to them can only improve the decisions made. Further, sensitivity analysis can lessen the potential fallout by showing over what ranges a particular alternative is optimal. Showing that a particular course of action is best over all reasonable ranges of the values at stake is a particularly vivid way of defending a decision.

Thankfully, the type of analysis illustrated here is becoming somewhat less controversial as many government decisions now require a cost-benefit analysis (environmental regulation, etc.). Decision analysis offers a particularly good way to tackle cost-benefit analysis because we include the uncertainty in both of those, rather than just calculating a single scenario and single number. An example will help make this process more concrete.[6, 7]

Suppose that you represent the government in an action against a company for violating a pollution statute. If you win, the company will be fined and enjoined from further pollution unless it installs remedial devices. If you lose (that is, you're unable to prove statutory violation), your agency may still have recourses, but you won't be able to bring the full statutory penalties and requirements to bear.

[6] For a very thorough discussion of applying decision analysis techniques to social problems, see Howard, R.A., Social Decision Analysis, *Proceedings of the IEEE* (Institute of Electrical and Electronics Engineers), Vol. 63, No. 3, March 1975. Although this article does get rather technical, it also contains excellent qualitative discussion of the concerns raised in this text. It is worth skipping the equations and reading the text if you practice in this area.

[7] Rather purposely, this example aims not to resemble a real statute or case, but, instead, to illustrate how such an analysis might be structured.

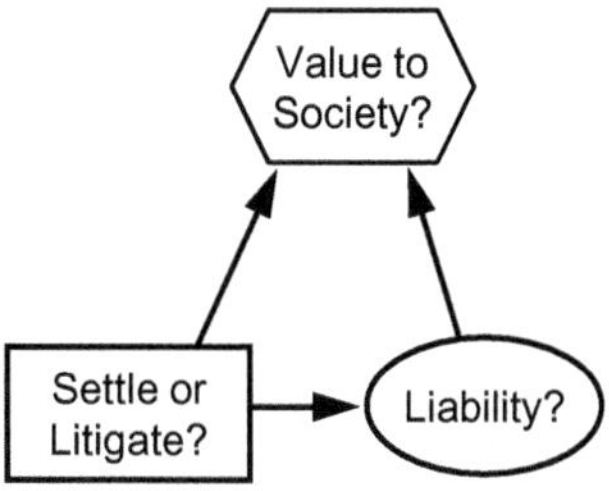

Figure 8.42 Start of relevance diagram for governmental enforcement action

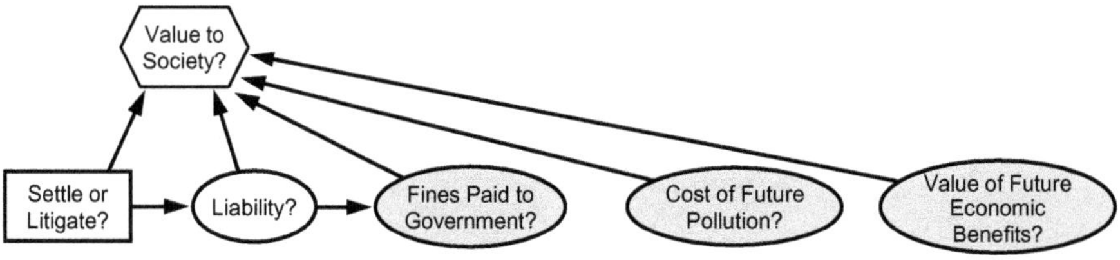

Figure 8.43 Adding the components of total value to society

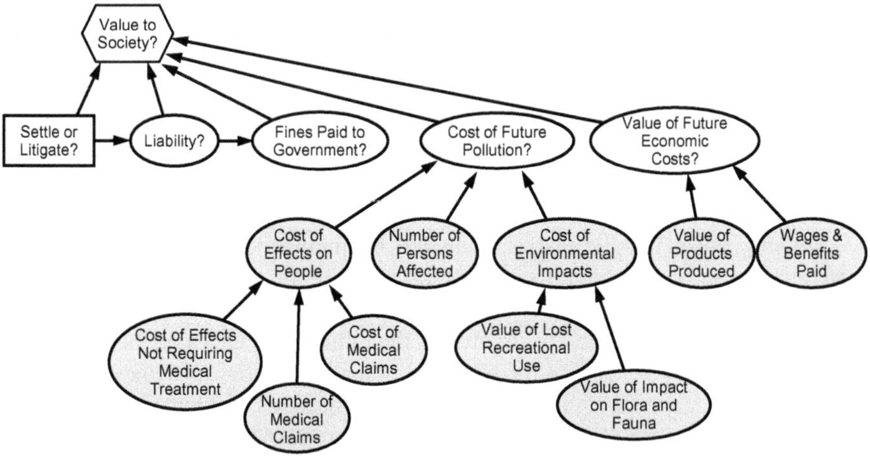

Figure 8.44 Adding the drivers of pollution costs versus economic costs

The company has offered to install certain remedial devices right now in return for dropping all charges. Should you accept the offer?

Although a full analysis is beyond the scope of the present discussion, a relevance diagram will illustrate how you can begin the process. Starting with the value to society instead of the value of the case, you would have the standard settlement and liability questions (Figure 8.42).

You might then begin considering some of the components of the value to society as whole, as shown below (Figure 8.43).

Adding another level of detail or two to the relevance diagram would then begin to get you to questions whose answers would allow assessments of the pollution effects and economic benefits (Figure 8.44).

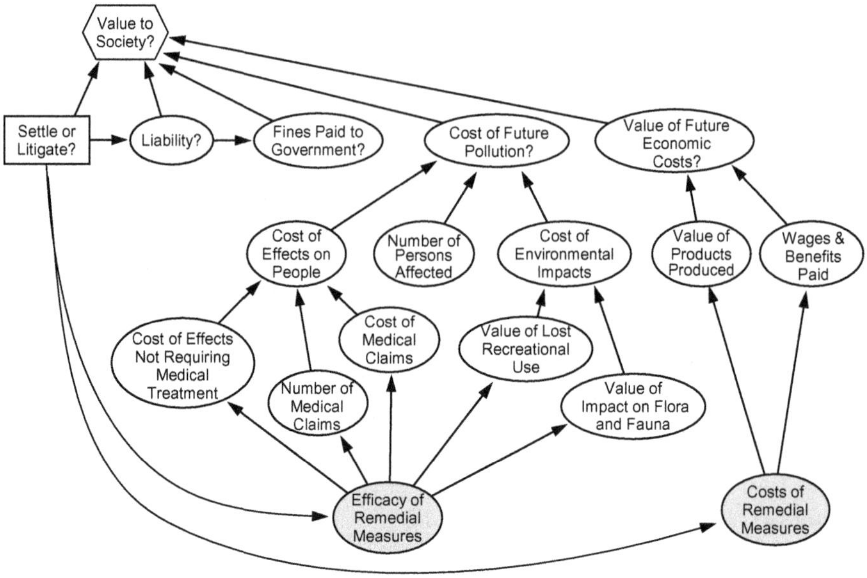

Figure 8.45 Impacts tied back to government decision

Finally, one more level of detail will show how your litigate/settle decision affects the other elements of value to society besides fines paid to the government (that is, pollution and economic benefits) (Figure 8.45).

You may object that the analysis is already beginning to look impossibly complicated. However, the problem itself is subtle and complicated: it involves trade-offs between parties involved and not-involved in the action and though indirect and long-term mechanisms. Merely structuring the influence diagram ought to help you think more clearly about exactly what's at stake, and, later, to demonstrate how carefully your agency weighed all the competing interests. Seldom (if ever) does a public hearing so clearly lay out the relationships between the potential government actions and the interests of the parties.

Further, measures exist and have been successfully employed to keep analysis like this tractable and lucid. The chief tool is relating the factors together through simple arithmetic models like the ones used above in the patent example. For example, the dollar value of the environmental effects might be expressed the value of recreational use minus the value of population reductions. Thus, (environmental effects) = (lost recreation) − (population reductions). Lost recreation, in turn, might be represented as shown in the graph below, where the value drops as areas are closed more frequently for pollution, till it drops to zero when the area must be closed altogether as too hazardous for any recreational use (Figure 8.46).

The cost value of the change in plant populations might be modeled as just the reverse: very little for small changes in population, and more rapidly increasing as the ecosystem incurs greater and greater damage (which becomes progressively harder to reverse) (Figure 8.47).

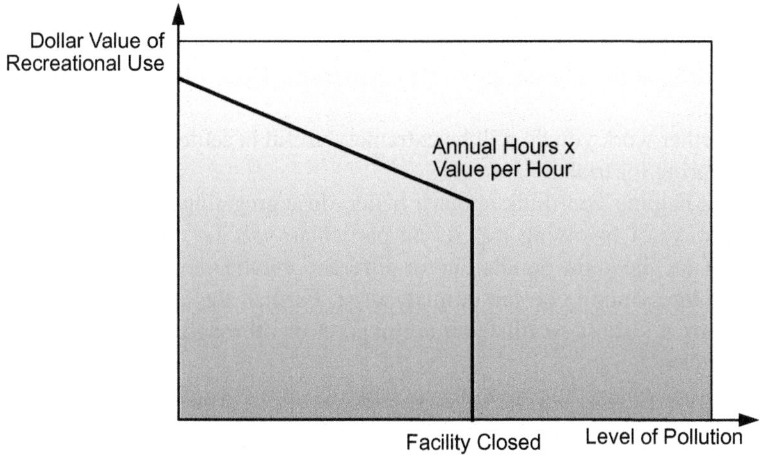

Figure 8.46 Simple model for value of recreational use

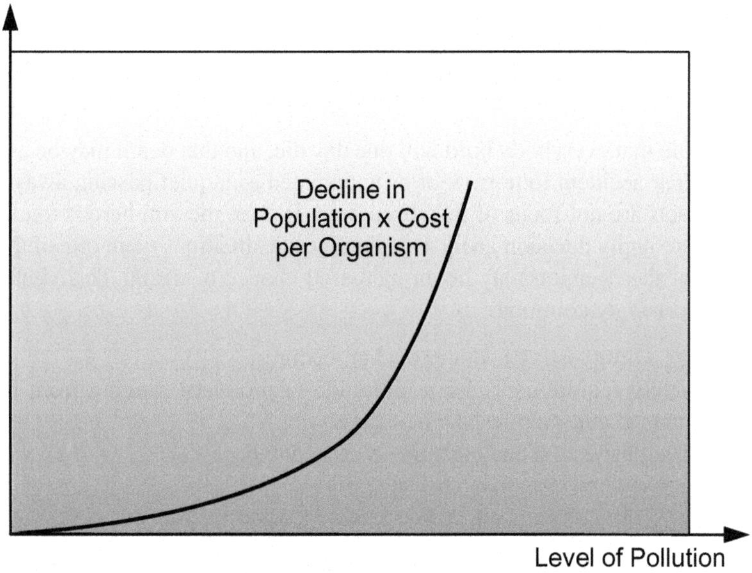

Figure 8.47 Simple model for cost of reduction in flora

Once the arrows in the relevance diagram have been expressed in simple relationships like these shown here, sensitivity analysis of the final value to answers to different questions (as opposed to sensitivity to their probabilities) helps to show which questions are important enough to have probabilities assessed and to go into the tree.

To sum up, then, even though the example given above may appear daunting at first, there are serious reasons why you should consider applying decision analysis to these kinds of problems.

- The problems are this complicated, and, if anything, even more in need of analysis to cope with the complexity and uncertainty than are simpler problems.
- You will understand your problem better even if you only draw the relevance diagram.
- Any further work you do will be extremely useful in settlement negotiations and in preparing for trial.
- Besides helping you think about it better, disaggregating the problem provides ready ways of involving experts on particular, specific questions (such as the percent decline in the population for different pollutant levels) without having to rely on their opinion on the ultimate issue. Further, the analysis will tell *which* questions it's most worth retaining experts on (through value of information calculations).
- The seriousness of the consequences won't go away by refusing to confront them explicitly and the trade-offs among them.
- The analysis will provide thorough and reasoned justifications if you later need to defend your actions.

8.5.3 Decision-Making Where Increased Chance of Death Is a Possibility

It is inevitable that everybody born will one day die, and that death may be as unexpected as a car accident tomorrow or as anticipated as a quiet passing away at age 89. These facts are not focus of this discussion. Rather, the aim here is to examine briefly how to apply decision analysis techniques in situations were one of the consequences of the decision may be an increased chance of death. This situation is perhaps surprisingly common.

- Deciding on drug testing and approval procedures;
- Setting safety requirements for a multitude of products, ranging from nuclear power plants to automobiles and hair dryers;
- Setting food inspection and cleanliness procedures;
- Deciding whether to become a private pilot.

In fact, every day each of us makes decisions that, when you think about it, cause small changes in your chances of dying that day. Driving a car is perhaps the most common example. Did you outfit your care with a five-point racing hardness and always wear a helmet when you drive? Those would reduce the already very small probabilities of dying in a collision, but at a very real cost. For similar reasons, cars aren't designed like army tanks even though that would virtually eliminate the chances of dying in a collision with another vehicle (hard on the pedestrians, though).

People also climb mountains, race cars, and even cross the street against the light. The point is that it is rational, even *necessary*, to balance very small probabilities of death against expected benefits in carrying on one's life. For instance, suppose you find a $100 bill on your doorstep with a note attached to it saying that the

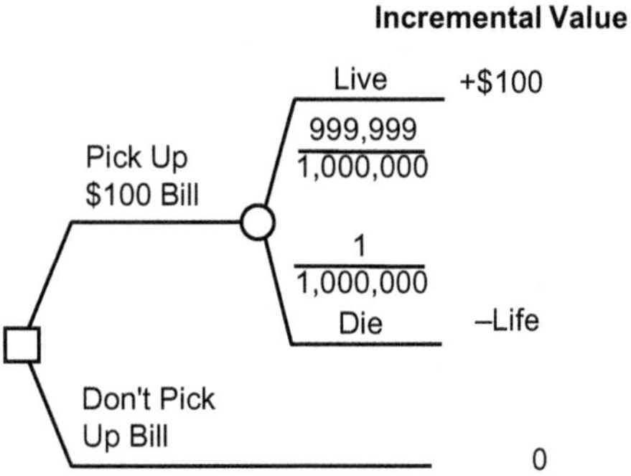

Figure 8.48 Decision with a very small probability of death resulting from it

bill is coated with a substance that's randomly fatal to one person in a million (to keep the example simple). You would face the following decision (Figure 8.48).

What would you do? For comparison, depending on who you ask, somewhere around 34 patients per million die from undergoing general anesthesia (the anesthesia itself—not any other causes). This risk is 34 times greater than the one posed above. Is your decision any different? Suppose the risk posed by the bill was only 1-in-10,000,000?

Most people would admit that there are some chances of death so small (or some rewards so great) that they are willing to undertake the chances of death for the expected reward. The way we live our lives bears this out. Further, continuing with the above example, even if you initially decided to pick up the $100 bill, there is some probability of death at which you would no longer be willing to do so. That probability is greater than 1-in-1,000,000 and likely very much smaller than 1-in-2 (50%). If you determined the probability of death (p) at which your decision switched, you could solve for the value you assigned to your life in this situation: $(p \times \text{Life}) = [(1-p) \times 100]$, or $\text{Life} = (1/p) \times [(1-p) \times 100]$. Thus, if the 1-in-1,000,000 probability were the point at which your decision switched, you would be acting as if you assigned a value to your life of $99,999,900.

Before proceeding, it is perhaps useful to pause on what this number means.

> While some might see this procedure as callous, it is important to observe that we are not attempting to value a human life or injury in a moral sense, we are rather assigning values to life and the other loss categories as a means of making societal decisions. The values are not measurements, but the results of decisions made by society [and individuals] concerning the balancing of social benefit and social risk.[8]

Thus, this number does not mean that you value your life at $99,999,900, and, in the extreme, would accept a payment of that amount in return for being murdered. All this number means is that, for situations involving very small probabilities of death, this is the number you want to use for rational decision-making. This number only applies where the chances of death are very small. If you thought about it, you

[8] Howard, R.A., Social Decision Analysis, *Proceedings of the IEEE* (Institute of Electrical and Electronics Engineers), Vol. 63, No. 3, March 1975.

might want to use larger and larger numbers as the odds of death increased, till the amount became infinite at some point (no one could pay you enough to undertake that risk of death). One analysis found that, with a probability of death of 1-in-10,000, a hypothetical individual might act as if the value of his life were $2.43 million, and that this particular individual would not accept any risks of death greater than .10 (that is, the value he used became infinite at that point).[9] The analysis also commented that, in one multimillion dollar product liability suit, the losing automobile company presented evidence at trial of small-probability values of life used in product design that were an order of magnitude below this figure (that is, around $100–$200,000).[10] Explicit use of a small-probability value of life would have produced a safer car and better legal defense.

The take-home message, then, is that it's *better*—rational, reasonable, justifiable, and moral—explicitly to use numerical dollar values of life when making decisions involving very small probabilities of death. People want neither automobiles that explode frequently, nor ones laden with $1 million worth of safety equipment. More safety is always possible. The critical question is whether the benefits justify the cost. In this determination, proper use of small-probability values of life is essential. As with the other nonmonetary consequences, they don't go away because you refuse to deal with them.

8.6 Uncertainty as to Which Forum

Uncertainty as to which forum your case will be heard in can make it very difficult to follow the Trier-of-Fact Rule in structuring and analyzing your case the way you think the trier will do. This is especially true where the different possible forums have different applicable law, such as when your case may end up in one of two different states or in federal court.

The solution (as laborious as it may sound) is to do a separate analysis for each jurisdiction and relate them together, as described in the following procedure.

- Conduct a separate analysis for the mean value of litigating the case in each individual forum. You should have a different mean value of litigating the case for each individual forum—otherwise, why is question relevant?
- Assess the probabilities of the case going to each forum.
- Compute the total mean value of litigating the case, as illustrated below (Figure 8.49).

The mean value of litigating the case, then, would be $(p_1 \times V_1) + (p_2 \times V_2) + (p_3 \times V_3)$. This value would figure into your settlement strategy as usual.

---

[9] Howard, R.A., *On Making Life and Death Decisions*, appearing in Societal Risk Assessment, R.C. Schwing and W.A. Albers, Jr., General Motors Research Laboratories, 1980.

[10] Ibid, p. 104.

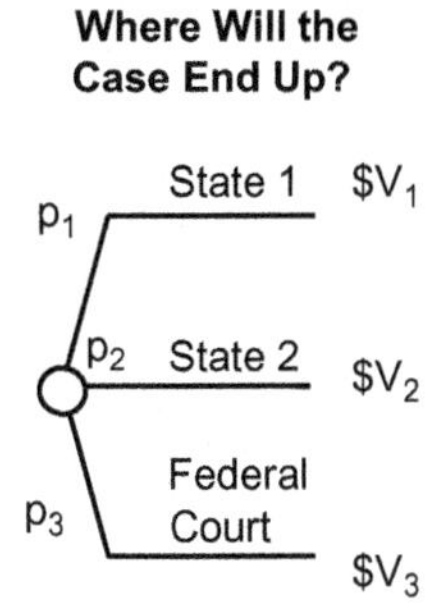

Figure 8.49 Uncertainty as to which forum

This way, you can effectively follow the Trier-of-Fact Rule and obtain a meaningful analysis of the case's value in each forum. Thus, you would have separate relevance diagrams for the value of the case in state 1, in state 2, and in federal court, and so on. Of course, there may be similarities that make the second and third analyses go faster, but don't be too quick to assume commonalities. As always, the differences and distinctions you draw (like the differences from adding more dependencies in the relevance diagram) are most valuable in planning your case.

If you face many cases that could end up in any one of those same three forums, then you might be able to avoid doing an analysis for each forum for each case. Do the multiple analyses for several cases and see if any consistent relationship between the forum values emerges. For instance, the value of litigating the case in state 2 (V_2) might consistently be 15% greater than the mean value of litigating the case in state 1, and the state 3 value (V_3) might be 20% lower. If you are lucky enough to find such a consistent pattern, then need only do the full analysis for one of the states, such as state 1, and then obtain the value for state 2 ($V_2 = V_1 \times 1.15$) and state 3 ($V_3 = V_1 \times .80$) by multiplication. Pick for the full analysis the forum that seems most critical or most influential to you.

8.7 Cases Where You Have Almost No Information

If you are faced with a great many similar claims (such as when product liability is at issue), you may find yourself with cases that you have almost no specific information about (For example, electric toothbrush claim #128.). If so, you may find it very difficult to value litigating that individual case and come up with some idea of settlement value. The following procedure will help you make best use of what information you have.

- If you have not already done so, pick out case you do have information on and analyze them.
- Compare those cases to the no-information case and try to determine whether the no-information case looks better or worse than each of them. Make adjustments as appropriate to value the no-information case (Of course, this method runs great risk of biasing from adjustment and anchoring.).

- If you cannot even estimate whether the no-information case looks better or worse than the other cases, calculate the average of (case value ÷ damages claimed) and apply this percentage to the no-information case: no-information value = (damages claimed × average ratio).

In my experience, I've heard the assertion "We have absolutely no information on this case!" many times. It's never yet turned out to be true. There is more information than people think; you just have to politely poke and prod a bit to get it out.

8.8 Problems

1. For the various difficult questions to structure presented in this chapter, are there differences in the decision analysis techniques applied to do so? Answer and give examples or reasons.
2. Are there any fundamental differences in the methods and tools for analyzing potential findings of liability at trial versus the amount of damages awarded? Why or why not?
3. Are there any limitations in applying the techniques illustrated in this chapter to other complex questions, such as which drug development projects a pharmaceutical company should fund?
4. What is the difference between structuring a difficult question and developing a settlement or trial strategy?
5. Besides the decision illustrated in this chapter, what other decisions might arise in the course of a dispute? Are any of them not susceptible to similar analysis? Why or why not?
6. Suppose you come across a difficult question in a legal dispute which is not illustrated in this book. List at least two examples. How would you tackle them and why?

8.8.1 Complicated Damages

7. What guides the level of detail and disaggregation you should use in analyzing complicated damages?
8. What are some of the ways different types of damages may affect each other?
9. How do you know when damages need year-by-year projections rather than assessing lump sums?
10. What is the relationship between modeling damages in a decision analysis and pleading them in a court filing?
11. You represent a large commercial real estate brokerage (Valley Realty). Your client recently sold a choice industrial tract in Silicon Valley, California, that had formerly been used by Galenz, Inc. to manufacture advanced non-silicon chips. The selling price was $1.8 million.

Prior to the sale, both your client and the buyer, Nebitz Computer, inspected the site and found it suitable. All California disclosure laws were at least facially complied with. However, upon beginning renovation of the facility, Nebitz found some 200-odd buried barrels of waste chemicals buried on the site, some of which were leaking. The cleanup bill will be in the millions of dollars, and Nebitz is suing both Valley Realty and Galenz for it. Nebitz is also contending that defendants knew or recklessly failed to discover and disclose the dump, and is therefore also suing for punitive damages.

Galenz, in turn, contends that the dump pre-dates its use of the facility, and was never discovered during Galenz's use of the facility for light manufacturing and assembly. Prior to Galenz's ownership, the site was used for some 10 years by another chip company that is now bankrupt, so it is plausible that they surreptitiously buried the barrels while they were going under. At that time, disposal of industrial wastes was less rigorously regulated in the Valley, so there is no record of when the barrels were buried. The barrels were found under a parking lot that Galenz claimed was paved over prior to its taking possession.

Galenz and Nebitz have come up with conflicting estimates for the cleanup, ranging from $1–4 million. The discrepancy stems from conflicting estimates of the number of barrels buried (100 to as many as 300), and how many of them are actually leaking (5–25%, which, in turn, determines the amount of soil removal, treatment, and special drainage installation required). These parameters can't be exactly determined until excavation actually begins. Further, though only a Galenz consultant has privately suggested it to you, it was common practice to bury such barrels in more than one layer. Thus, excavation may reveal a second layer of barrels underneath the first, potentially doubling the number of barrels. Whether this second layer doubles the cleanup cost depends on whether these barrels are also leaking, the type of chemical they contain, and its toxicity.

Discovery has thus far revealed that Galenz made a number of parking lot repairs during its tenancy for subsidence in the lot. Galenz contends that these were merely paved over, but Nebitz claims to have found a paving contractor who discovered a barrel during one such repaving. If this testimony is presented in court, it would bolster Nebitz's claim for punitive damages. You estimate that punitives could range from the amount of compensatory damages to triple them, depending on how damaging the evidence presented is. A recent survey of the lot shows no subsidence presently extant.

The claim to hold Valley Realty jointly liable hinges on a number of factors. First, Valley specializes in industry realty and in finding property for high-tech companies in particular. In this capacity, Valley has in two prior deals arranged site cleanup as part of a transaction. Second, the inspector used by Valley has been implicated in two other suits involving chemical dumping, one of which was settled recently for an undisclosed sum. Finally, Valley had a copy of a preliminary Superfund report that indicated the Galenz property as a possible cleanup site. It is unclear whether Galenz also had access to this report. However, the inference in the report was based only on the type of processes performed by Galenz's predecessor, and didn't indicate that any inspection of the property had actually been made. You therefore find it within the realm of possibility that Valley be held liable even if Galenz is not.

One final wrinkle is that Nebitz is unable to use the property until the cleanup is completed. Nebitz is therefore claiming a continuing injury for the $180,000 per year financing charges that it must pay on the property even though unable to use it.

11.1. Structure the relevance diagram for the value of the case. Include any key influencing facts.
11.2. What missing pieces are pointed out by the problem description so far? List any assumptions you find helpful or necessary to complete the problem description.
11.3. Draw the tree for the amount of damages awarded, and for Valley Realty's share of the liability. Do you think there is any spill-over between whether both the defendants are found liable and whether the jury awards punitive damages? How would this spill-over be reflected in the relevance diagram?
11.4. Assume that there will be such spill-over, and that you will therefore need three subtrees for liability: only Galenz liable; only Valley liable; and both liable. Would this spill-over be better represented by dependent probabilities or dependent outcomes?
11.5. Write the value formulas for the decision trees. Assume that the case will be settled within 6 months if settled, and that a trial will result in a verdict due in 2 years' time.
11.6. Assess all necessary probabilities. Make sure to develop your lists of reasons enough to make the probabilities meaningful to you. Feel free to make any assumptions to address questions raised by the problem statement, but make those assumptions explicit in your lists of reasons.
11.7. Calculate the mean damages award and Valley's expected share of the verdict. What should Valley's settlement strategy be?
11.8. Suppose that Valley could negotiate a payment to Galenz in return for a release from all further liability. How much should Valley be willing to pay?
11.9. Formulate a preliminary discovery plan listing the most important issues and the amounts Valley should be willing to spend on developing them. Which issue has the greatest potential for decreasing the value of the case to Valley?
11.10. Suppose that, even at this late date, Valley could purchase an insurance policy that would pay any judgment due from Valley over a $1.5 million deductible. How much should Valley be willing to pay for such a policy?

12. You represent Dr. Heinz Dill, a former research scientist at a large chemical company. Up until 5 years ago, Dr. Dill worked developing new varieties of nylon salts, an intermediate product that is then processed into the finished nylon of the desired properties, color, and so on. At that time, Dr. Dill was fired as a result of a dispute with the lab director, Dr. Paul Eihmer over the allocation of

resources among various research projects. Dr. Dill had an employment contract that required he only be fired "for good cause," and Dr. Eihmer (on the advice, you suspect, of the legal department), attempted facial compliance by charging Dr. Dill with interfering with the running of the department, with the research efforts, and with refusing to work on projects as assigned. Eihmer presented Dill with an ultimatum to accept a lower-level position in a different part of the country, and, when Dill refused, fired him. Besides losing his $100,000 per year salary, Dill lost half of the $200,000 allocated in his name to a 401(k) retirement plan because he was not yet fully vested, and lost stock options which, when they matured over the next 4 years, would have been worth $240,000.

Dill considered a wrongful termination suit at that time, but, decided instead to get even in the marketplace. On mostly his own time, he'd applied his chemical talents to developing a solution to quick-cure cucumbers, creating a pickle reminiscent of the so-called "deli-style" pickles that took up spice flavors in only one-fourth the usual time and only needed momentary heating: thus yielding a superior, crisp, and more flavorful pickle in less time, with less energy, and without any loss of shelf life. Part of his dispute with Dr. Eihmer concerned presenting his find to the food division for further development, which Eihmer responded to by forbidding any work on the process at all.

Dill sold his small investment portfolio, took out a second mortgage on his house, and went into the pickle business. Within a short time, he was supplying fresh pickles to many of the restaurants and delicatessens in his area. A friend then alerted him to a local bottling plant that could be bought from HUD for a song, and Dill cashed in his retirement savings to buy it. He then began supplying bottled pickles to stores and grocery distributors state-wide—beginning finally to cut into the sales of the food division of his old company. Over a 4-year period, his pickle sales had gone from only $100,000 the first year to $10 million per year, with a consistent net profit of 15% that Dill has been plowing back into the business.

Most recently, Dill has been approached by a national firm regarding a buy-out of his company, with the firm proposing to set up a new subsidiary that would produce and distribute his pickles nationwide, leaving Dill a 45% share of the subsidiary. The firm is a direct competitor with Dill's old company in all its markets.

Rumors in the trade of this last development got back to Dill's old company, which is now bringing a suit to claim the pickle formula. Three theories are advanced to support this claim. First, the company claims that the formula belongs to them under a contract in which Dill assigned in advance any inventions to the company. Second, they claim that, apart from the contract, they have a common law "shop right" to the invention under trade secret law; that is, an irrevocable license to use it because it was made on the employer's time. Finally, they claim that the time and money invested in Dr. Dill and his work give rise to a property right in the invention under the misappropriation theory of unfair competition, and that Dr. Dill's profits should be forfeited to avoid unjust enrichment.

Regarding the company's first claim (which you feel is their strongest), the assignment contract also has a clause that vests rights in the inventor for any invention the company refuses to develop. The contract does require that this refusal be in writing, but you feel there's a good case to be made that Dr. Eihmer made an effective refusal notwithstanding the lack of a writing. It was part of Dr. Eihmer's duties to handle assignment of inventions and he decided which prospects to develop. In a deposition, Dr. Eihmer has testified that he only instructed Dill to forego working on the pickle process so it could be transferred to the food division, but there is not yet any corroboration for this testimony.

Further, you feel that pickle process is not even properly subject to the assignment agreement, and this gets you into the second claim. Dr. Dill admits doing some work on it in the lab, but claims he mostly worked in his lab at home. He claims that he didn't do any nylon research at home for lack of the necessary equipment, but admits that some of the materials came from work, and that he did work on results and his lab notebook at home. However, Dill's position is bolstered by the fact that a pickle process was clearly beyond the scope of his duties and not what he was hired for. You therefore don't feel that the company has a legitimate shop right.

A threshold issue, of course, is whether the pickle formula is even a trade secret. The demonstrated commercial value of the formula and Dr. Dill's assurances that the formula cannot be determined from the final product because it is packed in a different solution after process do indicate a trade secret. Further, under the Uniform Trade Secrets Act, which this jurisdiction has adopted, the secret need not actually be used in the business (in contrast to the use requirement in § 757 of the first Restatement of Torts).

Under the misappropriation theory of unfair competition, there can be a property right in an invention where a party has invested labor, skill, and money in developing a salable product. Where a party has such a right, misappropriating that property for commercial advantage is actionable. Further, though actual competition between the parties is no longer a requirement, its presence (as in this case) bolsters the claim.

However, the same issue arises of whether the company did invest in developing the formula through its employment of Dr. Dill, or whether he developed it on his own. Thus, you feel the claim is again vulnerable on this ground.

Furthermore, you've convinced Dr. Dill to fight back by bringing his wrongful termination claim (Luckily, the statute of limitations hadn't yet run because, up until a year ago, Dill was required under the contract to perform occasionally consulting services for the projects he worked on.). The basic issue here is whether the company has documented "good cause" for which to fire Dr. Dill. Dill contends that he never interfered with the department or with the research, that he only voiced his opinions and opposition in meetings called to evaluate research projects, and that he never refused a work assignment. Eihmer, of course, claims the contrary on all these points, but his only corroboration is the memos he wrote himself. Dill's employment record is exemplary apart from Eihmer's comments. Dill further contends that Eihmer was jealous and vindictive of him because Dill's stock options were a coveted award for achievement that Eihmer had never received himself. Dill also commented that he had been very critical two major research proposals in particular, contributing largely to their being turned down, only to find later than Eihmer had authored them.

Accordingly, you feel there's a good case for finding Dill was wrongly discharged. If he prevailed on this claim, he could, at the most, receive 5 years worth of back wages. He would have to receive at least 2 years of back pay to achieve full vesting on his 401(k) funds and receive the balance (gaining an additional 25% vesting each year), and 4 years to vest his stock options (also vesting 25% each year).

A further interesting point to you is the questions of fact common to the two claims. Was Dr. Dill properly performing his duties and developing the pickle formula on his own time, so that he owns the formula and was wrongly discharged? Or did he develop the formula at work and raise such a ruckus over Eihmer's refusal to further it that there was good cause to fire him, and he was acting as a vengeful employee by absconding with it? You anticipated that Dill's former company would move for joinder on this basis, and accordingly, successfully moved for it first. The claims will therefore be tried simultaneously before a jury.

12.1. Structure the relevance diagram for the pickle formula claim. Be sure to include any important influencing facts. List any assumptions you make about issues or information to clarify the case.

12.2. Structure the relevance diagram for the wrongful termination claim, similarly listing any important facts or assumptions you make. What issues are common to the two claims? Do you see any spill-over effects besides?

12.3. Decide how you will handle common issues and spill-over effects between the two claims (Recall that you can't, for instance, simply subtract mean values where there is a dependency between the issues rolled up in each value—you'd be double-counting something.). You could, for instance, do one large tree, or do the common issues in a single subtree and use the summary mean value in a subtree for each respective claim.

12.4. Draw the decision tree for each claim and fill in the value formulas. Assume a 3% discount rate for any discounting calculations.

12.5. Assess the probabilities for the decision tree. For continuous variables, indicate why you chose a particular assessment method. Be sure to develop your lists of reasons, and add any assumptions you feel necessary to make the case realistic.

12.6. Solve for the mean value of the combined case. Can you solve for an independent mean value for each claim? Why or why not?

12.7. What should Dr. Dill's settlement strategy be?

12.8. Suppose you discover that Dill's old company has $400,000 of coverage for wrongful termination claims, with a $100,000 deductible (that is, the company must pay amounts over $500,000 as well). How does this affect Dr. Dill's settlement strategy?

12.9. Suppose you succeed in getting the pickle formula claim dismissed on partial summary judgment? What should Dr. Dill's settlement strategy be then? How does this differ from evaluating the wrongful termination claim while the pickle formula claim is still pending?

8.8.2 *Legal Fees and Costs*

13. List at least one purpose for which you would want to add together indemnity exposure and legal fees and costs to show the total exposure in a case, and one purpose for which it would be better to keep them separate.
14. What are the disadvantages of routinely settling cases for any sum less than the total exposure (indemnity and costs)? What are the advantages?
15. What are the differences in the decision analysis techniques applied to analyzing legal fees and costs versus other issues?
16. How would a case being retained on a contingency fee basis affect the case value from the defendant's perspective? From the plaintiff's perspective? From the perspective of the plaintiff attorney? How might this affect settlement and trial strategy?
17. You're pursuing a products liability suit for a client who lost several toes while using a power lawn mower. The special damages total $20,000, and you're asking for $40,000 in compensatory damages. You estimate that the court may award all or none of the compensatory damages, depending at least partly on what theory liability is found under.

    You're bringing the case under both strict product liability and breach of warranty theories. You can establish strict liability if there was a defect in the product itself or in the design (that is, even though the product was made to specifications, it was unsafe when used in a reasonably foreseeable manner). You can establish liability under the warranty theory if you can show that the implied warranty of merchantability was breached by the mower being unfit for its intended use.

    The mower in question is a heavy duty model, advertised as suitable for cutting tall grass. To this end, it had wheel height adjustments that allow it to be raised up enough to cut grass stalks without being bogged down in heavy clumps of grass. Your client was using the mower to cut tall grass in a swampy area when he slipped in some mud and his foot slid forward and into the mower, resulting in his injury. The mower had fixed safety guards all around it, but because these were raised up with the rest of the mower when the wheels were raised, there was 2-1/2 inches of clearance under the guards when the wheels were so raised. Client was wearing sneakers at the time.

    Under the strict liability theory, you contend that, because the mower was advertised as "heavy duty" and "for tall grasses," it was foreseeable that it would be used in a setting like the one your client used it in. Further, it was unsafe for use in such a setting because it did not have safety guards that adjusted downward as the wheel height was adjusted upward, and because it lacked a sufficiently sensitive safety disengagement feature on the mower drive for when the blade hit an obstacle. In response, the manufacturer contends that lower safety guards would bend the grass over as the mower was pushed over it, resulting in the grass being flattened rather than mowed, and that such low guards in heavy grass would make the mower prohibitively difficult to push.

The safety disengage is not set more sensitively because doing so would defeat the purpose of having a more powerful engine to cut tough grass. Further, the manufacturer points to its owner's manual, which advises wearing heavy boots when cutting difficult grass (as well as safety glasses).

Under the warranty theory, you contend that the mower was unsuited for heavy duty use because it lacks the larger wheels and power drive found in most other similar models, because it does not have adequate safety flaps for stones potentially thrown from the mower, and because it lacks the multiple blade design that most other heavy duty mowers use to achieve greater cutting capability without decreasing the sensitivity for blade disengagement. The manufacturer responds that most of these features are found in industrial models costing 20–50% more, and that its model is suitable for home use in all reasonable settings.

You have been considering which of the theories you have a stronger case under, and what the spill-over effect might be if you try both theories in front of the jury and lose on one theory before the jury considers the second one.

Your opponent has also moved for partial summary judgment on both theories. You estimate he has a 10% chance of getting partial summary judgment on the strict liability theory, and 40% on the warranty theory (Because these determinations are made by the judge, you don't think they'll affect your chances with the jury.).

Your client is paying you on an hourly basis and is concerned about the cost of trying the case. You estimate that it will cost between five and nine thousand dollars to try the strict liability claim only, between three and six thousand dollars to try the warranty claim alone, and between six and twelve thousand dollars to try them both together.

17.1. Structure the relevance diagram for the problem. Do you need one relevance diagram or two?
17.2. Structure the decision tree and fill in the value formulas. What more information do you need?
17.3. Assess any remaining probabilities. List any assumptions you make in assessing them.
17.4. Calculate the scenario values and mean values.
17.5. What is your client's best settlement strategy? What is the best settlement strategy if he loses under either of the motions for partial summary judgment?
17.6. What is the value of information on the amount of legal fees in each scenario? What is the value of information on the other uncertainties in the case? What does this tell you about your discovery strategy?
17.7. Suppose your client asked about hiring you on a contingency fee basis. Assuming a 1/3 fee, is it worthwhile for him to do so?
17.8. Do legal fees ever change the preferred settlement strategy? Explain and justify your answer.

8.8.3 Prejudgment Interest and Discounting

18. When is discounting required in analyzing case values?
19. How would you start with choosing the right discount rate to use?
20. List at least two ways of accomplishing a present value calculation.
21. When might the parties to a case have different time preferences for money? How might this affect settlement or trial strategy?
22. What is the relationship between discounting calculations and evaluating a structured settlement decision? (Note: if needed, you can Google "structured settlement.")
23. For certain types of damages, life expectancy of the plaintiff plays a role (e.g., how long the plaintiff might need medical care or otherwise be earning lost income). How would you include that uncertainty in a discounting calculation?
24. You represent General Utility Nuclear (GUN), an operating company whose main asset is a nuclear power plant, the power from which is sold under contract to neighboring utilities. GUN's plant has two reactors, both of which require refueling at 4-year intervals. The refueling times are staggered so that only one reactor is shut down at any one time. GUN thus requires a new shipment of fuel rods every 2 years.

    GUN currently has a contract with Atomic Yellowcake Corporation (AYEC), a company that supplies fuel rods both domestically and under government contract to approved utilities overseas. AYEC buys enriched uranium yellowcake and processes it into finished, metallic fuel rods. Difficulties in obtaining yellowcake some years back led AYEC to conclude a contract with the Department of Energy's Savannah River facility for the purchase of surplus yellowcake from their operations. However, revelations of insufficient safety at Savannah River led to a shutdown of the reactors there, leading to cancellations of planned shipments to AYEC. As a result, AYEC was forced to curtail production and is behind in its shipments.

    One such shipment to GUN is now 3 years overdue. Reactor #1 was shut down for a year pending refueling until GUN was finally able to obtain fuel rods elsewhere. At that time, a lawsuit was forestalled when AYEC agreed to the $2 million liquidated damages that GUN owed to its customers when forced to default on power sales. Reactor #2 has now also been shut down for a year awaiting refueling, and AYEC has refused to make a similar damages payment, citing impossibility and impracticability as excuses for nonperformance. GUN has now brought suit for breach of contract.

    GUN contends that performance is not objectively impossible because the requisite fuel rods could have been obtained from another source, such as the nationally owned French nuclear industry. Alternatively, Savannah river offered to irradiate spent reactor rods for AYEC. These rejuvenated reactor rods could be used in GUN's reactors in place of new rods.

    However, AYEC responds that it has been stymied in its attempts to obtain new rods from France both because of high demand for them in France and because of glacially slow federal procedures for approval. AYEC therefore contends that performance in this manner is impossible.

As for impracticability, irradiating spent rods would cost more than new ones normally would, but would only have a 2-year useful life instead of the normal 4-year life, thus requiring refueling and rods twice as often. Accordingly, AYEC contends that fulfilling its obligation in this manner would be an essential change in the nature of the required and would replace the already slim profit margin under the existing contract with a large loss. GUN responds that difficulty in obtaining yellowcake was a foreseeable event and that, as between the two parties, AYEC is responsible for the risk of it occurring.

GUN is currently paying damages for its continuing default under long-term power contracts in the amount of $4 million per year (paid since Reactor #2 shut down). The loss of revenues amounts to $20 million per year, and GUN is saving only slightly on operating costs because it is almost as expensive to maintain a dormant reactor as an operating one. Given the dim prospect of ever obtaining refueling from AYEC, GUN has arranged to obtain irradiated rods, but delivery of those rods won't be for another 1–3 years. When delivered, those rods will cost $3 million, versus the new rod cost under the AYEC contract of $2 million. By that time, Reactor #1 will also have needed refueling. At time of delivery, GUN has an option to purchase a second set of irradiated rods for the same cost. GUN expects similar financial consequences if Reactor #1 must be shut down again. GUN is asking for its increased fuel costs and lost revenues as damages, with prejudgment interest on both. You feel that a court would certainly award prejudgment interest if it awards the lost revenues, but that prejudgment interest is more doubtful on the increased fuel costs.

AYEC contends that, even if found liable for breach, it should not be liable for the loss of revenues to GUN, but, rather, only for increased fuel costs, because lost operating revenues are special damages that were not foreseeable at the time of the contract and not within the contemplation of the parties. GUN responds that it clearly indicated to AYEC that it was only undertaking one fuel supplier relationship, and that, therefore, it is clear that failure to deliver the fuel as promised would result in reactor shutdown. GUN further adds that its duty to mitigate does not extend to engaging in risky and expensive speculation via futures or option contracts for fuel rods.

Assume a 3% discount rate, that the statutory prejudgment and postjudgment interest rate is 10%, and that it would take 2 years to obtain a judgment if the case were tried. Assume also that AYEC would take a year to pay any judgment.

24.1. Structure the relevance diagram for whether AYEC is found liable. Include any key influencing facts and any additional assumptions you make.
24.2. Structure the relevance diagram for damages if AYEC is found liable. Is the damages question independent from liability? Why or why not?
24.3. Structure the decisions trees for the problem.
24.4. Assess any necessary probabilities. Indicate why you chose particular assessment methods for the continuous variables.

24.5. Calculate the values for each scenario (A computer spreadsheet may be helpful for doing this.). Do the value calculations themselves tell you anything about GUN's preferred settlement strategy?
24.6. What is the preferred settlement strategy for GUN?
24.7. Suppose AYEC proposed to limit its potential exposure by making a one-time payment today of $15 million in return for GUN dropping the claim for lost revenues. Should GUN accept the offer?
24.8. Suppose the court uses its equitable discretion to stipulate at the outset that, based on the facts so far, it will deny any prejudgment interest. What effect does this have on GUN's preferred settlement strategy? Can you draw any general conclusions from this on when you need to consider prejudgment interest?

8.8.4 Precedence

25. List two ways the impact of a case on future cases might occur.
26. How would you decide whether to include or exclude impact on precedence from an analysis?
27. What are the considerations for including or excluding legal fees and costs from an analysis of the effect on future cases?
28. From the plaintiff's perspective, are there particular plaintiffs or types of cases where the effect on future cases would be more or less important? From the defendant's perspective, are there similar types of defendants or cases where the effect on future cases would be more or less important? Give reasons for your answers.
29. You represent a pharmaceutical company (Krengle, Inc.) in a suit for strict liability and for negligence brought by a patient who used one of the company's birth control pills for a number of years, Marjorie Smith. While using the pill, Marjorie suffered acute kidney failure, eventually requiring a kidney transplant. Several years after that, Marjorie developed cervical dysplasia (a growth on the cervix), which her doctors thought might be caused by the immunosuppressant therapy she was receiving to minimize the risk of kidney rejection. Because this dysplasia indicated a higher risk of cancer, though not malignant itself, Marjorie elected a hysterectomy. Her specific damages total $300,000, and she seeks $2 million in general damages.

    Marjorie's suit alleges that her kidney failure was caused by malignant hypertension (high blood pressure) which, in turn, was induced by Krengle's birth control pill. She contends that there is strict liability in pharmaceutical cases for failure to warn of reasonably foreseeable side effects that physicians should monitor. This basis of liability has been established in a number of recent cases. However, the new ground in this case is Marjorie's attempt to introduce evidence of Krengle's subsequent remedial measures on the issue of strict liability. Krengle changed its package insert to warn doctors to monitor possible high blood pressure after Marjorie was taken off the pill, but well before her claim was brought.

The general rule, embodied in Federal Rules of Evidence 407, is that

> When, after an event, measures are taken which, if taken previously, would have made the event less likely to occur, evidence of the subsequent measures is not admissible to prove negligence or culpable conduct in connection with the event. This rule does not require the exclusion of evidence of subsequent measures when offered for another purpose, such as proving ownership, control, or feasibility of precautionary measures if controverted, or impeachment.

The rationales for this rule are to encourage subsequent remedial measures by ensuring that they won't affect liability for prior incidents, and that more wisdom in hindsight doesn't necessarily imply prior negligence. However, the landmark case of Ault v. International Harvester Co., 13 Cal. 3d 113, 528 P.2d 1148 (1974) established that this rule does not apply in strict liability cases because culpability is not at issue, and because the corporate mass-producer of goods will not forego making improvements in its products, thereby risking innumerable future lawsuits, simply to avoid such measures being admitted as evidence in strict liability cases.

All the same, the applicability of the rule is not settled. A number of cases have held that where a strict liability case involves a design defect or failure to warn, the Ault exception may not apply; first, because concepts of strict liability and negligence liability are often intertwined; and, second, because the duty to warn itself involves considerations very much like negligence. The inquiry focuses on whether there was adequate information and knowledge at the time of Smith's prescription such that the warning given later (the subsequent remedial measure) should have reasonably been given earlier—thus leading to the failure to warn.

Towards this issue of knowledge and reasonable warning, Smith plans to introduce a European report that Krengle had at the time of her prescription which warns of the danger of hypertension with the type and dosage of hormone used in the pill in question. The report concludes that doctors should monitor the blood pressure of patients on this particular variety of pill and switch to a different prescription if hypertension is noted. Krengle responds that, even though it had the report at the time of Smith's prescription, it took some time to evaluate its findings and to get label and insert changes approved by the FDA. A further wrinkle is that, after waiting in frustration for advance approval for a labeling change, Krengle made the change unilaterally and got FDA approval after the fact. Krengle thus contends that feasibility of making the labeling change is not at issue, and the exception built in to FRE 407 does not apply.

Aside from the consequences in this case, Krengle is worried about the consequences of subsequent remedial measures being admitted routinely in future lawsuits. Krengle is constantly changing its product information and, less often, making subtle changes in formulation, and dreads the prospect of every improvement being admitted in court as evidence of prior negligence. Thus, the holding in this case could affect future cases involving Krengle's entire line of products.

29.1. Draw the relevance diagram for the value of this case, then for the value of future cases. You can incorporate them into one diagram if you wish.
29.2. Draw the decision tree for the value of this case and write the value formulas.
29.3. Assess the probabilities for the uncertainties in this case. Be sure to include a realistic probability assessment on the award to Smith in the various scenarios. Her expected award will be important in examining future possible awards.
29.4. Assume that the number of future lawsuits have been assessed for the following scenarios as below. Discretize each distribution to obtain a chance node for the expected number of future lawsuits in each scenario.
- settle the case;
- litigate the case and win;
- litigate the case and lose, but win on excluding subsequent remedial measures;
- litigate the case and lose, but lose on excluding subsequent remedial measures.

29.5. Using the assessed damages for Smith in each scenario, assessing the expected cost of future suits for each scenario (Recall that, to avoid anchoring problems, you would be better off at least assessing a three-branch estimate instead of the expected cost directly.).
29.6. Calculate the expected cost of future suits in each scenario and plug the numbers into your main decision tree.
29.7. What is Krengle's settlement strategy without considering the cost of future lawsuits (the effect of precedence)? What is it with precedence considered?
29.8. What is the value of information on winning the exclusion of subsequent remedial measures? What is the value of information on winning the suit? What does this tell you about the relative importance of the precedence set?
29.9. Note that separate contingent distributions were not assessed for the evidentiary ruling if Krengle won the case. Explain why or why not you feel this is a realistic assumption.

8.8.5 Valuing Nonmonetary Consequences

30. You work for the Department of Transportation and are assigned the job of doing the cost/benefit analysis for proposed new regulations regarding roll bars in convertible automobiles. Roll bars are reinforced, roof-height extensions of the car's structure that can support the car's weight and protect the occupants in case the

car rolls over (Note that the windshield and frame cannot support the weight of the overturned car.). Roll bars have been standard in race cars for decades and are offered in some convertibles, but are the exception rather than the rule. The regulations you're evaluating propose to make roll bars mandatory.

To make matters more complicated, two types of roll bars are currently feasible. The standard variety is fixed in place whether the top is up or down. This type has not been adopted at least partly because it spoils the open-top look that is the whole point of a convertible. A new, more expensive and complicated type is spring-loaded and pops up into position in less than a second when triggered by sensors in the vehicle. The sensors detect a number of situations that might indicate an impending roll-over, such as sudden and hard deceleration, excessive body roll, or one wheel leaving the ground. This system is already been offered as standard equipment by one luxury automobile manufacturer.

To begin with you focus on the costs of imposing roll bars. There would be up-front design and development costs, and additional manufacturing costs. Further, any roll bar would add weight to the car, reducing gas mileage and increasing fuel costs. The actual increase in fuel costs will depend on the price of gasoline, the fuel efficiency of the cars, and the number of miles driven. The design and production costs and weight would likely vary by the type of roll bar required. You consider other possible costs, such as the esthetic penalty, but decide (for the first-pass analysis, at least) to limit your analysis to these more concrete varieties.

You then turn to the benefits of the regulation. You decide to focus on the amount that roll bars would reduce deaths from roll-over accidents. Thus, the benefits will depend first of all on the small-probability value of life you use in your analysis (see text for more on this). The number of deaths from roll overs, then, will depend on the number of new convertibles on the road, their frequency of roll over, and the frequency of death in roll-over accidents—both with and without roll bars (this breakdown corresponds to the way you and other experts have studied the problem). You only evaluate the convertibles sold in the future because only they are potentially affected by new regulations. The convertibles already on the road are a sort of "sunk cost" unless the regulations also require retrofitting—an exorbitantly expensive proposition. One other factor will be the average number of passengers in the car when a roll over occurs.

The other side to the benefits analysis is how auto company profits are affected. One simple way to handle this is to project convertible sales and figure an average profit margin on each car. The extra production and design costs would come out of these profits. You realize, of course, that car makers will likely be able to pass on part of their increased costs in the form of higher prices, but consumers would also have a dis-benefit of having to pay higher prices. It therefore seems good enough for a first-pass to just consider the profits reduced by higher costs because somebody will have to pay.

You then consider how the annual growth rate of convertible sales will be affected by requiring roll bars. The auto makers, of course, argue that the end of the truly "open car" will put a big dent in convertible sales. However, you feel it's at least possible that the extra safety would attract more buyers—sort of a Volvo-effect. You make a note to consider both increased and decreased growth in sales later on.

30.1. Structure the relevance diagram for the problem as given. Do you feel it gives a fair representation of the costs and benefits inherent in the proposed regulation? Explain why or why not, and indicate whether and why any changes you propose should be included in the first-pass analysis or held for later refinements.

After structuring the relevance diagram to get an idea of the numbers you'll need and their relationship to each other, you begin researching the data. Some numbers are directly available from past research, and others need to be estimated from a variety of related sources. You arrive at the following set of preliminary estimates. Note that 2020 will be the start of the analysis because it will take till then before any changes could be put into effect. You decide at the same time to make the first-pass analysis for a 10-year period to avoid getting into too many vehicle retirement questions.

Convertible Use and Efficiency

Average miles per gallon	22
Reference weight for miles per gallon	3500 pounds
Reduction in mileage for additional 100 pounds of weight	5%
Average cost of fuel	$4/gallon
Average annual miles driven	15,000
Average number of passengers	1.5
Percent of Vehicles Rolling Over Each Year	0.25%

Convertible Market

2020 estimated sales	300,000 cars
Number of models	25
Average profit per vehicle	$5,000

Roll Bar Data

	No roll bar	Fixed roll bar	Spring roll bar
Design cost ($mm) per model	0	150	300
Production cost ($ per vehicle)	0	300	800
Weight (lbs)	0	200	350
Percent of passengers killed when a rollover occurs	10%	1%	1.5%
Annual sales growth	2%	4%	5%

Economic Data

Small-probability life value	$2.43 million
Discount rate	3%

30.2. Put together a numerical model to calculate the net cost of each proposed alternative (no roll bar, fixed roll bar, spring-loaded roll bar). Use of a spreadsheet for this purpose is strongly recommended. One possible organization of such a spreadsheet follows.

- Organize a section for all the input data in the top of the spreadsheet.
- Organize the roll bar data into a table that lets you specify the type of roll bar selected (such as "none" or "spring") and picks the data for the remaining calculations. In Excel, the HLOOKUP function performs this purpose.

30.3. Calculate the number of new car sales each year, the total number of cars on the road, the number of roll-over accidents, the number of fatalities per accident, and the small-probability value of the loss of life.
30.4. Calculate the additional fuel costs.
30.5. Calculate the additional design and productions costs.
30.6. Calculate the total costs.
30.7. Calculate the total profits on convertible sales.
30.8. Calculate the net profits (total profits minus total costs).
30.9. Discount the 10-years of net profits to obtain a net present value.

Note that this calculation does not measure the total costs and benefits to society. It does not, for instance, include that value of the jobs provided by the auto manufacturers, and so on. However, by calculating a base figure for not requiring any roll bar, you can compare the additional costs and benefits of requiring a roll bar against it. Thus, you can evaluate the incremental costs and incremental benefits of the two roll bar configuration. Because only these incremental costs and benefits are relevant to the decision (the sunk costs point), you only need an accurate relation to a base figure. That base figure does not itself have to be a complete model of the value to society. It only has to be complete enough to see the differences made by the relevant alternatives (in this case, the roll bars).

30.10. Calculate the base net present values for each of the three alternatives (This step will be very simple if you've followed the above steps for constructing your spreadsheet.). On this basis, which alternative looks best so far? What does your intuition tell you about the strengths and weaknesses of the analysis so far?
30.11. Assume that you've gone through the three-branch shortcut for continuous variables to get the 10%, 50%, and 90% points for the input data. For some variables, it made sense to assess a plus or minus (±) variation on the historical data.

Convertible Use and Efficiency

	10% chance actual is lower	50/50 chance actual is higher or lower	10% chance actual is higher
Average miles per gallon	18	22	27
Reference weight for miles per gallon (pounds)	3300	3500	3900
Reduction in mileage for additional 100 pounds of weight	2%	5%	7%
Average cost of fuel ($/gallon)	$3.25	$3.75	$4.25
Average annual miles driven	11,000	15,000	20,000
Average number of passengers	1.1	1.4	1.8
Percent of vehicles rolling over each year	0.2%	0.3%	0.5%

Convertible Market

	10% chance actual is lower	50/50 chance actual is higher or lower	10% chance actual is higher
2020 estimated vehicle sales	200,000	300,000	450,000
Number of model	20	25	35
Average profit per vehicle	$3,000	$5,000	$8,000

Roll Bar Data

	No roll bar	Fixed roll bar	Spring roll bar	Variation
Design cost ($mm) per model	0	150	300	±30%
Production cost ($ per vehicle)	0	300	800	±20%
Weight (lbs)	0	200	350	±10%
Percent of passengers killed when a rollover occurs	10%	1%	1.5%	±50%
Annual sales growth	2%	4%	5%	±40%

Economic Data

	10% chance actual is lower	50/50 chance actual is higher or lower	10% chance actual is higher
Small probability value of life ($mm)	1	2.43	4

Discount Rate 3%

For each of the data inputs, set the input to its high and low values, recording the net present value each time. Be sure to reset each input back to its 50% value before proceeding to the next one.

This process is called deterministic sensitivity analysis (as opposed to the probabilistic sensitivity analysis you've seen earlier in value of information calculations), and its purpose is to identify those uncertainties that have the greatest effect on the value of the problem (in this case, the net present value).

To get a complete picture, you should do this analysis for each of the three roll bar alternatives. However, if you don't have a software package to assist you, try slogging through it for the fixed roll bar alternative.[11]

30.12. What are the four or five variables that cause the greatest variation in the net present value? How do these results square with your intuition about which data inputs and uncertainties should be most important? Can you verify that the changes are sensible, given the relationships expressed in your spreadsheet?

30.13. Select those three or four data inputs that had the greatest effect on the net present value and structure them into a probability tree. Note that the decision is which roll bar to require, so you will have an identical probability subtree for each alternative (If you have decision tree software, you can include more uncertainties.).

30.14. Calculate the scenario values and the mean values for the three alternatives. Which one looks best? Can you justify this in terms of the incremental costs and the lives saved?

30.15. Assess the analysis so far. What insights have you gained into the efficacy and cost-effectiveness of the three alternatives? Where do you feel the analysis is most deficient, and why? Would you add extra detail, or remove detail, and why?

30.16. Calculate the values of information and control on the uncertainties in your tree (Again, this is an arduous task without software.). What do these tell you about areas for further study? Does it have any implications for your recommendation about the best regulation?

30.17. Examine your three subtrees. Construct a table showing in which scenarios each alternative is best (That is, don't bother where a single alternative is best in each scenario.). How would this information help you to support your recommendation? How would it guide further development of your analysis?

30.18. Consider the small-probability value of life used in this analysis. Why is or isn't it appropriate? How would you justify to a congressional committee its use in this analysis?

31. See also problem 4 in Chapter 12 (the MoCo problem) for valuing nonmonetary consequences.

[11] Supertree has a free companion program called Sensitivity for performing deterministic sensitivity analysis.

Chapter 9
Special Probability Approaches

In Part II, you have learned about how subjective probabilities can be used to represent the relative likelihood of possible and uncertain answers to a question; how dependent probabilities can be used to represent the relationship between two questions; how to assess probabilities to avoid the cognitive and motivational problems that make their use necessary in the first place; and, finally, how probability arithmetic works to express the value of your case in a single measure (like the mean value).

This chapter expands on each of those topics. First, the section on Many Possible Arguments shows how, when there are many possible arguments you would win or lose your case by, you can use questions and probabilities to represent meaningfully the likelihood of winning or losing your case without having to analyze each individual cause of action or legal theory in the complaint.

Second, the section on spill-over shows how dependent probabilities can represent the subtle but significant effects different parts of your case can have on one another—effects separate from winning or losing on specific legal issues.

Third, the section on working with a group shows how the methods of avoiding cognitive and motivational problems are extended to group analyses (that is, any more than one person doing the analysis).

Fourth, the section on probability distributions shows how to actually construct these other representations of the value of your case. You've seen these forms before in Part II, but the procedure for constructing them was deferred to this chapter.

The final section on dependent probabilities returns to that subject once more, this time to explain finally how the direction of an arrow in a relevance diagram can be reversed (that is, the order of dependent probabilities reversed) and how this process is the key to explicitly calculating the value of *imperfect* information.

As mentioned in the introduction to Part III, these last two topics are advanced and may be omitted on a first reading.

J. Celona, *Winning at Litigation through Decision Analysis*,
Springer Series in Operations Research and Financial Engineering,
DOI 10.1007/978-3-319-30040-5_9

9.1 Many Possible Arguments

A common problem when you first begin to organize your case with relevance diagrams and decision trees is how to structure an analysis when the complaint that lists 20 different arguments why plaintiff should recover. A first reaction is to include all 20 arguments in your relevance diagram and tree because that's the way you have to go about preparing an answer to the complaint. If you do so, the relevance diagram may be manageable, but the tree especially will be impossibly large—not to mention the problem of assessing meaningful probabilities for all 20 chance nodes. Thus, although responding to each point is necessary for the answer you file, doing so in your case analysis is neither practical nor helpful.

The solution to this potential mess lies in applying the Trier-of-Fact Rule: how will the judge or jury decide the case? With experience, counsel will have a good sense of what the main issues and key arguments are. You need to separate the wheat from the chaff and focus your analysis on the key legal and evidentiary issues that will drive liability and damages.

If in doubt, start simple. You can always add more issues to the analysis.

The following sections give a number of ways that remembering this focus can keep your tree small and your analysis meaningful.

9.1.1 Focus on How the Judge or Jury Will Think About It

As mentioned above, the first approach to the 20-arguments problem is applying the Trier-of-Fact Rule: structure your problem the way the trier-of-fact (judge or jury) will think about it. Specifically, will the jury really consider separately each argument and make a separate determination of liability on the basis of that argument? Or, rather, will the jury mull them all over and make a finding not specifically tied to one of the theories?

If you really think the jury will consider multiple arguments separately (perhaps because of the substantive law or because of the way the judge will instruct them), then you should use the following procedure.

- For each argument, draw the relevance diagram for whether you win or lose on that argument, as below. We will do the analyses in separate pieces than show how to put them together (Figures 9.1 and 9.2).
- For each argument, draw the subtree and assess probabilities, as below. Do the arguments in the order you think the jury will, and make sure that each assessment is conditioned on the jury already having found you not liable under the preceding theory. This will help you to watch for any possible spill-over effect (such as synergies between the arguments) (Figures 9.3 and 9.4).

Regarding the tree for the second main argument, from a legal standpoint there cannot be any liability without causation. If for example, the jury form would require them to make a specific finding of causation in order to find liability, then a

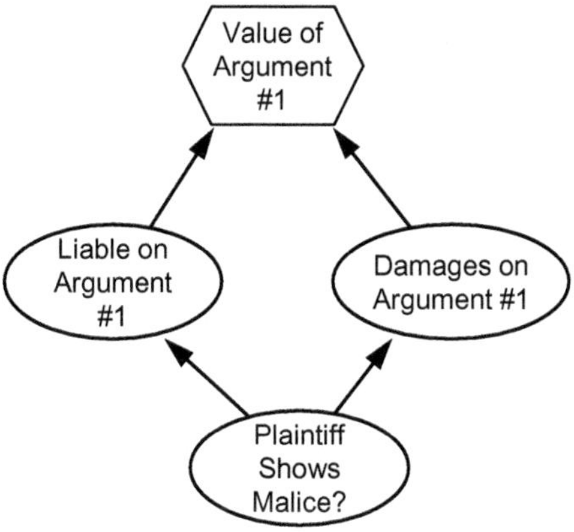

Figure 9.1 Relevance diagram for first main argument

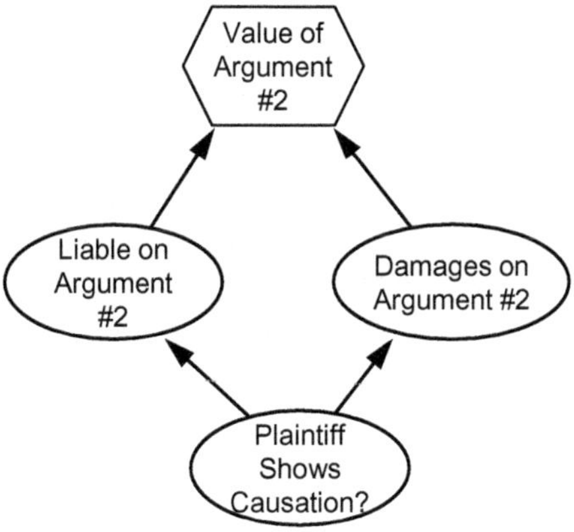

Figure 9.2 Relevance diagram for second main argument

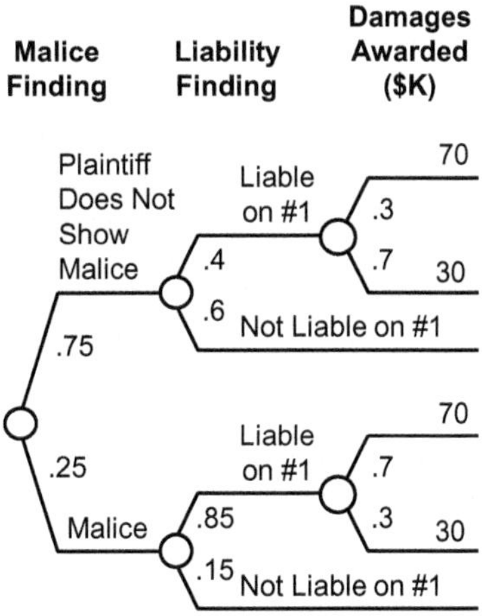

Figure 9.3 Assessed tree for first main argument

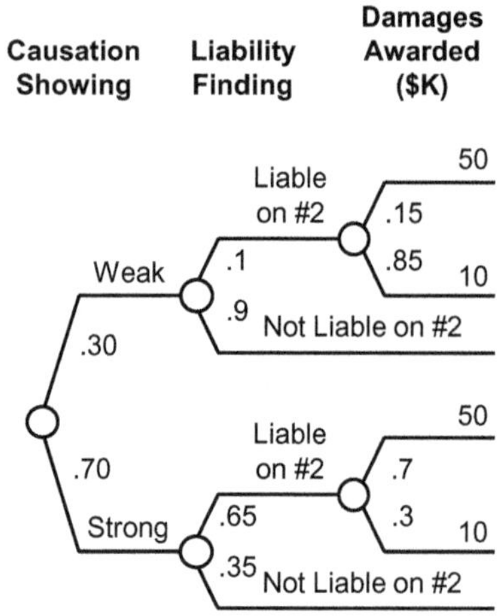

Figure 9.4 Assessed tree for second main argument

finding of no causation would mean a 0% probability of liability. This tree assumes that the jury will consider causation, but not have to make a specific finding on it.

- If you really think the jury is going to make an independent finding of liability or not on each possible argument, then they're probably going to consider damages based on which argument you lose by—and likely you'd have a damages node in each subtree, as illustrated here. If you think the jury is only going to make one damages determination, then do you really think they'll decide each argument separately?
- For each subtree, calculate the overall probability of winning by that argument, and the mean value of losing, as below (Figures 9.5 and 9.6).[1]
- So you can show the chance of winning and losing by each argument in the main tree and still get the mean values correct, compute the mean damages given liability for each argument as follows.
  - #1 mean damages given liability = 24.9 ÷ .51 = $48, 823.
  - #2 mean damages given liability = 17.535 ÷ .485 = $36,155.
- Draw the main decision tree with one chance node to represent the possibility of winning or losing by each individual argument. Remember to keep the nodes in the order you think the jury will consider them. Compute the overall mean value of the case, and the chance of losing or winning (Figure 9.7).

Alternatively, you could draw the two alternative arguments for recovering on a claim with a tree like the one below (Figure 9.8).

[1] Very advanced note: if you're considering the client's risk aversion and want to show her the full cumulative distribution, you'll need to summarize the subtrees with chance nodes from the discretized distributions, rather than with mean values. Otherwise, you'll lose the spread of possible outcomes from each subtree. See the section on subtrees for more details on how to do this.

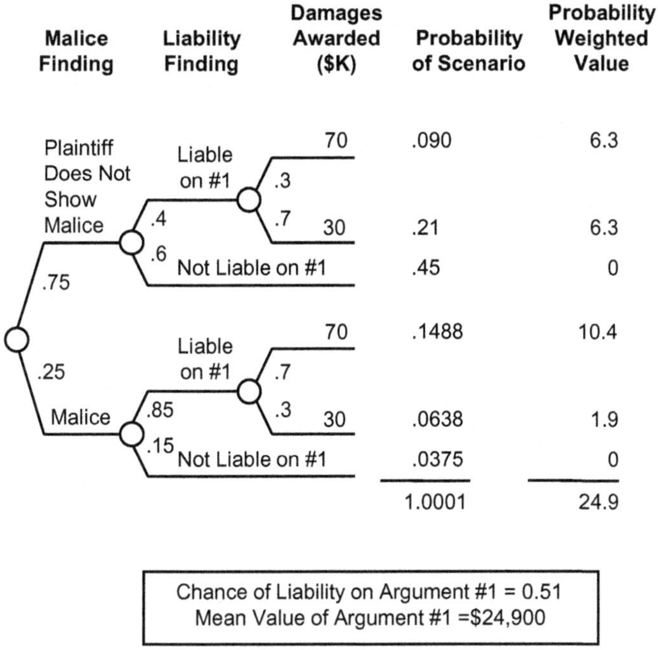

Figure 9.5 Roll forward mean value calculation for first argument

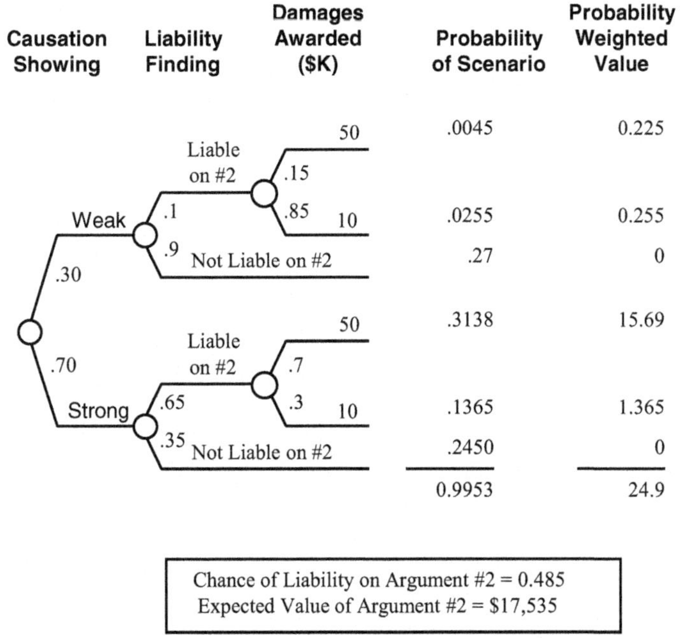

Figure 9.6 Roll forward mean value calculation for second argument

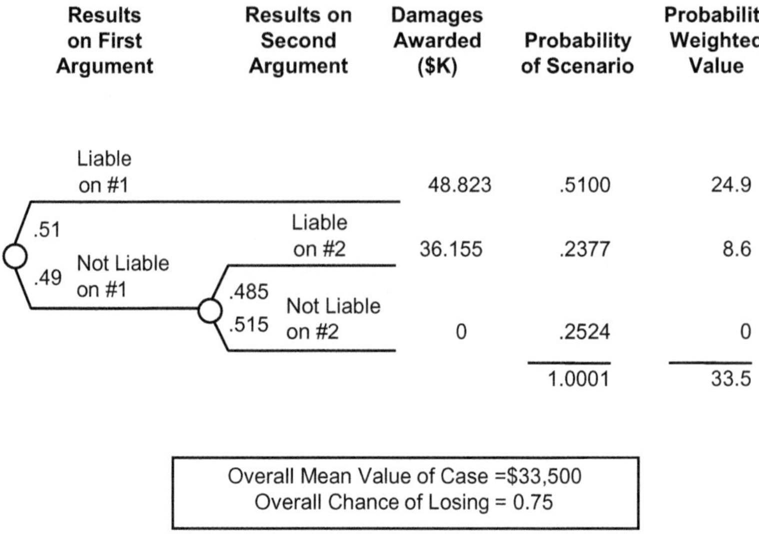

Figure 9.7 Overall tree for prospects for the case

Figure 9.8 Two alternative arguments done as a single tree

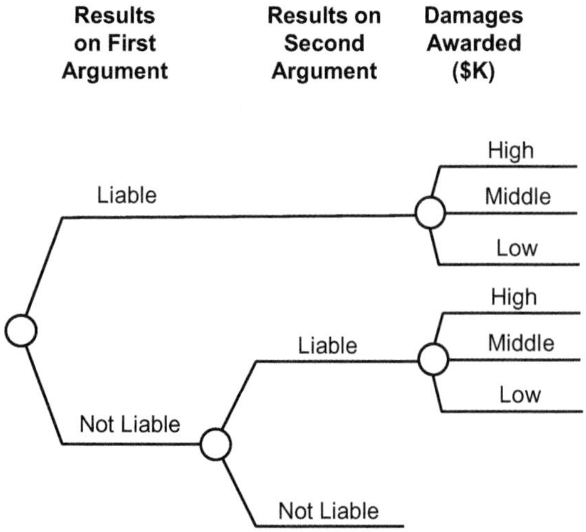

Either tree will work, and there are multiple correct ways to structure a problem. To make this formulation fully equivalent to the first formulation, you would need to add the issues of malice and causation.

This tree shows two alternative arguments to win on basically the same cause of action (with contract, tort, etc.). If the two arguments are actually independent causes of action with damages that would add together (e.g., a contract and a tort claim in the same case), you would need a tree like the one below (Figure 9.9).

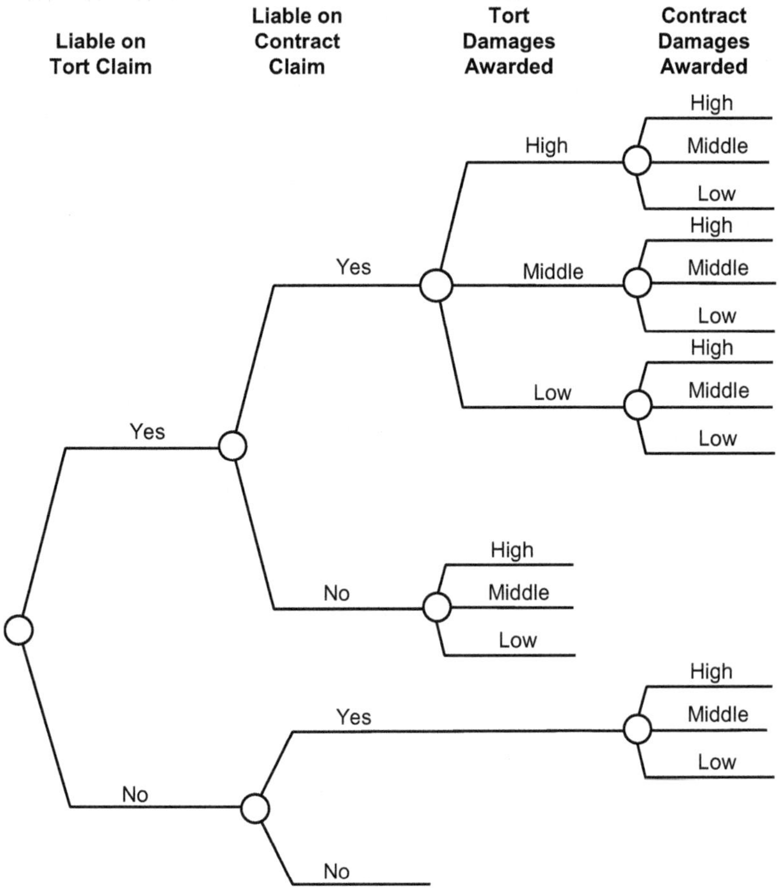

Figure 9.9 Two independent causes of action, rather than two arguments under the same claim

Part of the purpose of drawing the tree is to help you think more clearly about what is going on in the case and how it will play out. The structure should help you assess what could happen in various scenarios.

Of course, it's very easy (even easier with software) to make the tree so complicated neither you nor anyone else can make sense of it. Following are some suggestions for keeping your analysis as simple as possible, but no simpler.

9.1.2 Don't Use More Detail Than Is Helpful

The preceding section showed how to handle the 20-arguments problem with subtrees when you felt the jury really was going to separately argue and decide each argument. However, if you felt that the jury was not going to parse the problem this finely, then the extra detail actually interferes with your case analysis by giving you a different perspective than the jury will bring to bear—and thus degrading your estimation of the case value.

Another way to consider the effect of the 20 arguments is to list them in your tree as possible reasons why the jury came out one way or another. This method assumes that the jury will certainly listen to and consider each argument, but will kind-of roll them all together and make a decision without a separate finding on each argument. Thus, a node with reasons for the example shown in the preceding section might look like the following (Figure 9.10).

A similar situation is when an ultimate issue (such as "Insured Liable?") depends on more than three uncertain influencing factors, and these influencing factors all seem equally significant and are independent of each other. In this case, it may also be best just to directly assess the probability on the ultimate issue. You could also pick out one or two of the influencing factors that seemed far more significant than the rest and draw that factor as a chance node in the tree, leaving the others for your list of reasons.

For instance, suppose an insured was involved in an accident while making a turn in an intersection at dusk. Preliminary structuring produced the following relevance diagram on whether the insured would be found liable (Figure 9.11).

Figure 9.10 Listing reasons for various outcomes instead of separating factors

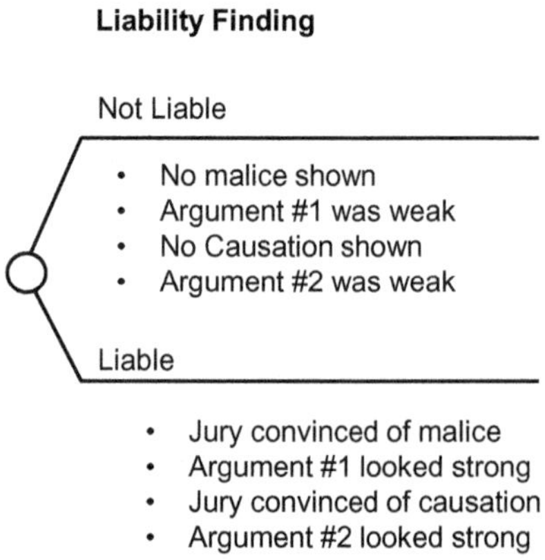

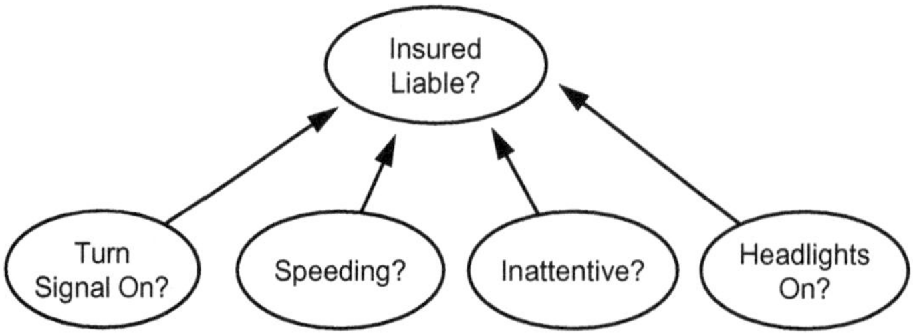

Figure 9.11 Relevance diagram for multiple factors bearing in liability

If you thought the jury was going to make a determination on each one of those issues before deciding liability, or if you felt most comfortable assessing probabilities on liability conditioned on a finding on each one of those issues, you could draw the full decision tree as below (Figure 9.12).

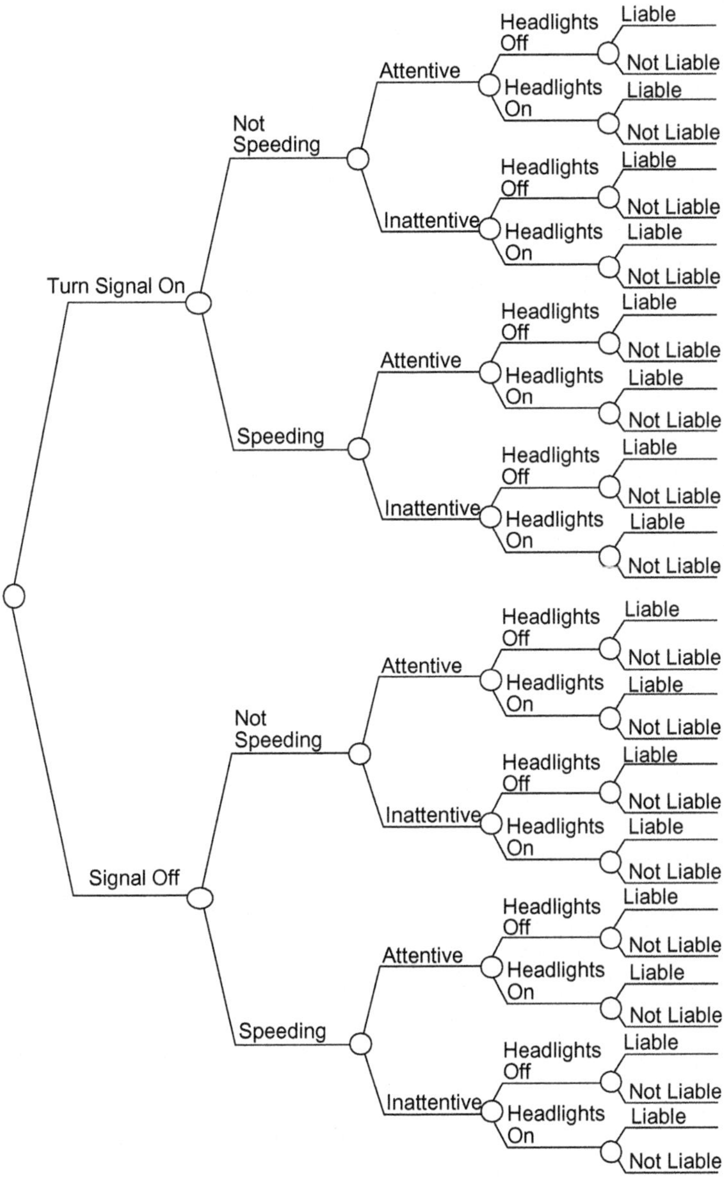

Figure 9.12 Too much breakdown to make sense of

This tree would require assessing 16 different sets of probabilities for liability. If it's shown that the insured had the turn signal on, was not speeding, and was attentive, how would the odds vary depending on whether the headlights were on or off? Breaking things down to a level of detail where people can't distinguish meaningfully between the scenarios is a sure sign you've gone too far.

The better approach is to start simple and only add as much detail as you need. For example, in a case like this, you could consider just a direct assessment considering all the possible factors (Figure 9.13).

However, this is probably not enough of a breakdown. Inattentiveness will be hard to show and headlights on or off at dusk may not matter much, but whether the insured was speeding and had a turn signal on would likely have a big effect on the chances for liability. A tree which breaks down just these factors is given below (Figure 9.14).

Figure 9.13 Not enough breakdown between key factors

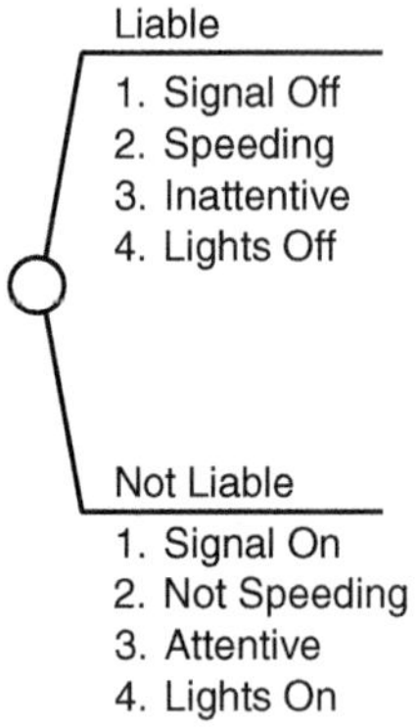

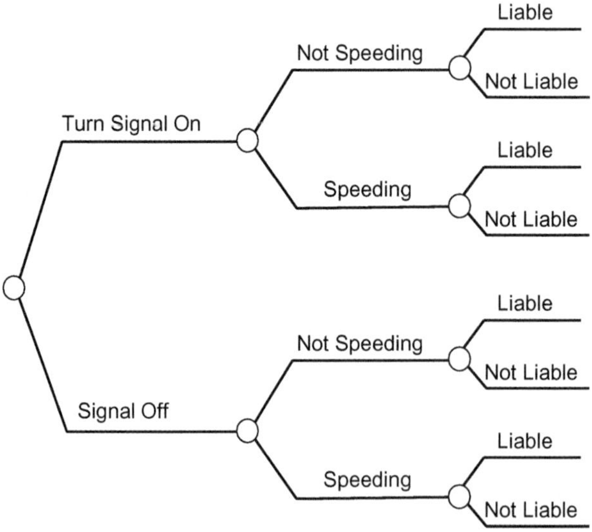

Figure 9.14 Possibly enough detail, but not too much

For this tree, it should jump out at you that, if the insured had his turn signal on and wasn't speeding, the odds of liability are very low, maybe around 5–10%. If the reverse is true (signal off and speeding), odds of liability are very high, maybe 90–95%. The two split scenarios in the middle of the tree might be like a 50–50. With adding the chances of the signal and speeding being demonstrated at trial, there'd be a complete tree to calculate the chances of liability.

As you can see, there is no hard and fast rule about what the right level of detail is. More detail that helps you think more clearly about the possibilities is helpful. Adding detail beyond that point simply muddies the waters.

Getting the level of detail right is part of the art of decision analysis. Like many arts, it is a skill learned with experience and practice. Hopefully, this example provides some guidance as to getting the level of detail right.

9.1.3 Look for "Clincher" Arguments

The third technique for dealing with the 20-arguments problem is a variation on putting the more significant influencing factor in the tree and leaving the rest for the list of reasons on the ultimate issue (discussed above). This variation is to look for a "clincher" argument; that is, any argument that could by itself determine the case. That one issue would then similarly be put into the tree as a separate chance node. All the other problems or issue (the ones not determinative individually), could then be put together in the list of reasons why you might lose the case even though you didn't lose on the clincher issue. The tree below illustrates this approach (Figure 9.15).

Of course, depending on whether you represent plaintiff or defendant, you might also possibly win on the clincher issue, and then win or lose on the whole combination of other issues.

For instance, consider the insured's traffic accident considered in the preceding section. You might think that speeding was a "clincher" issue, and that you could lose the case on speeding alone, and still lose the case on all the other issues even if you got by speeding. The tree for this representation appears below (Figure 9.16).

If you felt some issues stronger than others, but didn't think any of them a "clincher," or didn't think any of them really stood out enough to be singled out in the tree with a chance node, then you'd be back to the method of the preceding section, where you should directly assess the ultimate issue of liability and list the other issues as possible reasons for finding liable or not liable (Figure 9.17).

This approach also applies to cases where a series of events is at issue (such as a number of events leading up to an accident for which negligence is alleged), and there may or may not be one event (such as a particularly egregious action) that could fix liability on its own. Then, even if the court doesn't find liability on the basis of that one action, it could still find liability on the basis of the whole series of events considered together.

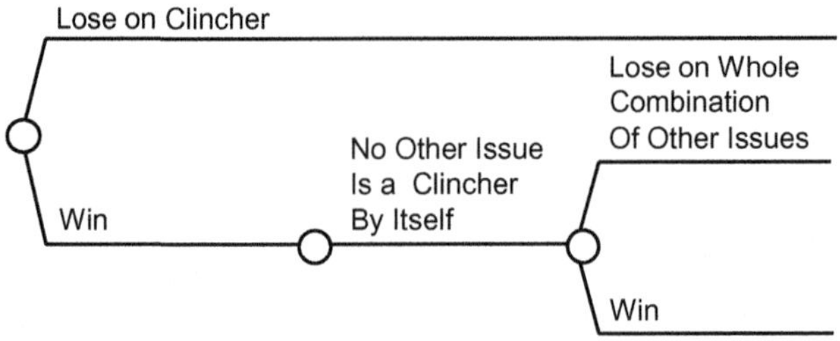

Figure 9.15 Tree where one issue is a "clincher"

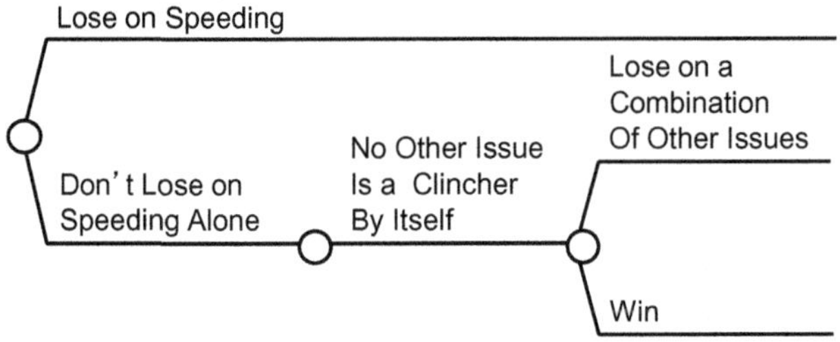

Figure 9.16 Tree where speeding along is a "clincher" argument

Figure 9.17 Tree with many contributing but non-determinative factors

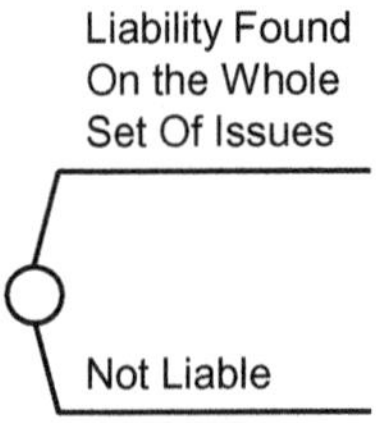

To summarize, then, you have three methods for dealing with the 20-arguments problem/many influencing issues.

- Do a separate subtree for each argument or each issue the trier-of-fact could find liability on.
- Directly assess the probabilities on the ultimate issue and list the arguments or other issues as reasons why the court could come out one way or the other.
- Pick out the one or two most important arguments or issues for their own chance nodes, and list the rest as reasons why the court might nonetheless come out for or against you (Of course, with this method you need to make sure your probabilities are properly conditioned on what's already happened in the tree.).

One of these methods should enable you to analyze the case the way you think the jury will, and thus help ensure a meaningful valuation of your case.

9.2 Handling "Spill-Over"

Throughout this book, you've heard the term spill-over used to refer to dependencies or effects between parts of your case that not formally related together in a legal sense (like the relevant standard of care is necessarily related to a finding of negligence). If you're a trial attorney, you've seen spill-over in various forms: on the law, such as when one legal claim makes another look better or worse; on the facts, such as when particular evidence hurts you client's credibility almost regardless of what other evidence you introduce; and on the emotion of the jury, such as when you know they're dead-set on a particular finding "come hell or high-water."

The spill-over effect is real; it affects your chances of winning or losing and the value of your case. Accordingly, throughout this text we have presented suggestions for when to look especially for spill-over, and how to incorporate it into your analysis. This section takes a closer look at the types of spill-over, and, in particular, suggests how dependent probabilities are particularly useful for handling it.

9.2.1 Argument Synergies and Order of Proof

One of the most commonly recognized types of spill-over is between alternative arguments and the order they're argued in. More specifically, two types of effects are possible.

- The mere fact of offering two legally independent and alternative arguments for finding liability may make each appear stronger (or weaker, if one argument is not credible).
- The order of proof may affect the overall chances of winning, such that presenting one argument first makes the second seem more or less likely. In this case, the order the jury considers the issues in may affect your overall chances of winning.

Accordingly, your probability assessments should always be conditioned on and reflect the influence of all arguments both sides will be making so you can properly account for these effects. An example will illustrate how you might use dependent probabilities to account for these effects.

Suppose you are defending a claim where plaintiff alleges there was a breach of the implied covenant of good faith and fair dealing arising from a fiduciary relationship. Plaintiff has a lot of evidence and testimony on the alleged breach and has been emphasizing that point in pleadings and arguments. However, you feel the evidence is much weaker that a fiduciary relationship existed in the first place, which would make your client's actions quite usual dealings in an arm's-length business relationship.

If the case went to trial that way, you might assess the chances on the case like this (Figure 9.18).

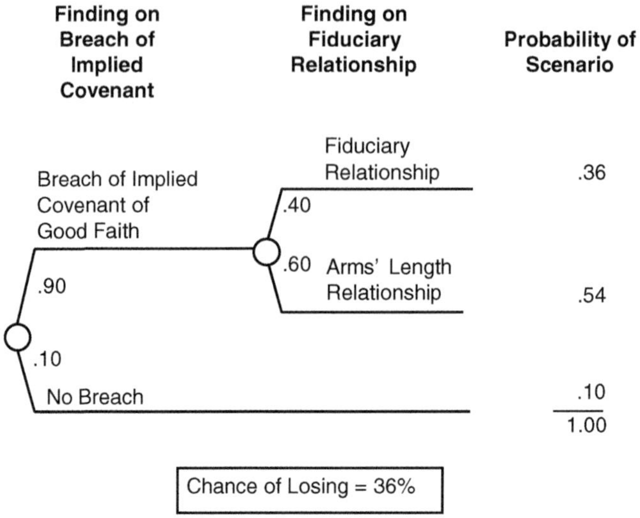

Figure 9.18 Assessment with plaintiff making his strongest argument first

However, suppose that as defense attorney you were able to reverse the order of argument and proof by contending that plaintiff must prove existence of a fiduciary relationship as a threshold requirement. Otherwise, the implied covenant is not even relevant. Further, you felt that, once having made your strongest case first, even if you lost on the issue plaintiff would look less credible on the implied breach. An assessment of the chances this way might look like the following (Figure 9.19).

With this order of proof, you feel your client's chances of losing dropped by almost two-thirds. Your client would have much better prospects at trial, and a stronger negotiating position for possible settlement.

Of course, there is no objective "proof" of the improved prospects for your client. But what you have done is quantified your experience, judgment, and intuition as to the prospects for changing the order of proof which you can use to drive trial and settlement strategy. Clarifying your thinking to make better decisions considering uncertain prospects is exactly the point.

9.2.2 Jury (or Judge) Synergy

In the previous example, the spill-over effect was between uncertainties which are formally related: there cannot be a breach of an implied covenant of good faith and fair dealing if there is no fiduciary relationship.

A second type of spill-over is when you think a jury or judge not accepting one of arguments means they're less likely to accept your next argument even though the two are not formally related. The converse may also be true: winning on one

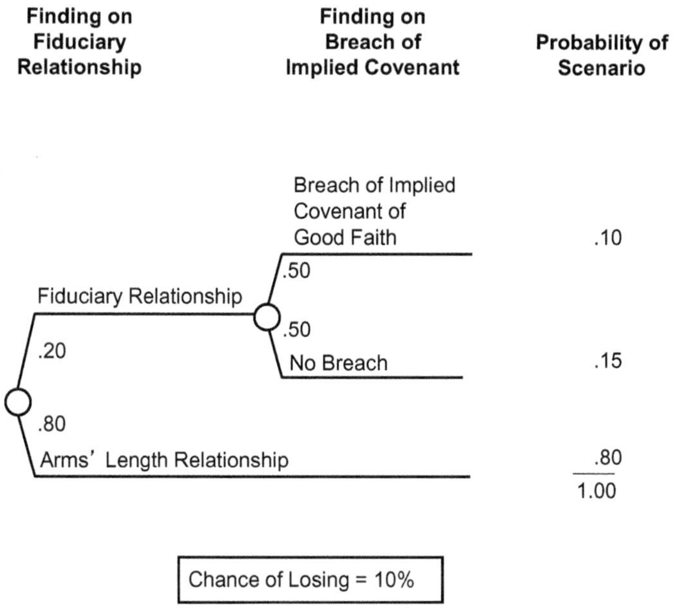

Figure 9.19 Defense attorney making his strongest argument first

argument may mean the trier is favoring your side and you have better chances of winning on the next count. And, of course, it may be that how the trier decides an early issue doesn't change your chances on later issues at all; that is, the resolution of the issues is independent and there are no connecting arrows in the relevance diagram (other than possible an "if no" or "if yes" arrow showing when the issue is not applicable).

If you feel this kind of synergy will affect your case, it should be reflected in dependent probability assessments, as illustrated in the tree below (Figure 9.20).

This tree shows a belief that how you do on your contract theory for recovering compensatory damages will affect how well you do on your tort theory for punitive damages, even though the two theories are not legally dependent on one another (as special and general tort damages would be).

This could happen for a number of reasons. For example, losing on one theory damages a party's credibility, which affects their prospects for other theories of liability. This happens frequently. Though counsel may feel professionally obliged to allege every possible theory of recover (as in a "shotgun" pleading), the weaker claims damage credibility on the stronger ones. The trier-of-fact will always be evaluating who to believe and whether they deserve to win—aside from the formal legal arguments.

As always, the Trier-of-Fact Rule applies to evaluating possible spill-over arguments: what order will the jury consider your arguments in, and how they view your second argument, given their result on your first?

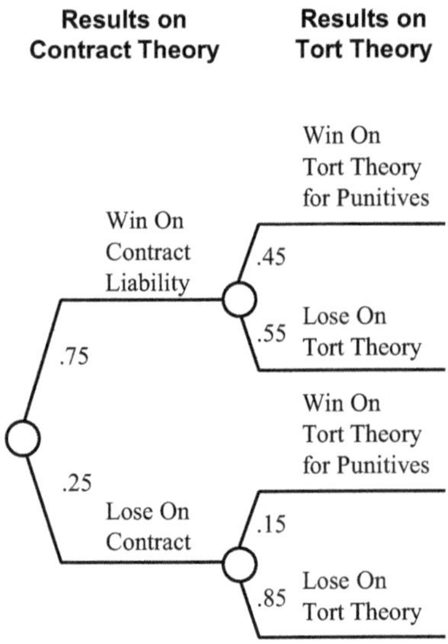

Figure 9.20 Spill-over between "independent" factors

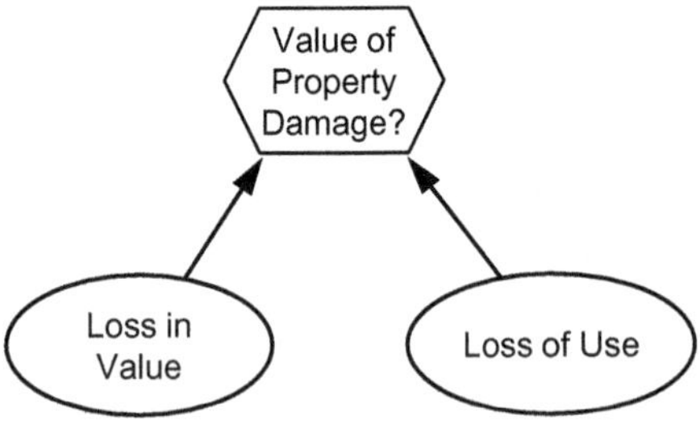

Figure 9.21 Relevance diagram for two elements to damages

9.2.3 Jury Consideration of an Issue in Stages

This third kind of spill-over arises when the judge or jury approaches an ultimate issue in stages: first, by making a preliminary determination on the ultimate issue, and then by considering the factors that influence the ultimate issue—rather as a way of testing their preliminary determination. The jury might then stand with their preliminary determination, or reverse themselves. An example will illustrate how this kind of spill-over might operate.

Suppose the jury were considering a claim for property damage, where possible components of that property damage include loss in property value and loss of use. The relevance diagram for these damages appears below (Figure 9.21).

However, you feel the jury will first consider whether there is any property damage at all before trying to parse the loss in value and loss in use components. You might be tempted to show that in a relevance diagram like this (Figure 9.22).

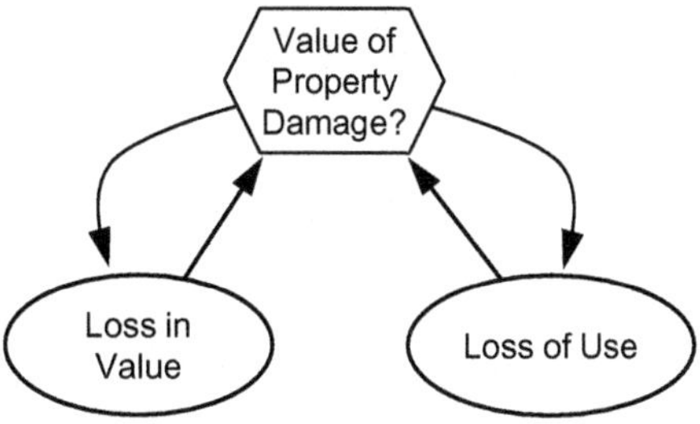

Figure 9.22 Mutual relevance?

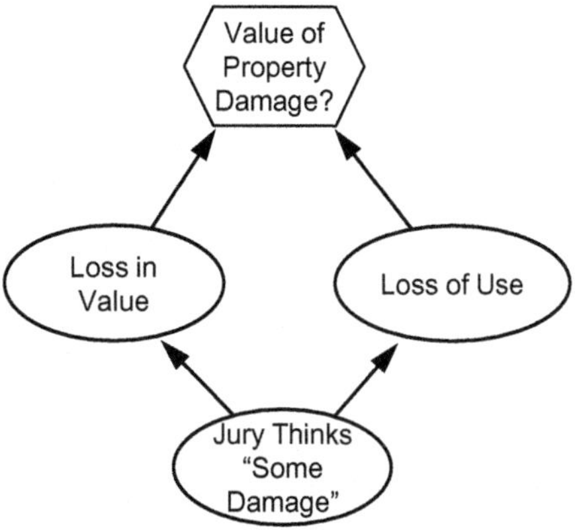

Figure 9.23 Adding a new issue to remove the loop

The error in this diagram becomes apparent when you try to draw the tree: where to start? The tree from left-to-right needs to follow the flow of the relevance diagram (the directions of the arrows). This diagram wouldn't work any better than household plumbing connected in a similar manner.

The plumbing or ordering rule for relevance diagram (no loops are allowed) helps you to sort this out. There needs to be a new issue at the bottom which flow upward along with the other issues.

In discussing this case with counsel, it was found that counsel was trying to express their belief that the jury would exhibit this kind of synergy by deciding the damage issue in two stages.

- 1st Pass: Did the jury think that there was any damage at all?
- 2nd Pass: If the jury thought that maybe there was some damage, then they would think harder about the issue and focus on the more specific kinds of property damage: loss in value or loss of use—with the possibility of the jury deciding they were wrong and there's no damage after all.

This kind of synergy can be more clearly seen in the corrected relevance diagram below (Figure 9.23).

This relevance diagram can be drawn as a tree following the usual ordering rule: start at the "source" and follow the arrows to the "sink." As in defining algorithms and data flows, loops tend to be errors (Figure 9.24).

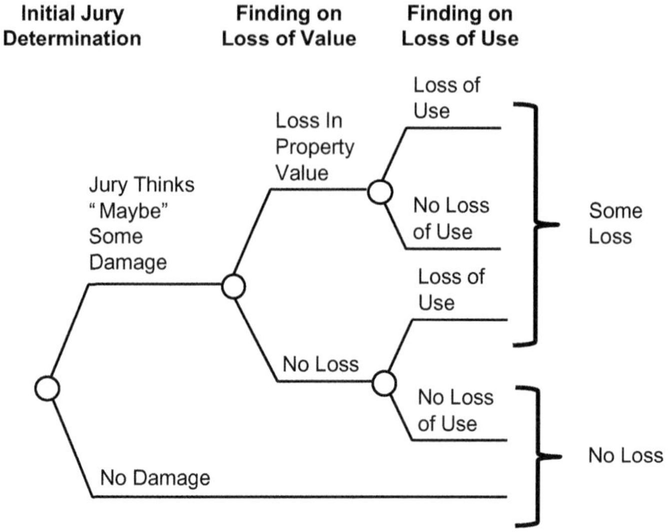

Figure 9.24 Tree for corrected relevance diagram

The final issue of issue of the value of the property damage is the value in each scenario, with the respective loss of use and loss of value damages (if any) added up.

This technique affords you a very insightful way of structuring the jury's consideration of your case. Remember, though, that to achieve meaningful results, the probability assessments on the jury finding loss in property or loss of use *must* be clearly conditioned on the jury having already found "maybe some damage." Otherwise, you end up with the same confusion (failing the clairvoyance test) evident in the first tree.

Using relevance diagrams, tree drawings, and dependent probabilities you can represent the very real phenomena of spill-over between parts of your case that are formally and legally unrelated. This gives you a clearer understanding of what's at issue and what's driving your case so you can make better decisions on how to proceed.

9.3 Working with a Group

The discussion in Part II gives general instructions on problem structuring and probability assessment than can apply equally to one or several persons doing an analysis. Indeed, when it comes time to make the litigation/settlement decision or to consider risk aversion, the client must be involved.

However, you may equally often work with other lawyers on a case, especially on larger cases. The following sections give further suggestions that will make working in a group easier and help to counteract some of the peer pressure that can introduce bias (and, thereby, inaccuracy) into the problem structure and into the probability assessments especially.

9.3.1 Problem Structuring in a Group

Whenever several lawyers are involved in the case, you will usually have a better and more insightful analysis if you involve all those lawyers, claims managers, etc. in structuring the problem. Several people are usually better at thinking of the possible issues, developing influencing factors, and arguing the strengths and weaknesses. Lawyers being in attendance, it's usually easy to find someone to play devil's advocate and argue against the prevailing sentiment. This is very helpful in testing your thinking.

Experience has shown that the several people do a better job of developing the real issues in the case (as opposed to what's pleaded) and thinking through the reasons the issues may turn out one way or another. Arguing about whether a probability is should be 20% or 60% is the most productive part of the discussion.

Accordingly, even if you're handling a case by yourself, you may find it useful at least to outline briefly the issues to a colleague and ask for his or her thoughts on what might influence the court to come out one way or the other.

The process for structuring the problem can be informal. You need no more definite agenda than the problem structuring steps given in Part II. Typically, we'll start off discussing the case for a bit, then at some point I'll start drawing possible structures on a white board to see if they work for the way people are thinking about the case. Sometimes, the first structure works. Other times, we need to go through several alternative structures before settling on for the first pass analysis.

A large white board is extremely useful for being able to draw out the entire tree and erase and revise as you go along. For a record, take a picture of the board with your cell phone. For very complex cases with many assessments, you can also use adhesive flip charts to write assessments large enough for everyone to read them, then post them on the wall as you go along. Likewise, when finished you can save the flip charts or photograph them.

As you go through the assessments, be sure to write out the reasons an issue may turn out one way or another. This is a very important part of documenting the state of understanding at the time of the assessment. This record is critical when taking a second look at a case and seeing if any new information has emerged, and whether any of the assessments should be updated.

9.3.2 Probability Assessment in a Group

Group probability assessments are best done with the same group and in the same session as problem structuring. Quantifying what people think in terms of numbers helps clarify their thinking, often leading to a problem reformulation along the way. With an expert facilitator, an initial problem structuring and assessment for a case can often be accomplished in a single 2–3 hour meeting.

The first time you work with a group on a decision analysis, I would recommend you begin with some materials explaining:

- What decision analysis is
- Why it's needed
- How it works
- What the results will be used for

In particular, the assessment and use of subjective Bayesian probabilities will likely be new to people. Having everyone complete a probability assessment exercise like the one in the problems at the end of Chapter 5 is highly recommended.

It is very difficult to bring other people in on this process part way through (e.g., for the assessment but not the structuring) because, having not been involved in the prior discussion, it will take some time for them to catch up with the group. Best to have everybody there for the entire process to build their understanding together. This also assures that the entire group understands, owns, and can support the analysis.

For certain issues, you may want to work with a particular expert to assess probabilities as discussed in the next section. It's still a good idea to get initial guesses from the group in the room, though it would be advisable not to disclose those guesses to the expert initially. You'll want to get the expert opinion unbiased by him or her knowing what the client thinks.

The process for assessing probabilities in a group is as follows.

- For each issue, discuss thoroughly all the reasons why an issue could be resolved favorably or unfavorably. This should involve identifying the legal and factual support for each side of the issue, and thinking about the biases of judge or jury (as well as possible synergies or spill-overs).
- For discrete issues, assess the discrete probabilities (negligence or not etc.).
- For each continuous issue, assess the 25%/50%/25% range.

If there is a difference of opinion on the assessments (as there often is), allow some time to discuss the divergent viewpoints and why they lead to different numbers. These differences are an important part of building understanding of the issues.

Most of the time, discussion will resolve these differences. Sometimes a person is convinced to change their initial opinion. Other times, the group will reach a consensus between the high and low viewpoints.

If there is a solid divergence on the probabilities, note both viewpoints and sets of numbers and move on to the next issue. You can run the analysis with both sets of numbers to see how much it changes the case value and whether it switches any decisions.

If the best alternative is the same regardless of which estimate it used, no further discussion is needed. If the best alternative switches or the settlement value of the case varies greatly, both perspectives need to be preserved and escalated to the ultimate decision-maker (client, senior claims manager, etc.).

As a case develops and further information emerges, the analysis can be revisited, preferably with the same group that originally did the structuring and assessments. Sometimes the group feels the numbers need to revised, sometimes not. The list of reasons for each assessment is especially important in helping people to recall what they were thinking previously and whether that has changed and assessments should be updated. Do date each set of assessments and record who was present for them.

9.3.3 *Working with Experts*

In your cases analyses, there may be specific questions that you don't feel comfortable assessing probabilities for, especially as the case involves more aspects of client's business or involves more technical questions (like the stress cracking of a particular metal alloy, for example). Accordingly, you'll want to consult an expert for these probabilities.

This is a different problem than working with a group because the expert won't have participated in structuring the problem and developing the reasons, and because the expert may have no training in Decision Analysis techniques. The process below therefore incorporates the aspects of the probability assessment procedure for counteracting cognitive and motivational problems, and other introductory steps that will help you obtain an accurate assessment with a minimum instruction.

- If possible, before consulting the expert, do a complete first-pass analysis and a probability sensitivity analysis on the questions you're considering consulting the expert about. There is no point in consulting the expert if the sensitivity shows that your preferred settlement or discovery strategy won't change over any reasonable range of probabilities or won't greatly impact case value.
- Explain to the expert what you are looking for, how it will be used, and why his role is important. Emphasize that, in most cases, the information provided will be confidential and privileged provided in preparation for possible litigation and not discoverable.
- You are looking for an accurate representation of his information, intuition, and judgment—emphasizing the uncertainty about what will actually happen—and not a "right" or "wrong" answer.
- The expert's information will be used in planning settlement or trial strategy etc. For this purpose, it is more important to get an accurate idea of the relative likelihood of the different possible answers, instead of emphasizing any one particular answer (In other words, you want the distribution and not the mean value. You'll calculate that later.).
- The expert is important because her information, intuition, and judgment are needed on this issue.
- Explain that the process you are using is important for obtaining a good estimate because of the traps people usually fall into when thinking about uncertain answers (that is, the cognitive and motivational problems).
- Go over the definition of the question and redo the clairvoyant test to make sure the question also passes according to the way the expert thinks about it.
- Develop some of the significant factors influencing how the answer will come out. You may want to draw a relevance diagram for just that question. If the

expert comes up with influencing factors that she can't integrate all together in a single assessment, you may need to do a different assessment for each scenario (that is, assess dependent probabilities).

- Work with the expert to develop her own list of reasons why particular answers might or might not happen. Then ask the expert about reasons found on your list but not on hers, and vice versa.
- Assess the answer with the procedure appropriate for your question (discrete or continuous uncertainty). You may made need to suggest your own initial guesses to help get the process rolling. If so, remember your own training in how people typically underestimate uncertainty and make guesses that will really get the expert thinking.
- Inquire about the possibility of a follow-up session, if necessary (Although, of course, your first-pass analysis and advance preparation is aimed at avoiding the time and expense of an additional session.).
- You may want to ask if she'd like further information about the Decision Analysis process.
- Although the expert may be curious, in some cases you won't be able to report back the results of the analysis. Advise what information you will be able to share.
- Thank the expert for her time and emphasize the value of her contribution.
- Make sure you have complete notes for the session and, if needed, create any additional documentation shortly thereafter while memory is fresh.

9.4 Tracking Estimates Over Time

Because assessing good subjective probabilities is a learned skill, tracking how good the guesses were over time is extremely helpful. With a single analysis, you can't tell much: assessing the uncertainty doesn't make it go away. You won't be able to tell with a single case how good the probabilities were; you'll just be able to see how the case turned out (if you're privy to that information).

However, with multiple analyses you can begin to see how good the assessments were. This is usually possible with multiple analyses done for a single organization (like the drug companies do with their drug development projects and Chevron does with its development projects). It was also possible with the work I did for Stanford University Medical Center.

Here are some of the questions to look at:

- For all the cases resolved with no indemnity payment, was there a low overall probability of liability if the case went to trial? This would indicate a good job of assessing low probability of liability at trial and successfully using that to drive case strategy. Some of the reasons a matter may be resolved without an indemnity payment include:
  - A matter never gave rise to a filed suit.
  - The statute of limitations ran before an attempt to file a suit.
  - The case was dismissed.

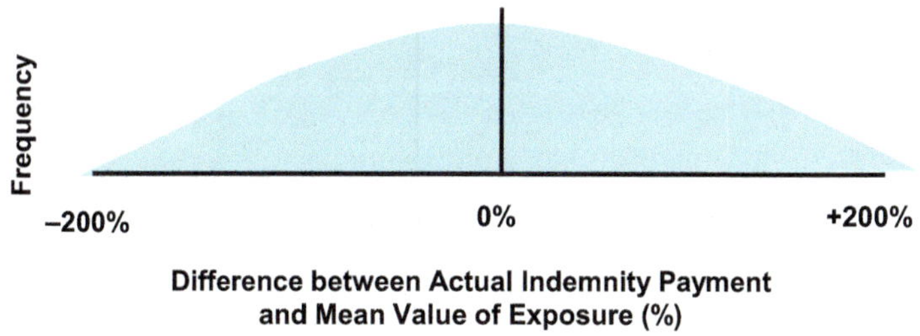

Figure 9.25 Analysis varied widely from actual payments

- The other party ceased pursuing the matter.
- The case went to trial with a verdict in your favor (or was resolved in your favor on appeal).
- A settlement was reached which did not involve an indemnity payment (e.g., remedial measures in lieu of a cash settlement).

- For the cases resolved with an indemnity payment, were there any indemnity payments outside of the range of the full distribution on indemnity exposure at trial? The low end of the distribution, of course, is always zero (a verdict in the defendant's favor). Since the full probability distribution is intended to capture the full range of possible outcomes (including very low probability scenarios), there should never be an indemnity payment greater than the maximum value of the distribution. This would indicate a failure to accurately capture the uncertainty in exposure in the case.
- How do the actual indemnity payments compare with the mean values of exposure at trial? If you plot the difference between the mean value of the analyzed exposure and the indemnity payment as a percentage (i.e., indemnity payment was 5% less or 15% more than the analyzed mean), here are some potential answers to this question.

The first graph shows that actual payments varied widely from the mean values of analyzed exposure at trial. This indicates that the analysis didn't go a good job of capturing the uncertainty in possible outcomes (Figure 9.25).

The next graph shows that actual payments on average were far greater than the means of the analyzed exposure at trial. This indicated that analysis systematically underestimated the exposure and that most cases cost far more to resolve than analysis would have predicted (Figure 9.26).

Next is what the data would show if the analyses systematically overestimated the exposure. In this graph, cases on average cost far less to resolve than predicted by case analysis. This would especially be a problem for insurance companies, indicating the reserves for claims and cases far exceeded the amounts necessary to set aside (Figure 9.27).

The first graph showed a failure to grapple with the uncertainty posed in a case. The next two show systematic biases in the analysis: either excessive pessimism or optimism (from the defendant's perspective).

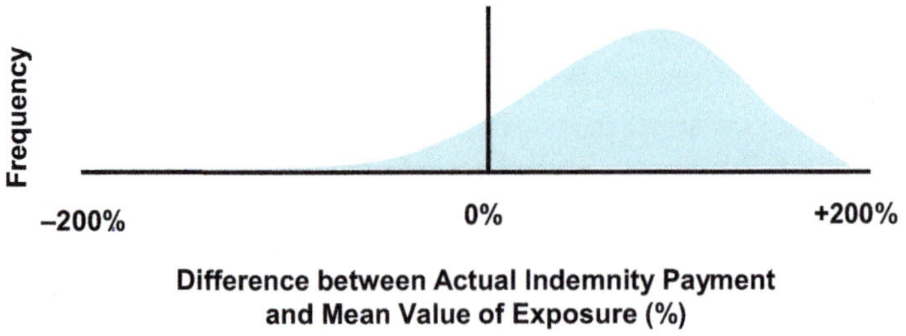

Figure 9.26 Analysis systematically underestimated exposure

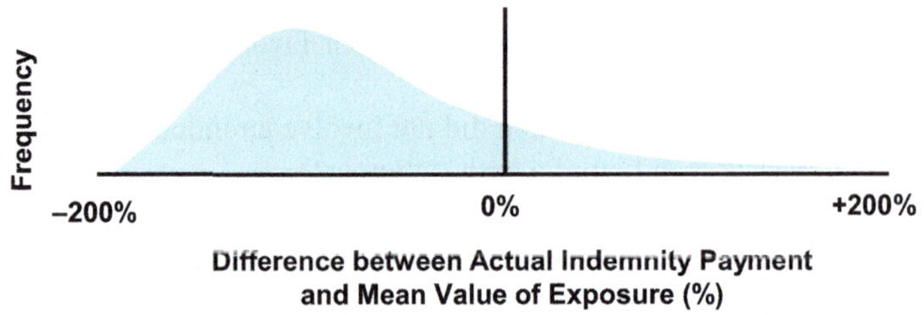

Figure 9.27 Analysis systematically overestimated exposure

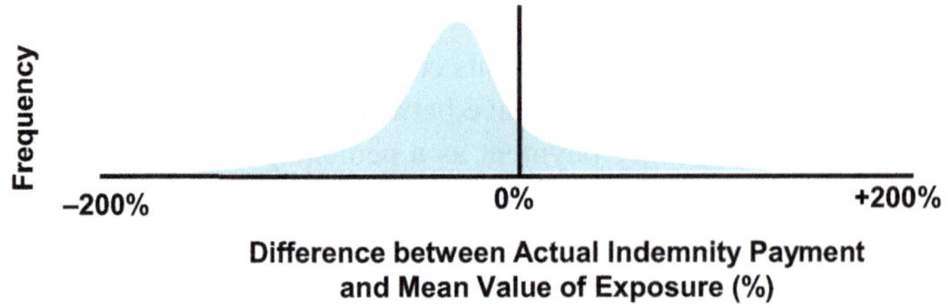

Figure 9.28 The indicator of good analysis and effective case strategy

The goal is to do a fairly good job at quantifying the exposure, recognizing that quantifying the uncertainty doesn't make it go away, and, from the defendant's perspective, resolve cases for less than the trial exposure. The last graph shows what that data for this would look like (Figure 9.28).

This graph indicates that, although there is variation in how individual cases were resolved (trials are inherently uncertain, as are negotiations), on average cases were resolved for less than the mean exposure of taking them to trial. From a defendant's perspective, this is a solid indicator that decision analysis has been effectively used to create winning legal strategies.

9.5 Constructing and Reading Probability Distributions

The mean value of a probability distribution is a convenient number for comparing alternatives and for reducing the value of a case to a single number. We still need to look at the full probability distributions for a number of reasons:

- To see if all the scenarios make sense and the calculations looks right. In particular, we look for the overall probability of no liability at trial and the possible values in extreme scenarios.
- To show the total uncertainty in the value of the case.
- To evaluate the effects of different risk preferences.

Accordingly, we'll now look at how to construct and read the full probability distributions. As with many other results, software can do this for you, but an understanding of what the software is doing is important for assuring the results are correct and make sense.

9.5.1 Histograms

The histogram is generally the most easily understood way of displaying and explaining the full probability distribution on the value of your case. It looks somewhat familiar because of its bar-chart format. Each bar tells you the probability that the value of your case will fall into the range displayed underneath that bar. Following is the histogram from the Howitzer case (Figure 9.29).

As you'll recall, because virtually all the probability for case outcomes from litigation fall in the range of zero to $1 million, we looked at a second plot on just the value between zero and $300K to get a better idea of the possible scenarios (Figure 9.30).

From the histogram, you can see there is about a .82 chance that the value of litigating the case will fall in the range $0 \leq \text{value} < \$50{,}000$. You know from the tree that this is the overall chance of winning the case and actually getting a value of zero. However, this result shows up in the histogram as the probability of a range. Is this result deceiving?

Answering this question requires thinking about where the individual numbers in the probability distribution come from. Remember that, for the tree, you had to divide the continuous questions (like damages) up into ranges, and pick a number that represented that range to go on the branch of a chance node. Accordingly, when you look at a full distribution at the end of a tree like the Howitzer distribution displayed below, getting a number like $220,000 (the second value from the top) really means getting a value in the range of values represented by $220,000. That's why it's correct to show ranges on the bottom axis of the histogram rather than the actual values from the end of the tree (Figure 9.31).

The one exception to this rule-of-ranges, of course, is the value of zero: it really is a zero value because it's associated with the chances of no liability at all.

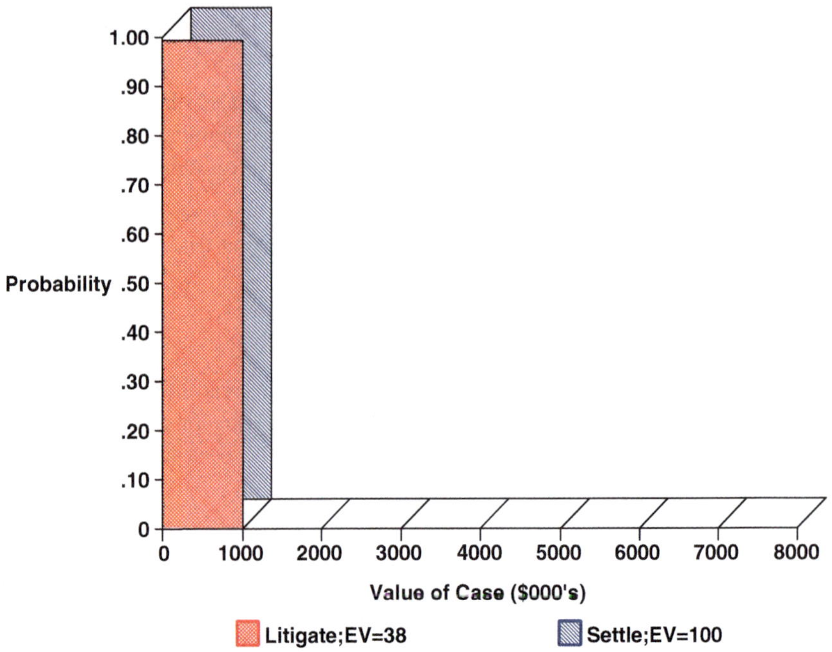

Figure 9.29 Overall histogram for Howitzer v. Crabtree

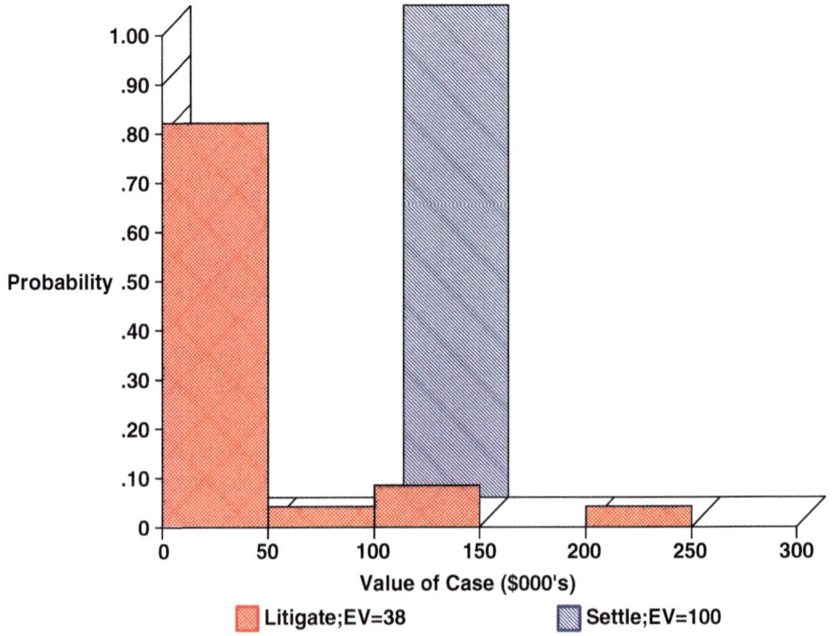

Figure 9.30 Histogram for values between zero and $300K

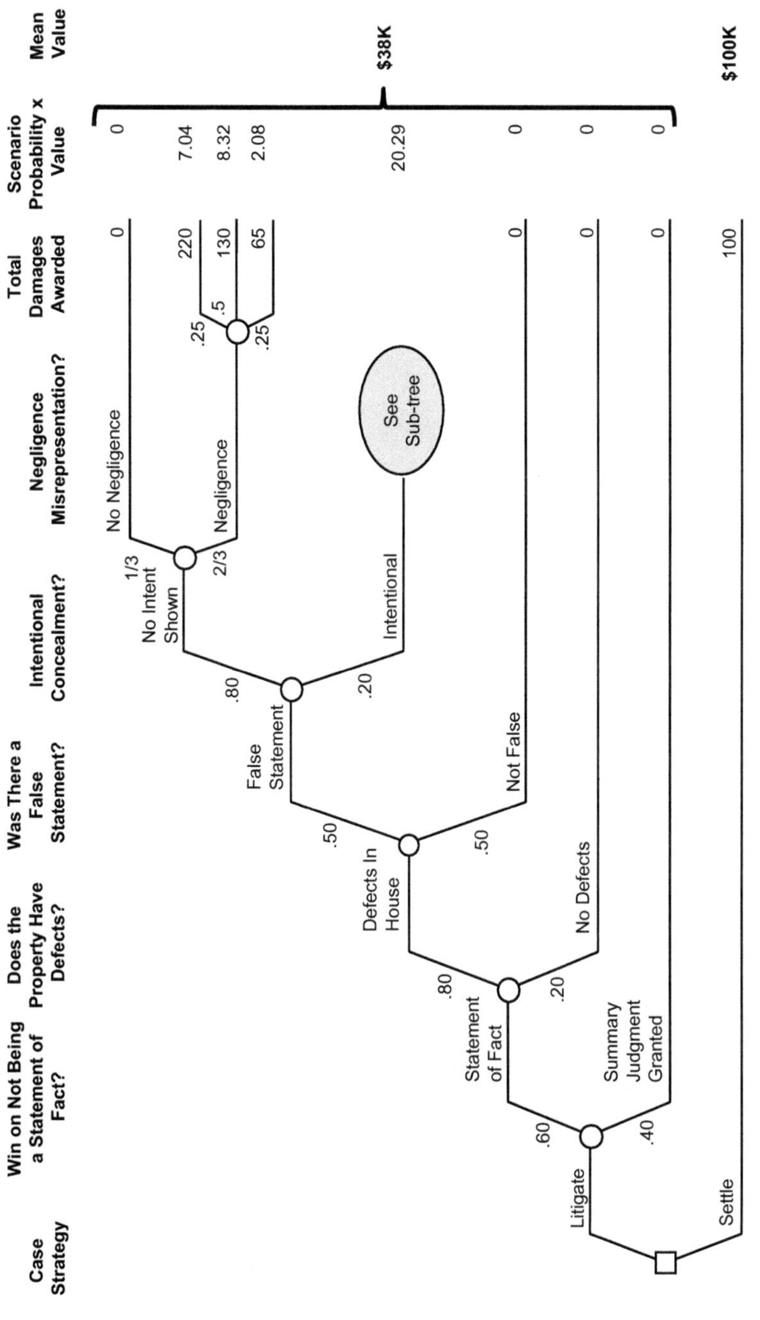

Figure 9.31 Main tree for Howitzer v. Crabtree

Table 9.1 All scenario probabilities and values for Litigate option from Howitzer tree

Probability	Scenario
0.064	0
0.24	0
0.12	0
0.4	0
0.0108	65
0.032	65
0.0216	130
0.064	130
0.0108	220
0.032	220
0.00006	565
0.00012	630
0.00006	720
0.00078	2065
0.00156	2130
0.00078	2220
0.00024	4265
0.00048	4330
0.00024	4420
0.00012	7065
0.00024	7130
0.00012	7220

In interpreting histograms, then, you'll have to remember that there's a real, range-less zero value in there for the chances of no liability (As you'll see later, the same is true of cumulative distributions.).

The procedure for plotting the histogram, then, is as follows.

- List all the values and probabilities for litigating the case from lowest to highest. Below are the values and probabilities for litigating the Howizter case.
- This includes all the values from the damages subtree and the main tree. We copied and pasted from the Excel worksheets shown earlier. To look at the full probability distribution, we need all the values (Table 9.1; Figure 9.32).
- The list above includes the numbers and probabilities from the subtree on punitive liability (repeated above), rather than the mean value of $422,750 from the subtree and its probability. For the full distribution, all the scenario probabilities from the damages subtree needed to be multiplied by the probability of getting to that subtree. From the main tree, this probability is $.60 \times .80 \times .50 \times .20 = 0.048$. If you're going to show the full probability distribution, you need *all* the numbers and values at the end of the tree—not just mean values from subtrees.
- Decide the number of ranges to divide the numbers into, then figure the ranges and write them down. Note that, for each range except the last one, the range will include all the numbers greater than or equal to the lower limit (≤) and less than the upper limit (<) (Table 9.2).

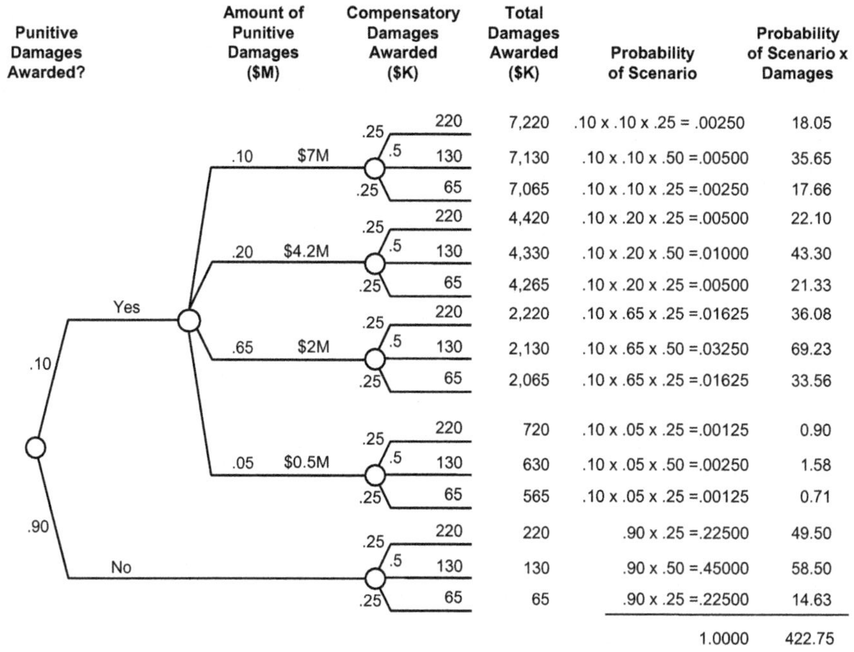

Figure 9.32 Damages subtree for Howitzer v. Crabtree

Table 9.2 Bin ranges for histogram

Lower limit ($K)	Values in this range	Upper limit ($K)
0	≤X<	100
100	≤X<	200
200	≤X<	300
300	≤X<	400
400	≤X<	500
500	≤X<	600
600	≤X<	700
700	≤X<	800
800	≤X<	900
900	≤X<	1000
1000	≤X<	8000

- When using software, values largest than the largest range are excluded. If doing it by hand, you can decide whether it makes sense to vary the range of the largest bin to include the low probability high value scenarios.
- Add up the probabilities for the values falling in each bin range (Table 9.3).

This produces the histogram below (Figure 9.33).

Note that the probability of a scenario in the first bin (87%) is higher than the probability of no liability (82%) because it includes the scenarios where liability is found but compensatory damages are only $65K and no punitive damages are awarded.

Table 9.3 Bins with probabilities of each bin

Lower limit ($K)	Values in this range	Upper limit ($K)	Probability of values in bin
0	≤X<	100	0.8668
100	≤X<	200	0.0856
200	≤X<	300	0.0428
300	≤X<	400	0
400	≤X<	500	0
500	≤X<	600	0.00006
600	≤X<	700	0.00012
700	≤X<	800	0.00006
800	≤X<	900	0
900	≤X<	1000	0
1000	≤X<	8000	0.00456

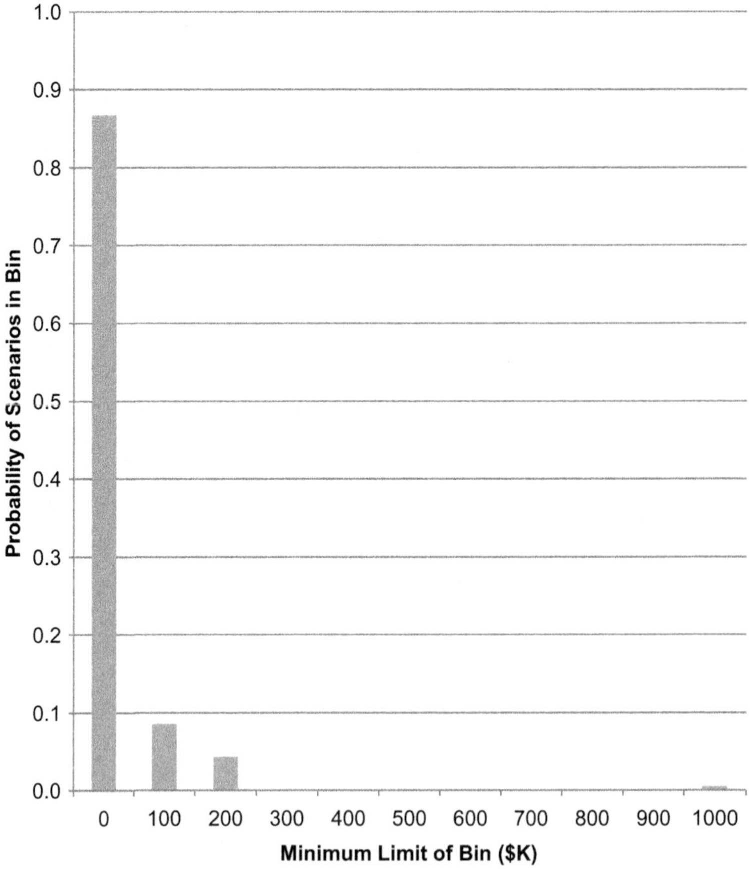

Figure 9.33 Histogram for Howitzer v. Crabtree

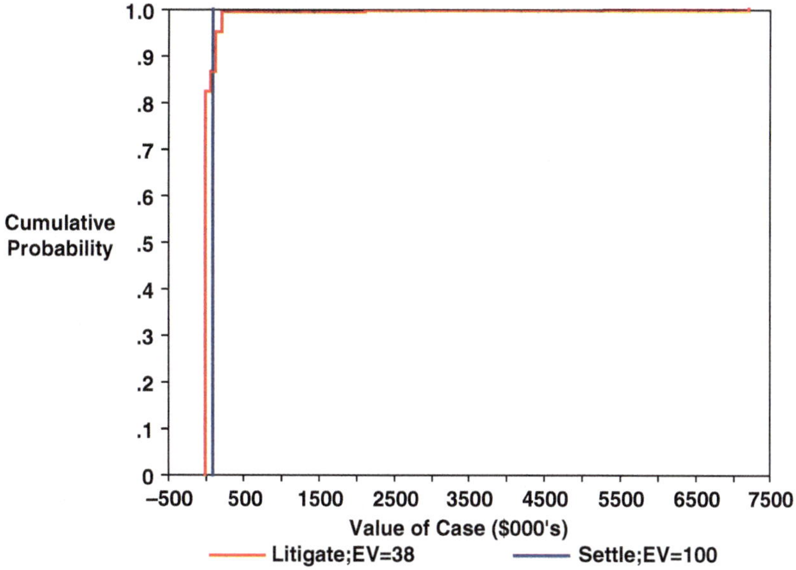

Figure 9.34 Cumulative probability distributions for Howitzer v. Crabtree

9.5.2 Cumulative and Excess Probability Distributions

The full probability distribution can also be displayed at a cumulative probability distribution. In contrast to the histogram, which gives you the probability of an answer within a particular range, the cumulative distribution gives you the probability that the answer will be less than a particular value. Thus, if you examine the Howitzer cumulative distribution (repeated below), you can see that, for instance, the probability that the value of litigating the case will be less than or equal to $100,000 is about .85 (Figure 9.34).

You've also seen cumulative distributions as a way to assess continuous variables. The cumulative is better suited than the histogram for this purpose because the cumulative only need not go down to assure that the probabilities will add up to 1.0, while a histogram assessment would require you to juggle the height of all the bars to make sure they add up to 1.0 (a much more difficult task).

Besides being assessed directly, a cumulative probability distribution can be constructed from your tree. The process for doing so follows.

- Order all the values and probabilities from highest to lowest, as was shown in Table 9.1 for the litigate option in the Howitzer tree.
- Calculate a running sum of the probabilities (Table 9.4).
- Plot these values the scenario values and running sum of probabilities. Connect them in stair step fashion: straight up from the first value to the probability of the first value, over horizontally to under the second value, up by the probability of the second value, over to the third value, etc. Plotting them as a scatter plot in Excel with connecting lines gives close to the same result (Figure 9.35).

Table 9.4 Scenarios with running sum of probabilities

Probability of scenario	Value of scenario ($K)	Running sum of probabilities
0.064	0	0.064
0.24	0	0.304
0.12	0	0.424
0.4	0	0.824
0.0108	65	0.8348
0.032	65	0.8668
0.0216	130	0.8884
0.064	130	0.9524
0.0108	220	0.9632
0.032	220	0.9952
0.00006	565	0.99526
0.00012	630	0.99538
0.00006	720	0.99544
0.00078	2065	0.99622
0.00156	2130	0.99778
0.00078	2220	0.99856
0.00024	4265	0.9988
0.00048	4330	0.99928
0.00024	4420	0.99952
0.00012	7065	0.99964
0.00024	7130	0.99988
0.00012	7220	1

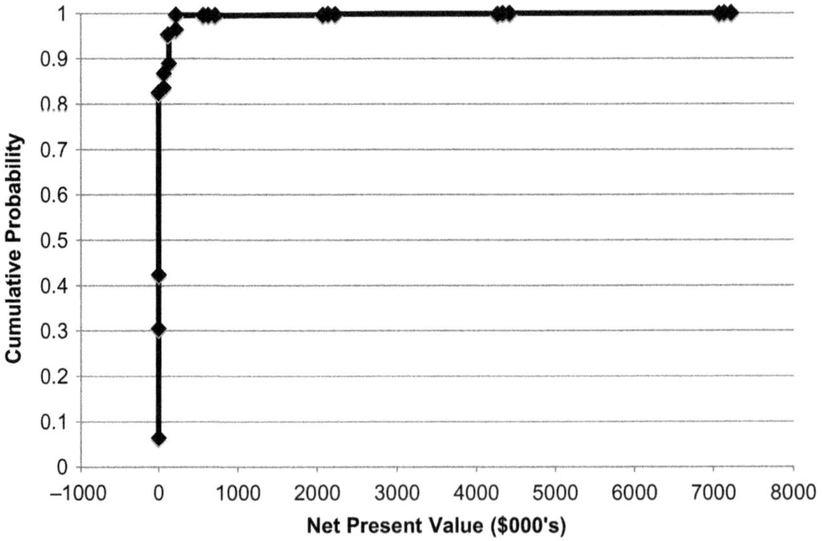

Figure 9.35 Rough approximation of a cumulative probability distribution with an Excel marked scatter plot

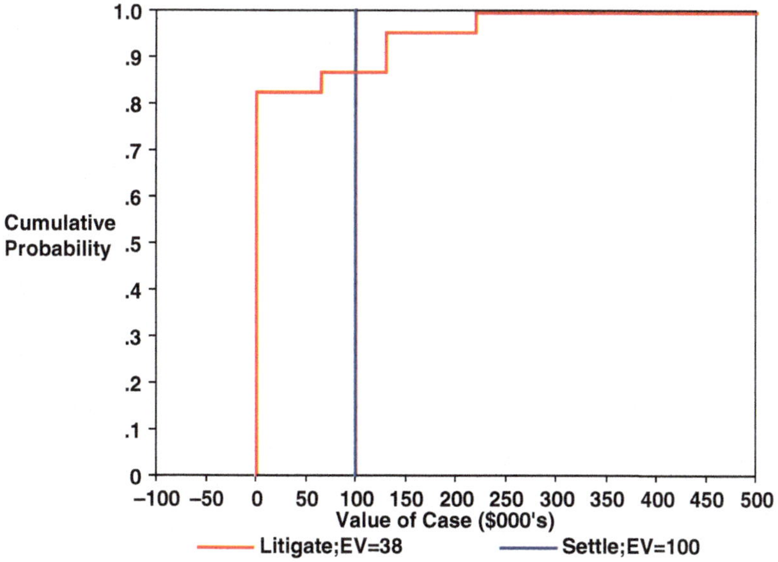

Figure 9.36 Properly drawn cumulative probability distribution drawn with Supertree

The proper stair step pattern is more apparent with the partial graph of the cumulative probability distribution we used Supertree to produce (Figure 9.36).

From this plot, you can read about the 80% probability of no liability and that all the probability is that the exposure at trial is less than about $220K.

The excess probability distribution tells you just the opposite of the cumulative: the excess tells you the probability that the value of the case will be greater than a certain value. You get an excess distribution just be adding a scale going from high to low on the other side of your cumulative, as shown below (Figure 9.37).

What is the probability that the value of litigating the case will be greater than $100,000? Greater than $200,000? Greater than $10,000?

Note that more than one cumulative probability distribution can be plotted on the same graph, as we have been doing to compare the litigate and settle alternatives.

It is also possible to smooth out the part of the distribution that has steps due to the discretization of the continuous variables rather than a change from one distinct scenario to another (as in litigate versus settle). A smoothed version of the litigate distribution appears below (Figure 9.38).

The procedure for smoothing is to draw a smooth curve that cuts through the middle of each step. This is the reverse of the procedure we used for discretizing a continuous distribution.

One has to be careful, though, that you're only taking the discretized continuous variables back to a continuous display, rather than omitting important information (such as a switch from only compensatory damages to compensatory plus punitive damages where the amounts are closer together than in this case). For these reasons, we prefer to stick with the stair steps. They describe distinct scenarios that can be traced back to the decision tree.

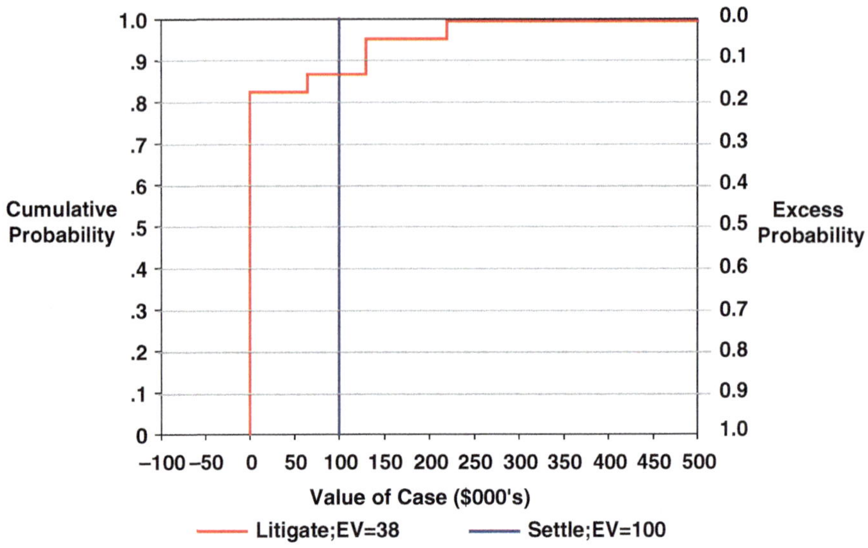

Figure 9.37 Adding the excess probability scale

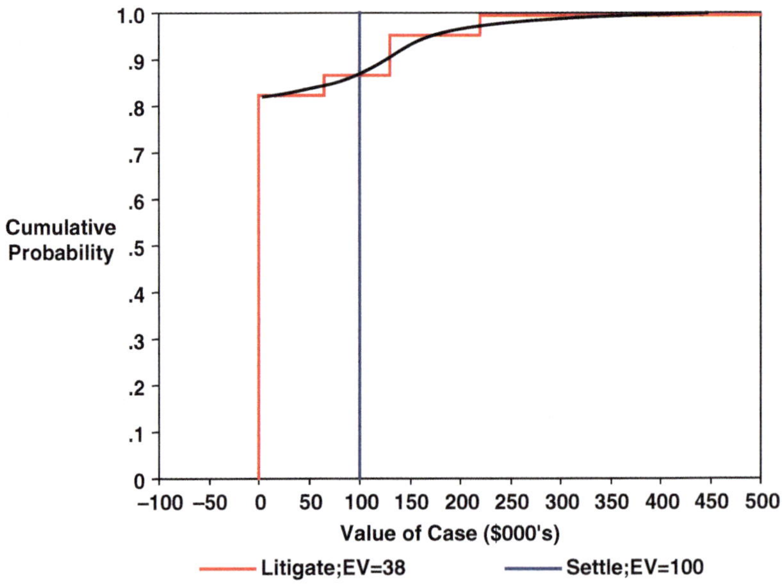

Figure 9.38 Upper portion for litigate distribution is smoothed out

As noted in the discussion of Monte Carlo simulations earlier, one of the issues with them is the display is always a smoothed curve. For litigation decision analysis in particular, we really need to see the steps to make sure we understand the calculations and that they make sense.

9.5.3 Piecewise-Continuous Distributions

Piecewise-continuous distributions are not a new kind of probability distribution. Rather, they are distributions that have vertical segments and corners like the Howitzer distribution shown above.[2] These kinds of distributions occur whenever there is a sudden jump in probabilities and outcomes, such as when you move from no liability to liability. They might also arise when you switch from one legal theory to another (such as from a contract theory to an anti-trust theory), or from one kind of damages scenario to another (such as from compensatory damages only to compensatory plus punitive damages).

Generally, you should not try to assess these kinds of distributions (like the total exposure at trial) directly. Doing so requires people to model and estimate the impacts of various factors in their heads, a process especially prone to errors and biases. Best to disaggregate, structure, and assess the multiple uncertainties and let the piecewise continuous distribution be a result of analysis rather than a directly assessed input.

Your analysis should reflect these sudden jumps because they genuinely reflect the effects of different influencing factors determining the value of your case. Up to now, you have done this almost without thinking by the way you structure your case. We can make use of the piecewise-continuous structure of the distribution to create simplified versions of the results that are useful for reviewing with clients and counsel.

Strictly speaking, we could discretize the smoothed version of the cumulative probability distribution the same way we've discretized other continuous distributions. However, it's better to work from the decision tree to ensure you obtain values that exactly match your other results. One of the objectives of these methods (and reasons for avoiding Monte Carlo simulations) is we want results that flow transparently from the inputs without introducing spurious variations.

The form I've found most useful is to show the overall probability of liability followed by the mean damages exposure if found liable (or a three-branch range on damages exposure).

For the Howitzer tree, we know that the probability of no liability at trial is 82% and that the mean exposure at trial is $38K (from previous calculations). So the start of the simplified tree is this (Figure 9.39).

The only number we're missing is X. The rollback mean value has to stay the same at

$$38 = (.82 \times 0) + (.18 \times X)$$

[2] Mathematically speaking, distributions with vertical segment or corners are discontinuous because the slope of the distribution is infinite for a vertical segment and there is no slope at a corner. Hence the name piecewise-continuous to describe these kinds of distributions.

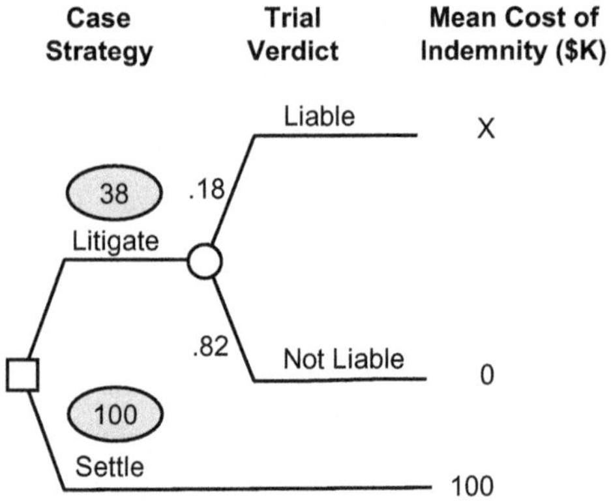

Figure 9.39 Start of simple abstract from Howitzer tree

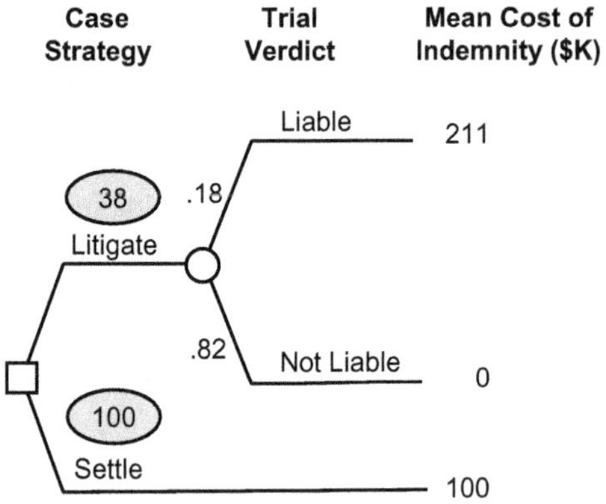

Figure 9.40 Completed abstracted tree for piecewise linear distribution

Solving for X:

$$38 = .18 \times X$$
$$X = 38 \div .18 = 211$$

This allows us to complete the abstracted tree (Figure 9.40).

For trees with simpler liability models than Howitzer, this operation will be easier to do. But the result you're aiming for will be roughly the same.

9.5.4 Contingent Probability Distributions

You've already seen how probability sensitivity analysis can be used to answer questions about different assessments and their impact on case value and strategy.

- Which issues have the greatest potential for improving the value of your case?
- Which issues could potentially change your decision, such as from litigate to settle?

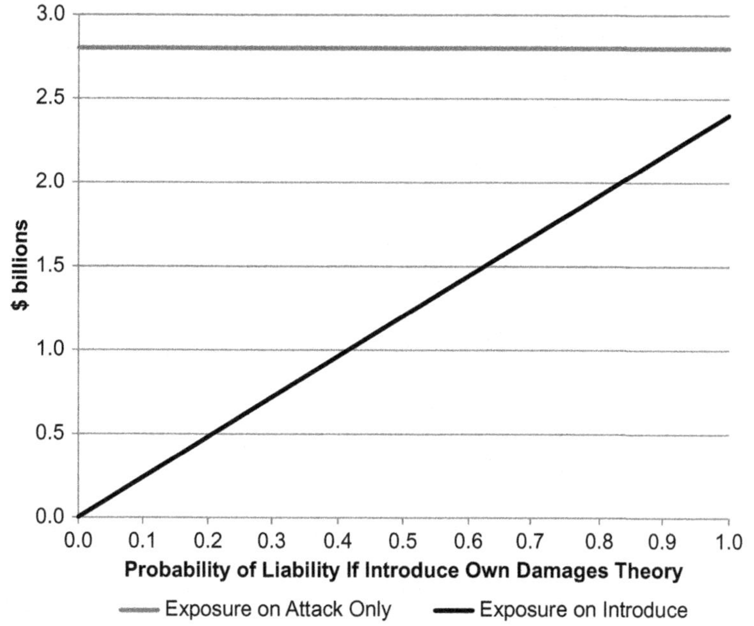

Figure 9.41 Sensitivity to probability to evaluate alternative trial strategies

- Which decisions should have differences of opinion on probabilities be resolved because those differences could change your preferred decision?

Thus, as you'll recall from the hypothetical Texaco trial strategy example, the probability sensitivity (repeated below) showed how the probability of losing affected the mean value of the case. However, for any probability of losing if Texaco introduced its own damages theory, it was still better (based on assessments of possible damages) to go ahead and do so (Figure 9.41).

This analysis relies on comparing mean values. In contrast, probability distributions allow you to see all the potential scenarios and probabilities behind the mean value.

We can also look to see how probability distributions change depending on how particular uncertainties turn out, rather than just looking at how the mean values change. These probability distributions are contingent or dependent on the result of another uncertainty. They give information on the full uncertainty remaining after one or more uncertainties has been resolved.

For example, if we revisit the Howitzer v. Crabtree main decision tree, each place where there is a rollback mean value shown in the grey oval is someplace where we could look at the full probability distribution rather than just the mean value (Figure 9.42).

For example, if we might want to look at the full uncertainty remaining if the decision is to litigate and we already knew how the motion for summary judgement

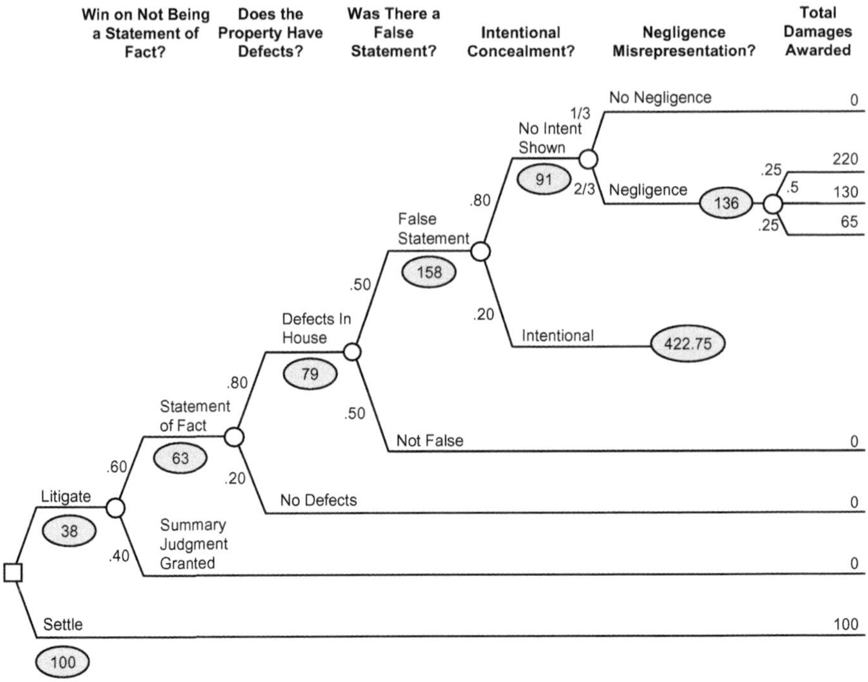

Figure 9.42 Main tree for Howitzer v. Crabtree

turned out. That would be a full probability distribution for the mean value of $63,000 at that point (lose the motion) and a vertical line for zero indemnity (win the motion).

To figure these distributions by hand, you would need to redo the roll forward scenario probability and value calculations starting only with the property defects question and proceeding to the end of the tree. Whatever that distribution turned out to be, there would be a 60% chance of that distribution and a 40% chance of zero, yielding the same overall mean value.

This operation is tedious by hand and easy with software. Below are these distributions plotted with Supertree (Figure 9.43).

This particular plot is not particularly informative. The distribution for litigating looks pretty much the same, just with mean value increased to 38 ÷ .60 = $63K for the uncertainty on summary judgment which it excludes. A look at any of the other uncertainties possibly leading to no liability (defects, false statement, and negligence) would be pretty much the same.

However, what might be interesting is looking at how the trial exposure changes depending on whether or not punitive damages were awarded. That's the uncertainty that dramatically changes the case value.

To look at these contingent probability distributions while leaving all the other uncertainties open, we need to change the order of nodes in the tree. We would like

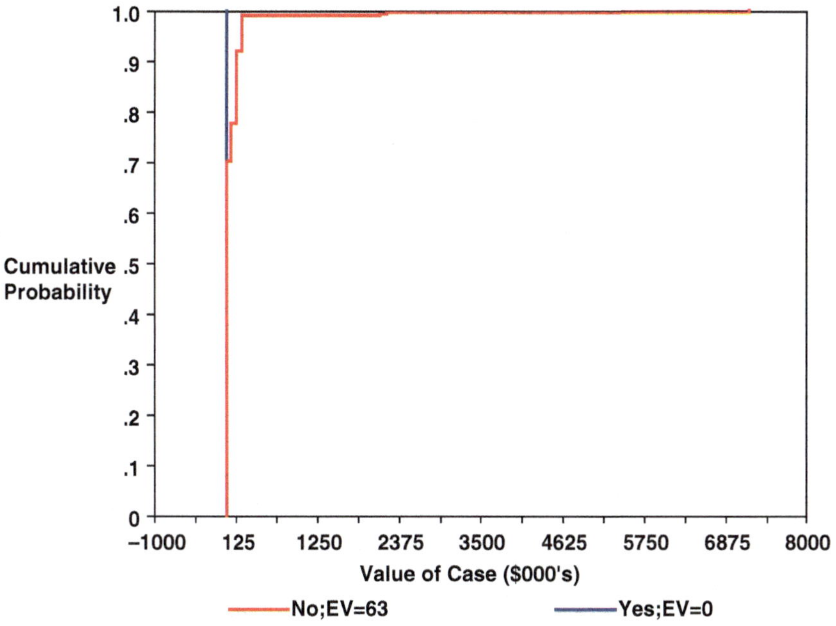

Figure 9.43 Probability distributions contingent on results for summary judgment motion

to move the uncertainty on whether punitive damages are awarded to right after the litigate or settle decision, with all the other uncertainties following it.

Changing the order of nodes in a tree is possibly the most complex decision analysis of all. Most software does not even attempt this operation. Accordingly, for interested readers we refer you to *Decision Analysis for the Professional* or *Foundations of Decision Analysis*.

For this discussion, we'll assume you're using software capable of performing this operation correctly, such as Supertree. The resulting display is below (Figure 9.44).

Punitive damages indeed drastically changes the uncertainty faced. In fact, we can see from the distributions that the worst case with no punitive damages is still better than the best case with punitive damages awarded—what you would expect given the relative magnitudes of the two damages assessments.

In fact, looking at the mean values of contingent probability distributions is what we were doing with the value of perfect information and value of perfect control calculations. Those likewise require a change of ordering in the tree. We can similarly draw the tree with the punitive damages uncertainty brought to the front (now before the litigate or settle decision) for values of perfect information and control (Figure 9.45).

If you knew that—if found liable—punitive damages would be awarded, you would choose to settle the case for $100K versus a mean exposure at trial of

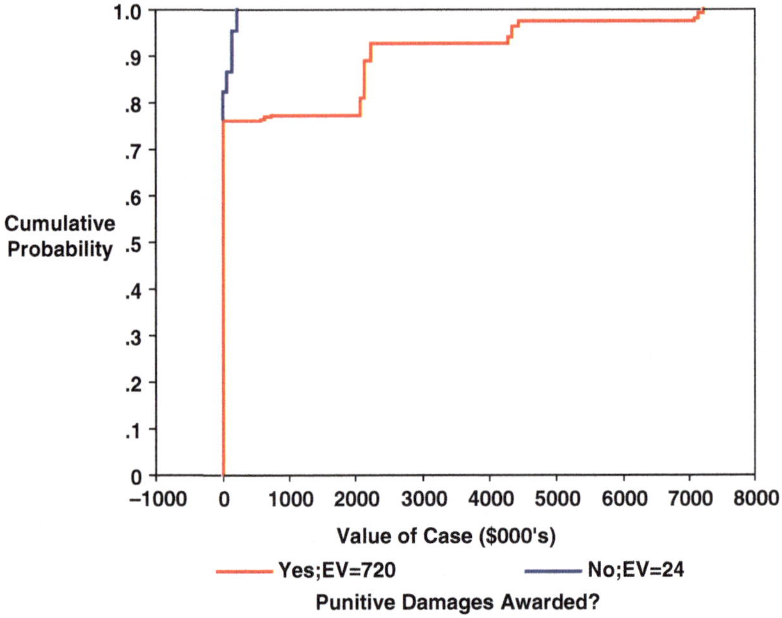

Figure 9.44 Probability distributions contingent on award of punitive damages

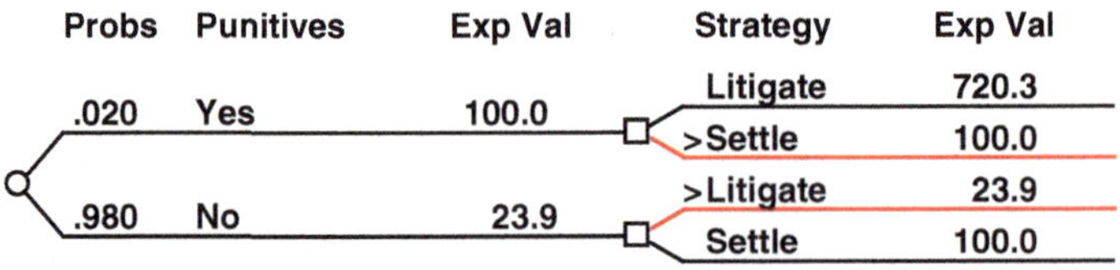

Figure 9.45 Contingent probability distribution mean values shown for values of perfect information and control

$720K. If you know they would not be awarded, litigating for mean exposure of $24K looks better.

Recall that information only has value if it changes a decision. Since the preferred decision switches, there must be a value for perfect information.

The overall mean value of this tree is $(.02\times 100.0)+(.98\times 23.9)=\$25.4$K. The overall mean value with the uncertainties and decisions in the original order was $38K. So the value of perfect information is $38-25=\$13$K.

If you could purchase perfect control, the value with that control would be 23.9 or $24K. So the value of perfect control is slightly higher at $38-24=\$14$K.

In summary, contingent probability distributions allow you to look at what the remaining uncertainty would be if you knew how just one uncertainty turned out. The mean values of the contingent distributions are the same numbers used in value of perfect information and control calculations to see if there is any value to learning how uncertainty turns out, or to controlling its outcome.

9.6 A Closer Look at Dependencies

Arrows in the relevance diagram (dependencies) can mean different things:

- Knowledge: at that point in the diagram, you know how the uncertainties leading to it turned out and what decisions were made.
- Determination: the nodes pointing to a double-oval determine the outcome of that uncertainty. Likewise, the arrows pointing to the value hexagon are inputs determining the total value.
- Different possible outcomes: the possible outcomes of an uncertainty depend on the uncertainties leading to it.
- Different probabilities: the probabilities of various outcomes vary according to the uncertainties pointing to it (same possible outcomes).

These last two methods, respectively, are known as dependent outcomes and dependent probabilities. This section takes a closer look at the implications of using these two methods, including why the direction of an arrow in a relevance diagram can be reversed with complete mathematical equivalency. The purpose of this discussion is to help you think more clearly about dependencies and to properly handle the more subtle forms of dependencies when they arise.

9.6.1 Dependent Probabilities and Dependent Outcomes

So far, you've likely not considered whether you were using dependent probabilities or dependent outcomes to represent a dependency. The choice flowed from the method used to assess the probabilities: dependent probabilities came from assessing the probabilities on a yes/no question; dependent outcomes from using one of the short-cut methods to assessing a continuous variable; and both probabilities and outcomes might be different if you assessed full dependent probability distributions for different tree scenarios (as illustrated in the discussion of precedence, for example).

However, when dealing with numerical answers, you do have a choice between using dependent probabilities and dependent outcomes. The two methods are equivalent to one another. For instance, the hypothetical Texaco trial strategy tree (repeated below) showed how the trial strategy changed the chances of liability (dependent probabilities) and the possible damages awarded (dependent outcomes) (Figure 9.46).

For an uncertainty that depends on another uncertainty, you should choose whichever method most naturally describes the relationship to the people you're assessing the probabilities with. For the verdict at trial (outcomes are fixed), the clear choice was to use dependent probabilities. For the potential indemnity, using the 25%/50%/25% approximation made choosing different possible outcomes (dependent outcomes) the natural choice.

For uncertainties with numerical outcomes (continuous variables), it is possible to use either dependent probabilities or dependent outcomes to describe the dependency.

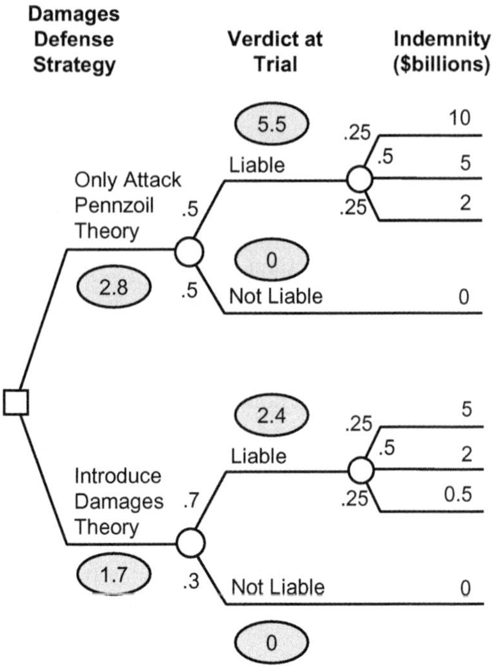

Figure 9.46 Dependent probabilities and dependent outcomes with trial strategy choice

The choice should be driven by what makes the most sense to the people assessing the probability.

Generally, for a particular dependency, you would choose either dependent outcomes or dependent uncertainties to represent the relevance. Using both is entirely possible but often confusing and usually indicates two different uncertainties being mixed together. Best to separate them and be clear what you're talking about.

9.6.2 Order of dependencies

One further issue with dependent probabilities is deciding how issues depend on each other; that is, which way should the arrow in the relevance diagram go? For instance, suppose you're considering the total damages awarded in a case where there could be compensatory and punitive damages awarded, as illustrated below (Figure 9.47).

This relevance diagram expresses a belief that the amount of punitive damages awarded will be affected by the amount of compensatory damages awarded, and that the jury will decide on the compensatory damages first (applying the trier-of-fact rule). This would be the case where possible punitive damages were computed as a multiple of compensatory damages (as is frequently done). That relationship is captured below (Figure 9.48).

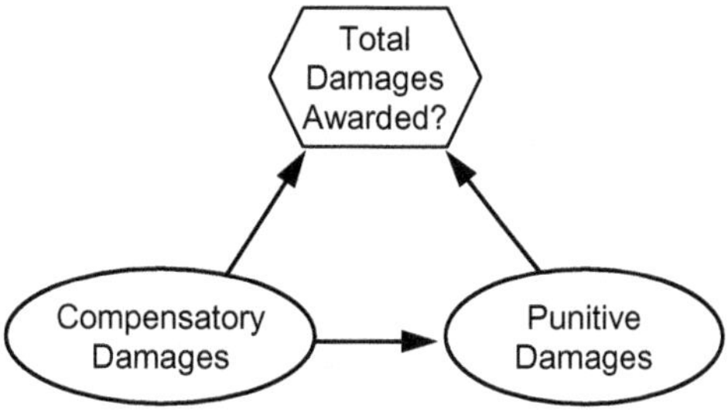

Figure 9.47 Punitive damages depend on compensatory damages

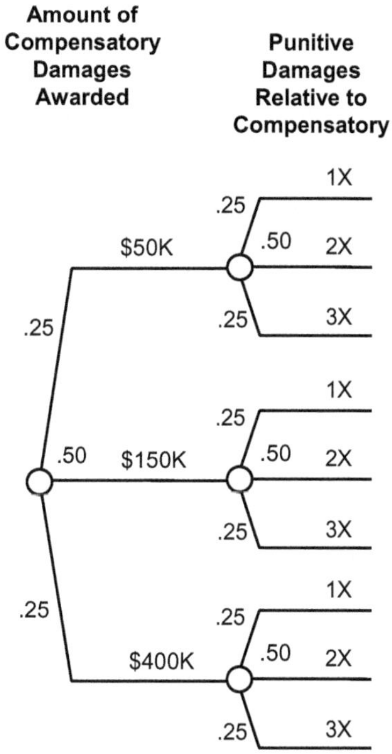

Figure 9.48 Dependent outcomes expressed with a calculation

You could multiply out the numbers for punitive damages awards and show those, but this form of display shows better how the people in the room were actually thinking about them. The objective is to come up with an assessment which captures the experience, intuition, and judgment of the people as best as you can.

Suppose the arrow were pointing the other way, as shown below. This meant that the jury would be considering whether it would award punitive damages before determining the amount of compensatory damages (Figure 9.49).

This formulation could lead to a very different assessment. It might indicate people feel that the jury, if only awarding 1X punitive damages, would "top up" the compensatory damages to increase the total award. On the other hand, if they were leaning toward 3X punitive damages, it might indicate they thought the compensatory damages were low and were trying to give the defendant more. This thought process is illustrated in the possible assessment below (Figure 9.50).

Figure 9.49 Relevance going the other way

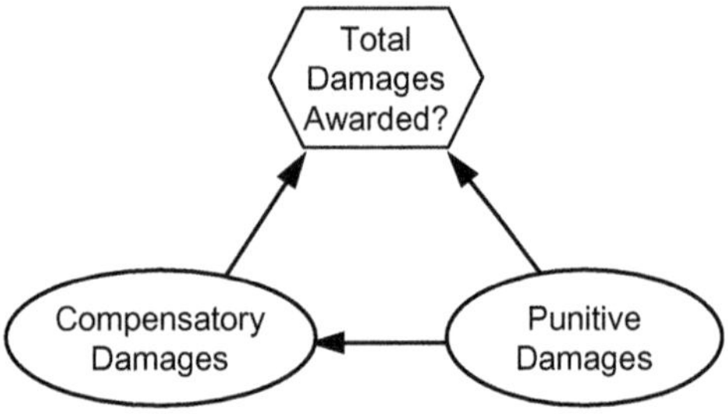

Figure 9.50 Different relevance leads to a different assessment

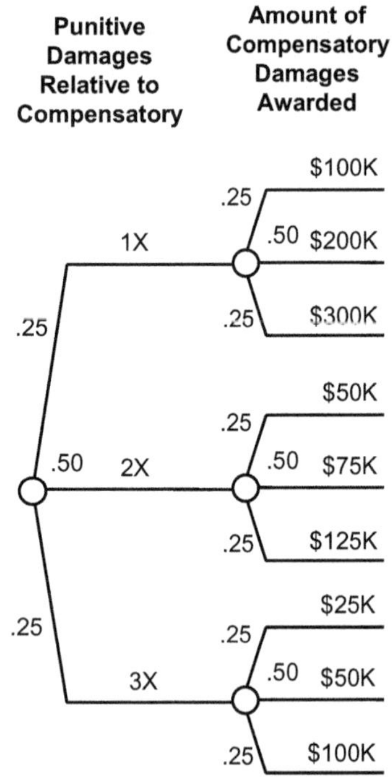

Either order and assessment is correct if it accurately captures the judgment, information, and intuition of the people at the time it is elicited. Part of the art of decision analysis is figuring the best structure and numbers to capture these.

Let's take another example, this time with dependent probabilities. Suppose you represent a large and famous hospital being sued for medical malpractice in connection with a surgery. The surgeon is a codefendant, but has a different insurer and attorney. The questions are whether the hospital or the surgeon or both are liable for malpractice.

The allegations regarding the surgeon concern whether he was negligent in the procedure itself, which resulted in damage to a nerve in the patient accompanied by loss of sensation, movement and strength in one arm. The condition is short of paralysis and may get better. It also may be a known complication of the procedure, which has a surgical site very close to the nerve.

The allegations regarding the hospital concern whether staff properly immobilized and stabilized the patient for surgery, and if failure to do so may have contributed to the injury.

As counsel for the hospital, you feel assessments on potential liability should start with the surgeon on the theory that his actions are most important and what the hospital staff did or didn't do are secondary. The patient was sedated and immobilized, just a question of whether the arm could have been strapped down a bit better and possibly in slightly different position. Your team assesses the probabilities of liability below (Figure 9.51).

However, counsel for the surgeon strongly feel that they need to start with the hospital's chances of liability. You suspect there is a strong motivational bias in this (the surgeon doesn't want to be on the hook alone), so work very hard in the assessment to get them to think broadly about the issues and what a jury could conclude. Working with them, you come up with the following assessment (Figure 9.52).

How to you compare the two assessments? The answer requires reversing the order for one of them. We'll work on the surgeon's assessment.

Start with multiplying the scenario probabilities, as you did for the roll forward mean value calculation (Figure 9.53).

Then draw the tree in the hospital's order and fill in the scenario probabilities from the surgeon's tree, matching the scenario and probabilities (Figure 9.54).

Add the scenario probabilities on the scenarios following the surgeon liable or not liable to find the overall probabilities of those (Figure 9.55).

Lastly, divide to find the intermediate probabilities. For example, $.59 \times X = .56$, on the Yes/Yes scenario, so $X = .56 \div .59 = .93$. Filling in all the numbers produces the tree below (Figure 9.56).

Figure 9.51 Hospital's assessments for respective chances of liability

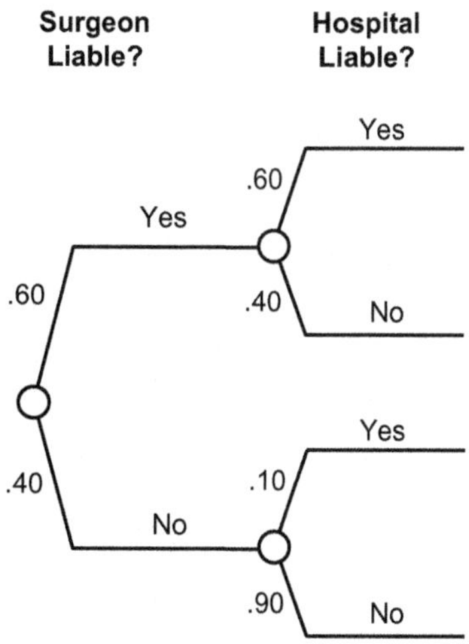

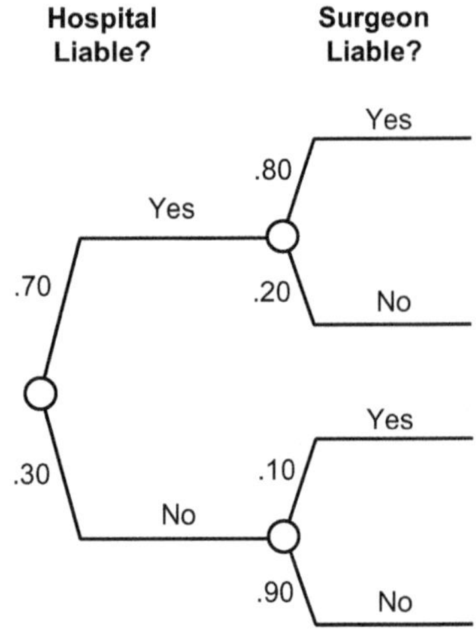

Figure 9.52 Surgeon's assessments for respective chances of liability

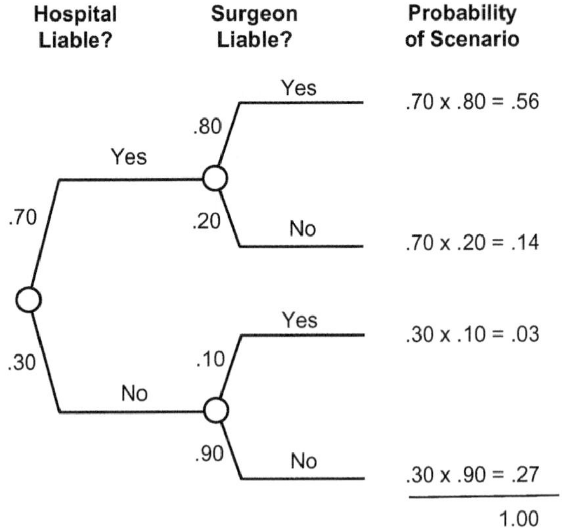

Figure 9.53 Scenario probabilities for surgeon's assessment

Comparing the two assessments, it turns out that the hospital's and surgeon's perspectives are not so far apart. You basically agree on the chances that the surgeon will be liable (60% versus 59%). The only difference is the chance the hospital will be liable. You feel there's a $(.60 \times .60) + (.40 \times .10) = 40\%$ chance the hospital will be liable, while the surgeon feels there is a 70% chance the hospital will be liable. This at least clarifies matters for discussion.

In practice, disagreements like this happen all the time, and are especially dangerous between codefendants. Discord among defendants is very good news for plaintiffs. With these methods, you can:

- Consider and assess dependent probabilities in various orders
- Translate between them to compare.

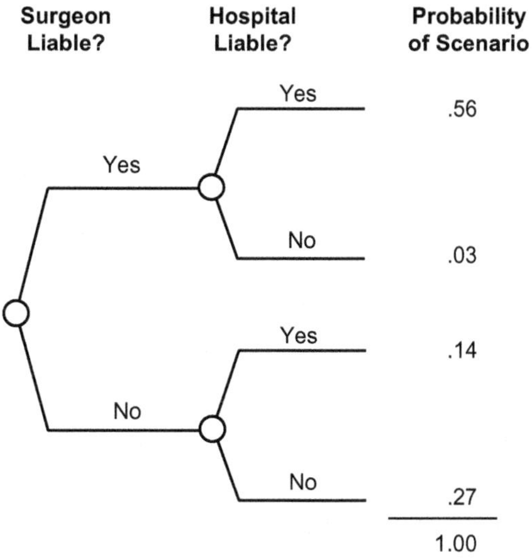

Figure 9.54 Surgeon's scenario probabilities on tree in hospital's order

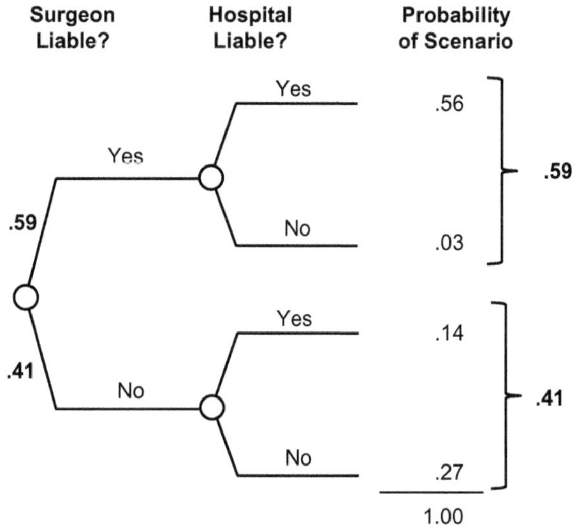

Figure 9.55 Overall probabilities of surgeon liable from his assessment

In addition, these methods are necessary to bring a node with dependent probabilities to the front of the tree for value of information and value of control calculations.

The calculation illustrated in the example is described by Bayes' Rule, which shows how to calculate the odds with a dependency reversed.

Suppose X is the question represented by the first chance node, with possible answers X_i ($i = 1, 2, \ldots, m$) and Y is the second chance nodes with possible answers Y_j ($j = 1, 2, \ldots, n$). The probability on branch Y_j following branch X_i can be written

Figure 9.56 Surgeon's assessment reversed to match hospital's order

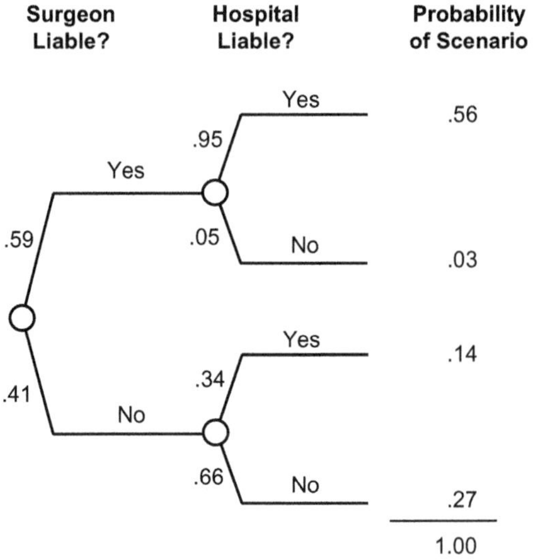

as $p(Y_j|X_i, \&)$ where Y_j means the odds of Y occurring given that X_i has already occurred and given your state of information &. The scenario probability we calculated is $p(X_i|\&) \times p(Y_j|X_i, \&)$.

Bayes' Rule gives the formula for reversing the order to find $p(X_i|Y_j, \&)$, that is, the odds of being on the X_i branch of the Y_j scenario given that Y_j has already occurred and your state of information. The formula is:

$$p(X_i \mid Y_j, \&) = \frac{p(Y_j \mid X_i, \&)\, p(X_i \mid \&)}{\sum_{k=1}^{m} p(Y_j \mid X_k, \&)\, p(X_k \mid \&)}$$

9.6.3 Jury Science and the Value of Imperfect Information

One last application of dependent probabilities is in calculating the value of imperfect information. First, though, a word on why you might like to think about the value of imperfect information.

An occasional objection to the value of perfect information calculation is that perfect information is almost never available: you won't be able to buy a perfectly accurate prediction of how most uncertainties will turn out. It is therefore more realistic to calculate the value of imperfect information.

However, this objection usually turns out to be less important for two reasons. First, the value of perfect information always gives you a *maximum* that you should be willing to spend for further information on any uncertainty. Even this maximum value lends a new perspective and discipline to case planning. Second, as you'll see

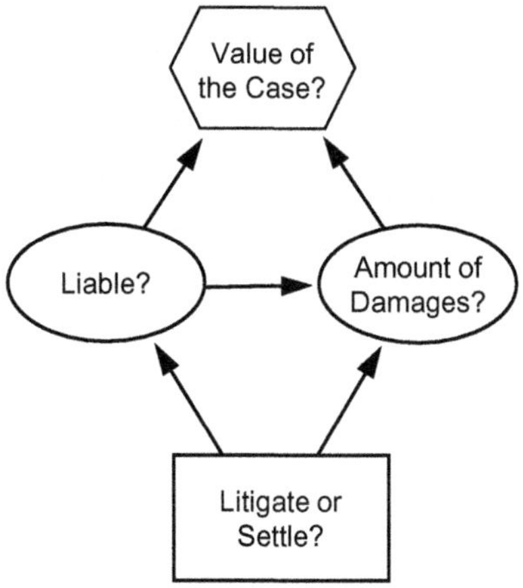

Figure 9.57 Basic structure for case analysis

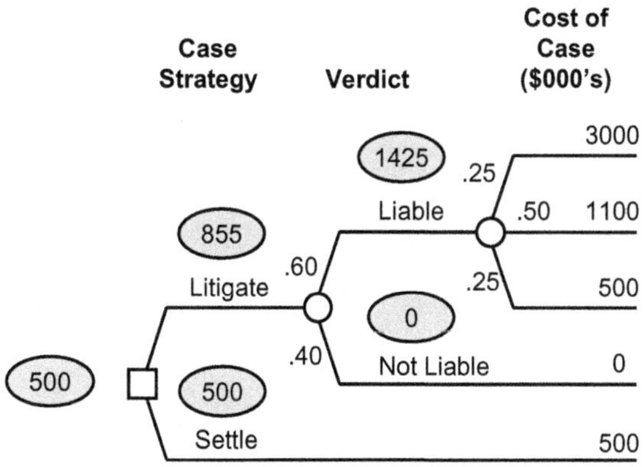

Figure 9.58 Rollback mean value calculations

below, calculating the value of imperfect information requires knowing *how* imperfect the information is; that is, how accurate past predictions have turned out. This kind of information is seldom available in a legal context. Accordingly, the applicability of the value of imperfect information is somewhat limited.

As for other reasons you might be interested in the value of information are best illustrated with an example (Figure 9.57).

Suppose you represent the defendant and estimate the chance of liability at trial is about 60%. Your 25%/50%/25% assessment on potential damages is $3 million/$1.1 million/$500K. A settlement offer of $500K is on the table. How does that look? (Figure 9.58).

Based on these assessments, the settlement offers looks good, but the client is uneasy both about writing a check for $500K and for taking the chance of a trial.

Suppose also you have an opportunity to relieve this anxiety by retaining a jury science firm to conduct a mock trial at a cost of $40,000. Should you do it? As a first step, you calculate the value of perfect information on liability to find the upper limit on the amount you should pay, as below (Figure 9.59).

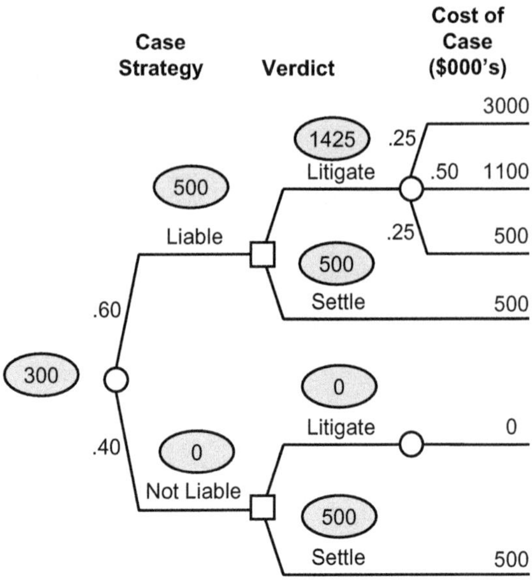

Figure 9.59 Value with perfect information calculation

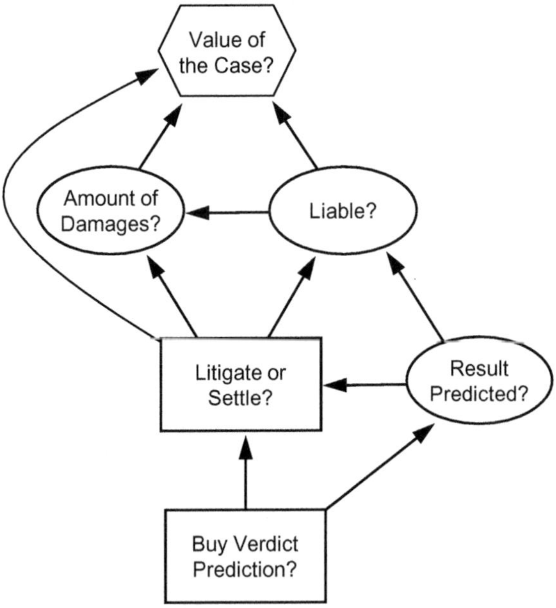

Figure 9.60 Verdict prediction: the Jury Science Decision

The value of perfect information on liability, then, is 500 − 300 = $200,000. Perfect information looks very valuable. But what's imperfect information from the jury science firm worth?

Start to structuring the relevance diagram for hiring a jury science firm (Figure 9.60).

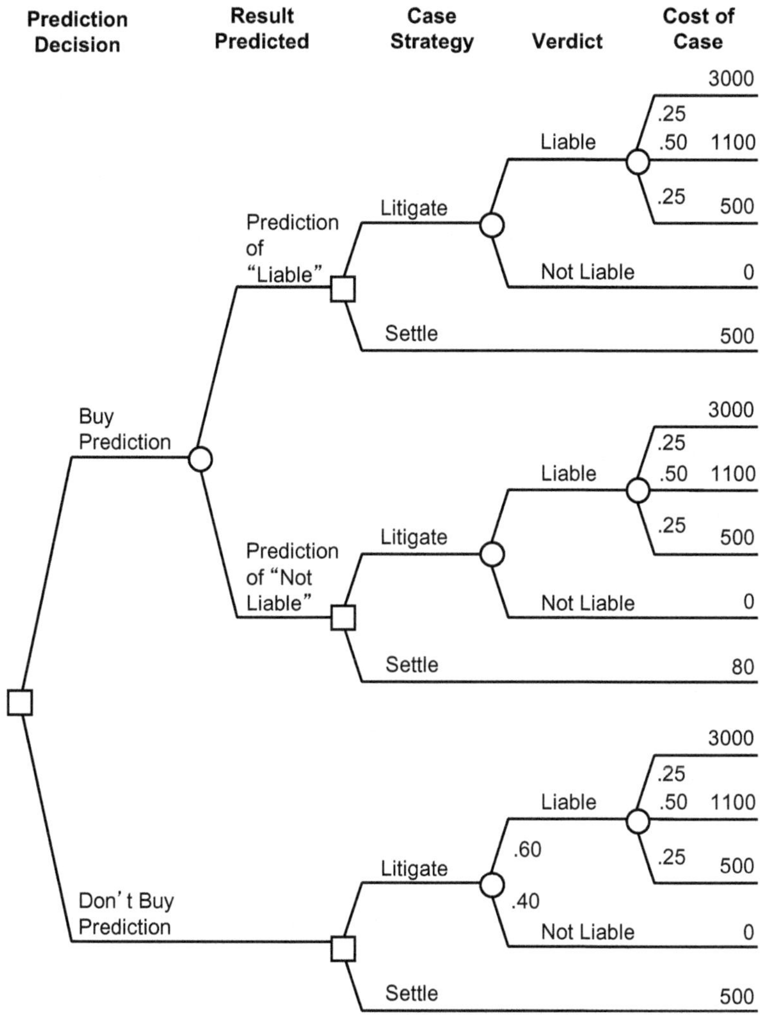

Figure 9.61 Tree for the Jury Science Decision

Thus, you would first have a decision on whether or not to purchase the prediction. Then you would either receive the prediction or not, and still have a decision on whether to settle or litigate. Further, you still face an uncertainty on liability because the prediction is imperfect (as opposed to perfect information, which eliminates uncertainty once you know the prediction). An initial tree drawing, then, might look like the following (Figure 9.61).

You reasoned (correctly) that, if you don't buy the prediction, you face the same situation as before (the bottom part of the tree). But what if you do buy the prediction? What probabilities should you put in for the prediction, and what for liability? You recall that there should be probabilities for the possible predictions because you

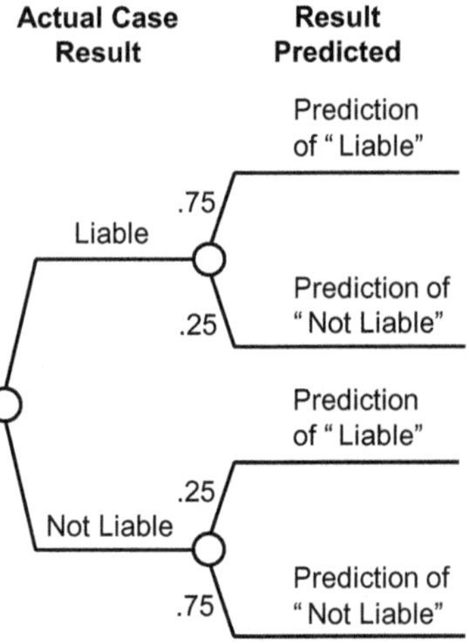

Figure 9.62 Record of prediction results

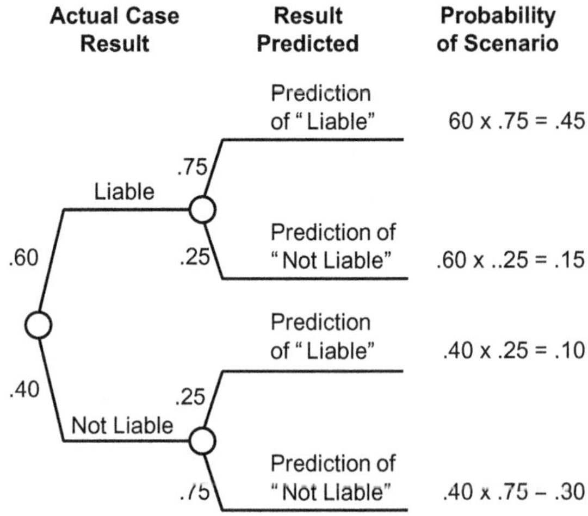

Figure 9.63 Scenario probabilities for Jury Science Decision

don't know what the jury scientist would predict, and if the prediction has any value at all, it should change your probabilities for the jury finding you liable or not liable.

The answers to these question rely on how accurate the predictions have been in the past. This information will tell you how a particular prediction should affect your probabilities for being held liable. For instance, suppose the jury science people tell you they correctly predict the result about 75% of the time. In tree form, this information would look like the following (Figure 9.62).

Notice that the event being predicted (the actual verdict) comes first, while the data on how accurate the predictions have been comes second. This order is the usual and correct way to think about the accuracy of a prediction.

You still need to fill in the probabilities for Liable and Not Liable. Your original probabilities go there. The tree ordering from left to right means that there's not yet any new information to change your probabilities. Accordingly, the full tree for the jury science track-record appears below (Figure 9.63).

Figure 9.64 Order of uncertainties reversed to find probabilities for the tree

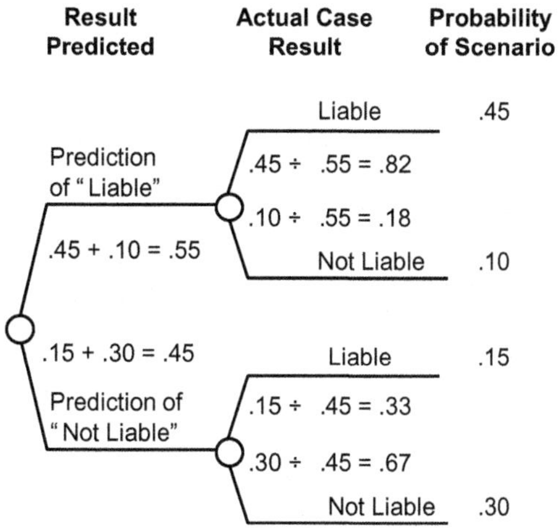

This order, of course, is the reverse of the order you need for the top part of your tree. Accordingly, you need to follow the procedure given above for reversing the order of the questions, as illustrated below (Figure 9.64).

Note how a prediction of "Liable" gives you an .82 chance of being held liable, while a prediction of "Not Liable" only gives you a .67 chance being held not liable. Would you have intuitively guessed that these odds were implied by the historical data? It is almost universally true that the reversed probabilities *cannot* be intuited, and, indeed, failing even to recognize that a reversal is needed is one of the most common errors people make in thinking about probabilities.

These probabilities are now in the right order for your decision tree, allowing you to calculate the value of the case with and without the prediction, as shown below (Figure 9.65).

The value of the case with imperfect information is $489K, so the value of imperfect information is 500−489=$11,000. This is quite a lot less than the $200K value of perfect information. As you can see, the value of imperfect information drops off quite rapidly as the information becomes less "perfect," and isn't proportional to the drop in accuracy from perfect information. Dropping from 100% accuracy to 75% accuracy is a 25% drop, while dropping in value from $200K to $11K is a 94% drop. This is one area where intuition definitely doesn't work. How good does the $40,000 mock trial look?

In summary, then, there are number of points that make it useful to know about the value of imperfect information.

- The value of imperfect information is always less than the value of perfect information, and, frequently very much less. The value of a prediction with only a 50% accuracy of predicting liable or not liable is zero! You might as well flip a coin (You should try plugging these probabilities into the above analysis and seeing if you can confirm this result.).

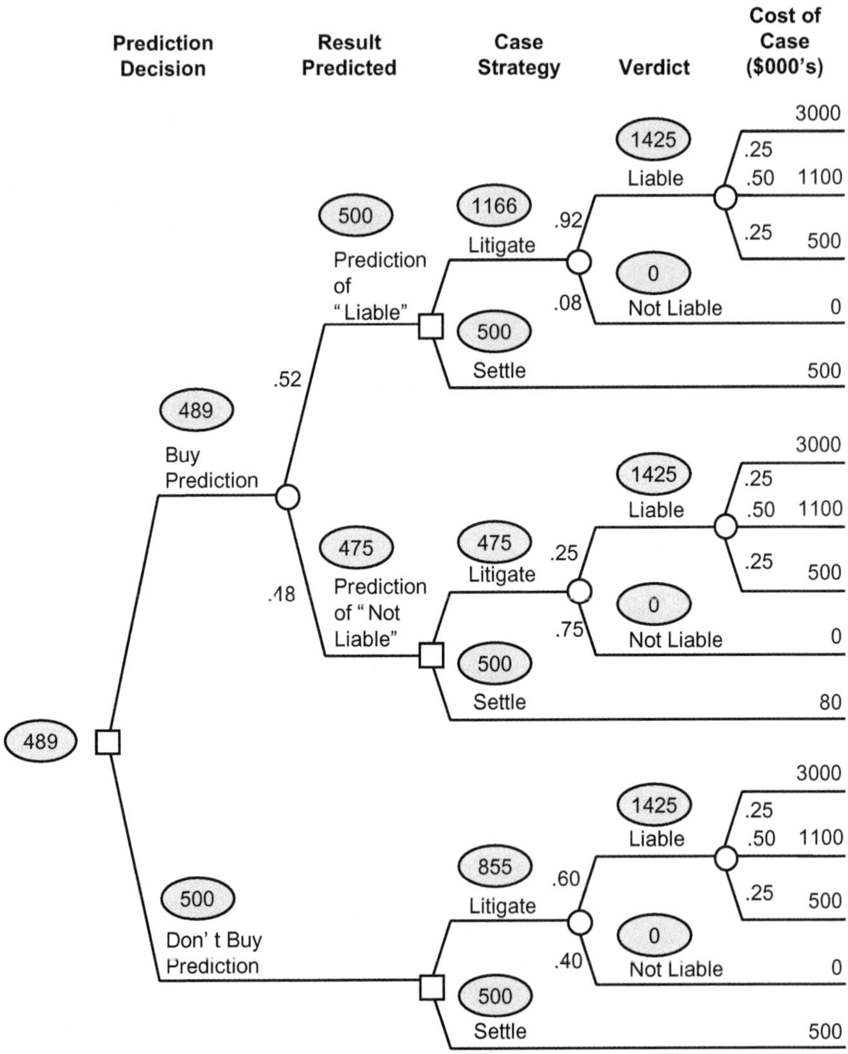

Figure 9.65 Rollback mean values for Jury Science Decision

- Even though you usually don't have the historical data to calculate a value of imperfect information, knowing how it's done will help you to think about the value of improving your assessment on particular questions.
- Knowing how the process of assessing and then reversing the probabilities for imperfect information will help you to think about dependencies and the order of probability assessments.
- Knowing how imperfect information calculations *should* work (or where to look it up) will help avoid the common trap people fall into when they deal with predictive information (that is, not drawing the tree in the right order and then reversing it).

9.7 Problems

9.7.1 Many Possible Arguments

1. There are several ways multiple arguments can arise. There could be multiple arguments for prevailing under the same cause of action, as in two different arguments that failure to perform under a contract was so material as to excuse the other party from further performance (such as the need to pay). There could also be two different legal theories for recovery, such as a contract theory and a tort theory. What are the differences in how you need to handle these alternative means in assessments and decision trees?
2. In a pleading including many arguments, how do you decide which ones to evaluate?
3. Draw the tree shown in Figure 9.8 for two alternative arguments with the malice and causation issues included in it (as in the preceding figures).
4. The web site of a contractor working on billing for a large hospital had their web site hacked, data stolen and posted on the Internet. The data included patient identifying information and diagnosis codes, which could be decoded to see what the patients were being treated for. Shortly after the posting, the hospital discovered this, had the posting removed and the contractor's system secured. However, the breach was discovered and the hospital is being sued. The contractor was also sued, but has since declared bankruptcy. You represent the defendant hospital.

   The hospital and contractor are being sucd on two grounds. First, the disclosure violated Federal Law. Depending on how the applicable federal law is interpreted, the statutory damages for disclosure could range from $1,000 to $10,000 per patient.

   They are also being sued for actual damages. Any patient who knew their confidential health information was disclosed could sue for negligent infliction of emotional distress. Those damages are very uncertain. For patients whose physical diagnoses were disclosed, you estimate the damages awarded could range from $10K to $50K. For those with psychological conditions disclosed, you estimate the damages could go from $10K to $200K.

   You guess there is a 60% chance the hospital could be liable for the disclosure, and a 90% chance the contractor will be liable. If both parties are liable, there is a 50% chance a court would award joint and several liability, which means either defendant is liable for the entire judgment. The contract won't pay anything because it's bankrupt. If it's not joint and several liability, the hospital would only be liable for its share of fault, which you estimate between 10% and 50%.

   Because some of the data was partial, it's hard to tell how many individual patients were in it versus multiple diagnoses for the same patient. You estimate that there were between 1,000 and 10,000 individual patients in the data, and that 5–20% of the diagnoses were for psychological conditions.

It's not yet known how many patient might have known their data was disclosed. It could be none, but could be as high as 20%.

4.1. Structure the relevance diagram for this case.
4.2. Draw the decision tree and fill in the numbers. Make estimates for any numbers you are missing.
4.3. Calculate the values for all scenarios and the rollback mean values.
4.4. What is the mean liability under the statutory penalties theory?
4.5. What is the mean liability under the emotional distress theory?
4.6. How does it impact defendant's exposure if the court decides that plaintiffs can recover under both theories of liability versus only under one?
4.7. Plot the probability distributions for exposure under each individual theory and total exposure.

9.7.2 Handling "Spill-Over"

5. Describe at least two types of "spill over" and why or why not they need to be handled differently in analysis.
6. Are there types of "spill over" which cannot be quantitatively analyzed? Explain why or why not?
7. Spill-over can be modeled with dependent probabilities or dependent outcomes. Are there other ways? If yes, describe it. If no, explain why not.
8. A driver was killed when the sports car he was driving went out of control and hit a tree on a street in an otherwise deserted industrial park on a Sunday afternoon. His widow is suing the manufacturer. You represent the manufacturer.

   Plaintiff alleges that the car crashed because of a defect, specifically, lack of a stability control system. The car was 10 years old. Though such systems are present on most vehicles today, they were not standard 10 years ago. In particular, they were not commonly included in high performance sports cars where they were thought to interfere with the driving experience. You feel there is a 10% chance of liability on this basis.

   However, exposure is still substantial because the deceased was very wealthy and can easily demonstrate $50–100 million in future lost income and loss of consortium damages.

   There is also the issue of speed. Police estimate the deceased was traveling at more than double the posted speed limit when he crashed. In addition, the original tires were on the car which, at 10 years old, would have become hard and lost much of their traction.

   You estimate there is some chance (come up with your own numbers) that a jury would make a preliminary determination that the tragedy was "pretty much" the deceased's fault and not find any liability. Further, if they do find liability, they might find that the deceased's contributory negligence made a significant contribution to the crash. They would then have to decide the percentage that the deceased was at fault and defendant would only be liable for its percent of the damages.

In the alternative, plaintiff has made a claim under the vehicle's comprehensive extended warranty for the full value of the car, which was around $1 million.

8.1. Structure the relevance diagram for the case.
8.2. Draw the decision tree and fill in the probabilities, including reasons for the probabilities you come up with yourself.
8.3. Calculate the rollback mean values to determine the manufacturer's mean exposure.
8.4. Plot the full probability distribution on the manufacturer's exposure.
8.5. Determine the manufacturer's maximum settlement value.
8.6. Which probability has the greatest effect on the value of the case? What are the values of perfect information and perfect control on this uncertainty?
8.7. In the interim since this car was made, the manufacturer has included some level of stability control system on all of its vehicles. You feel there is an 80% chance this fact can be excluded from the case as a subsequent remedial measure, and a 20% chance plaintiff will be able to introduce it. This could influence the jury's thinking on the defect question. Revise your analysis to include this question and calculate the manufacturer's new maximum settlement value.

9. Refer to problem 29 in Chapter 7 (the Cinelli problem). Do the first parts of the problem if you have not already done so, then go on to the following part.

   9.1. Suppose you reconsider the prospects for trying both theories together. You would then have to consider the "spillover" effects between the theories. Suppose further you believe these effects are as follows:

   - Your damages won't be affected by which theory you win on because it will all be muddled in the jury's mind, giving you a new damages distribution of .25/.50/.25 of $10/30/75 thousand regardless of how you win.
   - The jury will consider the consent theory first because it doesn't require as serious a finding of fault, and will only go on to consider malpractice if they don't find for you under the consent theory.
   - If the jury doesn't find for you on consent, their finding on whether the unwarned-of damages happened will be the same as their finding of causation under malpractice because they can't keep the distinction straight.

   Should you try both theories together?

10. Note that problems 6, 12, and 13 in Chapter 7 also involve evaluating spill-over.

9.7.3 *Working with a Group*

11. List the various roles for working in a group. What level of understanding of decision analysis is required for each role?
12. How does working with an expert on a particular assessment differ from including them in a group for an entire analysis?
13. For each of the following situations, do a group probability assessment according to the procedure in the text. If you do more than one exercise, designate a different person as the facilitator each time. Remember to make sure that the question satisfies the clairvoyance test for each member of the group before proceeding. Record any definitions you declare to make sure the question passes this test.

    13.1. What is the probability that a constitutional amendment prohibiting flag burning will be ratified within the next 5 years?
    13.2. What is the probability that at least one member of the group will be sued within the next 5 years? Do you need to differentiate the possible types of lawsuits for the question to pass the clairvoyant test?
    13.3. What is the probability that the US Supreme Court will grant certiorari to a case involving the permissible size of tort awards within the next 3 years?
    13.4. What is the probability that corporate tax rates will be reduced within the next 2 years?
    13.5. What is the probability that a court will award damages for emotional distress in a wrongful discharge case within the next 4 years?
    13.6. What is the probability of immigration reform within the next 2 years?
    13.7. For at least one of the questions above, apply the clairvoyant test to break the question into two separate questions and assess the probability for each.

9.7.4 *Tracking Estimates Over Time*

14. Why is it difficult to track how good subjective probability estimates are?
15. List at least one type of subjective probability estimate that can be tracked and one that can't. For the one that can be tracked, specify how you would do so.
16. In a presentation to an analysis, you are challenged to show the data proving the probabilities are correct. What is your answer?

9.7.5 *Constructing and Reading Probability Distributions*

17. What are the differences between a histogram, a cumulative probability distribution and an excess probability distribution?

18. What is the difference between a histogram and a probability density function, such as a bell curve?
19. When should you use a mean value for discussion and presentation purposes? When should you use a plot of the complete distribution?
20. How would you evaluate a cumulative probability distribution to see if it's correct? What is the definition of "correct" you would use?
21. When should you smooth a cumulative probability distribution?
22. What method should you use to smooth a cumulative probability distribution? Be sure to include a specification of which parts you should and shouldn't smooth.
23. Should you ever smooth a histogram? Why or why not?
24. What is the difference between discretizing a continuous probability distribution and assessing a 25%/50%/25% range?
25. Your client is considering introducing a new kind of combination microwave–convection oven that consists of three different microwave emitters, top and bottom gas burners, a convection fan, and is lined with bricks to deliver that true bakery taste and crispness. However, the client's director of development, Dr. Escoffier, is concerned that bricks pose a risk of severe burns to customers if they force the oven open during the self-cleaning cycle. The oven has a self-locking safety latch to prevent this, but it can be forced. Further, tests have shown that if customers physically break off the metal microwave safety latch (such as by striking it very hard with a heavy pan) and don't have the latch repaired, there will be about a half-second of microwave exposure before the secondary hinge sensor can detect the opening and shut the microwaves off. Dr. Escoffier is therefore worried about lawsuits should either of these occur.

    However, the vice president of marketing, Mr. Slickham, feels that the bricks and microwaves are the key features of the oven, enabling conventional-quality crispness and browning in microwave time. He doesn't want to give these features up if it can be avoided.

    You have therefore been asked to work with the planning department in evaluating the prospects for models with these features by focusing on the potential liability. The planning department analyzed three possible models: bricks + microwaves, bricks only, and microwaves only. To keep matters simple on the first run-through, they only estimated the sales for the year of introduction. If any of the models looks promising, the next step would be to analyze them over their product life cycle, looking to see if the liability profile and profitability change over time.

    As a result, they have given you a cumulative probability curve on first year sales for each of the three models.

    While you were waiting for the charts, you and Dr. Escoffier assessed the claims per 10,000 units sold (a measure both of you were familiar with) for each model, and the average present value settlement cost per claim. Dr. Escoffier discretized the average claim settlement cost for his own edification, but left you to do the curves for the number of claims (Table 9.5).

Table 9.5 Assessments for average present value settlement cost per claim ($000s)

	25% actual is greater than	50/50 chance actual is greater or less than	25% chance actual is less than
Bricks + microwave oven	500	80	10
Brick oven	200	40	15
Microwave oven	400	70	12

You note grimly that there are potentially very high settlement costs for those few cases where the client would contest the damages and go to trial.

Finally, the planning department informed you that their sales assessments were based on unit prices of $850, $1200, and $1800 for the microwave, brick, and brick + microwave ovens, respectively, which would yield an estimated profit per unit of $250, 500, and $950 (The microwave/gas oven had a much lower price and profit because there were already competing models, while the other two ovens would be the first home-models offered.).

25.1. Draw the relevance diagram. Do you need a different diagram for each stove model, or will one do for all the three? Draw the tree for each of the three stove models (three trees).
25.2. Discretize the probability distributions for sale into three-branch chance nodes. Fill all the numbers in for the probability trees.
25.3. Calculate the first-year profits or losses for each model for each scenario. Be careful to keep track of units: sales and settlement cost are given in thousands, claims in number be 10,000, profits in dollar per units. You will probably find it most convenient to end up with a dollar value in millions ($2.1 million etc.).
25.4. What is the mean value for each model?
25.5. What is the maximum amount you should be willing to pay to insure all the claims on each model? (Assuming full coverage.)
25.6. Pick the model with the highest mean value and plot the mean value of that model as a function of the probability of the highest and lowest percentage of sales producing claims (ignore the middle branch). Do the same for the average present value settlement cost. Which is the steeper line? What does this tell you?
25.7. Plot the cumulative probability distribution for each model. Which model has the highest potential profits? The greatest potential losses?
25.8. Compare the cumulative probability distributions to their respective mean values. How are the mean values reflected in the cumulative distribution plots?
25.9. Pick the model with the highest mean value and draw the histogram for it. Does this form of display tell you anything that the cumulative distribution did not? Would the histogram be easier to explain to someone not familiar with decision analysis than the cumulative distribution?

26. Refer to one of the cumulative probability distributions from the previous problem (the oven problem).

    26.1. Should it be smoothed or not for presentation or discussion with others, and why do you think so?
    26.2. Give at least one reason each why, in general, a cumulative probability distribution should or should not be smoothed.

27. Note that problem 29 in Chapter 7 (the Cinelli problem) offers good practice with cumulative probability distributions.

9.7.6 A Closer Look at Dependencies

28. Is the use of dependent probabilities or dependent outcomes equivalent? Why or why not?
29. Is it always possible to translate an assessment using dependent probabilities into one using dependent outcomes? Why or why not?
30. When should you use tree structure to show a dependency rather than a dependent probability or outcome assessment?
31. Is it always possible to change the order of a dependency? Why or why not?
32. Is it possible to have dependent probabilities or outcomes depend on a decision rather than another uncertainty? Why or why not?
33. Is it possible to reverse the order of a dependency when a dependency depends on a decision? Why or why not?
34. Calculate the mean damages awarded in both Figures 9.48 and 9.50. Are they the same? What does that tell you?
35. Assess one or more of the dependencies below. Be sure to apply the clairvoyant test to define exactly what you are assessing and include the definitions of the respective questions.

    35.1. The temperature outside given where you are exactly 1 year from now.
    35.2. The maximum speed of the vehicle which will be your primary daily driver 10 years from now.
    35.3. The number of your grandchildren who will be alive 30 years from now.
    35.4. The number of litigation decision analyses you will perform over the entire course of your career.
    35.5. The number of parts of this question an instructor will assign you to assess before the end of the course.

36. What information do you need to calculate the value of imperfect information?
37. What is the relationship between the value of perfect and imperfect information?
38. When do you need to calculate the value of imperfect information, and when will the value of perfect information suffice?

39. When is it not possible to calculate the value of perfect information?
40. When is it not possible to calculate the value of perfect control?
41. Why are the value of perfect information and perfect control useful even though perfect information and perfect control are seldom available?
42. Calculate the value imperfect information in the jury science example if the accuracy of the prediction is 50%, and again if the accuracy is 40%. What does this tell you?
43. Suppose you have been retained by Microsoft in connection with the lawsuit by Apple Computer regarding Microsoft's Windows program (Note: this problem was written before the actual litigation occurred. Consider it a look back.).

    As you are probably aware, the Windows program helps programs that run on MS-DOS personal computers (that is, IBM-compatible personal computers) interact with the user via windows on the screen, pull-down menus, dialogue boxes, etc.—very much like the Finder program on the Apple Macintosh computer. Suppose that Apple has brought suit on a number of different grounds, and Microsoft has hired you to help value and plan the case.

    You note, first of all, that many of the techniques used in the Finder and Window programs were actually developed at Xerox's Palo Alto Research Center (PARC) in the 1970s. However, Xerox never proceeded with commercial development and application of the user interface. You are unsure what effect this will have on Apple's claims. You also note that Apple has a patent application pending on its Finder program. You suspect it must be a process patent: an act, or series of steps performed upon the subject matter to be transformed and reduced to a new state or thing, with information being the subject matter in question. However, because patent applications are secret until approved, you don't know the contents or the type of the patent application—just that it has not yet been granted.

    Because patent infringement is not available to it, Apple has brought suit on a number of other grounds. First, it asserts trademark infringement. Although the Finder program (or, more properly, the appearance of its windows and dialogue boxes) is not registered on the federal principal register of trademarks, Apple can still assert common law trademark rights. The first question in establishing such rights is whether Apple has adopted and appropriated the Finder "look" and used it in its business. The answer seems clear.

    Next, the mark must be classified as to how it identifies and distinguishes the goods in question (Are the "goods" the Apple software environment? The physical Macintosh computer itself?). You could argue that the mark is descriptive; that is, the look of dialogue boxes, windows, and pull-down menus virtually is the Macintosh product, and therefore Apple needs to show secondary meaning: that consumers associate that look with Apple Computer, Inc. Alternatively, Apple could argue that the mark is arbitrary and fanciful: that the Finder features have no per se relationship to personal computers, but, rather, uniquely identify software on Apple Macintosh personal computers. You note how the classification of the mark determines the showing necessary to establish its validity.

If the mark is valid, then Apple still must show infringement, which courts establish by looking to a variety of factors. The sight, sound, and meaning test examines whether the marks look confusingly similar, sound the same, or evoke similar mental images. The Finder and Windows interfaces certainly look the same, but its hard to know how the sound aspect would apply. Would the mental image of a graphically oriented, mouse-driven interface with windows, pull-down menus, and dialogue boxes be the relevant one? Would the image of the computer associated with that look be the focus?

Other factors include:

- the strength of the mark: its tendency to identify the source of the good;
- the similarity of the competing mark, considered in light of the respective settings the marks are used in;
- the similarities of the products in form, composition, texture, quality, function, or orientation;
- the likelihood of direct competition between the prior owner and the alleged infringer;
- whether there is actual confusion as to the source of the goods;
- whether the defendant adopted in the mark in good faith and without knowing or intentional infringement;
- the sophistication of buyers; that is, whether the relevant purchasing public is discriminating enough to distinguish the source of the goods without any confusion.

The final question under the trademark claim is remedies. Normally, a prevailing plaintiff is at least entitled to an injunction against further infringement. You presume this would mean barring Microsoft from further sales of Windows or similar products. If Apple can establish that Microsoft infringed intentionally, they can also recover damages for past infringement. Those damages can be lost profits on lost sales if Apple can show that, but for the infringing sales by Microsoft, it would have made the sales. This requires showing that Apple had the capacity to make the sales, available product in all the relevant locations, lack of any other substitutes, and so forth. Alternatively, if it can make the showing for lost profits, Apple can still recover a reasonable royalty on Microsoft's past sales. Another way to calculate damages for intentional infringement is to prevent unjust enrichment by awarding Apple Microsoft's profits on its Windows sales.

Apple's second theory of action is based on a theory of unfair competition. In general, the law of unfair competition applies to a wide range of commercial conduct likely to confuse consumers about the source of goods or services. Under this heading, Apple brings a number of specific claims.

First, Apple asserts that Microsoft is passing off its goods as those of Apple's by imitating almost every visible aspect of the Finder program. To establish this claim, Apple must show that the imitation constitutes an indirect misrepresentation of the source of the goods. This can be shown by demonstrating that the Finder program is a common law trademark (with the same arbitrary and fanciful or secondary meaning analysis as under the trademark claim) and that there is a likelihood consumers will be confused, or, if the Finder is not entitled to trademark protection, that Microsoft is intentionally passing off its goods and there is actual consumer confusion.

The second unfair competition claim asserts dilution: the loss of the uniqueness or individuality of the Finder trademark by use without Apple's permission; or injury to the value of the mark caused by actual or potential confusion as to the source of goods. You note that this claim also hinges on the ability of the appearance of the Finder interface to function as a trademark in identifying the source of goods, and in the confusion analysis.

The final unfair competition claim is misappropriation: that Apple has developed a property right in the Finder program by investing labor, skill, and money in developing a salable product, and that Microsoft has taken commercial advantage by misappropriating the qualities, salability, and reputation of the original. This claim would therefore depend on how much Microsoft has marketed Windows by trying to take advantage of Apple's success, and in whether the court feels that the combination of original Xerox development and Apple's further development confers a property right on Apple.

Finally, you note that, under the unfair competition rules, the same remedies of injunctive relief and damages for willful infringement apply, but that courts can be much more innovative in crafting injunctive relief. Accordingly, a court could simply order Microsoft to change certain aspects of its program or marketing methods, rather than ordering sales stopped altogether.

Apple's last cause of action is under the law of undeveloped ideas. Under this area, a party can have a right in an idea itself if the idea is novel and concrete. The novelty requirement is met if the idea has never been previously thought of or put into practice. Thus, Apple's claim may founder because of the previous Xerox work. On the other hand, the Xerox work was never integrated into a functional computer operating system, so Apple may get by. Regarding concreteness, the idea must be fully fleshed out and ready to use without further work. Apple's commercial application of the Finder program would seem to meet this requirement.

If Apple succeeds in establishing a right in the idea of the interface itself, it still must establish a basis for recovery. The two grounds proposed are unjust enrichment, and a property right in the idea. Under the unjust enrichment ground, the idea is that the defendant would be unjustly enriched if he were allowed to use the idea without paying for it. Courts will commonly require that the idea be embodied in a concrete form attributable to plaintiff's own ingenuity, and that the concrete form be used by defendant. You are unsure how this requirement would be applied in this case. Certainly, Apple could argue that the computer code was a concrete form, and that Microsoft must have used this form by observing and imitating the appearance and operation of the Finder program. On the other hand, you could argue that Microsoft never actually used the text of the computer code itself, and that Microsoft expended considerable time and energy in developing its own code. The fact that the Finder code won't run in the MS-DOS environment that Windows runs in would bolster your argument. If Apple prevailed under unjust enrichment, it would likely be able to recover Microsoft's profits on Windows, though, again, the court has equitable discretion.

The property basis is less clear, with courts examining the same novelty and concreteness requirements to decide whether they will recognize a property right in an idea itself. This basis has most often been recognized in the context of professional services (such as be doctors, lawyers, and the like) where ideas are a large part of the service being rendered. Although the idea behind it (a graphic-oriented user interface) is certainly the most valuable part of the Finder interface, you are unsure how the court will apply this ground where there have been no explicit commercial relations between Apple and Microsoft. If Apple prevailed on the property basis, the hornbook remedy is the value of the property. You aren't sure how a court would value the appearance and function of the Windows interface, and suspect the court would resort to another method of determining the remedy as a surrogate.

43.1. Consider the possibility of spill-over between the claims in this case. Are there any common legal or factual issues? What kind of a spill-over effect (if any) do you think there will be? How will the spill-over affect the way a jury decides the case?

43.2. Assume that the case, if tried, will be tried before a jury in San Jose, California. You note that San Jose includes both parts of Silicon Valley and the Hispanic neighborhoods of East San Jose. Structure the relevance diagram for the value of the case, paying special attention to how you think the jury will approach the case (Of course, this is the usual rule, but this problem is designed to make the application of it crucial.). Your analysis may vary from a classmate's. In the real world, they usually do.

43.3. How and why is your relevance diagram different from the way the pleadings in the case lay out the claims? What are the strengths and weaknesses in the way your relevance diagram differs? How would you use these differences in arguing with a colleague that the relevance diagram shows a better way to approach the case?

Assume that Windows sales have steadily increased from \$1 million 4 years ago to \$20 million this past year, and are expected to remain steady at that level for the next 5 years (that is, look ahead 5 years in valuing any future relief.) Microsoft's net profit margin on those sales is 15%. Also assume that, because the Finder is bundled in with the sale of Macintosh computers, Apple's revenues on those same unit sales would have been between 10% and 20% higher, depending on how the court unbundles the revenues in a Macintosh sale, and that Apple's profit margin has been a consistent 20%.

Assume that an injunction against any future Window sales would cripple Microsoft because of the all the other products sold to run under Windows, and that Microsoft sales would likely plummet from $100 million to around $10 million. The company might be forced to declare bankruptcy.

Assume that if the case is tried, the trial will take 6 months, could commence immediately because of a quirk in the local court calendar, and that both sides would expend between $300–500 thousand in handling the case. Also assume that the court would award Apple's legal fees if it found willful or intentional misconduct by Microsoft on any of the possible bases.

Assume that a determination of reasonable royalties could range from 2% to 10% of gross sales.

43.4. Structure the decision tree for the value of the case. Use subtrees if you find it necessary or helpful, but pay careful attention to possible dependencies between the subtrees (which would prevent you from simply summing the resulting mean values).

43.5. Write the value formulas. State any additional assumptions you need to make in determining the value formulas.

43.6. Assess any necessary probabilities. Again, add any assumptions you find helping in filling in a gap or making the problem more realistic.

43.7. Write the probabilities into the decision tree and calculate the mean values. What should Microsoft's settlement strategy be?

43.8. Calculate the values of information and control on the uncertainties in the tree. What should Microsoft's discovery strategy be?

43.9. Which claims would Microsoft be best-advised to move for partial summary judgement on? For the most promising claim, add a decision and uncertainty to your tree to evaluate the decision on whether to move for partial summary judgment on that claim. How does making the motion affect the value of the case?

43.10. What insights have you gained that would held you in deciding what to file as an answer to Apple's brief? (Assume that the facts of the problem accurately summarize Apple's brief.)

43.11. Suppose that settlement attempts thus far have failed, and the case is progressing to trial. How could the way you've structured the problem thus far affect your plans for voir dire examination of jurors?

43.12. Suppose that jury selection is now complete, and you handled voir dire as stated in the previous part. List two ways each your challenge strategy could have succeeded or failed (that is, list two kinds of people you would and two kinds you would not want on the jury). For each type of person, how might you restructure your analysis in preparation for presenting your case?

44. Dr. Piltdown, a jury science expert, has offered to predict the jury's decision. His method is to take infra-red photographs of the jury's heads as they sit in the courtroom and then to compare those photos to a similar ones taken of your client and of the opposing party. His theory is that the jury's empathy with a

party is reflected in how similar the pattern of hot and cool regions on their heads are, and in whether the jurors have "plaintiff patterns" or a "defendant patterns." This information seems like it might be valuable to you because your research has shown that verdicts in this jurisdiction are almost evenly mixed, with 50 percent going for the plaintiff, 45 percent for the defendant, and 5 percent resulting in hung juries.

You've gone through Piltdown's records and compared his predictions with the actual results. He correctly predicted a verdict for the plaintiff 70% of the time, with 25% hung juries and 5% verdicts for the defendant. He correctly predicted a verdict for the defendant 60% of the time, with 30% hung juries and 10% verdicts for the plaintiff. However, most interestingly, he correctly predicted hung juries an astonishing 96 percent of the time, with 2 percent each verdicts for the plaintiff and for the defendant (An accuracy Piltdown attributes to "irreconcilable differences between left- and right-brain blood flow levels among the jury members.").

44.1. Given a prediction of a verdict for the plaintiff, what is the probability of a verdict for the plaintiff? Of a verdict for the defendant? Of a hung jury?
44.2. Given a prediction in your case of 50% plaintiff verdict, 40% defendant, and 10% hung jury, what is the probability of a verdict for the defendant? Of a verdict for the plaintiff? Of a hung jury?
44.3. Suppose your predictions were 80% defendant verdict, 15% plaintiff, and 5% hung jury. What are the chances of a hung jury? Of a verdict for the plaintiff? Of a verdict for the defendant?
44.4. Suppose your client is facing a potential \$10 million judgment, and Dr. Piltdown could increase his accuracy in predicting plaintiff verdicts to 80% (15% hung and 5% defendant) and in prediction defendant verdicts to 70% (20% hung and 10% plaintiff) by having your client shave his head before taking the infrared photo. How much is it worth to your client to shave his head?

Chapter 10
Multiple Parties and Claims

A number of complications typically arise when dealing with multiple parties and/or claims that make it difficult to analyze the value of litigating your case—and thereby to think about what a good settlement amount is. This chapter shows ways of handling a number of the most common problems. The approaches don't involve new concepts, and, indeed, you can (or may have) come up with them yourself by using standard problem structuring techniques. If you haven't, the discussion here will hopefully save your time for the more unique aspects of your case.

10.1 Codefendants

Properly considering a codefendant in valuing your case means, first, considering the overall value of litigating the case, and, second, accounting for how that verdict will be allocated between you and your codefendant(s). And, as usual, you need to analyze the value of litigating the case before you can consider a settlement value. Accordingly, the discussion will first hand the valuation and allocation problems for litigating a case, then show how to extend this result to considering settlements.

10.1.1 How to Determine Your Value of Litigating the Case with a Codefendant

When you are the sole defendant, your value of litigation is the mean damages award, discounted for the uncertainties in damages being awarded. When you are a codefendant, there are the additional uncertainties of whether your codefendant will

J. Celona, *Winning at Litigation through Decision Analysis*,
Springer Series in Operations Research and Financial Engineering,
DOI 10.1007/978-3-319-30040-5_10

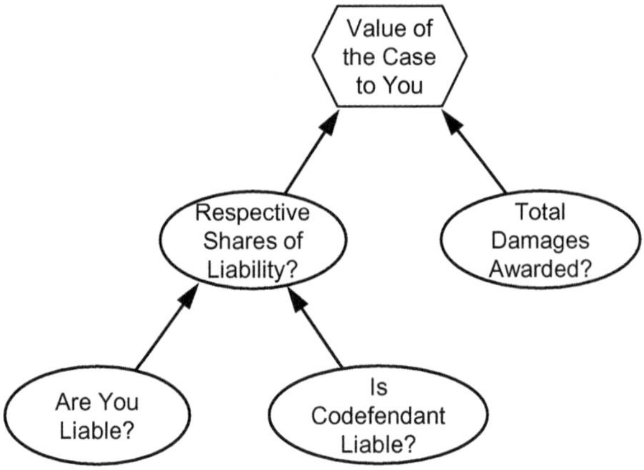

Figure 10.1 Value of the case to you with a codefendant

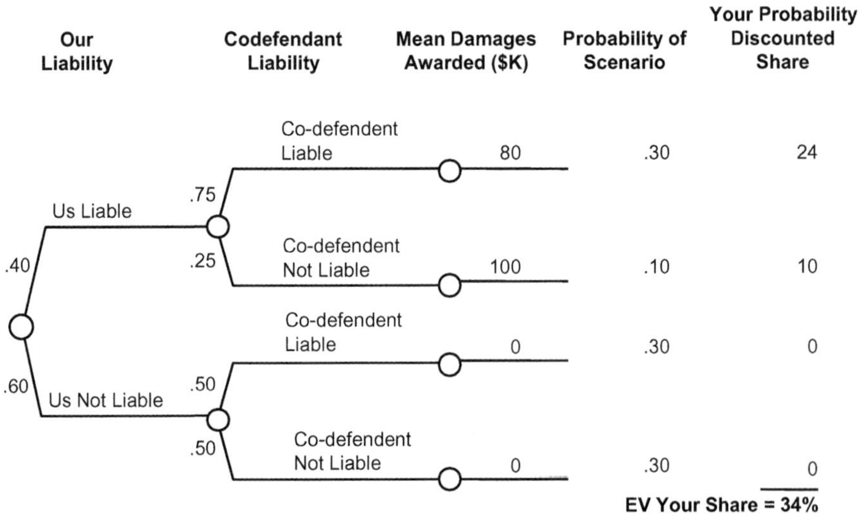

Figure 10.2 Assessed tree for your mean share of damages

be held liable, and what your respective shares of liability will be. The relevance diagram makes these relationships clear (Figure 10.1).

Of course, the usual question of spill-over should occur to you, such as if finding liability for you and the codefendant is related, and whether these questions also affect the damages awarded. Those possibilities can be incorporated into your analysis as well as the allocation problem discussed here.

Suppose, then, that, if you and the codefendant are both found liable, your share of the liability will be 80% (you could also put in a chance node for this question). One possible share for determining your share of the verdict appears below. This tree also omits the damages uncertainty (Figure 10.2).

Your value of litigating the case, then, should be .34 times the mean damages. If the mean damages were $75,000, your expected share would be $25,500, and you should be willing to settle for anything less than this amount. This figure is the number you should use when dealing with the plaintiff. See the next section for allocations between the codefendant and yourself.

This figure assumes, of course, that your codefendant has not yet settled or been found liable. See the section after the next one for when the codefendant has already settled. The same analysis would apply if the codefendant had already been found liable.

10.1.2 *How to Allocate a Settlement Between Yourself and a Codefendant*

If you are dealing with a codefendant, rather than with the plaintiff as in the previous section, a simple extension to the analysis is needed. The approach is to figure the expected shares of both yourself and the codefendant in your tree, and then to use the ratio of those expected shares to allocate any settlement between the two of you. The tree below illustrates this calculation (Figure 10.3).

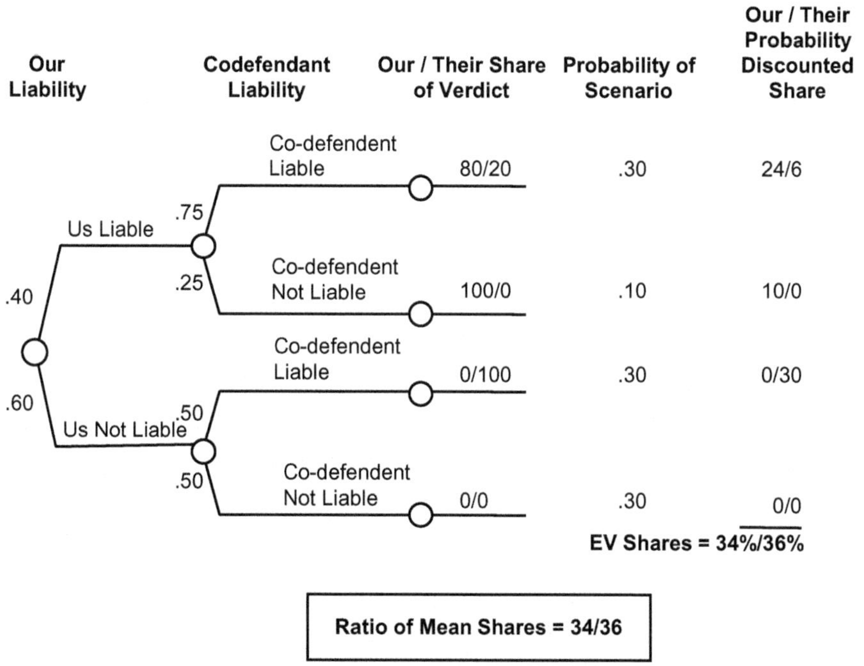

Figure 10.3 Allocating a settlement with a codefendant

Accordingly, because you and the codefendant have about the same mean value shares of the mean award, any settlement should be about split evenly between you. And remember that your prior analysis still applies: for $75,000 mean damages, you should not settle for any more than .34 × 75 = $25,500.

10.1.3 How to Figure a Settlement Value Where Codefendant Has Already Settled

The third variation is thinking about your settlement value where a codefendant has already settled or been held liable. For instance, suppose a suit is outstanding against you, and a codefendant has already settled for $15,000. You no longer have the problem of allocating a prospective judgment between you and the codefendant, and the liability and damages questions might look something like the following (Figure 10.4).

How should you think about a settlement value for the case? Should you consider the $15,000 the codefendant has already paid to calculate your maximum settlement value at 16 – 15 = $1,000?

As you might guess, simply subtracting the codefendant's payment from the mean value the way you subtract legal fees would be wrong. It would be assuming that your liability is 30 – 15 = $15,000 in the top scenario, 10 – 15 = –$5,000 in the middle scenario, and 0 – 15 = –$15,000 in the bottom one!

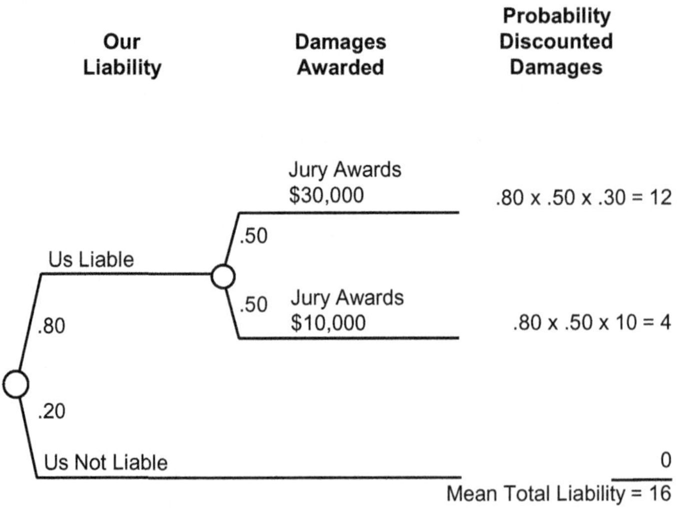

Figure 10.4 Total exposure remaining in the case

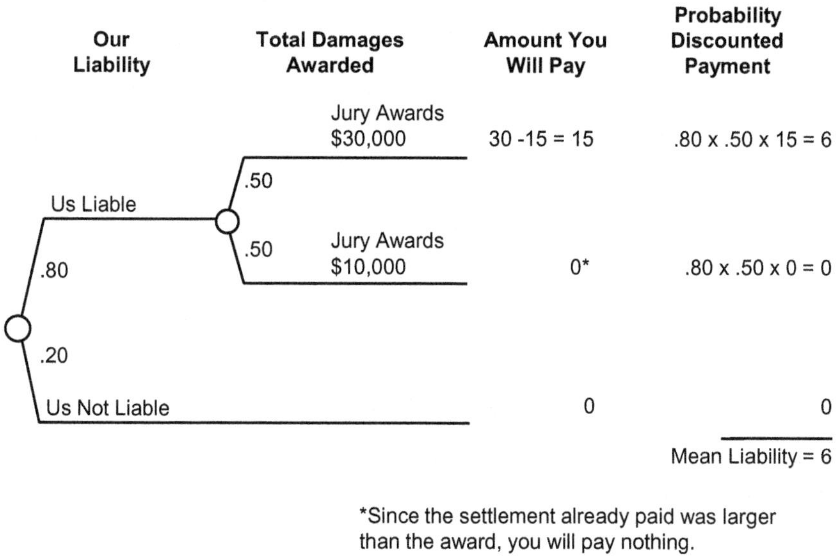

Figure 10.5 Remaining mean liability where the codefendant has already settled

The key is recognizing that the total expected liability of $16,000 is not the amount expected to be paid as a result of this second case. Rather, the value of *this* case to you is the amount of the award minus previous satisfaction of judgment from the codefendant's payment. Accordingly, the correct way to calculate your settlement value for this case appears below (Figure 10.5).

Thus, you should be willing to settle this case for any amount up to $6,000. This figure properly accounts for crediting the prior settlement payment toward satisfaction of your judgment.

The point of this tree, of course, is that you can't subtract a number from the overall mean value when that subtraction doesn't apply in every scenario. In this case, you can't subtract the codefendant's $15,000 payment where the judgment against you was only $10,000, or where the judgment was zero. Those excesses $(.4\times-5)+(.2\times-15)=-\$5{,}000$ are why the settlement value of the case came out $5,000 too low in the first calculation. Where the subtraction doesn't apply to all scenarios, you have to do the subtraction scenario by scenario.

10.1.4 Splitting a Settlement Between Insurer and Insured

An interesting parallel to the codefendant problem is that of splitting a settlement between insurer and insured. For instance, consider the following tree, which lumps insurer and insured together without differentiating policy payment terms (Figure 10.6).

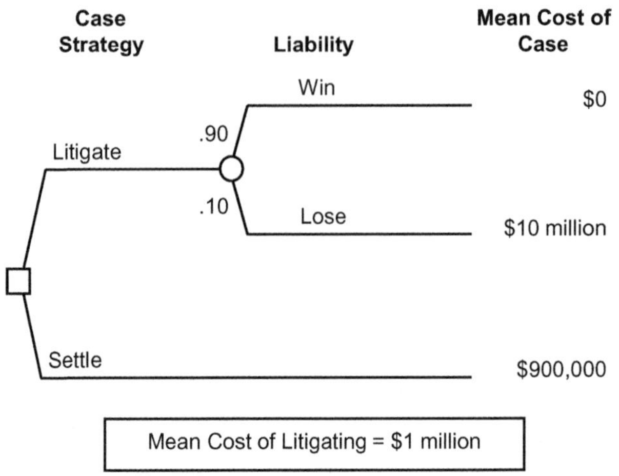

Figure 10.6 Total cost of case to insurer and insured

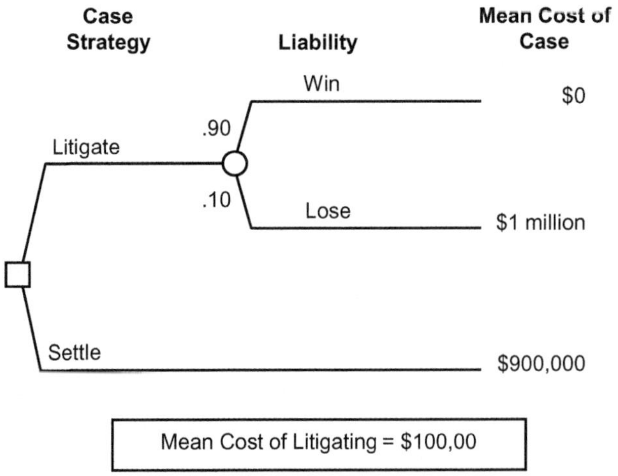

Figure 10.7 Tree for exposure to insured alone

The $900,000 is a good settlement for the insured plus the insurer together, since the mean value of litigating is $1 million (.10 × $10 million), as compared with the settlement demand of $900,000. As a total settlement amount, then, paying $900,000 makes sense.

However, suppose the insured has a $1 million self-insurance retention in its policy, and thus would only have to pay all amounts up to the first $1 million. Then,

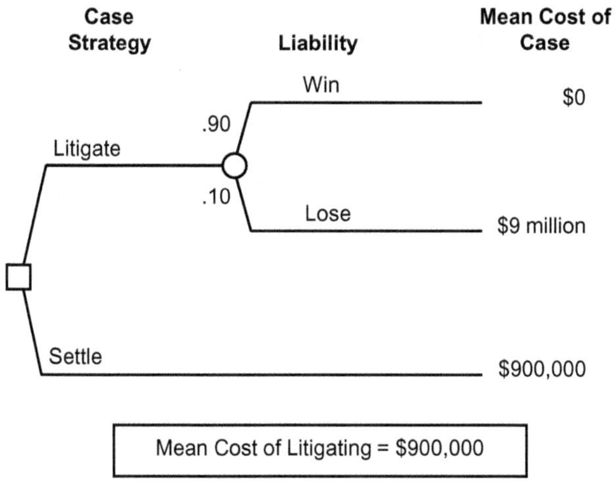

Figure 10.8 Tree for value to insurer alone

when the insured looks strictly at what he would have to pay, the settlement value looks very bad, as shown below (Figure 10.7).

The insured's mean cost of litigating is only $100,000! Thus, it would be a very bad business decision for the insured to pay all $900,000. In fact, anything over $100,000 would be too much for the insured to pay.

On a purely economic analysis, then, you could argue that insurer must pay at least $800,000 to make this settlement happen. Furthermore, drawing a parallel from splitting a settlement with a codefendant, the most equitable arrangement would be for the insured and the insurer to split the $900,000 demand according to their respective mean values, as was demonstrated above in the general case for allocating settlement with a codefendant.

As can be seen from the insurer's tree below, the insurer's mean cost of litigating is $900,000 (the insurer has the other $900,000 of the overall $1 million mean value of litigating), and thus the insurer should pay 9/10 of any settlement figure and the insured 1/10—even though the entire settlement demand may be less than the insured's self-insured retention (Figure 10.8).

You might argue that this analysis is irrelevant because policy language almost always requires the insured to pay any and all settlement amounts up to the self-insured retention amount. Perhaps that kind of standard policy language will change if insured's begin to follow this economic analysis and choose to litigate, ending up costing the insurer even more!

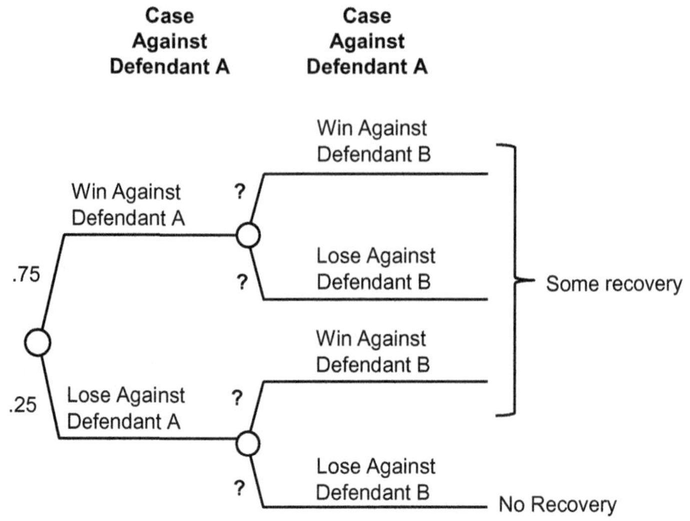

Figure 10.9 Main tree for multiple cases against multiple codefendants

10.2 Multiple Defendants

As opposed when you are one of several codefendants, the problem of multiple defendants arises when you are the plaintiff and are pursuing a single claim against multiple defendants whose respective liabilities are based on different and independent theories. Accordingly, the problems you need to consider may include handling separate, possibly related trials; problems of partial satisfaction of judgment credited to future defendants; and spill-over within a single trial—even when liability against the defendants is predicated on different legal theories. In this sense, this problem is similar to the problem of recovering a single loss on either of two independent arguments, which was discussed in Part 5 of the section on drawing decision trees.

These problems can be handled by standard problem structuring techniques. Likely you will want to use a separate subtree to handle the possibility of recovering from each defendant. However, to show how you might handle some of these problems, the following main tree for liability highlights some of the potential spill-over problems, followed by an example that shows several of these problems at work. To be more explicit, the spill-over problem in this case is that even though the defendants' liabilities are based on independent theories (indicating that you can use a different subtree for each defendant), there may be spillover between these nominally independent subtrees (Figure 10.9).

Here are some consideration in putting this analysis together.

- The probability for winning against defendant A would come from a detailed subtree showing the factual and legal uncertainties in your case against A.
- It may be unnecessary to consider the case against B if you win against A. For instance, if winning against A gets you all the damages you have a claim for, then you would not care whether you also win against B.
- You may need to consider how the chances of winning or losing against defendant B change depending on the result against defendant A.
- You may also need detailed subtrees to capture the uncertainties in damages in the various scenarios, such as this one of winning against B after having lost against A.

For example, if the chance of prevailing against defendant B were 60% regardless of the result against defendant A, the overall probability of winning (against either defendant) is $.75+(.25\times.60)=.90$, while the overall probability of losing against both the defendants is $.25\times.40=.10$.

Thus, careful structuring of the main tree is necessary to assure realistic appraisal of your chances of recovering your loss. It is rather like a "Win Either Way" problem, and capturing the right combinations in your main tree will ensure you don't over-value or under-value your case.

Let's see how this plays out in an example.

The plaintiff D-Zine, Inc. designed a special treatment building to be built by ConCorp. Work was to be completed in 1978. D-Zine was required by the building's owner to guarantee the work of Con-Corp because Con-Corp was some shaky financially. After completion, the owner had numerous complaints about the finished building, and, shortly following a minor earthquake in 1982, demanded that D-Zine make certain repairs. D-Zine finished those repairs in 1983.

There were two major areas of repair work. The first was extensive repairs to a peeling waterproofing membrane applied to certain areas of the building. The second was repairs to the heating, ventilating, and air-conditioning system (HVAC), which played a crucial role in the treatments at the plant.

In late 1985, D-Zine sued Concorp (more financially stable now), alleging that both of the repairs were required by ConCorp's negligent construction. Specifically, the plaintiff D-Zine alleges that the membrane was applied at the wrong temperature. D-Zine also sued Dry Co., the materials supplier, on the alternate theory that, if the membrane was properly applied, then it must have been either improperly manufactured or unsuitable for its intended use.

D-Zine further alleged that the HVAC system was improperly installed by ConCorp. The plaintiff counsel anticipates that ConCorp will answer this charge by claiming that plaintiff's design was faulty. D-Zine has begun a survey of industry experience with this same design, and is not sure how favorable the results will be. The counsel anticipates that a jury would not hold ConCorp liable if they found D-Zine's design faulty.

As mentioned above, D-Zine has long-since completed the repairs and is therefore suing both the defendants on a theory of indemnification. D-Zine contends that it was required to make the repairs because of its guarantee of ConCorp's work, and *is therefore entitled to recover the repair costs from the guilty parties. The actual costs were $3 million for the peeling waterproofing membrane and $2 million for the malfunctioning HVAC system and are not in dispute.*

D-Zine's right to indemnification depends on establishing it had a legal obligation to make the repairs when demanded by the owner. The defendants will argue, first, that the owner waited too many years before making the demand; and, second, that the alleged guarantee given by D-Zine wasn't legally sufficient—that is, it didn't legally obligate D-Zine to perform repairs. The defendants base this second contention on having found no more written evidence of the guarantee than a mention in correspondence between D-Zine and the owner that "D-Zine can be counted on to see that ConCorp performs." In fact, besides this correspondence, D-Zine plans to introduce testimony by its employees and the owner's supervisor on the clear extraction of a guarantee as a condition of the contract. The supervisor would make a more credible witness than D-Zine's employees, but, unfortunately, he retired 2 years ago and the plaintiff counsel has thus far been unable to locate him. A further possible obstacle to establishing the guarantee is the question of whether the law in the jurisdiction requires that guarantees be written to be enforceable.

The ultimate issues for the value of this case appear in the relevance diagram below. Note that the ultimate issues are whether D-Zine recovers at all for the repairs it's made—not what the specific results against ConCorp or Dry Co. are. The focus on what happens to you means that the questions of recovery are paramount, not which potential defendant you obtain indemnification from (Figure *10.10*).

Next, the counsel added questions on the whether the guarantee was a legal obligation until they felt comfortable assessing probabilities (Figure *10.11*).

Next, the counsel considered the question of recovery for repairs to the waterproofing membrane, considering that D-Zine could potentially recover from either ConCorp or Dry Co. for these repairs (Figure *10.12*).

Finally, the counsel completed the diagram by adding the influencing factors for whether D-Zine recovered for the repairs to the HVAC system (Figure *10.13*).

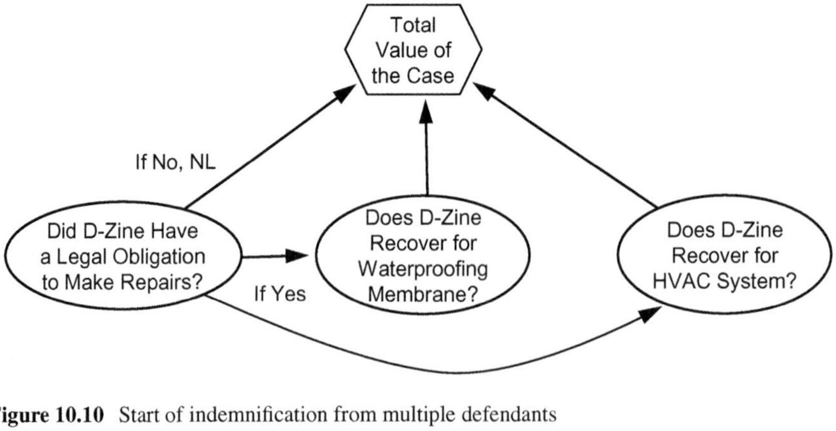

Figure 10.10 Start of indemnification from multiple defendants

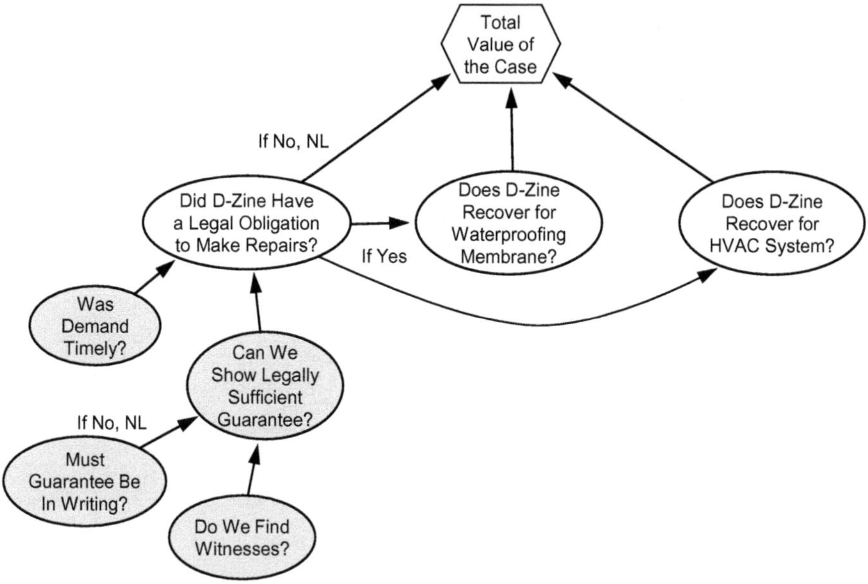

Figure 10.11 Adding the guarantee issues

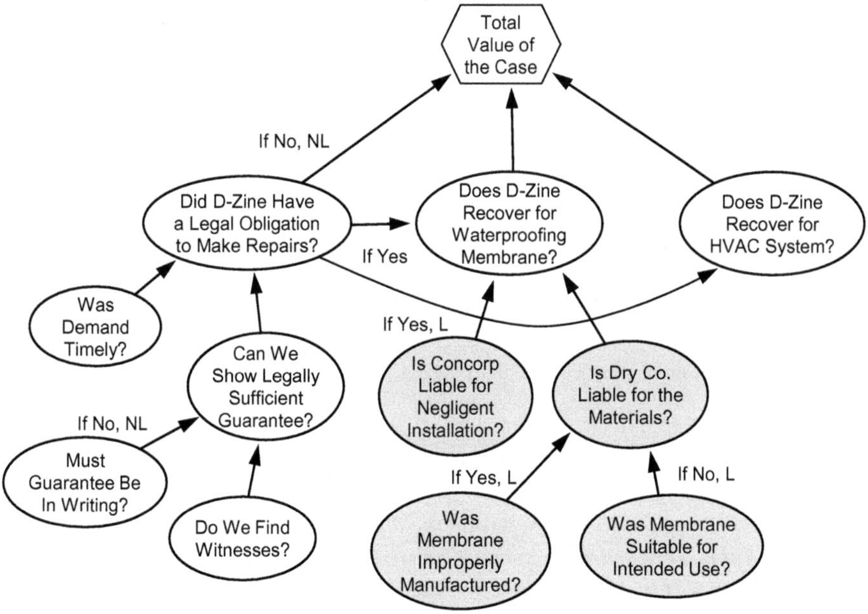

Figure 10.12 Adding the waterproofing membrane issues

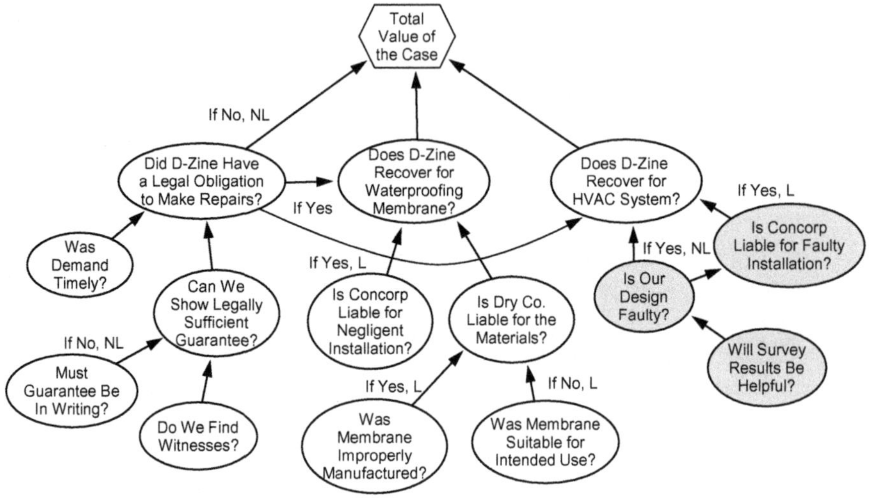

Figure 10.13 Completing diagram with HVAC issues

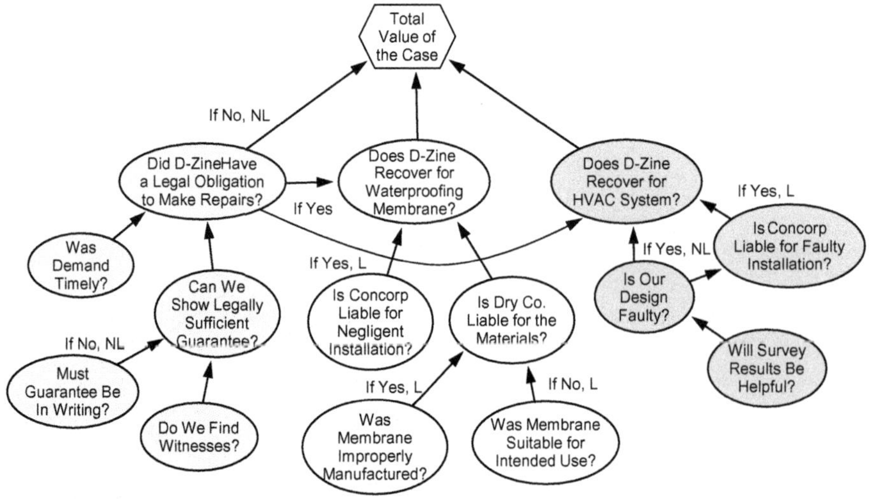

Figure 10.14 Questions grouped for subtree analysis

Lastly, the counsel indicated the subtrees that would be used to do the analysis in pieces (Figure *10.14*).

Note that it was possible to separate the diagram this way in spite of the arrows going into the circles drawn because those arrows indicated asymmetry in the tree—in other words, that whether D-Zine could recover on its respective claims

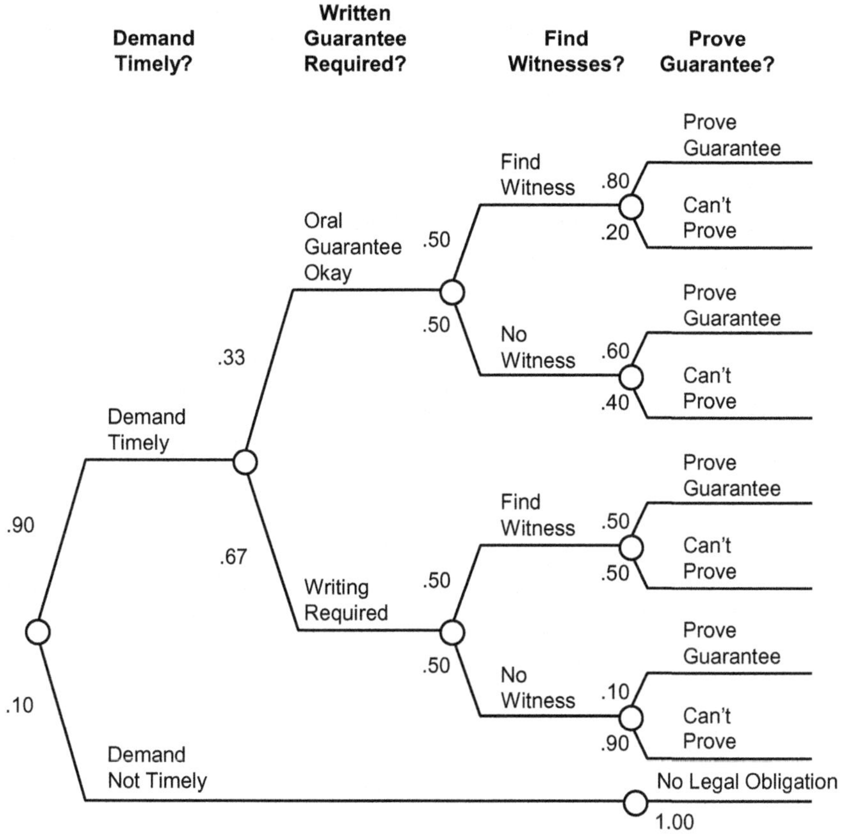

Figure 10.15 Subtree for guarantee issues

was irrelevant if they lost on the preliminary issue of whether the guarantee was legally binding.

The diagram done, the counsel drew the subtrees and assessed probabilities for them, as shown below. Though omitted there, the lists of reasons, as always, are the most important part of the process. For the legal obligation subtree, the chances of proving the guarantee depended on whether an oral obligation was sufficient and whether the plaintiff counsel found the owner's supervisor. As noted before, an untimely demand cut off liability (Figure *10.15*).

The subtree for recovering on the water-proofing repairs went through all the alternative theories that D-Zine could recover on from either defendant. A win on any one theory, of course, meant that the outcome of the remaining theories didn't matter because D-Zine would have its recovery (Figure *10.16*).

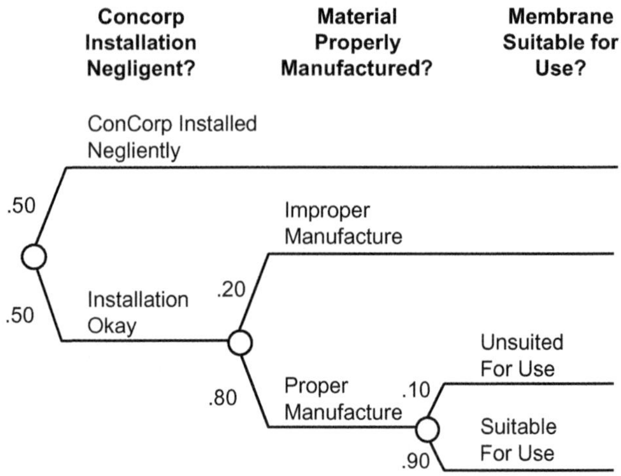

Figure 10.16 Subtree for waterproof membrane liability

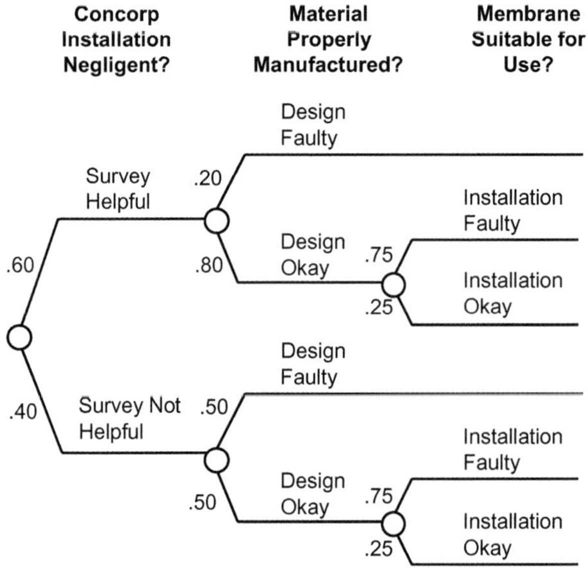

Figure 10.17 Subtree for HVAC system liability

For the HVAC system recovery subtree, it was the counsel's judgment (reflected by the lack of an arrow in the relevance diagram), that the survey results would not affect the finding on whether ConCorp's installation was faulty (Figure *10.17*).

Finally, solving all these subtrees yielded the probabilities for the main tree, shown below (You should try solving the subtrees yourself and see if you come up with the same numbers.), Note that the same probabilities appearing for the HVAC

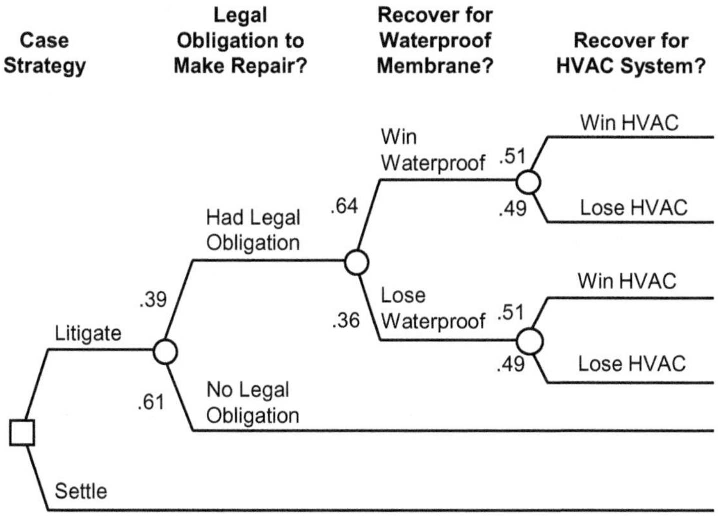

Figure 10.18 Completed main tree for codefendants' liability

win regardless of how D-Zine does on water-proofing indicates counsel's judgment that the court's finding on these two issues are independent of each other. In cases like this, you should consider whether there is spill-over between two legally independent theories and adjust probabilities accordingly. Thus, for instance, you might lower the probabilities of winning on HVAC after you've already lost on water-proofing (Figure *10.18*).

The settlement offer was $2.0 million (from both defendants, split somehow among themselves). Should D-Zine settle the case, or litigate? We will leave that for a problem at the end of the chapter.

10.3 Counter-Claims

Because counter-claims are also discussed in Chapter 8, part (A)(4) on offsets to an award, only a few more points will be mentioned here.

The basic method for handling counter-claims is to do a separate analysis for each claim and then to net out the mean values. The resulting number is the usual settlement value of your case. However, there are two reasons why this method may not be adequate.

- The two claims may not be independent of each other. They may be legally or factually related, or there may be spill-over between them.
- The client may be risk-averse.

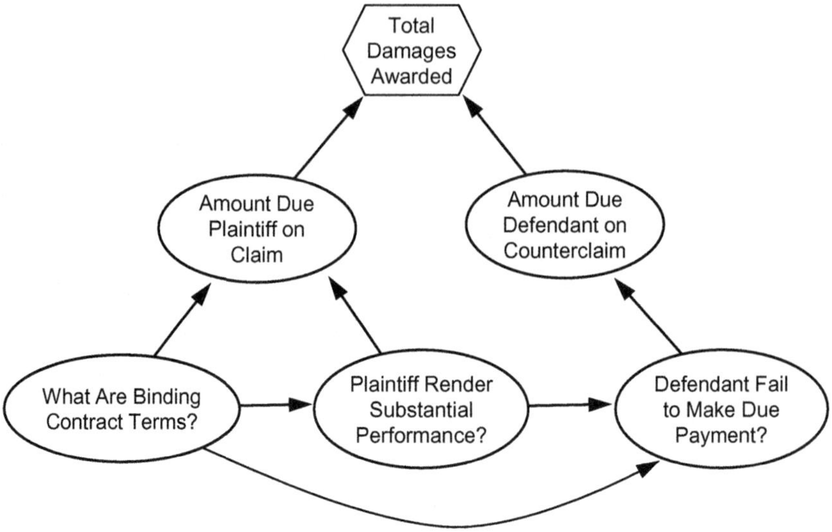

Figure 10.19 Relevance diagram for contract claim and counterclaim

Figure 10.20 Value calculation for claim and counterclaim

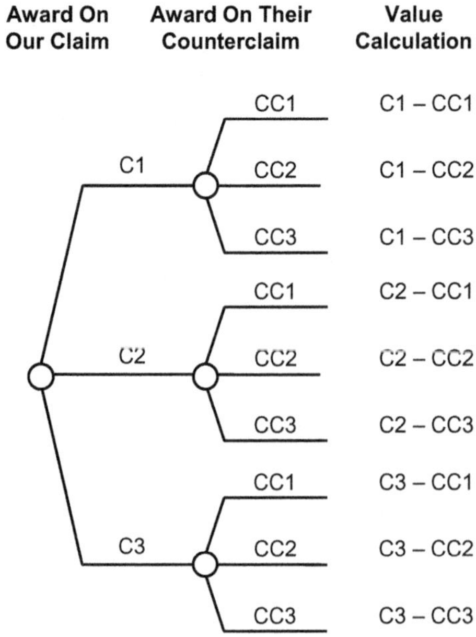

For the first reason, you'll need to use one main decision tree for the claim and counter-claim together to handle the interactions. Thus, in the contract claim and counter-claim below, the dependencies between the claim and counter-claim on the question of damages means that you should draw one tree for them all so you can properly assess different answers for each of the scenarios (Figure 10.19).

There may be similar dependencies on the question of liability, and even dependencies between the liability and damages questions. To avoid your tree getting unmanageably large, you'll need to follow the suggestions in Chapter 4 about limiting tree size. Pay particular attention to: first, building up your relevance diagram with levels of increasing detail to help you decide the level of detail necessary to assess answers for the tree; and, second, to looking for independent sections of your relevance diagram that can be handled with a subtree.

If the client is risk-averse, netting out the mean values also will not work because, as you'll recall, risk aversion depends on the total range of possible values, and the mean values don't convey this information. Your first alternative, then, might be to try summarizing the claim and counter-claim subtrees with chance nodes, and then netting out the two scenario-by-scenario, as shown below. Note that this distribution will likely have positive and negative numbers for when the plaintiff is a net recipient and net payor (Figure 10.20).

This method will give you a nine-value distribution on the value of the case. If the tree isn't too large or you're using software, you could just include these uncertainties in the main tree.

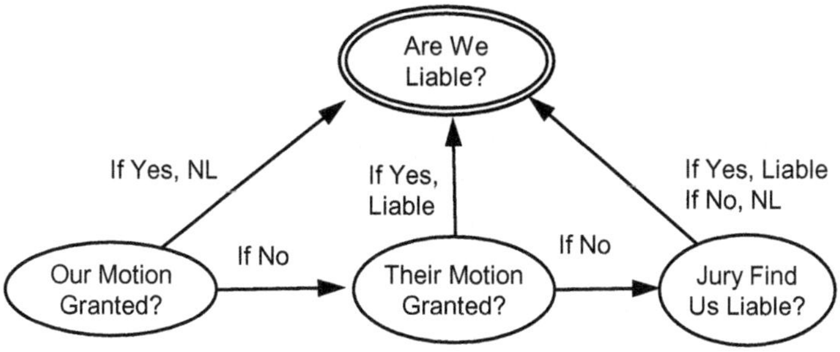

Figure 10.21 Two opposing motions for summary judgment

Figure 10.22 Tree for opposing motions for summary judgment

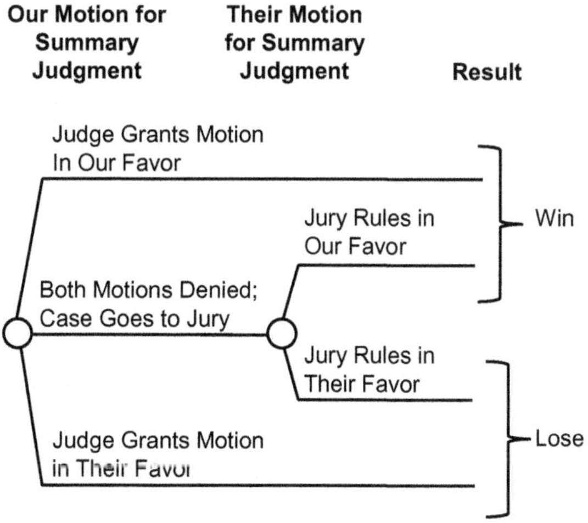

10.4 Two Opposing Motions for Summary Judgment

Handling two opposing motions for summary judgment is a matter of recognizing and correctly representing all the possible scenarios. Putting all the questions and the "if yes" and "if no" labels in the relevance diagram makes clear how the questions could be resolved (Figure 10.21).

Because all the influences to the liability question directly determine liability, this influence diagram can be drawn as a one-, two-, or three-node tree. The drawing for the two-node version appears below (Figure 10.22).

Which version of the tree your drew, of course, would depend on how you thought the trier-of-fact would decide the problem and how you felt most confident assessing the probabilities.

10.5 Problems

10.5.1 Codefendants

1. For the allocation with the codefendant shown in Figure 10.3, what is the maximum the codefendant should be willing to pay to settle?
2. You represent Mucho Mojo, a grower and retailer of marijuana in Colorado with a booming business. Your client also sells his own Ganja Din™ branded pipes, which he has specially made by a manufacturer in Boulder, Colorado. They constitute about 20% of his $15 million in annual sales, with a 70% profit margin on them.

   Although expert in breeding and growing plants, your client did not think to search the US Patent and Trademark Office database, where he would have discovered that another Ganja Din®, with a registered trademark and based in Amsterdam, has been selling hashish pipes for decades and exporting them to the USA as "novelty glass and ceramic art objects." Your client and the manufacturer are being sued by the original Ganja Din for trademark infringement. The plaintiff is suing for lost profits from both firms and an injunction against further infringement.

   You estimate that the manufacturer in Boulder has a 50% profit margin on his sales to your client, and those sales are 25% of your client's retail price.

   You raise two defenses for your client. First, not even the most stoned customer would mistake a real marijuana pipe for a "novelty glass and ceramic art object," much less a hash pipe. You estimate a 30% chance of winning on this argument (depending on jury selection and venue).

   Your second defense is that the original Ganja Din did not in fact lose any business to your client because any serious attempt to compete with your client would run afoul of federal law prohibiting the interstate transportation and sale of drug paraphernalia. You estimate a 50% chance of winning on this argument.

The manufacturer raises an additional defense, that he made the pipes and labels separately and that your client applied the labels to the products to creating the offending offering. You estimate they have a 30% chance of winning on this argument. You also would say that the manufacturer won't be liable if your client isn't and, even then, might still get off on the last argument.

2.1. Draw the relevance diagram.
2.2. Draw the trees for value to the plaintiff, exposure to your client, and exposure to the manufacturer.
2.3. Assume that lost profits would be awarded going back for 2 years, and that both would be joint and severally liable for their total profits on the product to the original Ganja Din. Calculate the total value of the case to the plaintiff.
2.4. Calculate the mean shares of the verdict you and the manufacturer would be liable for.
2.5. Calculate the maximum settlement values for your client and for the manufacturer.
2.6. You've just learned that the manufacturer received an offer from the original Ganja Din to settle for $1 million. Should you be worried about it and call the manufacturer? Explain and illustrate why or why not.
2.7. Suppose the manufacturer accepts the offer. What is your client's maximum settlement price now?

3. At a party on a panoramic deck overlooking San Francisco Bay, there was a partial collapse, injuring a number of the attendees. They are suing the owner and host. You were retained by the homeowner's insurance company to represent him. There is a $1 million policy limit with a $50K self-insured retention.

   The owner is only liable if negligent. Because the owner employed a licensed contractor to construct the deck, it was designed by a licensed architect and inspected and approved by the city, you estimate only a 20% chance of liability on this basis. However, some boards were somewhat loose and decayed and possibly should have been replaced. You estimate there is an additional 60% chance of liability on this basis.

   There are eight plaintiffs. Your 25%/50%/25% estimate for the total damages for each (medical and lost income) to each are $40K/$120K/$200K.

   The plaintiffs are also suing the contractor and architect. The contractor alleges the design is the architect's and wasn't strong enough. The architect alleges that the contractor used cedar rather than the redwood specified, which meant the wood wasn't as strong and the beam dimensions should have been increased. You estimate there is an 80% chance of recovering against the contractor, and 60% against the architect because he inspected the construction and known or should have known the contractor was using cedar.

3.1. Structure the relevance diagram for the cost of the is case to the owner and insured.
3.2. Draw the tree.
3.3. Calculate the mean cost of the case to the owner.
3.4. Calculate the mean cost of the case to the insurer.
3.5. Structure the influence diagrams and trees for the cost of the case to the architect and contractor.
3.6. Calculate the mean cost of the case to the architect.
3.7. Calculate the mean cost of the case to the contractor.
3.8. What is the maximum amount each of the parties (the owner, insurer, architect, and contractor) should pay to settle the case?
3.9. You've convened a strategy conference with the counsel for codefendants' architect and contractor. What would you propose and why?
3.10. Should the homeowner retain his own counsel to represent him separately from the counsel provided by the insurance company? Why or why not?
3.11. The court just dismissed the plaintiffs' cases against the contractor and architect for lack of privity. Assume the probabilities of recovery are the same if you pursue a subrogation and the insurer sues them directly. Now what are the maximum settlement values for the original plaintiffs, the insured, the insurance company, the architect, and the contractor?

4. For the D-Zine problem described in this chapter, should D-Zine accept a $2 million settlement offer from the codefendants?
5. For each uncertainty in the main tree, what probability would switch D-Zine's best alternative?
6. See also problem 11 in Chapter 8 (Valley Realty) and problem 9 in this chapter (the Irving Presto problem) for problems involving the codefendants.

10.5.2 Multiple Defendants

7. You represent PlayBall, Inc., a development corporation that owns a major league baseball stadium. The stadium was built as part of a rather complicated deal between Playball, the city of Salton, and the Salton Sliders baseball team. PlayBall bought the land for the stadium from the city and built the stadium, then leased the stadium to the Sliders for 5 years. The lease payments were set as a percentage of gate receipts. The city of Salton, in turn, suspended property tax payments on the stadium for as long as the Sliders were tenants at the stadium, and guaranteed a minimum monthly lease payment should the gate receipts fall below a certain minimum.

   Sliders started off slowly for the first 2 years of their lease, but, with a few good draft picks and some inspired managing, are now cruising towards the division title with a substantial lead and only 20% of the current season remaining. Games have been sold out and lease payments were looking to total $2 million this year (5% of projected gate receipts of \$40 million)—well above the \$1 million per year minimum.

However, a major scandal has broken involving the Sliders' owner, Sid Slinky. The commissioner of baseball charges that Sid has been doctoring the books for gate receipts to create a secret slush fund, and has been using that money to make payments to promising college ballplayers and lure them to the Sliders. In response, the commissioner has barred the team from any post-season play. Sid was furious, and responded by selling the entire team to a Japanese owner (at a substantial profit) who will play the team in Japan for the remainder of the season.

Sid is now attempting to get out of the remaining 2 years of the lease. The escape clause he is trying to invoke terminates obligations under the lease where "major league sanction and recognition has been withdrawn." Sid contends that, because pro baseball is a business, and because fees for post-season playoff coverage can be double the regular season earnings (and the Sliders were certainly headed for league playoffs, if not the World Series), that suspension from post-season play constitutes de facto non-recognition of the team. After heated negotiations regarding a severance payment broke down, Playball filed suit against the Sliders.

The Sliders responded by impleading the city of Salton for the lease payment guarantee. The city answered that the lease guarantee does not apply when there are in fact no gate receipts at all. You feel the city has a stronger chance of prevailing on this claim if the Sliders succeed in escaping from the lease, but that the guarantee should be binding in any event.

The city then counter-claimed that the departure of the Sliders makes them "non-tenants" under the terms of the property tax exemption and counter-claimed against PlayBall for property taxes, claiming that the exemption no longer applies. If the city were to prevail in this claim, it's not clear whether the entire \$1 million annual tax payment would be due, or only \$400,000 for the remainder of the year. Again, you feel that the city's argument is stronger if the application of the escape clause is upheld and the Sliders are not liable for lease payments. However, because the tax exemption was granted separately from the lease guarantee, you feel it's possible that a court could find the departed Sliders "non-tenants" even if they are found still liable for the lease payments.

7.1. Structure the relevance diagram for the value of the case to PlayBall. Can you handle the city's property tax counter-claim separately?
7.2. Structure the tree for the value of the case and write in the value formulas for each scenario. Be sure to use discounting where payments (such as possible lease, guarantee, and tax payments) would not be due for some years in the future.
7.3. Assess any necessary probabilities. State any assumptions you make. Justify any differences of opinion with the preliminary indications given in the problem statement.
7.4. What is PlayBall's settlement strategy? Can you formulate distinct and separate settlement strategies for dealing with the Sliders and with the city of Salton?
7.5. Suppose the city attorney proposed foregoing the property taxes claim in return for PlayBall not holding them to the lease payment guarantee. Should

PlayBall accept the offer? Assume that this deal can be made in spite of the affect on the Slider's rights, then explain why you think this assumption does or does not hold.

7.6. Which probability has the greatest effect on the value of the case to PlayBall? (That is, which uncertainty has the greatest value of information?)

7.7. Assume that PlayBall settles with the Sliders for the mean value of the recovery from the Sliders. Now what is PlayBall's settlement strategy with the city?

10.5.3 Counter-Claims

8. See problems 12 in Chapter 8 (Dr. Dill) and problem 6 in this chapter (the PlayBall problem) for problems involving counter-claims.

10.5.4 Two Opposing Motions for Summary Judgment

9. For the two opposing motions for summary judgment shown in Figure 10.22:

   9.1. Draw the one-node version of this tree

   9.2. Draw the three node version of this tree

   9.3. If you had to assess these probabilities, which version would you use and why?

10. You represent Irving Presto, a delivery driver who's truck plowed into the back of a horse carrier, killing a champion thoroughbred.

    Irving works for Mercury Express, a delivery company. On Saturday, March 4, Irving went to the Mercury garage and signed out a truck. He was not working that day, nor was he scheduled to work on Sunday. He drove the truck to his apartment, where he met some friends who helped him pack the truck for Irving's move to a new apartment. Irving drove the truck to his new apartment, where the friends also helped him unload. It was a warm day and they were drinking beer. After unloading the truck, Irving drove it back towards his apartment at around 8 p.m. He somehow lost control of the vehicle and hit the back of a horse carrier belonging to Pegasus Farms, killing the thoroughbred Donatello in the back. In the previous racing season, Donatello was generally regarded as one of the top five thoroughbreds in the country. He was being taken to a lab for fertility testing before being put out to stud.

    Pegasus Farms is suing Irving and Mercury Express. Regarding Irving's liability, the site of the accident was a poorly lit rural road, a light mist was falling at the time so the road surface was wet, and it was dark out. Irving hit the truck as it was proceeding through an intersection across Irving's line of travel. The Pegasus truck had a stop sign in its direction, and the driver claims to have come to a full stop. There was no stop sign in Irving's direction. The Pegasus truck had all the requisite back and side lighting on the trailer. However, a blood alcohol test shortly after the accident showed Irving at 0.05%—well below the legal drunkenness level of 0.1% but still significant. Pegasus need only show Irving was negligent to establish liability.

Mercury Express' liability is more problematic. Pegasus holds Mercury responsible because it was their truck, and under the doctrine of respondeat superior. Mercury disputes whether they are liable for Irving's actions. They assert that it was Irving's day off, that he was not working or performing any service for them, and that he improperly signed out the truck. Irving claims that, although he used the truck that day to move, he thought he was scheduled to make a distant pick-up early Sunday morning and would therefore pick up the truck on Saturday to get an early start. The Mercury schedule shows Sunday as a day off for Irving. Further, the Mercury policy is not to allow truck pick-ups more than 12 hours before a job, but the attendants at the garage generally do not check the schedule before allowing a driver to sign out a truck.

Mercury has filed a motion for partial summary judgment on the issue of their vicarious liability. Pegasus, in turn, filed two motions for partial summary judgment: one for Irving's negligence, and the other for Mercury's vicarious liability.

Further, Pegasus has offered to stipulate to the value of Donatello at $1 million, an offer you consider tempting. One million dollars is a rather conservative figure for a champion race horse, and Pegasus seems to want an early settlement of the case (You suspect the farm has cash flow problems.). If you choose to dispute Pegasus' value, you might be able to establish a figure as low as $300–400,000. However, if Pegasus wants to throw in Donatello's as-yet untested stud value, the figure could go as high as $3–4 million.

10.1. Structure the relevance diagram(s) for the value of the case. Can you do a separate diagram for the damages issues? Explain why or why not.

10.2. Draw the decision tree for the case and fill in the value formulas.

10.3. Assess any necessary probabilities. State any assumptions you make to clarify the case or make it more realistic.

10.4. Before knowing how any of the motions turn out, what should Irving's settlement strategy be? Assume for now that a rich aunt would help with any judgment against him.

10.5. Construct a table showing Irving's settlement strategy given the various possible outcomes of the motions for summary judgment. Should Irving try to settle now, or wait to see how the motions turn out? Explain your answer.

10.6. Suppose the rich aunt phoned to say she'd only pay up to $300,000 for Irving, and that Irving is essentially judgment proof beyond that. Now what should Irving's settlement strategy be before and after finding out the results of the various motions?

10.7. Which issue has the greatest value of control for Irving? Does this suggest any possible motions you might make?

Chapter 11
Pretrial and Trial Strategy Planning

The examples in Chapter 7 showed the methods of Decision Analysis can be used to plan pretrial strategy (such as by using values of information control to guide discovery) and to plan trial strategy (such as by analyzing how various possible decisions could affect the value of your case). This chapter adds a few more points to these areas, first, by discussing how your analysis should develop with your trial preparation, and, second, by showing a number of additional techniques you may find helpful in planning trial strategy.

11.1 Updating Your Probabilities and Analysis

Subjective probabilities are based on a person's information, intuition, and judgment at the time they're assessed. The point of the assessment techniques is to obtain an accurate representation of that information without the distortion of biases. The resulting probability, then, is like a snapshot of that person's information at that time.

It is no surprise, then, that a person's probability can change over time as that person receives new information or reevaluates her existing information (perhaps as a result of seeing what effects her information has on the value of the case).

To keep track of this vital and changing information base and to help keep your analysis current and meaningful, it is therefore important that every analysis be dated. It is similarly a good idea to date probabilities as they're assessed and note who was involved. Then, for instance, if your cocounsel has just completed a very revealing series of depositions, you can review the analysis to date for possible updating.

Generally, as the case progresses, you should expect to learn new information that could change the probabilities and possibly even the relevance diagrams and decision trees themselves. It is important that you update your analysis to reflect these changes. You no more want to be working from an outdated analysis than

J. Celona, *Winning at Litigation through Decision Analysis*,
Springer Series in Operations Research and Financial Engineering,
DOI 10.1007/978-3-319-30040-5_11

driving with an outdated directions. Further, it generally takes very little time to update an analysis you've already gone through once. Look especially for changes in the value of the case that affect your settlement or trial strategy, and for changes in the values of information that affect your discovery and research plans.

These changes do not mean that the earlier analysis was wrong in any way. It simply points out that analyses (and, you may recall, good decisions) depend on the information available when they're made. Indeed, it is the lack of complete information from uncertainty and the cost of discovery and legal research that makes analysis necessary in the first place.

In summary, then, you should use the following procedure to review periodically the analyses for your pending cases.

- For each analysis, set a review interval. Depending on the case, weekly, monthly, etc., may be appropriate.
- Review the analysis, checking, perhaps, with questions like the following.
  - Have the issues changed? Are there new ones, or ones that are no longer important because of new information (e.g., a cause of action was dismissed in partial summary judgment).
  - Has the problem structure changed? Do the relevance diagram and decision tree still reflect your thinking on what influences the ultimate question in the analysis?
  - Are the probabilities still current? Has there been new information since the probabilities were assessed that would indicate a new assessment is advised? (Of course, proper dating is crucial for this step.)
  - Do you know the answer yet to any of the uncertainties? (This is the most dramatic form of updating.) Do these answers change your preferred settlement or trial strategy?
  - Can you think of any new alternatives that could dramatically affect the ultimate question (like the value of your case)?
  - Does the best course of action still make sense, and can you explain it persuasively and succinctly?
- Look ahead to the next analysis review.
  - What steps will you need to take before your next analysis review? Do have all the information and analysis you need to decide what those steps should be?
  - Will you need to present your analysis (or, more likely, your recommendations) to anyone before the next review, like the client? The entire Decision Analysis process aims to make you feel comfortable explaining and defending your recommendations, but do you need to set aside time to prepare drawings, slides, etc. for presentation?
  - What else could change your analysis between now and the next review? Should you make a note of any particular changes to look for at your next analysis review?
- Finally, confirm that the next scheduled review date is still appropriate.

These steps are well worthwhile to assure your analysis reflects the clearest possible thinking and the best information. The aim of the analysis is to help you better understand and plan your case. A Decision Analysis is a better way to think about the value of your case—not a one-time exercise to be done and then filed away. You won't get the maximum value out of your analysis unless you use it as a thinking tool continually to develop and test your thinking. Consider Decision Analysis as a way of planning and continually updating your case strategy to assure it's a winning one.

11.2 Contingent Strategy Tables

Besides updating your analysis by reviewing your structuring and information, you can also use your analysis to look ahead to what future decisions you should make, allowing you to anticipate and to plan for them before you actually have to make them. You can do this by applying the problem structuring techniques you've learned to future decisions, and using the tree to determine what the best decision is in different scenarios. When you put the decisions and scenarios in table form, it's sometimes called contingent strategy tables. An example will help make this process more concrete.

Suppose you represent the defendant in a traffic accident. Your client insists he was not at fault, the plaintiff insists he was, and the physical evidence is not definitive but tends to favor the plaintiff. On this basis, you feel there's about a .60 chance the defendant will be found liable.

However, purely by chance, you've discovered the name of a witness who was at the scene. You're not sure if you should depose the witness because the physical evidence leads you to believe it more likely that his testimony would hurt your client than help him. If you don't depose the witness, you don't think the plaintiff will find out about him unless you yourself are deposed (an unlikely possibility). The question, then, is whether you should depose this witness or just forget about him.

One possible relevance diagram for this situation appears below. It shows that you must first make your decision on the deposition before any of the other questions are resolved, and expresses your belief that the witness' testimony has the potential of affecting liability, but not damages. Note also the arrow indicating that you will know the effect of the witness before you have to make your decision on whether to litigate or settle the case (Figure 11.1).

A possible decision tree for this relevance diagram appears below, reflecting the judgments that there's a .65 chance the testimony of the witness will be bad for your client, and that bad testimony will increase your chances of being held liable to .90, while good testimony would cut your chances of liability in half, to .30. Note also the \$100,000 settlement offer, and the assessment on damages (Figure 11.2).

The roll-back calculation for the mean value follows below (A roll-forward wouldn't work to calculate the overall mean value for this tree. Do you understand why?) (Figure 11.3).

Figure 11.1 Whether or not to depose a witness

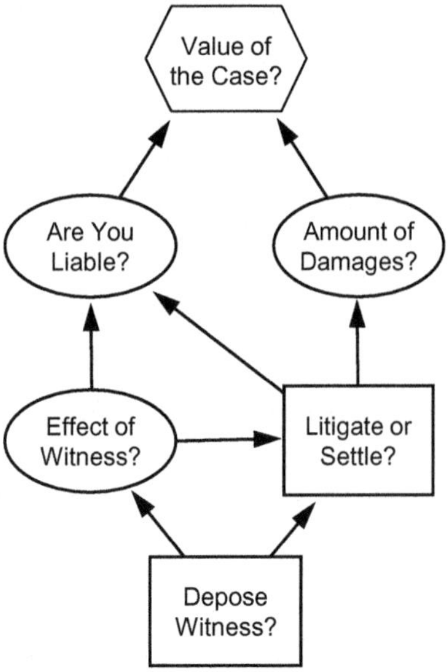

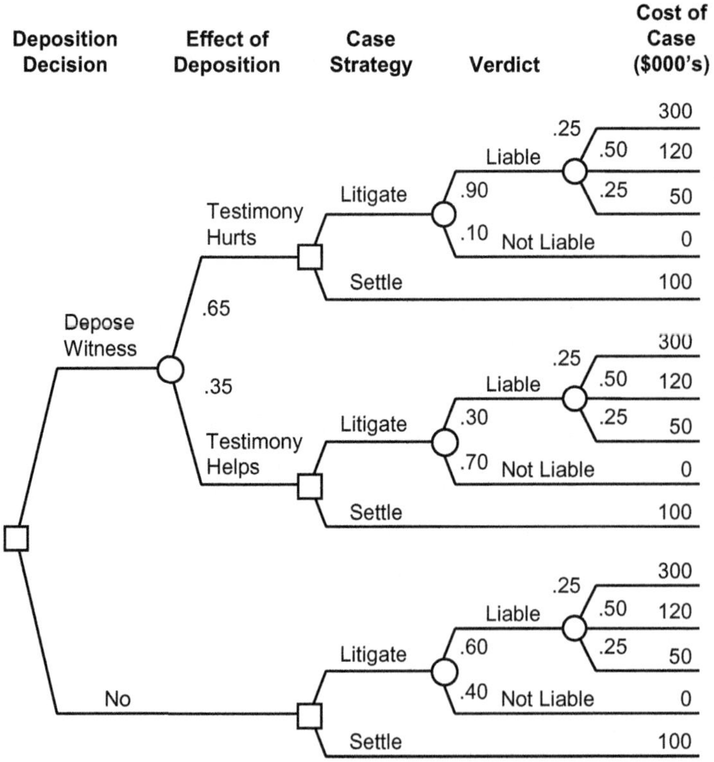

Figure 11.2 Assessed tree for deposing a witness

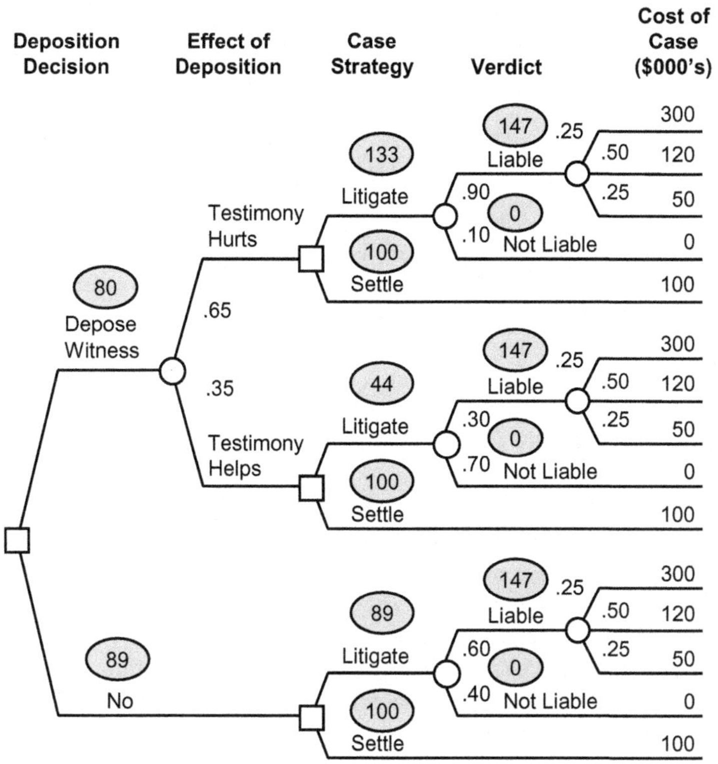

Figure 11.3 Rollback mean values for decision to depose witness

In spite of the .65 chance that the witness' testimony will hurt you, the preferred decision is still to depose the witness, with an $8,000 mean value advantage over not deposing the witness. If you'll examine the tree, you'll see that this is because there's a .65 chance you'll be $11,500 worse off and have to settle for $100,000, and a .35 chance you'll be $44,250 better off by reducing the value of the litigation by this amount (because of the reduced chance of losing)—thus producing the $8,000 advantage. Thus, deposing the witness has greater potential for reducing the mean value of the case than for increasing it, and you should do it. Even if your client is risk-averse, he should prefer to depose the witness because, since damages are unaffected, the deposition improves the mean value without changing the spread of possible outcomes (upon which risk aversion depends).[1]

[1]Indeed, a sensitivity analysis to risk aversion—a procedure that's a lot of work without software—shows that increasing risk aversion merely causes the certain equivalent values of deposing and not deposing to converge on $100,000 because the client becomes too risk-averse to litigate and always chooses to settle. However, the values just become the same; not deposing never looks better than deposing because of risk aversion.

Table 11.1 Contingent strategy table for decision to depose witness

Decision on deposing witness	Outcome of deposition	Preferred case strategy	Maximum settlement value
Depose witness	Hurts case	Settle	$133K
	Helps case	Litigate	$44K
Don't depose witness	(not applicable)	Litigate	$89K

The contingent strategy table, then, is just a matter of taking the information from the tree and putting it into tabular form and is easier to read than the tree. A possible contingent strategy table for this tree appears below (Table 11.1).

Note that the maximum settlement value changes according to the scenario. Accordingly, even if the plaintiff changes his settlement offer with the results of the deposition, your strategy won't change unless a bad deposition for you causes him to raise his settlement offer above $133K, or unless a good deposition for you causes a reduction to $44K.

Other kinds of tables are possible, and you can extract other sorts of numbers from your tree. The same kind of settlement strategy and preferred decision analysis you've done for your case overall is possible for future decision points. The common structure is finding out the results of one or more uncertainties and then having to make a decision (usually with more uncertainties pending). The relevance diagram and tree will help you put together the relationships and time sequences.

In summary, then, contingent strategy analysis helps you do the following.

- Analyze what the best decision is under possible future scenarios.
- Anticipate and plan for those future decisions.
- Analyze and plan sophisticated strategies that involve sequences of decisions and uncertainties.

11.3 Claim Inclusion or Exclusion

Claim inclusion or exclusion is another variation on trial strategy planning. It uses the same techniques to analyze the consequences of including or excluding a claim in your case, and is appropriate for those cases where including an additional claim might actually hurt you. This might happen, for instance, in the following circumstances.

- When introducing the claim opens up lines of inquiry that, overall, hurt you more than the additional theory of recovery helps you—rather like introducing good character evidence opens the floodgates for bad character evidence;
- When a claim sounds so incredible to the jury that it hurts your credibility on more solidly grounded theories (a kind of spill-over effect); or
- When is troublesome to introduce and argue and makes you look bad in front of the jury and, as a result, adds almost nothing to the value of your case—so you're out the extra time it took to prepare the claim, and maybe suffered on damages to boot.

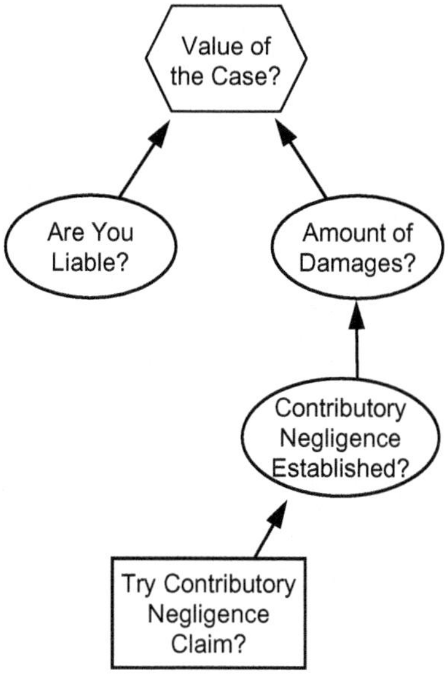

Figure 11.4 Start of relevance diagram for contributory negligence claim

The last possibility was discerned in an actual, tragic case in which an 8-year-old child was killed while playing on a railroad crossing. There had been other accidents at the crossing over the years, the parents knew about the accidents and the hazards, the child had been warned not to play there. The defendant was presented with the painful choice of whether or not to introduce a claim for contributory negligence on the part of the child (Figure 11.4).

However, the counsel thought that, even if the contributory negligence claim were established, its only effect would be on the damages awarded—not on the question of the railroad's liability, which would be determined solely on comparative negligence (who was more at fault). It was inconceivable that maximum fault would be a young child.

The only benefit, then, of introducing the claim was the possibility that establishing contributory negligence on the part of the child would reduce the damages awarded if the railroad was found liable.

When the assessments on damages were done, it was found that, even if the defendant were able to establish contributory negligence, the jury's expected subtraction from damages for the negligence of a deceased 8-year-old was so slight that it simply wasn't worth the trouble to include the claim. Counsel, who was torn between serving his client and feeling badly about trying the claim, was considerably relieved.

In summary, explicitly analyzing claim inclusion or exclusion can help you plan trial strategy in the following ways.

- It helps you to identify alternative legal theories and to analyze the contribution and spill-over effects of each.

- It allows you to value explicitly including or excluding a theory so you can make confident and defensible decisions about doing so.
- It helps to analyze how you should shape your case for the trier-of-fact.

11.4 Analyzing from Your Opponent's Point-of-View

In the section in Chapter 7 on whether you'd left out anything important, a mention was made of using your tree as a basis for discussing the value of your case with your opponent, hopefully thereby obtaining enough of a meeting-of-minds to facilitate settlement. You might, for instance, invite your opponent to fill in her own probabilities, see what the resulting case value is, and argue about probabilities and values. This kind of communication aims at resolving informational differences about probability that result in divergent case values. Information and judgment differences are an inevitable consequence of subjective probabilities, and, in this sense, using your opponent's probabilities could be called analyzing from your opponent's point-of-view.

However, perhaps a more fundamental difference between your view and your opponent's occurs when there is a *structural* difference in the case value (rather than an *informational* difference). A structural difference would be when there different questions in you opponent's relevance diagram than your's, or they're related differently.

For instance, recall the insurance analysis from the section on codefendants in Chapter 9. If the defendant in your case is insolvent and insured (such as when the incident occurred when the defendant was still solvent and insured), the value of the case to the defendant is zero because it's all going into the bankruptcy court's allocation anyway, and the value to the plaintiff would look like the following (Assuming, of course, that the insurer would pay amounts over the deductible even if insured didn't pay it.) (Figure 11.5).

In contrast, if the defendant is insured and solvent, the value of the case to the plaintiff is the amount of the award, and the value to defendant looks like the following. This defendant will be indifferent to all settlements and judgements that are between its deductible and policy limit (unless, for instance, the size of the insurer payment affects future premiums) (Figure 11.6).

Another possible asymmetry is when there are codefendants. Assuming codefendants are solvent, the value of the case to the plaintiff will be the amount of the award. In contrast, the value to a particular defendant will depend in whether he is held liable for that award and to what percentage. The tree from the earlier discussion of this problem is repeated below (Figure 11.7).

A plaintiff faced with this situation would have to be very clear about her case values in settlement discussions. She might not only use something like the above analysis to consider what a particular codefendant might settle for, but also she might use analysis like the one repeated below to consider what her future expected settlement with a subsequent codefendant could be, given this settlement (Figure 11.8).

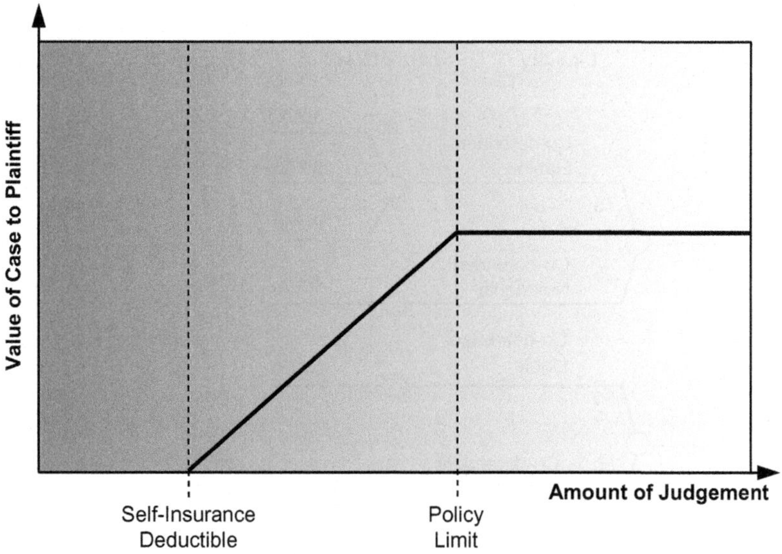

Figure 11.5 Value of case to the plaintiff with the insolvent, insured defendant

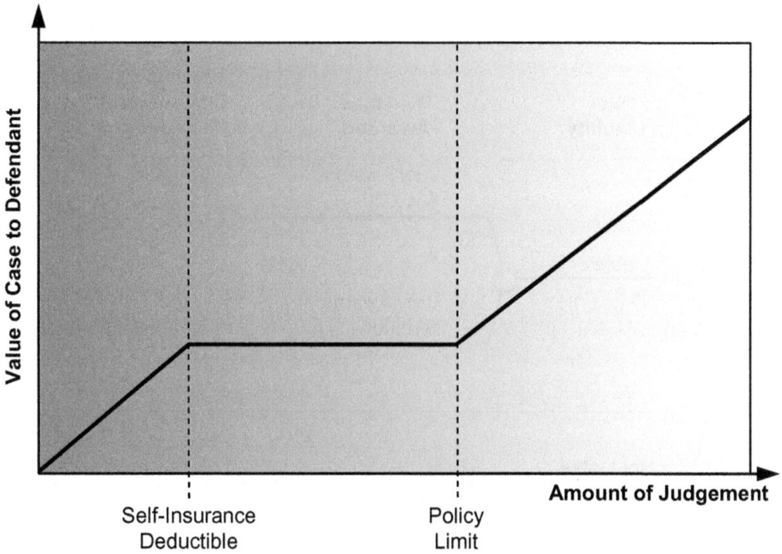

Figure 11.6 Value of case to the solvent, insured defendant

Yet another possibility is where the defendant could obtain an indemnification from another party, perhaps a potential codefendant not joined in the instant action. Depending on how the defendant estimated her chances of obtaining indemnification from the other party, the case values might diverge as illustrated below (Figure 11.9).

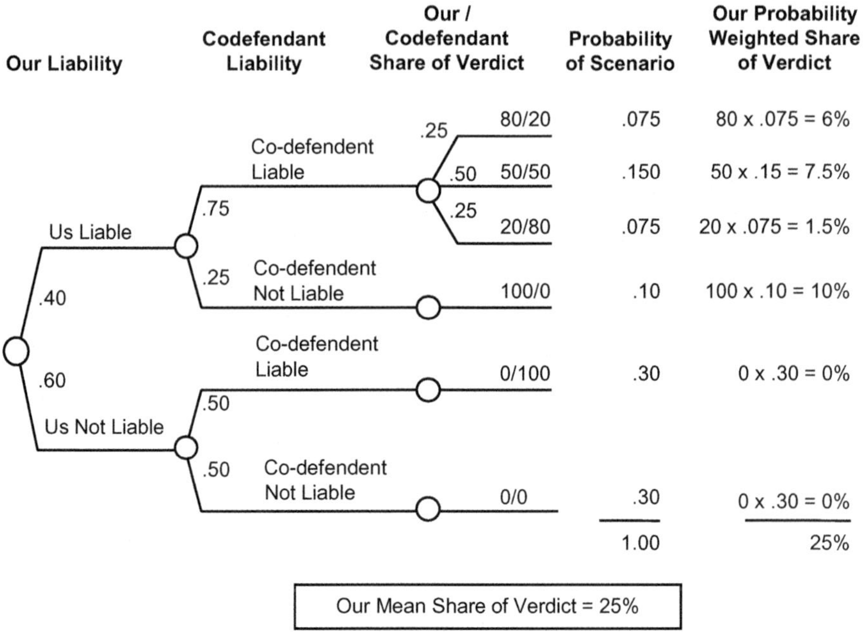

Figure 11.7 Codefendants value case differently

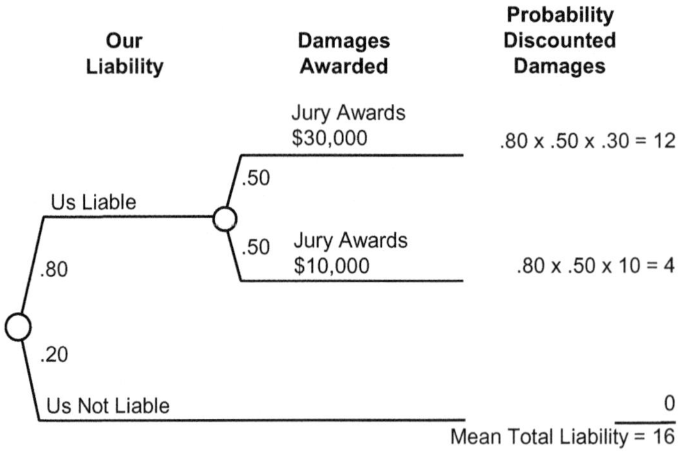

Figure 11.8 Remaining liability where the codefendant has already settled

A possible further complication is when the plaintiff counsel is working on a contingent fee. In that case, the amount the plaintiff client receives will be less the contingent fee. However, unless the contingent fee is different before trial than after a trial date is set (as is often the case), the contingent fee won't change plaintiff's settlement *strategy*. Can you write the value equations and explain why this is the case?

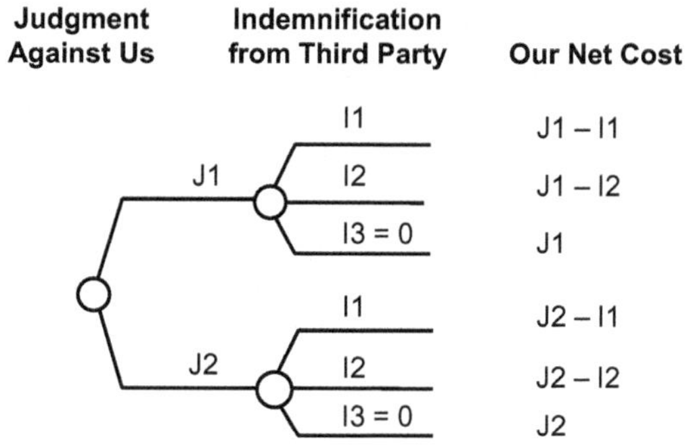

Figure 11.9 Cost of the case to use with indemnification from a third party

In summary, then, structural analysis from your opponent's point-of-view requires the following from the plaintiff's perspective.

- Complete your own usual analysis to value your case and plan your settlement strategy.
- Look for questions that don't affect the amount of the judgement you obtain, but, rather, affect the amount that the defendant will actually have to pay.

From the defendant's perspective:

- Likewise complete the case valuation and strategy from your client's perspective.
- Revise the analysis to reflect what the plaintiff will actually receive.

11.5 Applying Decision Analysis to Alternative Dispute Resolution (ADR)

ADR is enjoying tremendous popularity today, both in binding and in non-binding forms. Legal conferences and publications are devoting increasing time to discussing the various options and reporting on successful outcomes. Law schools teach it. This popularity is easily understood when you consider the tremendous financial and human costs involved in the long process of litigation. Less expensive and disruptive alternatives are clearly welcome. However, even if you opt for ADR, the potential trial risks must always be evaluated.

11.5.1 Why Do Trial Risks Need To Be Evaluated?

You might ask why trial risks need to be evaluated at all if you choose an ADR technique. After all, haven't you opted out of the possibility of a trial?

Consider first non-binding ADR. Certainly, during the process you may learn a lot about the strengths and weaknesses of your case through hearing and testing witnesses, contentions, and the opposing lawyer, and through learning the reactions and opinions of a neutral third party to all of these. However, at the conclusion of the process, the counsel and the client still must decide under which terms a settlement would be preferred to full litigation.

The non-binding ADR results are not a substitute for trial results, nor is the opinion of "one" neutral third party (be it one person or a jury) a clairvoyant prediction of how a judge or jury would rule. Because the results are neither binding like trial results nor infallibly predictive of trial results, the counsel must still evaluate trials risks to weigh against possible settlements. If anything, evaluating trial risks is even more urgent because each side will have an opportunity to remedy shortcomings revealed in the ADR process.

In contrast, binding ADR is a trial substitute, but the counsel still must decide a priori whether binding ADR would be better than a trial. Will the odds of winning and the magnitude of the award be almost as good as or better than a trial? The differences between binding ADR and trial may make this conclusion very difficult to draw.

- The arbitrator is not the judge or jury, and will view the case differently.
- The amount of information discovered may be much less.
- The length of the proceeding may be much shorter.
- The method of presentation may be very different than at trial.

The difficulty of comparing binding ADR and trial make the rigorous, quantitative analysis of Decision Analysis very valuable for accomplishing the comparison. Accordingly, regardless of whether you choose non-binding or binding ADR, the prospects and risks of trial are still the alternative you must compare against to decide if a proposed settlement (or settlement procedure) is the best choice.

Decision Analysis also handles a further (at least potential) problem common to non-binding and binding ADR. ADR may appear easier and safer than settlement negotiations because a third party tells you what to do, rather than the counsel have to make difficult evaluations of a case's prospects early on when she may feel least comfortable doing so.

However, reliance on a third party means foregoing the potentially even cheaper, faster, and far more flexible alternative of traditional settlement negotiations. Decision Analysis is aimed at exactly this problem of valuing cases in the face of uncertainties and incomplete information. Using Decision Analysis can prepare you to create and to take advantage of settlement opportunities potentially more beneficial to both sides than what might be obtained through the intervention of a third party.

In summary, then, Decision Analysis is not only a necessary step for deciding whether to accept non-binding ADR results or to enter into binding ADR, it can also facilitate potentially more advantageous settlements when the counsel would otherwise feel uncomfortable settling the case.

11.5.2 How Does Decision Analysis Improve the Traditional Settlement Process?

From the discussion of the benefits of Decision Analysis in Chapter 2 and from the discussion throughout this book, you've already seen how Decision Analysis can improve the settlement process. However, it is perhaps useful at this point to recap those benefits in the context of ADR.

11.5.2.1 Decision Analysis Encourages Earlier and Better Settlements

Using relevance diagrams and decision trees to structure the value of your case helps overcome the natural tendency to resist thinking about settlement amounts in face of multiple uncertainties and unfinished discovery. These structuring methods also help bring out the considerable information the counsel has even before formal discovery begins, all of which enable the counsel to think about quantitative measures and thereby get the settlement process off the ground far sooner than otherwise would have happened.

11.5.2.2 Probability Sensitivity Analysis Shows Important Questions and "Safe" Settlement Amounts

Once an initial tree has been completed, probability sensitivity analysis can show which information is most critical to determining the client's best settlement amount (you'll recall, by which changes in probabilities change the value of litigating the most). Probability sensitivity can also show you that a particular settlement is a "safe" amount because it's a good deal for your client over the entire range of reasonable probabilities for the most sensitive issues.

11.5.2.3 Decision Analysis Improves Estimates of Settlement Amounts

Even if the counsel feels comfortable thinking about settlement amounts without a decision tree, she will usually feel more confident after undertaking a decision analysis because doing so forces her to think harder and more realistically about the important issues, their relationships, and the odds of prevailing on each issue; and because the analysis uses formal logic to combine those judgments. Final, the tree makes explicit the consequences of alternative assumptions. All these combine to yield a settlement value that is more carefully thought-out, more broadly and solidly grounded, and thereby more defensible and arguable.

11.5.2.4 Decision Analysis Facilitates Communication and Agreement on the Value of the Case

Each side in a lawsuit typically involves several parties, from the minimum of a single counsel and the client to the possible cocounsel (both in the law firm and in a corporate law department) and the executives or groups of executives who must make ultimate case strategy decisions.

Lack of a clear, explicit, and comprehensive picture of the risks of a lawsuit can be a major stumbling block to developing consensus about the value of the case, especially early in the case when uncertainty, complexity, disorganization, and confusion may be at their greatest.

The relevance diagram and decision furnish a clear, unambiguous picture of the risks and value of a lawsuit, enabling the discussion and agreement necessary to a settlement strategy. The analysis can also help to focus client attention on the merits and value of the case—away from the emotions that, especially early in the case, can cloud judgment and interfere with a rational and economic decision.

11.5.2.5 Decision Analysis Facilitates Communication with Opposing Counsel and Client

Just as the clear and explicit valuation from a Decision Analysis enables your side to understand the risks of proceeding to trial, it can also be persuasive in explaining your view to the other side. A decision tree can be a very effective device in educating your opponent about the risks of a suit and agreeing on a settlement.

Occasionally, the counsel may be concerned about revealing too much to an opponent by showing her a complete tree with all the arguments and numbers filled in. Chances are, if you really want to settle and feel your opponent would perceive all the critical issues anyway, there is little risk in discussing your analysis in detail. Nonetheless, if you are concerned, try showing a tree that only has the obvious damages and liability issues and ask your opponent to fill in her own probabilities. Solving for the resulting mean value should provide a powerful new starting point for settlement discussions.

11.5.2.6 Decision Analysis Creates an Atmosphere Conducive to Settlement

Besides allowing explicit and communicable valuations of your case, Decision Analysis also acts in a unique fashion to create a climate conducive to settlement. The decision tree illustrates to your opponent your belief that not all scenarios end in your victory and their defeat, defusing adversarial tendencies. The probabilities recognize that there are no sure things in litigation, and that the risk and uncertainties are considerable and critical. The analysis itself demonstrates your seriousness and rigor in evaluating these risks. This seriousness is likely to impress your opponent and make her more attentive and involved.

Finally, disaggregating the value of the case into the influencing factors avoids stalemated discussions. Instead of coming to an impasse on a difference in settlement figures, the respective counsel can see where their judgements differ and see the valuation consequences of those differences—thereby pinpointing and facilitating potential compromises.

11.5.2.7 Decision Analysis Allows Proper Evaluation of an Opponent's Responses

This same explicit representation of the issues affecting the value of your case allows you to react properly to an opponent's responses. For instance, you may no longer believe a settlement demand too high when the opposing counsel examines your relevance diagram and points out weaknesses or omissions in your case. This experience, although unpleasant, can save you from undertaking costly discovery (or, far worse, losing at trial), only to discover the weaknesses later on when the price tag is higher.

Similarly, the lists of reasons you do along with your relevance diagram and probability assessments can prevent you from over-reacting to new information from opposing counsel. Reviewing your list of existing reasons and integrating the new information into it will help you response appropriately.

11.5.2.8 Decision Analysis Has Tactical Advantages

Although perhaps not in the spirit of benefitting both sides, Decision Analysis offers a number of tactical advantages. For instance, having done an analysis should give a bargaining advantage over a less well-prepared opponent. You'll know from your sensitivity analysis which issues are most critical, and can afford to give ground on less critical issues that won't affect your case value as much. Similarly, doing an analysis from your opponent's point of view (discussed in Chapter 10) will help you understand how your opponent reasonably should react to different settlement offers.

Finally, an analysis can be a very convincing means of educating a settlement judge and showing him how reasonable and rigorous you've been, convincing him (in the words of one attorney who had this experience) "to beat on the other side."

These effects should all make your a more confident and effective settlement negotiator after having performed an analysis.

11.5.3 What Benefits Does Decision Analysis Bring to ADR?

The preceding discussion outlined why conducting a Decision Analysis should put the counsel in a better position to settle a case easily, perhaps sooner, and maybe even on more favorable terms. However, even if settlement is not possible and cannot be considered an alternative to ADR, a Decision Analysis is still necessary to compare ADR against the prospects of trial. Further, even if you elect ADR, a Decision Analysis brings specific benefits to the ADR process.

11.5.3.1 Decision Analysis Can Limit the Scope of ADR

If the two sides used decision trees for an issue-by-issue, probability-by-probability comparison during their unsuccessful settlement negotiations (or, one side introduced trees to develop their differences of opinion), then both sides should have a fair idea of which issues and probabilities they have substantial disagreement on, and which of those cause a significant differences in the respective mean values (As you recall, some differences in probabilities may cause little or no difference in the mean value of litigating the case.), It may be possible, then, to limit the scope of ADR to just those differences of opinion that have a significant effect the mean value of the case. Doing so would reduce the complexity and cost of ADR.

11.5.3.2 Decision Analysis Helps Cope with the Time Urgency of ADR

Many forms of ADR take place sooner and over a shorter period of time than trial typically does (This, of course, is one of ADR's chief attractions.). However, with the time for preparation and the time to present witness and evidence both limited, an enormous premium is placed on identifying the issues and arguments that can most improve the value of your case, and, therefore, should merit more of the limited preparation and presentation time. Through value of information and probability sensitivity analysis, you can prepare just such a prioritized list of issues and arguments for allocating research and evidence.

11.5.3.3 Decision Analysis Can Help You to React Appropriately to New Information Gained in the ADR Process

Just as their is a danger of over-reacting to new information from your opponent (discussed above), there is a danger of over-reacting to new information gained in the ADR process and to the opinions of the neutral third party.

Having carefully thought through your case in conducting a Decision Analysis and in preparing your lists of reasons are the best preparation for reacting appropriately to new information and opinions (neither under-reacting nor over-reacting). In particular, the preparing the lists will help you to think in advance about how surprised you'd be by the various possible results of ADR. And, as known by statisticians and "jury scientists," too much predictive weight should not be given to a single observation—like the opinion of one neutral third party (one person or one jury).

11.5.3.4 Decision Analysis Brings to ADR the Same Case Development and Opportunity It Brings to Settlement

As discussed above, non-binding ADR may be a cost-effective way of getting more information about the strengths and weaknesses of your case. If you've done an analysis of your case in advance, you will not only be better prepared for the ADR, you will also find it much easier to revise appropriately your settlement position afterward by updating your analysis with the new information. Thus, an analysis can help you better develop your case for ADR and to take full advantage of the ADR for further development.

This same early preparation and develop can ready you to take advantage of ADR opportunities much sooner than you otherwise would be. Early and explicit valuations of the worths of ADR and of going to trial can be a powerful tool for deciding whether to seize on an ADR opportunity or not waste your time on it.

In summary, then, many of the settlement benefits of ADR (and traditional settlement negotiations)—their being cheaper, faster, and easier than trial—depend on the counsel and the clients being ready and willing to talk about the values of their cases. These benefits can be frustrated when, as one attorney experienced, the opposing counsel "wasn't ready to talk dollars or specifics yet." Thus, although many attorneys (and, especially, their clients) like the idea of shutting down litigation early, meetings of any sort will be fruitless unless at least one side has done the thorough preparation necessary for discussing case values (both sides would be ideal). An analysis of the litigation alternative will usually be a critical step before any form of dispute resolution can proceed.

11.6 Problems

11.6.1 Updating Your Probabilities and Analysis

1. Refer to problem 11 in Chapter 8 (Valley Realty) and do the analysis as requested there. Then go on to the following parts (Note that you won't see the updating effects unless you first do the problem as originally stated. Try not to look ahead!).
    1.1. Suppose you depose a Nebitz witness and discover she was present at a meeting where the Valley Realty realtor and the president of Galenz specifically discussed the barrels buried under the parking lot and whether they could be concealed from Nebitz, at least until after the sale. Suppose also you discover that a Superfund cleanup is currently in progress at another Nebitz plant, and that it would be a relatively simple matter to get the Feds to pick up half the cost at this site also if Nebitz accepted responsibility.
    1.2. Revise the analysis to reflect these changes. Now what should Valley's settlement strategy with Galenz and Nebitz be?
    1.3. How are the results from part (b) of this problem different from merely examining the contingent results in the original decision tree?

2. Refer to problem 24 in Chapter 8 (the GUN problem) and do the analysis as requested there. Then go on to the following parts (Note that you won't see the updating effects unless you first do the problem as originally stated. Try not to look ahead!).
   - 2.1. Suppose you discover that, at the same time it's been defaulting on its shipments to GUN, AYEC has been making full shipments to another customer—one who contracted at twice the GUN contract price. AYEC contends that it only could make enough of the particular rod to satisfy one customer, and it shipped to the other one for obvious reasons. Suppose you also discover that France has a cache of inferior, older fuel rods which will last for 4 years, but will only produce 70% of the power of the newer rods. France is willing to sell the rods at GUN's original contract price. AYEC was unaware of these rods.
   - 2.2. Revise your analysis, including structure and probabilities. What is the preferred settlement strategy for GUN?
   - 2.3. Should GUN now accept the $15 million payment from AYEC in return for dropping the claim for lost revenues?
   - 2.4. How have the values of information changed as a result of the new information? How do you explain this change?
   - 2.5. Redo your analysis from AYEC's perspective. What is their preferred settlement strategy, given the new information? (Be creative.) How should this affect your settlement strategy?

11.6.2 Contingent Strategy Tables

3. Refer to problem 28 in Chapter 7 (the Helixir problem). Do the problem if you have not already done so, then go on to the next part.

   Construct a table showing under which circumstances it would be worthwhile to apply for the patent on the entire molecule, and for which circumstances an application on the partial molecule would be better. Include the compound probabilities for each scenario. List the three or four most crucial points you would make in presenting your conclusions from this analysis to the board of directors of Helixir, Inc.
4. Problem 10 in Chapter 11 (the Irving Presto problem) also illustrates the use of contingent strategy tables.

11.6.3 Claim Inclusion or Exclusion

5. (Note that this problem also involves multiple defendants.) You represent Herschel, Rotenberg, Inc., a large brokerage firm. About a year ago, Herschel moved into a new office complex in the high-rise district of San Francisco, occupying four floors of the main tower. The complex was a vast and complicated

project, including one smaller office tower besides the main one, a hotel, a five-level underground parking facility, and four floors of shops and restaurants linked together on the lower levels of the two towers. The complex was a show-piece effort by the world-renowned architectural firm of U.K. Fang, and is owned and run by the firm Montgomery Properties, a firm organized for the sole purpose of developing and running the center.

Besides the attractive location and facilities, Montgomery Center offered Herschel and chance to consolidate and reorganize its west coast computer center. The computers had previously been split between smaller facilities in San Francisco and Los Angeles. As part of the move and reorganization, Herschel invested in a system of IBM mainframes. The computer center became Herschel's largest facility outside of New York, and handled the firm's Pacific business as well as the western US.

About a year after the center opened and Herschel moved in, San Francisco was hit with an unusual heat wave. Normally, summer temperatures in San Francisco rarely climb out of the 70s before ocean breezes and fog move in. However, this time the mercury climbed into the 90s and stayed there for a week before cooling down.

Cooling problems started as soon as the temperature went above 90°. Besides being uncomfortably warm for the workers, the high temperatures were devastating for the computer center. Cooling units were unable to cope with the outside heat and the heat from the computers, and the mainframes began to go down as temperatures climbed out of their operating limits. Herschel first complained to Montgomery Properties, then, as their efforts failed improve cooling operation, tried to install temporary cooling units in the computer rooms. However, lack of sufficiently high-voltage electrical power and the inability to physically move units large enough into the rooms doomed Herschel's efforts. After 24 hours of computer failure that had Herschel's customers over half the globe screaming, Herschel desperately looked for alternative machines. The finally located a suitable center at the Lockheed research facility, and persuaded Lockheed to allow them sole use of a mainframe cluster for 2 weeks. After a frantic transfer of tapes and memory units, Herschel was back on-line a day later. The cost for this 2-week rental was $2 million.

Herschel is now suing Montgomery Properties for damages, contending that Montgomery misrepresented the suitability of the facility and that Herschel suffered the damages as a result. Besides the computer cost, Herschel's damages include $500,000 for temporary employee relocation expenses, and $10 million for damage to Herschel's business and reputation. Herschel also suspects it may have incurred substantial trading losses through the disruption at the center, but cannot quantify those losses, so is including them in the $10 million claim.

Montgomery contends that the facilities were adequate for use as a computer center, and that Herschel must have drastically overloaded them to cause the breakdown. Herschel responds that they made their requirements quite clear in negociating the lease, and Herschel should have known the loads the system would have been placed under. You believe that, besides showing the computer center was a reasonable and foreseen use of the facility, there is evidence that

Montgomery's maintenance of the complicated cooling system was inadequate. You feel that if you can convince the jury of these two issues, you have a very good chance of prevailing against Montgomery.

Montgomery further contends that the $10 million damages were not foreseeable or within the contemplation of the parties, and thus should be barred as special damages. You feel it is quite foreseeable that Herschel would suffer wide ranging damages if its second largest computer center went down, especially if you establish the use of the facility as a prime computer center was reasonable and foreseen. It also occurred to you that the issue of reasonableness and foreseeability may impact your chances of establishing the misrepresentation.

Montgomery has also made some rumblings that inadequate design may be responsible for the failure, rather than Herschel's over-burdening the system or any lack of maintenance. Accordingly, you are considering adding U.K. Fang as a defendant in the suit. However, you realize that it would be very complicated to establish Fang's liability because you would have to establish that use was also reasonable and foreseeable vis-a-vis Fang's design, and that Fang's design was defect because it was not adequate as reasonably intended to be used. You have made preliminary estimates that it might cost as much as $500,000 to develop and try the design defect claim.

As a final wrinkle, you've discovered that Montgomery seriously overextended itself in building the center, and is very close to insolvency (That may help account for the inadequate maintenance.). You are therefore unsure how much you could actually collect from Montgomery. You've also learned that Montgomery's liability policy has a $500,000 self-insurance deductible, and a coverage limit of $4,000,000 after the deductible is paid.

5.1. Structure the relevance diagram for the value of the case? Can you do separate diagrams for the damages issues, or do you need to use subtrees?
5.2. Draw the decision tree for the value of the case, assuming that Herschel doesn't bring a claim against U.K. Fang. Further, for a variety of reasons, Montgomery can't bring the claim themselves and will be responsible for any insured judgment (assume they can pay it). Write in the value formulas.
5.3. Assess any necessary probabilities, being sure to write down your reasons and to list any assumptions.
5.4. Calculate the mean value of the case. What is Herschel's settlement strategy? What is the settlement strategy from Montgomery's point of view?
5.5. Suppose Herschel won't be able to collect anything from Montgomery without levying judgment against the center itself, a process that could take years. Now what is Herschel's settlement strategy? What is the settlement strategy from Montgomery's point of view?
5.6. Still assuming that Montgomery is insolvent, draw the decision tree for whether or not to bring the claim against U.K. Fang. How does this tree relate to your earlier tree? Write in the revised value formulas.
5.7. Assess any additional probabilities, and calculate the mean values. Should Herschel include the claim against Fang?

5.8. Draw up a contingent strategy table for settlement strategy according to whether the claim against Fang is brought, and whether Montgomery is solvent. Do you need additional information and analysis to construct this table? (need a probability of insolvency).

11.6.4 Analyzing from Your Opponent's Point-of-View

6. You represent Nikola Motors, Inc., makers of a low-volume but high-popularity electric vehicle. One of the over 100,000 Nikola Model T sedans sold had its battery catch on fire while the owner was participating in a track day at Thunderhill Motor Raceway in central California on a day where the temperature reached 104°F on the track at the time of the fire.

   Nikola is being sued by the owner of the vehicle in question, Mario Treehugger. The firm representing Mr. Treehugger is attempting to obtain class certification for a class action lawsuit against Nikola on behalf of all Model T owners.

   Plaintiff attorneys may or may not be successful in obtaining class certification. If they are not, Mr. Treehugger will be the only plaintiff. If they are successful, the court decide that all approximately 110,000 Model T owners have been damaged due to the diminution in vehicle value and certify all these owners as members of the class. Alternatively, the court could decide that only the approximately 30,000 owners of model T+ versions (the version Mr. Treehugger drove) have been damaged are eligible to be class members.

   If Mr. Treehugger is the only plaintiff, there is a 50–50 chance the court will decide the Model T is defective for the use Mr. Treehugger put it to. If the vehicle is found to be defective, Mr. Treehugger's damages will be the $150,000 cost of the vehicle plus approximately 1–2 times that amount in compensatory damages.

   If there is a class certification (which requires the court to certify the plaintiffs have sufficiently similar claims and damages), the chances are lower the court will find a defect. If found liable, Nikola may have to modify the vehicles or compensate the owners for lower market value.

   The compensation for lower market value may range from 5 to 10% of the vehicle purchase price. The cost to modify the vehicle depends on which remedy the court finds acceptable. One option is a software upgrade, which is essentially free to Nikola Motors (covered by an already scheduled upgrade). As an alternative, the court may find that Nikola needs to modify the vehicle's cooling system, which would cost Nikola $15,000–20,000 per vehicle.

   The chances of class certification for all Model T owners is even less than for Model T+ owners. The chances for the court finding a defect for all Model T owners is also less than if the class consists only of Model T+ owners (which have a higher capacity battery that can get hotter).

   In case of a finding of a defect, the bigger the class, the more likely the court would allow a fix rather than compensation, and the more likely the fix allowed

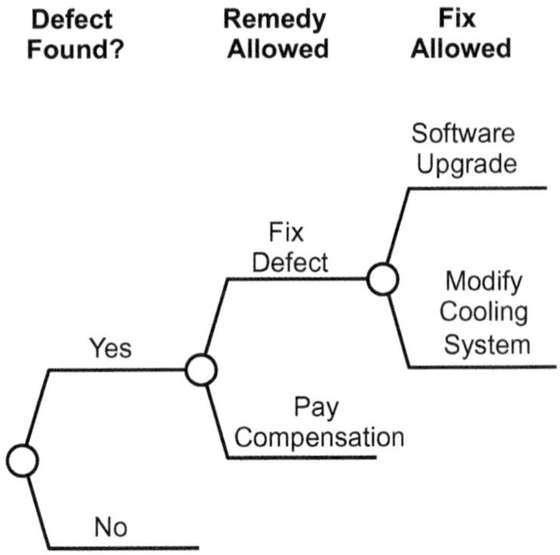

Figure 11.10 Partial tree for Treehugger v. Nikola Motors, Inc.

will be a software upgrade rather than cooling system modifications. This sequence is illustrated in Fig. 11.10.

Mr. Treehugger is represented by the team at Floogle &Bogg, Attorneys at Law. Floogle and Bog took the case on a contingent fee arrangement in which they will be paid 40% of any cash damages paid to plaintiffs.

6.1. Draw the relevance diagram for the value of the case.
6.2. Draw the tree for the value of the case.
6.3. Assess the probabilities.
6.4. Write the value formulas and calculate values in all scenarios.
6.5. What is the value of the exposure to Nikola Motors?
6.6. What is the value of the case to Mr. Treehugger?
6.7. What is the value of the case to Floogle & Bogg?
6.8. What is the maximum amount Nikola should offer to settle the case? Should Mr. Treehugger accept the offer? Is the offer attractive to Floogle and Bogg and why or why not?
6.9. Suppose Nikola offers to stipulate that the Model T+ is defective for track use, though not for road use. Should Mr. Treehugger accept this offer? Why or why not is this offer attractive to Floogle & Bogg?

7. See problems 12 in Chapter 8, part h (the Dr. Dill problem), problem 24 in Chapter 8 (the GUN problem), and problem 6 in this chapter (the Herschel, Rotenburg problem) for analyzing from your opponent's point-of-view.

11.6.5 Pretrial and Trial Strategy

8. You represent a group of aggrieved shareholders of B.Bank, a US subsidiary of a London-based commercial bank (Blakely Bank, P.L.C.). Forty percent of Blakely's stock is publicly owned and trades on the New York Stock Exchange.

Unfortunately for the public stockholders, B.Bank has been in trouble for some time, both as result of foreign loans which have proven uncollectible and because rising interest rates reduced the value of the bank's outstanding loans while increasing the cost of its own debt. When on the verge of insolvency, Blakely was taken over by federal regulators and is in the process of being liquidated. Shareholders expect to receive nothing out of the proceedings.

Accordingly, the shareholders you represent are trying to recover some of their losses by suing the solvent parent corporation, Blakely Bank, P.L.C., for fraud under Rule 10b-5 of the 1934 Securities Act. The contention is that the bank had a duty to reveal the condition of the uncollectible loans 1 year prior to when it actually did so, and that failure to reveal those material facts constitutes misrepresentation of a material fact under Rule 10b-5. If you can establish this, then all actual buyers and sellers of the stock during the time period when the bank had the duty to reveal the information and did not do so will be able to recover any actual damages.

However, establishing liability is proving to be very problematic. You feel your chances will be best if you can subpoena the loan officer of Blakely Bank, P.L.C., who previously gave some very incriminating testimony before the Securities & Exchange Commission hearings. However, he is now in England, and you don't think you'll be able to obtain his presence at trial unless he's extradited for criminal proceedings in connection with the insolvency of the US subsidiary (Assume that the testimony you require won't be protected under the Fifth Amendment.). You've learned that the local district attorney plans to do apply for extradition, but there are a number of problems with an extradition request (which you'll get back to in a minute).

If you the loan officer is not extradited and you can't obtain his testimony, you may be able to introduce his former testimony under Rule 804(b)(1) of the Federal Rules of Evidence, which reads as follows.

(b) Hearsay Exceptions. The following are not excluded by the hearsay rule if the declarant is unavailable as a witness:

 (1) Former Testimony. Testimony given as a witness at another hearing of the same or a different proceeding, or in a deposition taken in compliance with law in the course of the same or another proceeding, if the party against whom the testimony is now offered, or, in a civil action or proceeding, a predecessor in interest, has an opportunity and similar motive to develop the testimony by direct, cross, or redirect examination.

 Whether you can establish that B.Bank was a predecessor in interest to Blakely Bank depends on whether the court is willing to look to the fraud charges and invoke its equitable powers to pierce the separation between the entities. The fact that Blakely Bank held an controlling interest in B.Bank bolsters your position. As for the similar motive and opportunity, because the S.E.C. hearings involved the same subject matter of the insolvent loans as your claim depends on (as well as other matters), you feel the chances are pretty good of establishing that B.

Bank had a similar motive and opportunity to develop and cross the loan officer's testimony. The court might also go against you on this point if it found that, at the earlier hearing, the loan officer and B.Bank was cooperating was cooperating with the S.E.C. in an attempt to settle the matter without further proceedings.

However, none of this will be necessary if the district attorney is able to extradite the loan officer so you can subpoena him. Whether the extradition request succeeds depends on three factors. First, the extradition treaty between the USA and the UK is currently in technical lapse because it expired and has not yet been renewed. There is no obligation to extradite unless a valid treaty is in force. However, the President has extended the treaty under an executive agreement until the Senate formally ratifies an extension. Because such executive treaties have been interpreted as having legal "treaty" status in the past, you feel the treaty validity will be upheld by the US judge (and, thereby, honored in the UK) unless the Senate formally rejects extension of the treaty—a real possibility as a result of the disputes over extraditing Irish Republican Army (IRA) activists to the UK.

The second problem with the treaty looks to be the most difficult. Under the doctrine of dual criminality, a requested state (the UK) need not extradite a person under a treaty unless the offense is a crime in both nations. The UK does not have the kind of statute-based securities regulation that the US does, nor has it policed securities fraud as vigorously as the S.E.C. and private litigants have done under §10(b) of the 1934 Act and Rule 10b-5 promulgate thereunder. However, the UK does have a strong common law tradition of prosecuting fraud in all its forms. Accordingly, you're very unsure of whether the dual criminality requirement will be met.

The third and final obstacle to extradition is that there must be sufficient evidence to try the accused under the law of the requesting nation. US courts have interpreted this as requiring probable cause, and you feel the UK courts will take a roughly equivalent approach, though they don't have the same constitutional probable cause jurisprudence. You would be very surprised if the request foundered for lack of probable cause after getting over the other two hurdles.

You note that even if extradition fails and you can't introduce the former testimony, you feel you'll still get to the jury on the liability question. However, determining damages poses a final problem.

As noted previously, B.Bank made a press release regarding its loan collection problems in September. Exactly 1 year prior to that, an internal report to the board of B.Bank delivered a pessimistic assessment of the loans. The best case for your clients would be for a court to find the duty to disclose arising upon receipt of that initial report. However, the court may find no duty to disclose so speculative a report at that time. Your second position is that the duty arose in January, when a board meeting approved the report and initiated a review of all outstanding loans. Your final fallback date is June, when a number of division heads report ominous delays in collecting interest payments. B.Bank actually made its announcement in September when a number of important loans were clearly in default.

The important of the date the duty arose is that it establishes the date at which losses from the plunging stock price become part of the damages. Those losses total $300 million if the court finds the duty arose on September a year ago, $250 million for January, $100 million for June, and, of course, nothing if the court finds no duty to disclose before B. Bank actually did so (that is, no liability).

Blakely Bank, P.L.C. is likely to argue that, even if liability is established, much of the decrease in the stock price stems from the exposure to changing interest rates, which, as noted earlier, contributed to B. Bank's insolvency. You feel that, entirely apart from the issue of the date the duty arose, the court could find 10%, 50%, or 70% of the losses resulting from interest rate changes (and therefore not recoverable), with discretized probabilities of .25, .50, and .25.

8.1. Structure the relevance diagram for the value of the case.
8.2. Indicate which parts of the relevance diagram can be handled with subtrees (Subtrees are optional, but will make the analysis much easier.).
8.3. Draw the decision trees for the main tree and any subtrees. Fill in the value formulas.
8.4. Assess any necessary probabilities. Include any important reasons as important influencing factors in your relevance diagram if you like.
8.5. Calculate the mean values for the subtrees. What quantities will be plugged into the main decision tree?
8.6. Calculate the mean value for the main decision tree. What should the settlement strategy for your clients be?
8.7. Suppose that Blakely Bank, P.L.C. offered $20 million to settle all claims. Which uncertainties in the main tree have the greatest prospects of improving the value of your case? Over what ranges of probability should you accept this offer?
8.8. What is the value of information on when the court finds the duty arose? Why does or doesn't this value necessary indicate that this uncertainty should be in the main decision tree? Answer the same questions for the percentage of damages caused by interest rate fluctuations.

9. You represent GyroVision, Inc., a designer and manufacture of highly sophisticated gyroscope systems for a variety of applications, including military guidance systems. Six months ago, GyroVision completed delivery 250 specially designed gyroscope units to another Navy contractor for installation in the Navy's newest generation of submarine-launched missiles. At that time, the Navy made the final payment on the contract total of $50 million.

However, operational testing of the missiles has not gone well. Four successive test missiles have gone off course during the crucial boost phase where the gyroscope guides its course, and have had to be destroyed in mid-flight—a greatly embarrassing result for a new showcase Navy technology. The Navy has conclude that the fault lies with the GyroVision units, and has sued for refund of the entire contract price, as well as the cost of redesigning and replacing the units. The Navy would also return the units, but the salvage value on these purpose-built units is negligible.

The Navy contends that the return should be allowed because the units did not meet contract technical specifications, and because they failed to meet a contract requirement that the units meet certain performance requirements—entirely apart from the contract specifications. Both of these claims are highly problematic.

Regarding the technical specifications, they range from sensitivity to changes in axis orientation to response time to orientation changes and modulation of response voltage. All of these requirements are critical for a correct interface with the booster modulation and direction systems that actually change the course of the rocket's flight. GyroVision has been working closely with the Navy contractor to identify and meet the technical specifications, which have been subject to innumerable revision as the Navy's requirements were modified and the contractor's booster designs changed.

The specific problem seems to be associated with the mass of water that is ejected into the air with the missile when the compressed air system shoots it out of the submarine, at which time the booster engines ignite. This mass of water has increased more than was expected from the increased length and width of the missile, and the guidance and boost system as whole seems to never recover from the initial turbulence.

However, GyroVision experts note that the water problem could also be associated with hydrodynamic and aerodynamic problems with the design of the missile's exterior skin, or with the response of the exhaust nozzles that swivel to actually change the direction of the missile. You therefore feel it's highly questionable that the gyroscope unit is solely at fault.

The satisfactory performance requirement is even more problematic. The final test of a system's performance is, of course, whether it works in its final setting, but the GyroVision unit performed flawlessly in bench tests and in field tests in the previous version of the Navy's missile. You therefore feel that a court could find the performance requirements met even if it finds shortcomings in the technical specifications, though such a finding would be less likely in light of a technical specification shortfall. Further, even if a court found GyroVision at fault on the specifications and performance requirements, you're not convinced that they would allow the return. They could very well find that the Navy got what it bargained for, and that the fault lies with the incessant revisions (Naturally, national security poses a problem with any court proceedings, but secret tribunals have been set up in the past to handle such proceedings.).

Entirely apart from the issue of the $50 million return, the claims for damages for replacement costs are particularly irksome. The Navy contends that its damages should include the redesign and installation cost on replacement units. Again, you feel that the court's findings on whether GyroVision met Navy specifications would influence whether it awarded replacement costs if GyroVision were found liable, but it's not a foregone conclusion. You don't feel the performance issue would influence whether replacement costs were allowed because it's simply too nebulous.

If a court did award replacement costs, their determination poses a further obstacle. Redesign has not been completed, much less installation. Redesign costs could range from $10–20 million, and installation from $5–10 million. A court could award these costs on an estimated basis, or order them billed to GyroVision as actually incurred, but you're not sure this variation affects your assessment of them for the present case.

9.1. Structure the relevance diagram for the value of the case.
9.2. Can the replacement damages be structured as a separate subtree? Indicate graphically why or why not with the relevance diagram and justify your result verbally.
9.3. Structure the decision tree (and subtree, if any) for the value of the case. Write in the value formulas.
9.4. Assess any necessary probabilities. If you used a subtree for the replacement damages, does it depend on the main tree at all? Does this prevent you from using a subtree?
9.5. Calculate the mean values for any subtrees.
9.6. If you used a subtree, assume that the board of directors of GyroVision will want to see the full distribution of potential outcomes, and thus you need to summarize and subtrees with chance nodes.
9.7. Calculate the mean values for the main tree, including chance nodes for any subtrees.
9.8. What is the settlement strategy for GyroVision?
9.9. Calculate the values of information for the uncertainties in the main tree. How should GyroVision allocate its discovery efforts and use of expert witnesses?
9.10. Can you calculate the values of information on the uncertainties in any subtrees? Do so, or explain why not. If so, what is the maximum cost GyroVision should incur in studies to determine redesign cost?
9.11. Suppose that, in spite of the past difficulties, GyroVision would consider designing the replacement units. What is the maximum cost they could incur in doing so before coming out behind?

11.6.6 Applying Litigation Risk Analysis to Alternative Dispute Resolution (ADR)

10. List at least three ways that applying Litigation Risk Analysis to ADR differs from applying it to pretrial work. Do these differences make Litigation Risk Analysis more or less valuable? Can you think of any procedural changes that would take better account of the differences in ADR?

11. You're a lawyer for the United Auto Workers at a local component assembly plant where Motor Company (MoCo) has single parts assembled into devices like compressors, alternators, heat and ventilation control units, and so on. MoCo is now shifting those operations overseas, and proposes closing part of the plant and shifting the remainder over to assembling the components from overseas into subassemblies like engines and suspension subframes, which will then be shipped to a nearby final assembly plant for incorporation into finished automobiles. A dispute has arisen over how those new subassembly jobs will be filled. The union wants the plant workers with the most seniority shifted over to those jobs. MoCo would rather hire new workers at lower pay scales, contending that the subassembly jobs are less skilled than the component assembly ones because tolerances are generally greater and no skills like soldering or fine part adjustment are required (the engine components are already machined to the necessary tolerances).

    The dispute will soon be submitted to binding arbitration. Three contract clauses look to apply most directly to it. These clauses are as follows.

    "Management has the right to set skill requirements and pay scales in new skill areas, consistent with the skill and pay guidelines in appendices B and C, respectively."

    "Workers with the greatest seniority and skill level shall be preferred for new positions in their areas of employ."

    "For any continuing job or skill position, management may not lay off workers and replace them with workers of a lower skill grade or seniority."

    For short, you refer to the provisions as the new skill provision, the preference provision, and the layoff provision, respectively. Management has made a final offer to fill half of the new positions from the top of the seniority ladder, with discretion to hire the rest according to new skill and pay standards. Your problem is whether to advise the union to accept this offer, or to go into binding arbitration.

    As you think about how the arbitrator will decide the problem, you decide the first and most important issue he'll tackle will be whether the new skill provision applies; that is, whether the jobs will be a new skill area, or similar enough to the previous component assembly jobs that it doesn't qualify for new pay standards. You're not sure whether the arbitrator would then go to the preference provision or the layoff provision, but you feel his finding on the skill provision will affect his finding on both of these other issues, and, of the two, the preference issue will be affected most strongly. The arbitrator will likely "split the baby" by coming up with a number for what percentage of the new positions management must fill according to existing seniority, leaving it free to hire the remaining percentage as it chooses.

    11.1. Structure the relevance diagram for this problem. What is the relevant value measure, and how does it fit into the relevance diagram?
    11.2. Draw the decision tree for the problem. What's the problem with filling in the value equations at this point?

11.3. Is it meaningful and possible to assess probabilities at this point? Explain why or why not.

Suppose you struggled long and hard with how to relate the arbitrator's findings together to get a handle on the union's prospects. You thought that the new skill clause would clearly weigh the heaviest, with the preference clause less so, and the layoff clause even less. You then hit upon the following way of expressing the relationship.

Suppose this arbitration was like a football game, and you scored every time you won on the application of a clause. Then maybe winning on the new skill clause would be like scoring a touchdown (7 points), winning on the preference clause would be like a field goal (3 points), and the layoff clause would only rate a conversion (1 point). You figured that if you won on the application of all three clauses, the arbitrator would say that all the positions had to be filled according to seniority. On the other hand, if you lost on all three, then you might get nothing: MoCo could hire as it pleased. In between, you weren't sure exactly how things might come out, but a simple weighting scheme seemed appropriate enough: add up all the points you got in any scenario, divide by the maximum number of points ($7+3+1=11$), and call this the percentage the arbitrator awarded you. This scheme would at least allow you to get a handle on your chances in arbitration, and might even be simple enough to explain to other people and get their opinions.

11.4. What are the strengths and weaknesses of this method of valuing the outcomes? Do you think it would make any difference if management got the points (instead of zero) if you lost on a clause? (A numerical answer will be requested below.) Does a better valuing method occur to you? How would you decide a particular valuation method was good or not?

11.5. Fill in the value formulas in the decision tree.

11.6. Assess any necessary probabilities.

11.7. Calculate the mean values. Is the union better off accepting management's offer, or going into arbitration?

11.8. Plot the probability sensitivities. Over what ranges of probabilities does the preferred course of action differ?

11.9. Calculate the values of information on how the arbitrator applies the clauses. What does this tell you about how to split your limited presentation time?

11.10. Change the valuation method so that management gets the points if you don't. You'll have to do a little figuring with the formulas to get sensible results out of this. Does the preferred alternative change? What qualitative statements can you make about the relative merits of the two methods?

Chapter 12
Review and Communication

With almost every Decision Analysis, you will at some point need to discuss your results with someone else: the client, opposing counsel, co-counsel, etc. At that point, your ability to explain and communicate your results will be as important in establishing their credibility as your creativity and rigor in reaching them. Your analysis must be believed before it will become a basis for taking action.

Before then, you'll want to take a hard look them to test your own understanding. We'll start with that step before moving on to documenting and communicating the results.

12.1 Reviewing Your Analysis

Recall our overall perspective that decision analysis frames the interchange between intuitive and logical thinking, providing a framework for constructing thought experiments as to which course of action might be best and why (Figure 12.1).

The objective is a brief, cogent, persuasive recommendation as to what to do and why, with the detailed analysis furnishing an explanation of how you got there.

We'll go through each of the pieces of a sample analysis and look at how you might poke and prod at it to test your understanding.

12.1.1 Relevance Diagrams

The relevance diagram captures the definitions of factors and their relationships to each other without the adding complexity of numbers. Throughout your analysis, you should be reviewing those definitions of factors and their relationships to each other, including which are uncertainties and which are decisions. We'll use a sample case to poke at.

J. Celona, *Winning at Litigation through Decision Analysis*,
Springer Series in Operations Research and Financial Engineering,
DOI 10.1007/978-3-319-30040-5_12

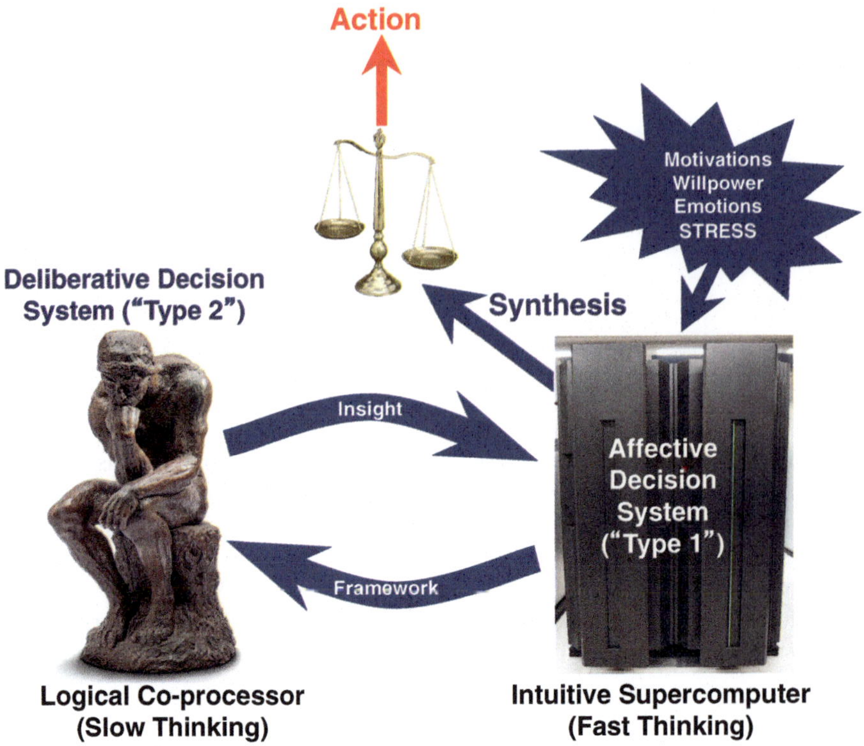

Figure 12.1 Decision analysis frames the interchange between intuitive and logical thought

Suppose you represent the manufacturer of a drug eluting stent. The stent is a metal mesh tube inserted via angioplasty to keep a blocked artery open. The drug is embedded in a polymer material coating the stent and releases slowly to prevent the inside of the stent from blocking from scar tissue formation.

The manufacturer is being sued by four plaintiffs alleging that something in their stents caused a severe autoimmune reaction. Two patients were successfully treated with epinephrine followed with antihistamines. Their symptoms subsided and their stents remain in place. One patient had his stent surgically removed and a bypass operation in lieu of it. One patient died. In all cases, doctors weren't sure of the origin of the symptoms and speculated their stents (all the same exact type from your client) were the cause. Hundreds of patients have been successfully treated with the stents without similar reactions.

The plaintiffs allege your client's product caused their injury. After researching the matter, you're not sure and have been developing possible defenses. The ultimate issues the jury will need to determine are whether you clients' product caused the injuries complained of and the amount of damages.

On the causation issue, you note the extremely low incidence given the use of the product. Many other factors can cause anaphylactic shock (bee stings, allergy to peanuts, etc.). You obtained the plaintiffs' medical records as part of discovery, but there wasn't any prior allergen testing in them. You don't think you will be able to compel new testing to see if there are any other allergies which might be the cause. You're looking for other ways to defend on causation.

One defense is that the product conformed to the state-of-the-art when developed and approved by the FDA. Subsequent remedial measures or changes to the state-of-the-art are not admissible to prove liability. If introduced and accepted, this defense would relieve your client of liability. However, you're not sure if the judge will allow this defense to be raised. Partly this depends on whether the plaintiff will be able to find a competent witness to testify that the current FDA standard should be admissible.

If you are able to raise it, a further issue is that the current FDA standard for drug eluting stents requires immune reaction testing and data. At the time the stent was submitted for FDA approval, it conformed to the FDA standard for testing. That standard required data on many different factors (corrosion resistance, radial stiffness and strength, kink and crush resistance, etc.) but didn't require any testing or data on possible autoimmune effects.

If the state-of-the-art defense fails, the issue of causation remains. The plaintiff has retained an M.D. as an expert witness who will testify that he believes the stent caused the symptoms and injuries complained of. You're not sure the jury will find the M.D. credible given that he is general surgeon (not a cardiologist nor an immunologist), has not practiced in a number of years and makes a living testifying for the plaintiffs as an expert witness. You think the credibility of this witness will affect whether the product caused the injuries to the plaintiffs and the amount of damages awarded.

After some effort, you put all this together in a relevance diagram (Figure 12.2).

At a broad level, the questions throughout the review process are:

- Do you understand the results and do they make sense to you?
- Can you summarize and explain them to others?
- Does anything in there look wrong somehow?
- Is anything missing?

Here are some more specific questions for relevance diagrams.

- **Is the Final Question Correct?** For most Decision Analyses, the final question (the one at the top of the relevance diagram) will simply be the value of litigating the case, as in the example above. However, when an analysis has sub-trees, there may be separate relevance diagrams for the subtrees, in which case you should ask whether a number or the probabilities of a yes/no answer is properly being sought for plugging into the main decision tree.
- **What Are the Ultimate Issues?** The ultimate issues directly determine the answer to the final question, such as the ultimate issues of liability and damages determining the value of litigating a case.
  - On the liability side, there may be more than one ultimate issue is there's more than one legal theory that could establish or cut off liability. Thus, in the example above, both the state-of-the-art defense and causation are ultimate issues because they could cut off liability.

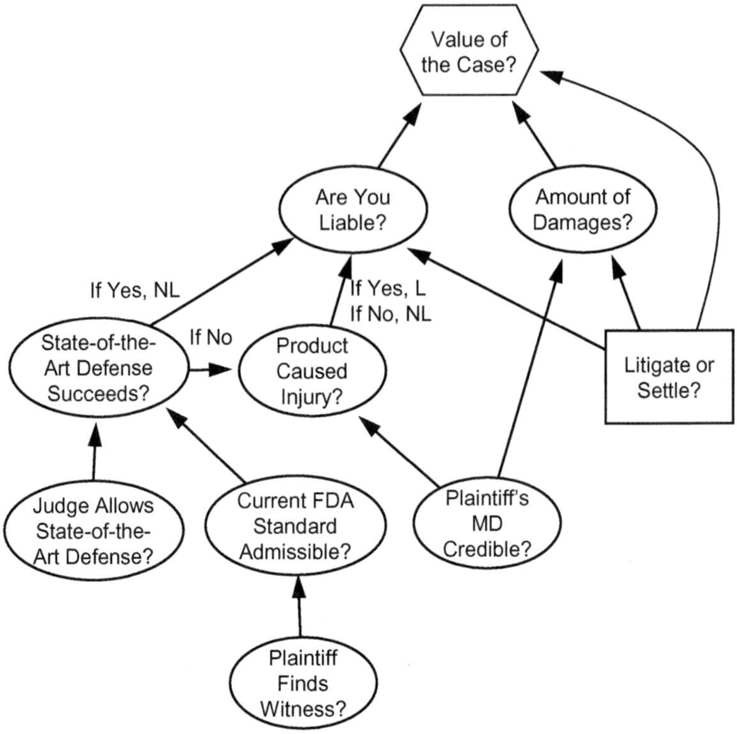

Figure 12.2 Relevance diagram for product liability case

- On the damages side, the ultimate issue is how much damages the trier-of-fact will award. There may be more than one ultimate damages issue if different types of damages are possible, like compensatory and punitive damages.

- **What Are the Key Influencing Factors?** The influencing factors are questions that affect how the ultimate issues turn out. For example, in the diagram above, both whether the judge allows the state-of-the-art defense and whether the current FDA standard is admissible are factors influencing the ultimate issue of whether the state-of-the-art defense succeeds.
- **Have Any Ultimate Issues or Important Influencing Factors Been Left Out?** This question is another application of the clairvoyant test: given the answers supplied to each question by the arrows into it, can you answer each question with yes, no, or a number, or do you need to know something more?

On the liability side, ultimate issues include any other legal theories that the plaintiff might be able to recover on. You should also examine whether the stated ultimate issues are truly so; that is, for example, does success of the state-of-the-art defense really guarantee a defense verdict?

Similarly, on the damages side, you should ask whether the stated ultimate liability issues are sufficient to ensure damages will be awarded. Or, for example, could the plaintiff's own conduct be a bar to damages in spite of the findings on the stated ultimate liability issues?

As for the influencing factors, you should ask yourself whether there are any others that would either greatly aid in assessing probabilities on one of the questions, or would greatly affect how a question turns out (these are really the same thing.) The object here is to think about whether ***important*** factors have been left out; you can always add another factor, but will it be really helpful to do so?

- **Do the Relationships Among the Factors Make Sense?** Remembering that there are different, equally correct ways of diagramming any one case, ask yourself whether the relationships in this diagram make sense to you. Do the arrows correctly describe which question(s) you'd like to know the answers to before assessing a particular question? Is the ordering correct? (Will the trier-of-fact consider the state-of-the-art theory before the causation issue?) Are the "if yes" and "if no" labels correct? Are there subtle influences or spill-over effects that have been omitted?
- **Do You Feel Comfortable that the Diagram Is a Good Representation of the Value of the Case?** With this last step, you should make an overall appraisal of the relevance diagram. Does it seem a reasonably complete and accurate way of thinking about the value of the case? Would you feel comfortable discussing this diagram with someone else, such as the client? If not, go back through the preceding steps to try to find what you disagree with or is lacking.

12.1.2 Assessments and Decision Trees

Next you review the assessments and the tree. First off, you did a small sub-tree to handle the question of whether the plaintiff would be able to find a witness to testify as the admissibility of the current FDA standard and the chances the standard would be admitted (Figure 12.3).

In this case, you just needed to add up the scenario probabilities to get the overall admissibility of the current FDA standard rather than reversing them (as you would for the value of imperfect information). Those probabilities plus your other assessments are shown in the main tree, along with a $5 million settlement offer currently on the table (Figure 12.4).

Here's a look at similar questions applied to the decision trees.

- **Is the Final Question Correct?** Rather than looking for the top oval in a relevance diagram, you'll need to look to the far right-hand side of the tree. Are there numbers there for the value of the case in each scenario? Do they make sense and can you explain how they were calculated?
- **What Are the Ultimate Issues?** To look for the ultimate liability issues in the tree, you'll need to look for those nodes that lead directly to the end of the tree (settlement or no liability), or continue on to the next question. For the ultimate damages issues, look for dollar damages amounts on the branches. Can you find all the ultimate liability and damages issues in the example tree?

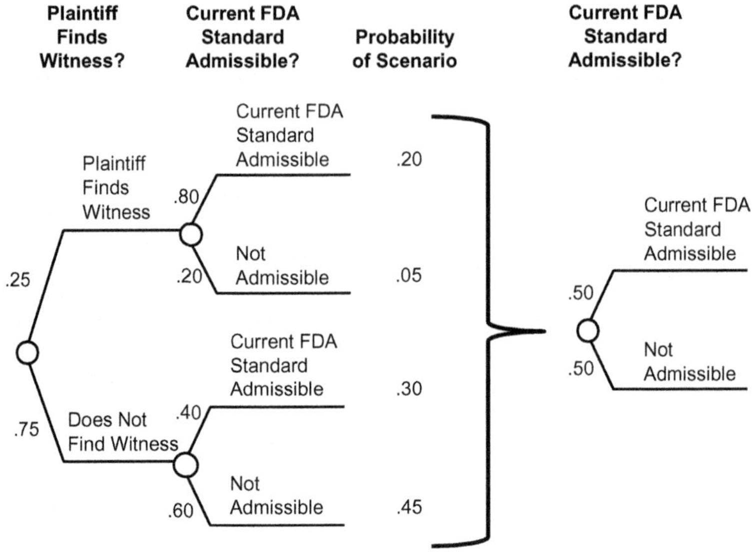

Figure 12.3 Subtree for admissibility of current FDA standard

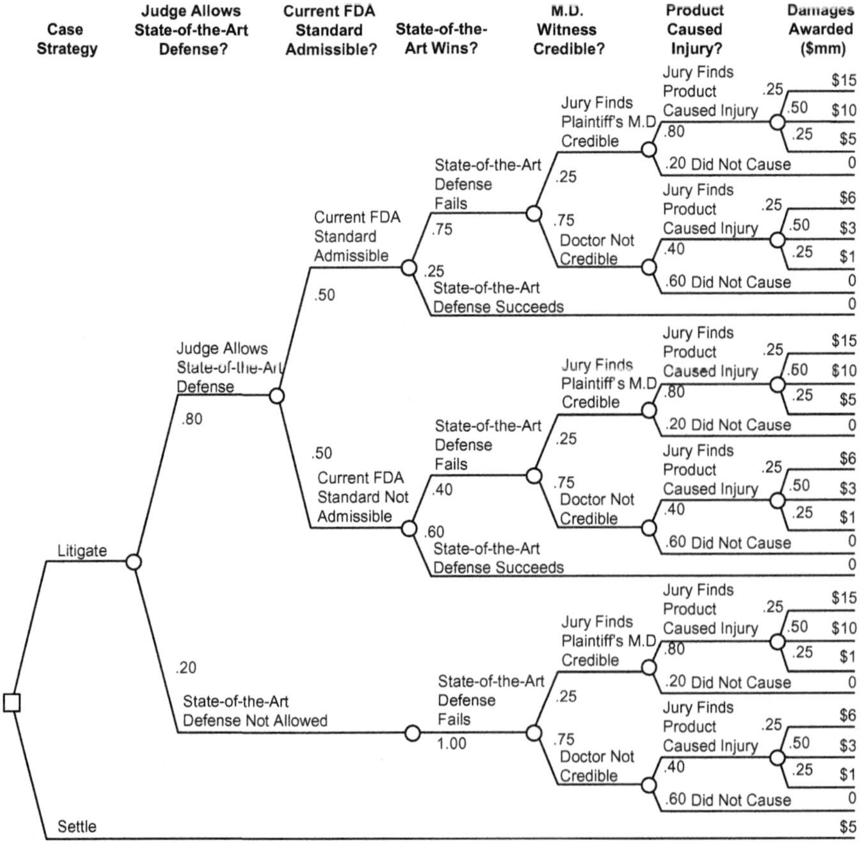

Figure 12.4 Main tree with assessments filled in

- **What Are the Influencing Factors?** The influencing factors in the tree are those that neither determine liability nor the amount of damages, but affect either one of these. On the liability side, an effect will be indicated by different probabilities for an ultimate liability issue, or for another influencing factor. On the damages side, the effect will be indicated again by different probabilities for another influencing factor, or by different probabilities or outcomes for an ultimate damages issue. In the end, each arrow in the relevance diagram should be represented in the decision tree by one of these effects (including the asymmetry effect where a question no longer applies at all).

For instance, on the liability side of the example tree, whether the current FDA standard is admissible changes the probabilities of success for the state-of-the-art defense. On the damages side, the credibility of the plaintiff's M.D. affects the possible outcomes for damages. Can you find the liability effect of the plaintiff's M.D.?

- **Have Any Ultimate Issues or Important Influencing Factors Been Left Out?** The same considerations apply here as for relevance diagrams. Another way you can use the tree to approach this step is follow each scenario through the tree from left-to-right and ask yourself if that scenario tells a complete story for how the case turned out in that instance. Have all the important questions been answered? Does it make sense the way each scenario plays out?

For instance, in the example tree you'll note that the question of whether the plaintiff finds a witness to support admission of the current FDA standard has been left out. There are three possible reasons for such an omission, and three corresponding ways to deal with it.

- The issue may have been unintentionally (or, perhaps, unwisely) omitted, in which case it should be included in the decision tree. It should then, accordingly, cause a downstream change in the tree.
- The issue may have been omitted because the person doing the analysis didn't think it important enough to include separately in the tree. In this case, it likely at least should have been included in the list of reasons for the questions in the tree it influences.
- The issue may have been handled in a sub-tree and only the summary answer was plugged into the main tree. For example, the witness question was handled in the sub-tree shown to produce the .50/.50 probabilities for admissibility of the current FDA standard.
- **Review the probability assessments at each node.** For each node in the tree (decision or chance), you should examine whether the alternatives/outcomes are mutually exclusive and collectively exhaustive (See Chapter 4 for a review of these concepts.). Be alert for omitted alternatives at decision nodes, and recall that, at numerical chance nodes, the numerical outcomes each represent a range of the full continuous distributions.

To review the probabilities, you will need to go behind the numbers and examine the reasons that support the counsel's judgment on both sides of the issue? What reasons might you enjoy a good outcome, or a bad one? How do witnesses, evidence, precedence, or biases come into play? How *else* might a good or bad outcome come about? You won't really know whether or not you agree with someone's probabilities until you examine the reasons for it. Initial agreement or disagreement may be deceptive.

- **Look for Downstream Variations in the Tree.** In general, every node in the tree should make a difference downstream in the tree. Three types of effects are possible.
  - Asymmetry, where other questions do or don't apply according to how the node turns out.
  - Different probabilities at other nodes according to how a node turns out.
  - Different outcomes at other nodes according to how a node turns out.

If there is no such variation, then the question shouldn't be in the tree because it won't make any difference in the value of the case. For instance, the only difference the admissibility of the current FDA standard makes in the tree is changing the probabilities of the state-of-the-art defense succeeding.

To see some of the downstream effects, you'll need to complete the rollback mean value calculations (Figure 12.5).

For example, if you litigate and the judge allows the state-of-the-art defense, the exposure in the case is $1.7 million if the standard is admissible and $0.9 million if it's not. The difference is accounted for by the differing probabilities on whether the state-of-the-art defense will win.

On the other hand, the mean exposure is the same if the state-of-the-art defense fails regardless of whether the FDA standard was admitted or not. At that point, the remaining issues are credibility of the witness, causation, and damages awarded. This leads to the next questions.

- **Examine Whether the Downstream Variations Make Sense.** After finding all the variations and the nodes they vary by, you should question whether those variations make intuitive sense. For instance, in the example tree, does it make sense that finding the plaintiff's doctor credible would boost damages by that much? That the doctor would also have the stated effect on the chances of the jury finding causation? Would you have put an arrow in the relevance diagram between the two questions, and is the variation in the tree a reasonable representation of the effect of the influencing factor?
- **Look for Repeated Issues In the Tree.** Whenever an issue is repeated in the tree, look to see if its probabilities or outcomes vary. If so, you should have examined the variation in the previous step. If not, ask yourself if it makes sense that they don't. The same probabilities for an issue appearing in different places in the tree means that issue is independent of everything the trier has considered before.

For instance, in the example tree, the admissibility of the current FDA standard affects the success of the state-of-the-art defense, but not the chances of the trier finding causation. Whether there is such an effect is a very detail question that turns on the content of the FDA standard, but these are the kind of subtle and important

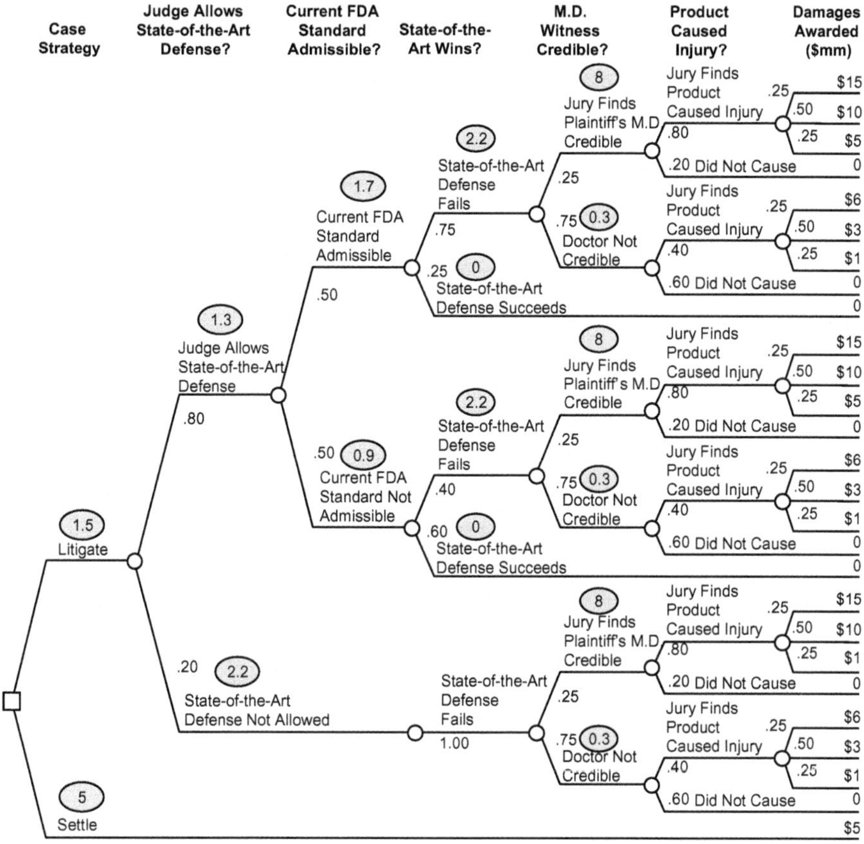

Figure 12.5 Rollback mean value calculations

effects you should be looking for at this point. Another potential effect is spill-over. Do the questions in the tree affect one another even though they're not formally related by legally theory or common facts?

- **Check the Value Calculations.** Check to see whether the values shown at the end of the tree for each scenario make sense. You should be able to trace through exactly how you got there. The values are directly assessed in this tree, but in others will result from sometimes complex value calculations.
- **Look at the Recommended Decisions**. For this example, the only decision in the tree is to settle or litigate the case. Other trees may have one or more additional distributions. Can you explain why the best alternative looks best, and justify how you came up with the numbers?

For this example, the mean exposure from litigating the case is $1.5 million in indemnity (not counting legal costs, which can be quite large in product liability cases). That looks far preferable to paying the $5 million settlement offer on the table. Even with maybe $1 million in costs to litigate a complex case, it's still better to refuse the current settlement offer.

In actual practice, for the defendant the strategy would be to continue down the road towards litigation and see if it's possible to get a more reasonable settlement offer, keeping in mind the total exposure in costs and indemnity to litigate the case. As you get closer to trial, more of those costs will be sunk costs, bringing an economic settlement offer closer to the pure indemnity exposure.

When you're satisfied that the relevance diagram, assessments, tree, and rollback mean values make sense, it's time to look at other kinds of results from the analysis.

12.1.3 *Probability Distributions*

Typically, you'll want to start with the cumulative probability distributions for the alternatives. In this example, it's just settle or litigate (Figure 12.6).

The mean values for the alternatives match the rollback mean values from the tree. The new information is that the overall probability of no liability at trial is about 80%. This graph is also useful for showing the best and worst outcomes if liability is found. The best case scenario (for the defendant) is about $1 million, and the worst is $15 million. These can also be read from the tree.

The steps in the distribution represent all the possible scenarios represented in the tree from the discretized probability distributions. In this example, we can look at the tree itself to make sure the scenario values make sense. In more complex

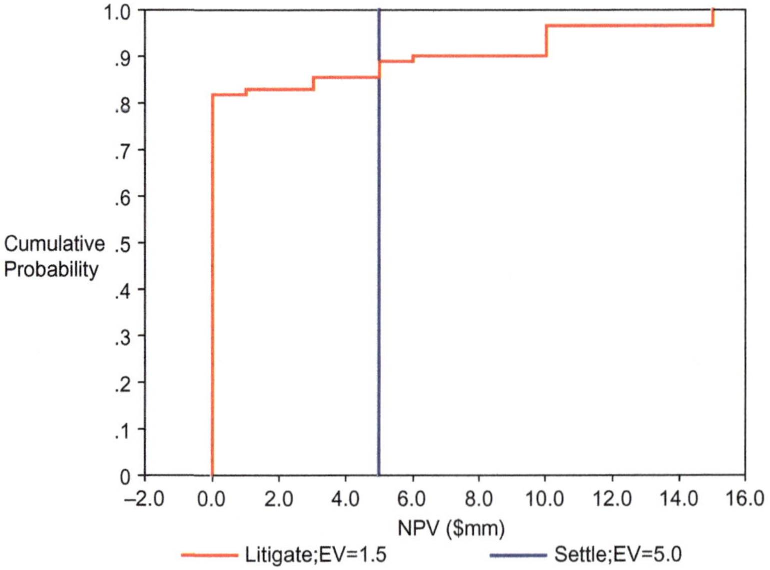

Figure 12.6 Cumulative probability distributions for litigate and settle alternatives

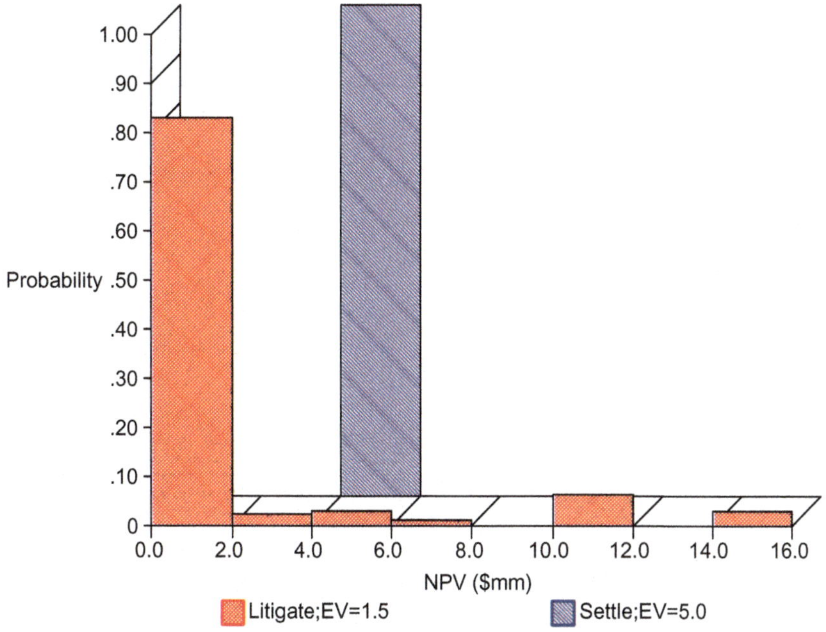

Figure 12.7 Histogram for litigate and settle alternatives

analyses, it's not practical to look through a tree with thousands of scenarios, so you need to look at the steps on the distribution to make sure all the individual scenario values make sense. For any particular value, you should be able to construct how the particular value was arrived at and why it makes sense.

You can also view the distribution in the form of a histogram, which is easier for some people to interpret (Figure 12.7).

The histogram is the exact same information as in the cumulative probability distribution, just shown in different form. The probability of the bin for values between zero and $2 million is slightly higher than the chance of no liability from the cumulative distribution because it includes the possibilities of the expert witness not being found credible and the jury award the lowest scenario of $1 million in damages.

It's generally also a good idea to calculate at least some of the sensitivities to probability to how changes in the odds on key questions change the case exposure and recommendation.

12.1.4 Sensitivity to Probability

Looking back at the tree, the key questions you might be interested in are whether the judge will allow the state-of-the-art defense, whether the current FDA standard will be admissible, and whether the jury will find the plaintiff's M.D. expert witness credible. You could also look at the ultimate questions of whether the state-of-the-art

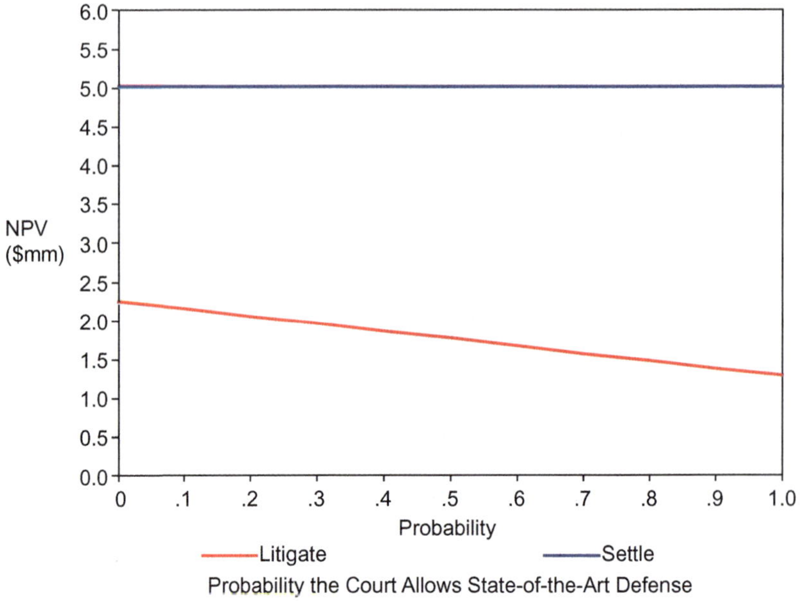

Figure 12.8 Sensitivity to probability the state-of-the-art defense is allowed

defense wins and whether the jury finds causation, but those are conditional probability assessments in the tree. That conditionality goes away when calculating sensitivity to probability, making them not as useful to look at.

Begin with the question of whether the judge will allow the state-of-the-art defense (Figure 12.8).

As you might expect, the mean exposure at trial drops as the probability rises that this defense is allowed. The drop might be less than you would guess because there are so many other factors at play. This uncertainty by itself only varies indemnity exposure in the case from about $2.2 million to around $1.5 million—not enough to dramatically swing settlement value.

However, this graph does show that, whatever the probability is, it doesn't make a huge impact in case exposure or strategy. At the present time and given current information, this is not an issue worth expending a lot of time and expense on.

Next we look at sensitivity to the current FDA standard being admissible (Figure 12.9).

As the probability of admissibility of the current FDA standard increases, mean indemnity exposure increases. This make sense. However, the potential variation in exposure is even less on this question alone (only from around $1.2 million to just under $2 million), and no value of this probability make the current settlement offer attractive. This question is likewise not a candidate for a lot of additional research or discovery at this point.

Lastly, we turn to the credibility of the plaintiff's M.D. expert witness (Figure 12.10).

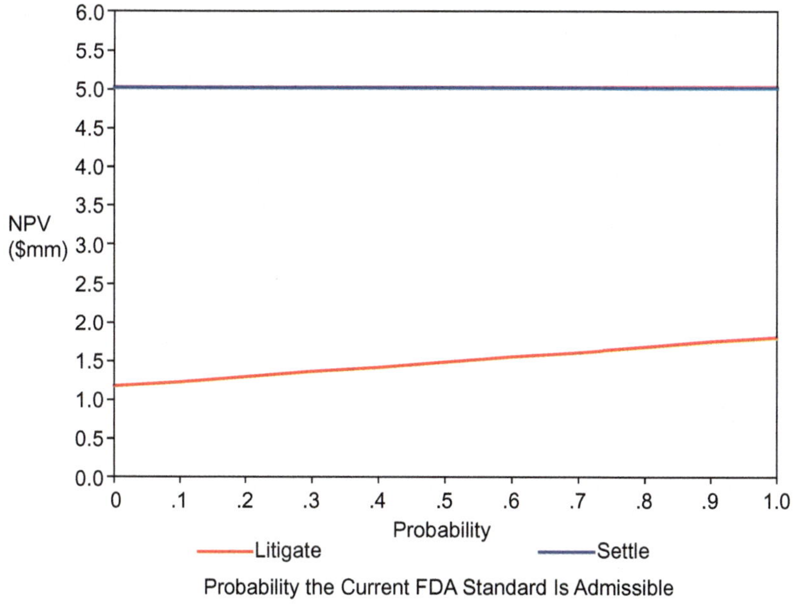

Figure 12.9 Sensitivity to probability the current FDA standard is admissible

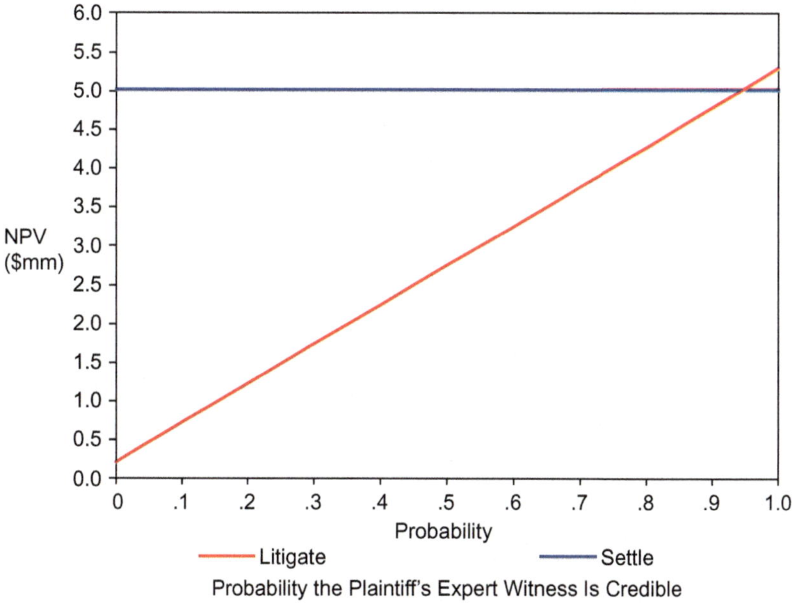

Figure 12.10 Sensitivity to probability the plaintiff's expert witness is credible

This uncertainty makes a big difference in indemnity exposure. If you know for sure the witness would not be found credible, the mean indemnity exposure in the case is only around $250,000. If you knew for sure the witness would be credible, mean indemnity exposure is almost $5.5 million.

This question, then, is one bearing a hard second look. Anything above a probability the witness will be credible of 80% or more makes the settlement offer look good. The initial assessment was only a 25% chance the witness would be credible. Do you really believe it, and can you explain why?

12.1.5 *What to Do and Why*

As we said in the beginning, the purpose of the decision analysis is to come with a cogent, clear, and defensible recommendation as to what to do and why.

For this case, the recommendations at this point are:

- Refuse the $5 million settlement offer on the table, or try to negotiate it down to $1.5 million or less (maybe a bit more with the cost of trial).
- Continue preparing the case for trial.
- Take a hard look at the credibility of the plaintiff's expert witness, or whether the plaintiff might be able to find another expert witness who would be more credible.

The reason is that the settlement offer is far above what the exposure at trial looks like at this point. You can go into details of why you think this is so with any of the materials presented.

In summary, then, as Decision Analysis becomes more widespread, you may be called on more and more often to review the analyses of the cocounsel or of the opponents. Thus, even though you may not regularly perform such analyses as a part of your work, you will need to know how to interpret and critique what others have done. This section provides a guide to doing just that for the most important analysis results.

12.2 Discussing the Case

After the hard thinking required to complete the analysis and clarify your thinking, one is often tempted to discuss the analysis—it was so much work to do! This is not recommended.

Best to start with our objective: a clear recommendation of what to do and why. Everything else is backup to describe how you get there or to answer "what if" questions.

People will want to know that you have all the backup and will frequently want to see enough of it to satisfy themselves that your thinking and methods are sound, but seldom want to go through the step-by-step details. The cocounsel may want to see more detail. Hopefully, if they are working on the case with you, they likewise worked with you on the analysis.

What clients want to see varies. Clients familiar with numbers and analysis (especially business clients) may want to dive in at some depth. If the analysis involved projection of company or market information, hopefully the experts at the client were likewise closely involved in putting those parts together.

My usual plan is to prepare the entire presentation documenting the results, present only the executive summary in the front section and cover other material as the audience requests.

12.2.1 Purposes of Discussion with Client or Co-counsel

When discussing your analysis with the client or co-counsel, a number of different objectives are possible.

- You may want merely to inform them of the risks of litigation in a particular case, without requiring a specific response.
- You may want to obtain other information from them, such as probabilities or what the potential effects of the litigation on client's business are (related to the problem of valuing nonmonetary consequences discussed in Chapter 8).
- You may want to assess the client's risk attitude, either formally or informally, to help you compare a settlement offer with the risks of litigation.
- You may need to present your recommendation and to obtain a decision from the client on whether to settle or to litigate.

From reading this book and from your own experience, you likely have some ideas for how to go about each of these objectives, and the method you use may vary from client to client. To augment your own ideas, following are some suggestions on how to approach each of these objectives.

- To inform a client or co-counsel about the risks of litigation, the relevance diagram works very well. It shows the major questions and their relationships without all the quantitative detail of the decision tree. Most people seem to grasp the meaning of a relevance diagram rather rapidly (though, as you've seen, translating one into a decision tree is a rather more difficult matter).
- To display the risks in terms of the possible dollar values of litigating the case, I prefer the cumulative probability distribution. It gives a clear vertical line at zero value for the probability of no liability at trial in addition to showing the best and worst case scenarios if found liable. Unless there are a lot of bins, the histogram mixes low-value liable outcomes in with the zero no liability outcome.
- To obtain more information from a client or co-counsel, you will need to involve them in problem structuring, or, at least, will have to formally assess probabilities from them. See the discussion in Chapter 8 for more on problem structuring and probability assessment with a group (which is, of course, applicable to a group of two).

- You can informally tackle the topic of client's attitude towards risk by discussing their comfort level with the possible outcomes of litigation. You may want to include in your presentation a sensitivity to risk attitude in case it's needed. Possibly some discussion of how to handle attitude towards risk will be needed. For many people, the approach presented in this book is a change from how they commonly think of risk.
- To obtain a decision from the client on the settlement strategy, you will want to present your recommendation and to support it. In addition to a verbal recommendation and rationale, the relevance diagram and decision tree work well for this purpose. The relevance diagram introduces the key questions and their relationships, and the decision tree shows the scenarios, the probabilities of various outcomes, and how they relate to produce the mean value. These lead to the recommendation of settling for anything less than the mean value of litigating the case. You will also recall that factors normally omitted from a first pass analysis, like avoidable legal fees, may argue for accepting a settlement offer somewhat above the mean value of litigating.

There will almost always be a question of how much depth you should go into. Should you try to stick to your conclusions, or should you get into the ultimate issues and the influencing factors? Should you go into the probabilities and lists of reasons at all?

The answers to these questions depend on who you're dealing with and what kind of response you need from them. The more a person will be involved in and contributing to an analysis, the more she'll need to be appraised of its details—perhaps to the level of structuring the relevance diagram with you. In contrast, a client's board of directors will typically only want to see conclusions, recommendations, and qualitative insights—maybe no numbers at all. Your experience and judgment will be the best guide to the appropriate level of detail.

12.2.2 Discussing the Analysis with an Opponent

Discussing your analysis with an opponent presents a slightly different problem than discussing it with a client or co-counsel. The counsel sometimes express concern about telling your opponent too much about your case, or about informing them of a legal theory they hadn't considered. You may perhaps want to protect your client's interests by only educating your opponent enough about the main issues in the case to facilitate settlement.

These concerns are perhaps less urgent than they may seem. If your opponent is reasonably competent and diligent, they will likely perceive all the most important issues on their own. Presenting your analysis will help to facilitate communication and settlement (as discussed in the section in Chapter 12 on ADR). Your judgment and that of senior counsel should guide you on this.

If you're concerned about giving too much away, you can limit your disclosure and still use your analysis to facilitate settlement with the following procedure.

- Make sure you've conducted thorough probability sensitivity analyses and know which uncertainties make the most difference in the mean value of litigating the case.
- Present a decision tree to your opponent without any probabilities filled in and, perhaps, with a theory excluded that you don't think she's considered. Use the tree to discuss the main issues with your opponent until the two of you have agreed on a tree that represents them.
- In some cases, it's possible to set probabilities in discussion with the opposing counsel and agree to a settlement value on the case. This won't be possible in highly adversarial cases.

There's no standard formula for how to integrate decision analysis in your negotiating strategy. But, whatever your negotiating strategy is, the analysis will make you vastly better prepared and effective in executing it.

12.3 Documenting Your Work

I will usually document my work in a presentation. Here's the usual outline.

- Executive Summary
  - Background and objectives
  - Summary of alternatives evaluated
  - Summary of work completed (scope, who was involved, etc.)
  - Data sources
  - Key financial results (your choice of analysis)
  - Recommendation and rationale
- Appendices
  - Background on decision analysis methodology
  - Detailed description of alternatives
  - Key probability assessments
  - Deterministic model results
  - Deterministic sensitivity analysis (tornado diagram), if any
  - Probabilistic analysis: probability distributions and contingent probability distributions
  - Sensitivity to probability
  - Sensitivity to risk attitude

The package will be very extensive for complex business litigation, and much simpler when liability or damages issues don't require that level of complexity.

These results can also be written up in report form following about the same headings.

Usually, the analysis and results are separate from other formal documents prepared by the counsel (formal facts, law, analysis-type documents). They are usually marked confidential and privileged and prepared in preparation for litigation.

12.4 Problems

1. For any of the case studies in this book:
    - 1.1. Complete the analysis for the problem.
    - 1.2. Follow the steps in this chapter to review the analysis and revise it if indicated.
    - 1.3. Prepare materials as described in this chapter documenting your analysis (recommendation, rationale for it, supporting analysis, etc.).
    - 1.4. Designate one or more persons to present to and specify their roles (co-counsel, client, or opponent).
    - 1.5. Specify your objective in having the discussion.
    - 1.6. Have the discussion.

Bibliography

Allen, M.S.: Business portfolio management: valuation, risk assessment, and EVA strategies. John Wiley & Sons, New York (2000)

Amran, M., Kulatilaka, N.: Real options: managing strategic investment in an uncertain world. Harvard Business School Press, Boston (1999)

Barabba, V.P.: Meeting of the minds: creating the market-based enterprise. Harvard Business School Press, Boston (1995)

Behn, R.D., Vaupel, J.W.: Quick analysis for busy decision makers. Basic Books, Inc., New York (1982)

Bodily, S.: Modern decision making: a guide to modeling with decision support systems. McGraw-Hill, New York (1984)

Brown, R.V., Kahn, A.S., Peterson, C.R.: Decision analysis: an overview. Holt, Rinehart, and Winston, New York (1974)

Bunn, D.W.: Applied decision analysis. McGraw-Hill, New York (1984)

Clemen, R.T.: Making hard decisions: an introduction to decision analysis. Duxbury Press, Boston (1996)

Cotovello, V.T., Merkhofer, M.W.: Risk assessment methods: approaches for assessing health and environmental risks. Plenum Press, New York (1993)

Golub, A.L.: Decision analysis: an integrated approach. John Wiley & Sons, New York (1996)

Hammond, J., Keeney, R.L., Raiffa, H.: Smart choices: a practical guide to making better decisions. Harvard Business School Press, Boston (1998)

Holloway, C.A.: Decision making under uncertainty: models and choices. Prentice-Hall, Englewood Cliffs, NJ (1979)

Howard, R.A., Abbas, E.E.: Foundations of Decision Analysis. Pearson, Boston (2016)

Howard, R.A., Matheson, J.E. (eds.): Readings on the principles and applications of decision analysis, 2 volumes. Strategic Decisions Group, Menlo Park, CA (1984)

Hubbard, D.W.:The failure of risk management: why it's broken and how to fix it. Wiley, Hoboken (2009)

Kahneman, D.: Thinking, fast and slow. Farrar, Straus and Giroux, New York (2011)

Keeney, R.L., Raiffa, H.: Decisions with multiple objectives: preferences and value tradeoffs. Cambridge University Press, New York (1993)

Keeney, R.L.: Value focused thinking: a path to creative decisionmaking. Harvard University Press, Boston (1996)

Luenberger, D.G.: Investment science. Oxford University Press, Oxford (1997)

J. Celona, *Winning at Litigation through Decision Analysis*,
Springer Series in Operations Research and Financial Engineering,
DOI 10.1007/978-3-319-30040-5

Matheson, D., Matheson, J.: The smart organization: creating value through strategic R&D. Harvard Business School Press, Boston (1998)

McNamee, P., Celona, J.: Decision analysis for the professional, 4th edn. SmartOrg, Inc., Menlo Park (2007)

Merkhofer, M.W.: Decision science and social risk management. Reidel, Boston (1987)

Newendorp, P.D.: Decision analysis for petroleum exploration. Planning Press, Aurora, CO (1998)

Raiffa, H.: Decision Analysis. McGraw-Hill College Division, New York (1997)

Raiffa, H.: Decision analysis: introductory lectures on choices under uncertainty. Random House, Inc., New York (1986)

Raiffa, H., Richardson, J., Metcalf, D.: Negotiation analysis: the science and art of collaborative decision making. Belknap, Cambridge (2002)

Samson, D.: Managerial decision analysis. Richard D. Irwin, Inc., Homewood, IL (1988)

Schlaifer, R.: Analysis of decisions under uncertainty. Robert E. Krieger Publishing Co., Inc., Melbourne, FL (1978)

Skinner, D.C.: Introduction to decision analysis. Probabilistic Press, Gainesville, FL (1999)

Spurr, W.A., Bonini, C.P.: Statistical analysis for business decisions. Richard D. Irwin, Inc., Homewood, IL (1973)

Thompson, M.S.: Decision analysis for program evaluation. Ballinger Publishing Co., Cambridge, MA (1982)

Tummala, V.M., Henshaw, R.C. (eds.): Concepts and applications of modern decision models. Michigan State University Press, East Lansing, MI (1976) (various): Making smart decisions. Harvard Business School Press, Boston (2006)

von Winterfeldt, D.: Decision analysis and behavioral research. Cambridge University Press, New York (1988)

Index

A
Accident and Safety Program tree, 84
Activity frustrated value of, 268
Adjustment and anchoring, 16–17
Alternative dispute resolution (ADR)
  benefits of analysis, 411, 413–415
  differences with trial, 410, 425
Alternatives
  arguments, 98–102, 310, 312, 319
  element of a good decision, 280
Analyzing from your opponent's point of view, 245, 406–409, 419–420. *See also* Opposing counsel
Analyzing results, 106, 171, 185. *See also* Interpreting analysis
Appropriate level of investment in discovery (from probability sensitivity), 217. *See also* Discovery strategy
Arrows. *See also* Relevance diagrams
  choice of direction, 33, 79, 307, 323, 347
  direction for assessment, 33, 79, 323
  use in relevance diagrams, 33, 79, 307, 347
Assessing continuous variables
  procedure to assess the cumulative distribution, 141–147
  why to assess the cumulative distribution, 139, 140
Assessing probabilities
  direction of arrows, 307, 347
  with experts, 327
  for forum uncertainty, 286–287
  with a group, 137
  procedures, 37, 130–149, 157–159
  selecting a method, 137–138
  updating probability assessments, 37, 130
  use of dependent outcomes or probabilities, 159
  use of probability wheel, 132–134
  value of improving assessment (value of information), 200
  Yes/No variables, 131–134
Availability bias, 18

B
Bad publicity value of, 268
Bayes' Rule (for reversing dependencies), 353, 354
Biases
  adjustment and anchoring, 17, 287
  availability bias, 18
  avoiding with probabilities, 37
  missing issues, 16
  mis-valuing combinations of uncertainties, 20–21
  representativeness and avaialability, 17
Boxes, 32, 57, 368, 369. *See also* Relevance diagrams
  in relevance diagrams, 32
Burden of proof, 69, 71, 76, 79
  role in framing issues, 62

C
Certain equivalent
  calculation, 193–195, 198, 257
  graph versus mean value, 195
  variation with probability of winning, 195

J. Celona, *Winning at Litigation through Decision Analysis*,
Springer Series in Operations Research and Financial Engineering,
DOI 10.1007/978-3-319-30040-5

Claim inclusion/exclusion, 416–419
as aid to trial strategy, 404–406
Clairvoyance, 59, 85, 128, 134, 150, 151, 157, 175, 201, 203–206, 245, 246, 268, 269, 324, 364
tree illustrating, 202
Clairvoyant, role in value of information, 56, 59, 64, 65, 121, 129, 139, 185, 200–202, 206, 268, 410
Clairvoyant test
applied to sub-trees, 176
in looking for factors left out, 430–431
Clincher argument, 317–318
Cocounsel, 211, 399, 412, 429, 442–444, 446
verifying analysis, 442
Co-defendants, 57, 375–381, 392–394
Cognitive problems, 15–20, 86
Communication
benefits of analysis, 47
with decision analysis, 429, 430
discussing analysis results, 429, 442
discussing analysis with opponent, 444–445
with probabilities, 438–442
Compensatory and punitive damages, 91, 211, 240, 348, 432
Complexity, 2, 5, 7, 13–15, 22, 31–33, 41, 46, 55, 72, 117, 185, 211, 229, 233, 269, 284, 412, 414, 429, 445
general discussion, 14, 23, 33–35
Complicated damages
insurance or reimbursement, 246–248
multiplier on damages, 243–248
an offset to the award, 244–246
share of the verdict, 241–243
Contingent fees, 255, 256
Contingent probability distributions
procedure for construction, 342–343
reading graphs, 343
Contingent strategy tables, 401–404, 416, 419
reasons to use, 401
Continuing injury (in present value calculation), 258–261, 290
Continuous variables. *See also* Cumulative probability distributions
discretization, 149, 215, 339
discretizing cumulative distributions, 337
"nice" and "skewed" distributions, 137
procedures for assessing, 139–147
ranges, 139
Contract breach tree, 61, 127
Contract liability
decision tree, 89, 90
relevance diagram, 90
Control, value of. *See* Value of control
Costs
legal fees and costs, 106, 248–256, 294–298
sunk costs, 106, 249–253, 256, 301, 303, 438
Counter-claims
contract claim and counterclaim, 390
effect of client risk aversion, 192
Cumulative probability distributions
effect on risk aversion, 169, 186
importance with risk-aversion
procedure for constructing, 331–346, 364–367
reading a distribution, 331–346, 364–367
reading graphs, 170, 171
smooth distributions, 341

D

Damages
complicated damages, 106, 239–248, 288–290
damages not correctly calculated, 185
Death
decision-making with increase chance of, 284–286
small probability value of life, 286, 301
Decisions, in relevance diagrams, 38, 47, 280, 308, 399, 411
Decision trees
for additional oil drilling, 249
for calculating a multiplier on damages, 243–248
with chance of death, 284–286
for clincher argument, 317–318
for codefendant, 243, 382, 389
contingent fees in tree, 256
for contract claim and counterclaim, 390
for defendant's indemnification, 245
for deposing a witness, 402
drawing one node or two (structure equivalency), 104–105
to handle large and uncertain legal fees, 254
introduction, 33, 35
for an offset or counterclaim, 244
opposing summary judgment motions, 392, 396–397
ordering the tree, 78–105
for order of arguments, 320
for settlement with codefendant, 190, 377
for a share of the verdict, 241–243
procedure for drawing, 77–78
pruning the tree, 90
reading decision trees, 6
significant and uncertain factor, 85

steps if too large, 88–98
value formulas, 116–120, 181, 224, 293, 295, 300, 395, 397, 418, 422, 424, 426
Defendant's indemnification tree, 245
Dependencies. *See also* Relevance diagrams
drawing dependencies in the tree, 56–62, 80–84
finding dependencies, 58–59
from asymmetry, 80–83
special dependencies, 59–62
testing Dependencies, 59
Dependent outcomes, 347–349, 367
reasons to use, 355, 357
Dependent probabilities
correlations between sub-trees, 255
in the decision tree, 234
direction of arrows, 307, 347
discussion of reversing dependencies, 351–352
relation to dependent outcomes, 290, 347–348, 362, 367
to represent spill-over, 290, 307
reversing dependencies, 351–352
when to use dependent outcomes, 109
Devil's advocate
playing devil's advocate, 71–72, 91, 92, 325
pruning the tree, 91
Discounted cash flow. *See* Present value
Discounting patterns, 260
Discovery strategy
appropriate level of investment, 217
introduction, 199
probability sensitivity, 213–218
safe settlement level, 218
value from probability sensitivity, 214–218
value of control, 206–211
values of clairvoyance (Information), 199–212
Discretizing probability distributions. *See* Piecewise-continuous distributions/continuous probability distributions
Dispositive issues (in relevance diagram), 68
Double ovals
in decision tree, 77
in relevance diagram, 86
D-Zine case, 383, 384, 386, 387, 389, 394

E

Equivalent trees, 105. *See also* Structure equivalency
Excess probability distributions, 337–340
reading graphs, 340

Expected utility. *See* Risk aversion. *See* Utility function
Expected value
calculation, 164–167
with roll-back, 164, 181, 183, 193, 194, 201, 208, 401
with roll-forward, 39, 40, 164–167, 181, 183, 194, 201, 311, 351, 401
introduction, 161
intuitive reasons for using, 41
reading the mean value, 162
relation to good decisions, 42
variation with probability of winning, 195, 214, 217
wealth maximization, 41
Experts, working with, 122, 131, 327–328

F

First pass analysis
considering issues handled grossly, 189–190
considering omitted issues, 89, 188
considering risk aversion, 188
discussion of level of detail, 65
evaluating the level of detail, 89
evaluating the results, 89, 188, 189, 191
Focusing on What's Most Important, 56, 88–89
framing questions, 61
handling legal fees and costs, 249
hints for drawing relevance diagrams, 56
for influence on precedence, 261–267
limiting your analysis, 89
other strategy considerations, 190
selecting a probability assessment method, 252
updating the analysis, 399–401, 415–416
where you have little information, 85
which factors should go into the tree, 77, 84–88, 177
Formulas, 4, 5, 162, 197, 198, 247, 256, 291–293, 354, 427, 445. *See also* Scenario values
to calculate scenario values, 162
Forum, uncertainty as to which, 286–287
Framing questions, 61
applying trier-of-fact rule, 62

G

Global issues (in relevance diagram, 68
Good decisions
definition, 42
elements of, 43

Good decisions (*cont.*)
relation to mean value, 42
Good outcomes, 42–43, 45, 71, 206, 252, 433
definition, 42
Governmental problems
modeling non-monetary consequences, 269–277, 279–284
sample relevance diagram, 281
use of LRA to clarify and to communicate, 280

H

Harm foreseeable tree, 72, 91
Harm reasonably foreseeable tree, 20, 72, 131, 294
Histogram
constructing histograms, 253, 254
reading the histogram, 439
use in handling large legal fees and risk aversion, 188
Howitzer v. Crabgrass et al. case, 62–65, 186, 191

I

Imperfect information value of
procedure for calculating, 288
reasons to calculate, 288
Important influencing facts (in relevance diagram), 69
Incremental costs, 252, 303. *See also* Sunk costs
Influencing factors, 65, 77, 84–88, 174, 177, 254, 314, 317, 325, 328, 341, 384, 413, 422, 430, 431, 433, 434, 442
which should go into the tree, 84–88
Information
element of a good decision, 43–45
value of (*see* Value of information)
Injunctive relief, value of, 271, 274, 275
Insolvent, insured defendant, 247, 248, 406, 407, 420
Insurance/reimbursement, 246–248
splitting settlement between insurer and insured, 379–381
Insurer tree, 105, 381
Interpreting analysis
reading a cumulative distribution, 169, 201
reading a probability sensitivity, 342, 343
reading decision trees, 276
reading numerical results, 207
reading relevance diagrams, 217
Intuitive sense
assessing probabilities, 44
evaluating analysis results, 47
expected value, 41
mutually exclusive and collectively exhaustive events, 102
probabilities, 36, 128
results, 45
"selling" others, 46
value of control, 207
Intuitive test (for dependencies), 59
Issue identification. *See* Relevance diagrams
Issues handled grossly, 189–190
Issues omitted from the analysis, 188–189
Issues unintentionally omitted, 185

J

Juries. *See* Trier-of-Fact Rule
Jury science, 354, 372. *See also* Imperfect information value of

K

Key influencing facts, 63, 65, 66, 89, 121
Key influencing uncertainties (in relevance diagram), 68

L

Labels, 60, 61, 66, 71, 77, 94, 109, 299, 392, 393, 433
on dependency arrows, 60
Legal fees and costs
contingent fees, 255–256
sunk costs, 249–253, 256
use of histogram with large fees, 253
using a best guess, 253
using separate trees, 243
varying by scenarios, 180–182
Lists of reasons
applied to handling details, 131
in assessing probabilities, 131–132
Logic, 4, 43, 45–47, 161, 162, 411
element of a good decision, 43–45
Lose either way
contract/tort relevance diagram, 127
contract/tort tree, 127
relevance diagram, 127
tree, 61, 127

M

Many possible arguments, 98, 307
"clincher" argument, 317–318
Mean of a continuous variable, 135–137

Miscommunication, 19–20
Missing issues, 15–16
Mis-valuing combinations of uncertainties, 20
Modeling non-monetary consequences, 269–279
Motivational problems, 21, 24, 32, 37, 55, 145, 185, 307, 327
Multiple arguments, 98–104, 308, 361. *See also* Many possible arguments
Multiple claims, 416–419
  inclusion/exclusion, 404–406
Multiple defendants, 384, 394–396
Multiple liability theories, 34
  sample tree, 34, 35
Multiplier on damages, 243–244
Mutually exclusive and collectively exhaustive events, 128–130, 152

N

"Nice" probability distributions, 137
No continuing injury, 258–260
Non-monetary consequences
  in business cases, 268
  direct assessment of, 268–269
  in governmental problems, 279–284
  modeling, 269–279
Numerical test (for dependencies), 59

O

Offset to the award, 244, 389
One argument only, 103–104
Only lose if both, 127
  relevance diagram, 127
  tree, 61
Opponent's point-of-view. *See* Opposing counsel
Opposing counsel
  analysis from their viewpoint, 412
  benefits of analysis, 413
  defendant's indemnification tree, 245
  different case value, 245
  different issues, 131, 412, 413
  discussing analysis, 412
  effect of insurance, 248
  steps to analyze their value, 431
Outside experts, 140
  need for cumulative probability assessment, 139, 140
Ovals (in relevance diagrams), 33, 56, 241, 433

P

Picking the "average" shortcut for assessing continuous variables, 204–206
Piecewise-continuous distributions/continuous probability distributions, 341–342
  assessing, 341
  discretization, 341
Precedence
  in relevance diagram, 262
  steps to analyze, 261
  with a first pass analysis, 261
Predictions. *See* Imperfect Information value of
Preferences, 4, 43–45, 257, 296, 331
  element of a good decision, 43–45
Pre-judgment interest, 69, 106, 257, 259, 260, 296–298
Present value
  choice of discount rate, 224–225
  discounting patterns, 255
  pre-judgment interest, 258–261
  in scenario values, 162–163
  in value formulas, 105–107
Pre-trial strategy, 399–426
  deposing a witness (contingent strategies), 188, 402
Probabilities
  dependent probabilties in the tree, 82–83
  improving assessment (value of information), 200–203
  not the best, 186
  right/wrong probabilities, 122, 186
Probability arithmetic rule, 128
Probability assessments. *See also* Assessing probabilities
  background, 121–130
  improving assessments, 130 (*see also* Value of information)
  introduction, 135–138
  procedures, 130–149
  scenarios, 128
Probability distribution. *See also* Cumulative probability distributions. *See also* Excess probability distributions
  contingent probability distributions, 342–346
  histogram, 331–335
  piecewise-continuous distributions, 341–342
Probability sensitivity
  benefits for settlement, 343
  discovery strategy, 213–218
  effect on value of the case, 213
  procedure, 213
  reading a graph, 198, 199

Probability sensitivity (*cont.*)
reading a probability sensitivity, 411
sample graph, 198
summary for discovery strategy, 218
Probability theory
arithmetic rule, 126–128
assessing probabilities, 130–149
background, 150–155
continuous variable, 135–136
"and" events and "or" events, 126, 127
mutually exclusive and collectively exhaustive events, 128–130
Probability wheel, 132, 133
procedure for use, 132
Procedure for cumulative probability distribution, 169–172, 182, 186, 196
Product liability, 20, 72, 91, 286, 287, 294, 432
decision tree, 21
Punitive and compensatory damages, 240

R

Ranges of continuous variables, 136
Reading a probability sensitivity analysis, 411
Reading decision tree, 433–438
Reading relevance diagrams, 350
Reading numerical results, 207
Reading the cumulative distribution, 201
Reading the expected value, 39, 167, 191, 193, 194, 256
Reading the histogram, 169–180
Relevance diagrams
for alternate arguments, 98
for claim inclusion or exclusion, 404–406
for codefendant, 57, 94
for compensatory and punitive damages, 240
contract claim and counterclaim, 390
for deposing a witness, 402
hints for drawing, 67–77
for insolvent, insured defendant, 247, 248, 407
insurance or reimbursement, 247
introduction, 56–62
key influencing facts, 62, 86, 114, 290, 297
for many possible arguments, 309
for opposing summary judgment motions, 130
procedure to draw, 64–67
for product liability, 432
reading relevance diagram, 350
symbols, 56
use to clarify your case, 72–77
Representativeness and availability, 17–18
Representativeness bias, 17

Results of analysis, 45–47. *See also* Interpreting analysis
Risk-adjusted discount rate, 41, 224, 225, 258
Risk aversion
attitudes toward risk-taking, 192
behavior, 194, 197
calculating a certain equivalent, 193–195
certain equivalent, 193
with counter-claims, 389–391
definition, 192
double-counting for client's risk aversion, 102
importance of cumulative probability distribution, 195–196
incorporating legal fees, 253
risk penalty, 193, 253
risk premium, 193
settlement strategy for risk-averse decisions-maker, 192–199
settlement strategy for risk-neutral decision maker, 191
use of utility function, 194
variation by probability distribution, 169
Risk penalty, 193, 194, 196, 197, 253
graph, 195
Risk premium, 193
Roll-forward calculations, 37

S

Safe settlement level (from probability sensitivity), 217, 218
Scenario values. *See also* Value of the case. *See also* Non-monetary consequences
calculation, 49, 161, 162
choice of discount rate, 257–258
with codefendant, 375–377
with counterclaim, 244
different value structure for opponent, 406
discounting patterns, 260
discounting patters, 260
offsets to award, 244–246
pre-judgment interest, 260
varying legal fees by scenario, 253–254
Settlement strategy
benefits of analysis, 190
with codefendant, 190, 378–379
with insured and insurer, 379–381
with insured defendants, 248
introduction, 190
risk-averse decision-maker, 192, 194
risk-neutral decision-maker, 191, 194
Share of liability, 136

Share of the verdict, 189, 241–243, 290, 376
Significant factors in the tree, 317
Skewed probability distributions, 137
Small-probability value of life, 305
Smoothing probability distributions, 339
Solvent, insured defendant, 248
SoMoLo, Inc. case, 270–272, 274, 275
Spill-over
 argument synergies and order of proof, 319–320
 between damages and damages offset, 239–241
 between main tree and sub-trees, 382
 between multiple arguments, 98–104
 between multiple defendants, 382–389
 between sub-trees, 372
 caution when tree pruning, 90–93
 in choice of arguments and dependent probabilities, 104
 claim inclusion or exclusion, 404–406, 416–419
 with codefendant, 375–381
 in drawing equivalent trees, 105
 evaluating many possible arguments, 308–318
 in evaluating probabilities, 319–324
 from order of issues in the tree, 319
 general discussion, 319
 jury or judgy synergy, 320–324
 in legal fees, 253–254
 in lists of reasons, 89–90
 using devil's advocate to find it, 71–72
State of information
 dating and updating analysis, 399–401, 415–416
 versus difference in value structure, 354
State of knowledge. *See* Probability theory
State-of-the-art defense, 20, 72, 431, 432, 435, 436, 439–440
Strategy. *See also* Discovery strategy. *See also* Trial strategy
 contingent strategy tables, 401–404, 416
Structure equivalency, 104–105
Subjective probability, 37, 38, 121–126, 130, 131, 139, 152, 307, 328, 399, 406. *See also* Probability theory
Substantive law, 6, 15, 187, 308
Sub-tree
 applied to many possible arguments, 308–318
 clairvoyant test, 56, 64, 128–131, 139, 141, 200, 327, 364, 367, 432
 for compensatory and punitive damages, 94, 240, 348
 for contract liability, 89, 90
 correlations between sub-trees, 255
 definition and purposes, 173–177
 for disability compensation, 18, 177
 for future medical costs, 174, 176–178
 to handle large and uncertain legal fees, 254–255
 list of reasons, 89–90
 with multiple defendants, 382–389
 procedure, 177–180
 purposes, 173–177
 to refine analysis, 189
 requirements, 174–177
 vicarious liability, 81
Sum of two kinds of damages, 239–241
Summary judgement, 24, 50, 59, 66, 75, 77, 94, 129, 151, 188, 206, 208, 211, 253, 293, 295, 344, 345, 372, 397, 400
 opposing motions for, 391, 392, 396–397
Sunk costs, 106, 249–253, 256, 301, 303
 examples, 249
Symmetry, 81
 in relevance diagram and tree, 81

T

Texaco hypothetical
 illustration of dependent outcomes and dependent probabilities, 347–348
 trial strategy example, 218, 343
Three-branch shortcut for assessing continuous variables, 145, 241, 303
Trial strategy
 claim inclusion or exclusion, 404–406
 Texaco hypothetical example, 343, 347
Trier-of-Fact Rule
 applied in framing issues, 61
 applied to a multiplier on damages, 243–248
 applied to compensatory and punitive damages, 211, 432
 applied to consideration in stages, 322–324
 applied to contract claim and counterclaim, 389–391
 applied to forum uncertainty, 286–287
 applied to judge or jury synergy, 7, 320–324
 applied to many possible arguments, 308–318
 applied to opposing summary judgment motions, 392, 396–397
 applied to order of arguments, 320
 applied to picking the level of detail, 56

Two Arguments to Recover a Single Loss, 5, 128
Two-branch shortcut for assessing continuous variables, 241

U

Ultimate issues (in relevance diagram), 68
Uncertain factors in the tree, 174
Uncertainty(ies). *See also* Ovals in relevance diagrams
  determined by other questions, 83–84
  general discussion, 14
Updating your probabilities and analysis, 399–441
  contingent strategy tables, 399–401, 415–416
Utility function, 194, 195, 257. *See also* Risk aversion
  definition, 194

V

Value of clairvoyance. *See* Value of information
Value of control
  calculation, 206–211
  intuitive answer, 207
  value with control, 346
Value of imperfect information. *See* Imperfect information value of
Value of information
  calculation shortcut, 204–206
  definition, 200
  derivation, 4
  value from changed decision, 208
  value with information, 356
Value of the case. *See also* Scenario values. *See also* Opposing counsel
  cumulative probability distribution for, 182
  effect of decisions and uncertainties, 38, 93
  effect of probability sensitivity, 213–218
  effect of value of control
  effect of value of information, 203
  for insolvent insured defendant, 247, 407
  intuitive sense of, 191
  mean value calculation, 167
  with risk aversion, 196–199
  for solvent insured defendant, 248
  as value measure, 58
Value of wizardry. *See* Value of control
Values. *See also* Scenario values. *See also* Mean values. *See also* Certain equivalent values. *See also* Value of the case
  in decision tree
  formulas, 55, 105–106, 244, 263, 290, 372
Venn diagram, 129
  mutually exclusive and collectively exhaustive events, 130
Vicarious liability
  decision tree, 82
  relevance diagram, 83

W

Win if either, 61
  relevance diagram, 61
Wizardry, value of. *See* Value of control
Wrong ordering in the tree, 185

Printed by Printforce, the Netherlands